Ion Exchange in Environmental Processes

Ion Exchange in Environmental Processes

Fundamentals, Applications and Sustainable Technology

Arup K. SenGupta

Professor
Lehigh University
Bethlehem, USA

WILEY

This edition first published 2017

Registered Office
John Wiley & Sons, Inc., 111 River Street, Hoboken, NJ 07030, USA

Editorial Office
111 River Street, Hoboken, NJ 07030, USA

For details of our global editorial offices, customer services, and more information about Wiley products visit us at www.wiley.com.

Wiley also publishes its books in a variety of electronic formats and by print-on-demand. Some content that appears in standard print versions of this book may not be available in other formats.

Library of Congress Cataloging-in-Publication Data

Names: SenGupta, Arup K. author.
Title: Ion exchange in environmental processes: fundamentals, applications and sustainable technology / by Arup K. SenGupta, Professor, Lehigh University, Bethlehem, USA.
Description: First edition. | Hoboken, NJ, USA : Wiley, [2017] | Includes bibliographical references and index. |
Identifiers: LCCN 2017016090 (print) | LCCN 2017016885 (ebook) | ISBN 9781119421283 (pdf) | ISBN 9781119421290 (epub) | ISBN 9781119157397 (cloth)
Subjects: LCSH: Ion exchange–Industrial applications.
Classification: LCC TP156.I6 (ebook) | LCC TP156.I6 S45 2017 (print) | DDC 660/.29723–dc23
LC record available at https://lccn.loc.gov/2017016090

Cover image: Foreground: Courtesy of Arup SenGupta and Michael German; Background: © MirageC/GettyImages
Cover design by Wiley

Set in 11/13pt WarnockPro by SPi Global, Chennai, India

Printed in the United States of America

10 9 8 7 6 5 4 3 2 1

To
Susmita, Neal and Soham for their love and support
and
Mother Nature for Her infinite tolerance

कर्मण्येवाधिकारस्ते मा फलेषु कदाचन।

"Thy right is to the work only, but never to the fruits thereof"

Bhagvad Gita: Verse II:47

Contents

Preface

Ion exchange is a fascinating scientific field, as central to natural and biological systems, as to the engineered processes. Historically, application of ion exchange always stayed far ahead of theory and the design approaches for ion exchange systems were mostly empirical. The intrinsic complexity of the field was poorly understood and the science of ion exchange was accepted as mere exchange of ions. After the Second World War, ion exchange theory took root, progressed gradually on a scientific foundation and new applications were conceived and implemented. The intrinsic complexity of the field of ion exchange and its many seemingly eccentric behaviors were unraveled. Understandably, learning the subject requires revealing its scientific core in appropriate sequence, interjected with key scientific inquiries of "*why*" and "*how.*"

It was during the fall of 1996 when I was in England on a sabbatical leave at the invitation of long-time friend and colleague, Prof. Michael Streat, that the thought of writing a book on Ion Exchange dawned on me and I initiated the process. While there, I was informally giving a series of lectures to a group of senior graduate students and young faculty members on topics related to fundamentals and recent developments in ion exchange. Some difficulties arose. I struggled to communicate some experimental observations of others that are seemingly counter-intuitive. So I started preparing notes of my own and that was the modest beginning. Needless to say, the effort went back and forth, the book project proceeded at a snail's pace and turned dormant. Finally, 3 years ago, I undertook the assignment as a mission that needs to be brought to a closure. However, the key questions or motivating factors – *Is such a book necessary and whom is this book for* – remained unchanged throughout.

No specialty grows in isolation. Ion exchange is not a recent invention, but over the last five decades, the science of ion exchange has permeated into a myriad of other growing fields – from decontamination to deionization, from mining to microelectronics, from gas separation to green processes, from novel synthesis to nanotechnology, from drug delivery to desalination, to name a few. The following figure from the Google patent search includes the number of ion exchange-related US patents issued during the last three decades, illustrating continued inventions of new products and processes.

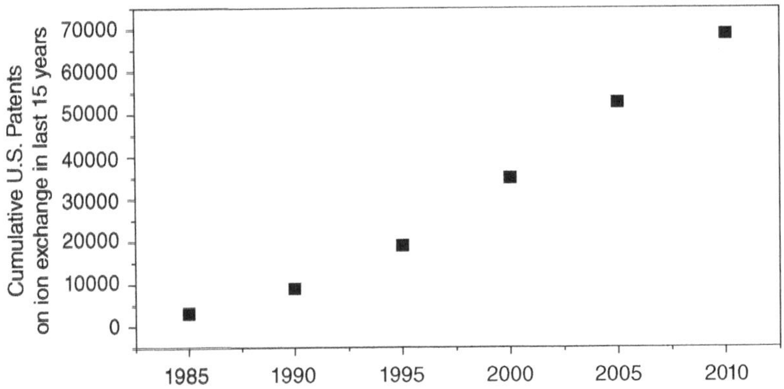

Very high US patent numbers only reinforce the dynamics of the field and its blending with many other seemingly disjointed scientific areas. It is only appropriate to mention that the worldwide push for sustainability and stringent environmental regulations has seen ion exchange technology as a major player in the development of the next generation of environmental processes and efficient materials. Such a move has demanded a need to revisit the fundamentals of ion exchange with a renewed perspective. As expected, this book presents the "why" and the "how" of multiple ion exchange phenomena with varying degrees of complexity. However, a conscious attempt has been made to present *physical realities* of every ion exchange phenomenon of interest right up front. Only then, underlying theories and quantitative approaches have been discussed to validate observed physical realities.

Presentation of theoretical tools that might help the reader in solving or addressing specific problems were given due importance. At the same time, overemphasis on mathematical models and abstract theories has been avoided. Even when mathematical deductions and related equations have been adequately presented, qualitative explanations and interpretations have not been ignored. Thus, a mathematically or a thermodynamically disinclined reader, with deep understanding of the subject through experience or other means, may comfortably navigate through the entire book and gain new knowledge or identify areas warranting further innovation.

Writing or introducing a new book on *Ion Exchange* will always remain incomplete unless an honest discussion is made about how it complements or adds to the existing title on Ion Exchange written by Fred Helfferich over 50 years ago. Helfferich's book is a seminal contribution in the field and will continue to remain so. I take pride in stating that I knew Fred Helfferich. He was an esteemed professional colleague and we interacted in several ways. I personally keep a copy of his book both at home and in the office, consulting it whenever necessary. Nevertheless, it is also my finding that people always refer to Helfferich's book when confronted with a question or uncertainty, but rarely do they read it for learning the subject of ion exchange. Classical step-by-step learning through Helfferich's book and applying the knowledge appropriately pose some genuine challenges. The book was not really written to serve that purpose. Also, during the last few decades, new ion exchangers, namely, macroporous, fibrous, hybrid and biomaterials have emerged with distinctive attributes; novel use of the Donnan membrane principle has opened up new opportunities to produce sustainable materials and processes. Further, gas- and solid-phase ion exchange may soon provide new platforms for novel, environmentally benign processes. More and more, ion exchange is being used synergistically with other known processes resulting in key breakthroughs in processes with enhanced sustainability. This new book will substantially complement the existing body of knowledge in the public domain and serve as a major learning tool for young scientists and engineers.

Readers with a moderate knowledge of physical chemistry, chemical/environmental engineering principles and mathematics, should be able to progress through individual chapters on their own. For academic teaching, the book is suitable as a text or a reference for an undergraduate senior or first- year graduate level chemical or environmental engineering course in separation, environmental processes or ion exchange. Attempts have been made so that a potential reader, while gradually assimilating the content, will be prepared to apply the acquired knowledge for real-life scenarios, improve existing processes and develop an instinct for innovation through use of fundamentals. From that perspective, the content of the book will be useful also for polymer chemists, consulting engineers and technology companies seeking long-term holistic solutions. To facilitate the use of this book as a text or a handout in a short course, several numerical exercises have been included.

The book has altogether eight chapters that unfold connecting ion exchange processes and materials with fundamentals:

Chapter 1. Ion Exchange and Ion Exchangers: An Introduction
Chapter 2. Ion Exchange Fundamentals
Chapter 3. Trace Ion Exchange
Chapter 4. Ion Exchange Kinetics: Intraparticle Diffusion
Chapter 5. Solid- and Gas-Phase Ion Exchange
Chapter 6. Hybrid Ion Exchange Nanotechnology (HIX-Nanotech)
Chapter 7. Heavy Metal Chelation and Polymeric Ligand Exchange
Chapter 8. Synergy and Sustainability

A reader with prior exposure to the field of ion exchange, does not need to be deterred from jumping into any chapter of choice out of sequence and still comprehending the materials. Over the decades, widely used softening and deionization processes have been tailored to be more sustainable from chemical usage point of view and the subject has been discussed in both Chapters 1 and 2. Along the same vein, the ion exchange fundamentals have been appropriately harnessed to produce selective sorbents for nitrate, arsenic, fluoride, phosphate, boron and others. A relatively new field of hybrid ion exchange nanotechnology or HIX-Nanotech has emerged and the Donnan membrane principle plays a crucial role in expanding its application potential. Solid and gas-phase separations show promise for recovery of valuable materials with minimum chemical usage. In every such discussion presented in Chapters 5–8, the role of scientific fundamentals has been adequately articulated. Chapter 8 includes a new route to a simple-to-apply softening process without using an excessive amount of brine, often causing major disposal problems in arid regions.

It is generally agreed that the solutions to challenging problems of our time will not so much occur through evolution of new fundamental knowledge, but through synergistic integration of knowledge from seemingly disconnected fields. As the author of this book, I am quite optimistic that the science, technology and materials related to ion exchange, as presented here, will help fill some void and create new synergy for the next generation of innovators and inventors in the field.

Arup K. SenGupta
November, 2016
Lehigh University
Bethlehem, USA

Acknowledgment

During my first job as a process chemical engineer, my then supervisor in early seventies, N. K. Chowdhury, introduced me to the complexity and excitement of water science and technology. The excitement is yet to cease and my professional world during the last four decades has truly revolved around water in so many ways. In the same period, I was also exposed to the field of producing ultra-pure water for electric power generating utilities using ion exchange processes. Subsequently, I worked with Professor Dennis Clifford for my PhD; my graduate student life in the University of Houston was truly eventful and intellectually stimulating. The concept of gradual breakthrough during fixed-bed column runs was solidly confirmed through my doctoral work on chromate ion exchange. Dennis and I have remained friends and professional colleagues for nearly four decades and I am thankful to him in so many ways.

During the eighties and nineties, I had the opportunity and privilege to meet, chat, befriend and discuss matters of mutual professional interest related to different separation processes including ion exchange with many personalities around the world during Gordon Conferences on Reactive Polymers, IEX conferences at Cambridge (UK), and various ACS and AIChE conferences. I have very fond and rewarding memories of meeting and interacting with George Boyd, Robert Kunin, Fred Helfferich, Jacob Marinsky, Mike Streat, Charlie O'Melia, Wolfgang Hoell, David Sherrington, Spiro Alexandratos, Robert Albright, Steve Cramer, Menachem Elimelech, Ruslan Khamizov, Zdenek Matezka, Mimo Petruzzelli, Nalan Kabay, Kesava Rao, Gary Foutch and others. I am thankful to Jacob Brodie and Francis Boodoo for their continued cooperation with material support pertaining to our research efforts in environmental separation. The electron microscopy work of Debra Phillips for Hybrid Ion Exchanger-Nanotechnology is gratefully acknowledged. I sincerely acknowledge the US Department of State, US Fulbright Program, the Department of Science and Technology of the Government of India, WIST, Inc., Rite Water Solutions (I) Ltd. and Technology with a Human Face (NGO) for their support and assistance toward field-level implementation of ion exchange technologies invented in Lehigh University.

However, more than anything, I am most grateful to my graduate students and post-docs with whom I have worked closely for over three decades. Since I may not have many more opportunities, I would like to recognize them by name who have made meaningful contributions to push the frontiers of ion exchange science and technology inch by inch through their research: Yuewei Zhu, Sukalyan Sengupta, Anu Ramana, Yi-min Gao, Ping Li, Indra Mitra, Dongye Zhao, Esmeralda Millan, Matthew

DeMarco, David Leun, Luis Cumbal, Arthur Kney, John Greenleaf, Parna Mukherje, Sudipta Sarkar, Prakhar Prakash, Lee Blaney, Prasun Chatterjee, Surapol Padungthon, Ryan Smith, Mike German, Yu Tian, Jinze Li, Chelsey Shepsko, Robert Creighton and Hang Dong. Most of them started as students, but down the stretch, most of them became mature, thoughtful and innovative in their own rights. I sincerely believe that the knowledge acquisition has truly been a two way process and the students have enriched my professional life. It is likely that some names may have been omitted but that is unintentional and I offer my sincere apology in advance.

During the last four years, Beth Yen, the department secretary, unfailingly responded to my every request – be it copying, typing, scanning, editing or even running an errand, and often with time constraints due to poor planning on my part. I am immensely thankful for her cooperation and continued service.

For my education, from the second grade in the elementary school in India to my PhD in the US, I never paid any tuition. It was *gratis* all the way for my entire student career. Now I know that ordinary people, who pay taxes or are undercompensated, truly funded my education. I consider myself immensely fortunate and blessed.

I acknowledge continued cooperation from Wiley, the publisher of the book, and I am thankful to Saleem Hameed, Beryl Mesiadhas and Michael Leventhal for attending to necessary details and bringing the book project to a successful closure.

Last but by no means the least, without the incessant help and involvement of Michael German, this book could not be brought to a successful completion. In addition to carrying out his regular duties as a senior PhD student, Mike relentlessly responded to various details about the book project – from completing figures to collecting copyright permissions and many other associated pieces of work in between. Mike helped me overcome the activation energy barrier with his unselfish effort and I am indeed indebted to him.

1

Ion Exchange and Ion Exchangers: An Introduction

1.1 Historical Perspective

Evolution is traditionally viewed to occur in a slow but continuous manner for living organisms and creatures gradually acquiring new traits. To the contrary, many areas of "science" undergo periods of rapid bursts of fast development separated by virtual standstill with no significant activity. The first historically recorded use of ion exchange phenomenon is from the Old Testament of the Holy Bible in Exodus 15:22–25 describing how Moses rendered the bitter water potable by apparently using the process of ion exchange and/or sorption. Another often quoted ancient reference is to Aristotle's observation that the salt content of water is diminished or altered upon percolation through certain sand granules. From a scientific viewpoint, however, the credit for recognition of the phenomenon of ion exchange is attributed to the English agriculture and soil chemists, J.T. Way and H.S. Thompson. In 1850, these two soil scientists formulated a remarkably accurate description of ion exchange processes in regard to removal of ammonium ions from manure by cation exchanging soil [1,2]. They essentially simulated the following naturally occurring cation exchange reactions as follows:

$$NH_4^+(aq) + Na^+(soil) \leftrightarrow NH_4^+(soil) + Na^+(aq) \tag{1.1}$$

$$2NH_4^+(aq) + Ca^{2+}(soil) \leftrightarrow (NH_4^+)_2(soil) + Ca^{2+}(aq) \tag{1.2}$$

Some of the fundamental tenets of ion exchange resulted from this work: first, the exchange of ions differed from true physical adsorption; second, the exchange of ions involved the exchange in equivalent amounts; third, the process is reversible and fourth, some ions were exchanged more favorably than others.

As often with many groundbreaking inventions, the findings of Way and Thompson cast doubts, disbeliefs and discouragement from their peers. In the following years, these two soil scientists discontinued persistent research in this field. As a result, the evolution of ion exchange process progressed rather slowly due to the difficulties in modifying or manipulating naturally occurring inorganic clayey materials with low cation exchange capacities.

Inorganic zeolites (synthetic or naturally occurring aluminosilicates) later found wide applications in softening hard waters, that is, removal of dissolved calcium and magnesium through cation exchange. However, the anion-exchange processes remained unexplored and practically unobserved. Even at that time, it was not difficult to conceptualize that the availability of both cation exchangers and anion exchangers

Ion Exchange in Environmental Processes: Fundamentals, Applications and Sustainable Technology, First Edition. Arup K. SenGupta.

in the ionic forms of hydrogen and hydroxyl ions, respectively, would create a new non-thermal way to produce water free of dissolved solids as indicated below:

$$\begin{aligned} &H^+(\text{solid}) + OH^-(\text{solid}) + Na^+(\text{aq}) + Cl^-(\text{aq}) \\ &\quad \leftrightarrow H_2O(\text{aq}) + Na^+(\text{solid}) + Cl^-(\text{solid}) \end{aligned} \tag{1.3}$$

The biggest obstacle to realize this concept was to identify and/or synthesize ion exchangers which will be chemically stable and durable under the chemically harsh environments at very high and low pH. The immense potential of ion exchange technology scaled a new height when the first organic-based (polymeric) cation exchanger was synthesized by Adams and Holmes [3]. In less than ten years, D'Alelio prepared the first polymeric, strong/weak cation and anion exchangers [4–6]. Since then, synthesis of new ion exchangers never seemed to slow down and application of ion exchange technology in industries as diverse as power utilities, biotechnology, agriculture, pharmaceuticals, pure chemicals, microelectronics, etc. are continually growing. No specialty grows in isolation; ion exchange fundamentals, ion exchange resins and ion exchange membranes continue to find new and innovative applications globally. Figure 1.1 includes the number of ion exchange related US patents issued during the last three decades, illustrating continued inventions in new products and processes.

Ironically, the Second World War and, more specifically, the race for nuclear technology helped catalyze the growth and maturity of the field of ion exchange at an accelerated pace. Ion exchange was found to be a viable process for separating some of the transuranium elements and, for understandable reasons, its application aroused a great deal of interest. In fact, some of the most fundamental works on ion exchange equilibria and kinetics were carried out during the Second World War period by Boyd *et al.* and reported afterwards in the open literature [9–11]. All along, the scientific understanding of ion exchange fundamentals consistently lagged well behind its applications. Table 1.1 attempts to summarize milestones in regard to the development and application of ion exchange technology over time.

Figure 1.1 Number of patents per year for "anion exchange" and "cation exchange" per a Google Patents search. *Source:* Data taken with permission from Google [7,8].

Table 1.1 Historical milestones in ion exchange.

Year	Description	Patent #	Authors
1850	Discovery of ion exchange properties of soil	N/A	Thompson and Way [1,2]
1876	Zeolites or aluminosilicates recognized for base exchange and equivalence of exchange is proved	N/A	Lemberg [12,13]
1906–1915	Industrial manufacture of sodium permutit for hardness removal	914,405; 943,535; 1,131,503	Gans [14]
1934	Invention of sulfonated condensation polymers as cation exchangers	2198378A	Ellis
1935	First synthetic organic ion exchangers	2104501A, 2151883A	Adams and Holmes [15]
1938	Mixed-bed ion exchange process or duplex ion exchanger	2275210A	Stemen, Urbain, and Lewis
1939	Invention of sulfonated polystyrene polymerization as cation exchangers	2283236A	Soday
	Invention of aminated polystyrene polymerization as anion exchangers	2304637A	Vernal
1942	Cation exchange resin beads made from polymerized acrylic acids	2340110A,	D'Alelio
	Cation exchange resins with sulfonated, polymerized poly-vinyl aryl parent resin	2340111A 2366007A	
	Anion exchange resins with aminated, polymerized poly-vinyl aryl parent resin	2366008A	
1947	Element 61 (Promethium) was discovered by ion exchange of the by-products of fission	N/A	Marinsky, Glendenin, and Coryell [16]
1953	Use of zeolites as molecular sieves	2882243A	Milton
	Magnetic ion exchange resin for NOM removal (MIEX process)	2642514A	Herkenhoff
	Invention of weak acid cation exchangers	2838440A	Thurmon
	First countercurrent ion exchange using suspended/agitated beds of resin	N/A	Swinton and Weiss [17]
1954	Higgins countercurrent ion exchange contactor invented	2815322A	Higgins [18]
1955	Ligand exchange	2839241A	Albisetti
1956	Pellicular ion exchange resin	2933460A	Richter and McBurney
1958	Agitated bed contactor for semicontinuous ion exchange	N/A	Arden, Davis, and Herwig [19]
	Ion exchange in drug delivery	2990332A	Keating

(Continued)

Table 1.1 (Continued)

Year	Description	Patent #	Authors
1958 (publicly released)	Uranium separation, intraparticle diffusion (Manhattan Project)	2956858A	Powell
1959–1960	The book on "Ion Exchange" by Friedrich Helfferich was printed and laid the theoretical foundations for the field of ion exchange	N/A	Helfferich [20]
1962–1971	Cloete–Streat countercurrent contactor invented	3551118A (1962) 3738814A (1969) 3957635A (1971)	Cloete and Streat [21]
1964	Cellulosic ion exchange fibers synthesized	3379719A	Rulison
1965	Sirotherm process – thermally regenerable ion exchange resins	274-029; 59,441/65 (Australia)	Bolto, Weiss, and Willis
	Partially functionalized cation exchange (shallow-shell technology)	3252921A	Hansen and McMahon
1966	Macroporous ion exchange resin	3418262A	Grammont and Werotte
1968	Boron selective resin	20110108488A1	Chemtob
1969	Development of poly(methyl methacrylate) anion exchange resins or macroreticular polymers that reduced fouling by natural organic manner	N/A	Kressman and Kunin [22,23]
1971	Continuous moving bed ion exchange	3751362A	Probstein, Schwartz, and Sonin
1972	Phenolic ion exchange fibers	3835072A	Economy and Wohrer
1973	Iminodiacetic acid chelating resin	3936399A	Hirai, Fujimara, and Kazigase
	Metal-selective biosorbents	CA1036719A1	Stamberg, Prochazka, and Jilek
1973	Iminodiacetic acid chelating resin	3936399A	Hirai, Fujimara, and Kazigase
	Metal-selective biosorbents	CA1036719A1	Stamberg, Prochazka, and Jilek
1975	"Himsley contactor" multistage fluidized bed continuous counter-current ion exchange contactor	CA980467A1	Himsley
1976	Solvent impregnated resins	4220726A	Warshawsky *et al.* [24,25]

1980	Monosphere ion exchange resin (Dow Chemical Co.)	4444961A	Timm
1979	Ion exchange induced supersaturation (IXISS)	N/A	Muraviev [26,27]
1981	Radium selective resins	EP0071810 A1	Hatch
1983	Nitrate selective resin	4479877A	Guter
	CARIX (carbon dioxide regenerated ion exchange) process for brackish water desalination	EP0056850 B1	Kiehling and Wolfgang
1985	Short bed ion exchanger	EP0201640 B2	Brown
1990	Selective anion exchange for gold from cyanide solution with a simple and straightforward chemical regeneration	N/A	Schwellnus and Green [28]
1991	Bifunctional ion exchange resins (Diphonix)	EP0618843 A1	Alexandratos, Chiarizia, and Gatrone
1997	Polymeric ligand exchange	6136199A	SenGupta and Zhao
2003	Fluoride selective resins: strong acid cation exchange resin in aluminum form	WO2005065265A2	Jangbarwala and Krulik
2004	Donnan principle-based hybrid ion exchanger	7291578 B2	SenGupta and Cumbal
	(Ion Exchange) Membrane capacitive deionization (MCDI)	6709560 B2	Andelman and Walker
2007	Macroporous copolymers with large pores (0.5–200 μm)	20080237133A1	Dale, Sochilin, and Froment
2008	Rapid sensing of toxic metals with hybrid inorganic materials	WO2008151208A1	Chatterjee and SenGupta
2009	Removal of alkyl iodides by strong acid cation exchange resin loaded in Ag^+-form	7588690B1	Tsao
2010	Separation of ionic aqueous mixtures with ion exchange materials in an immiscible organic phase	8940175B2	Khamizov
2013	Hybrid ion exchange-reverse osmosis processes	WO2014193955A1	SenGupta and Smith
	Fluoride-selective resins: hybrid anion exchange resin with zirconium oxide nanoparticles	20130274357A1	SenGupta and Padungthon

Note: Patents are issued from the USA, unless mentioned otherwise.

1.2 Water and Ion Exchange: An Eternal Kinship

Ion exchange is a heterogeneous process where water, the most abundant polar solvent in our planet, is inevitably present. Even the ion exchange processes involving gases or solids require the presence of water. It is imperative that we understand the fundamental properties of water in order to follow the science of ion exchange. Oxygen is present in Group VIA of the periodic table and water (H_2O) is essentially a dihydride of oxygen. Note that sulfur (S) and selenium (Se) are also in the same group with oxygen but their dihydride, namely H_2S and H_2Se are volatile at room temperature. In contrast, water is liquid and an excellent solvent for salts with ionic bonds. In the electronegativity scale, hydrogen and oxygen are far apart. While hydrogen is electropositive, oxygen is strongly electronegative. Thus, covalent O—H bonds in water molecules are polar due to unequal sharing of bonding electrons with residual negative and positive charges on oxygen and hydrogen atoms, respectively. Hence, water molecules are essentially dipoles (dipole moment = 1.85 D), as shown in Figure 1.2a. The electronic structure of the water molecule corresponds to the tetrahedral arrangement with the oxygen atom having two lone pairs of electrons as presented in Figure 1.2b. The dipolar water molecules experience a torque when placed in an electric field and this torque is called a dipole moment. When molecules have dipole moments, their intermolecular forces are significantly greater, especially when dipole–dipole interactions or hydrogen bonding is possible. Water molecules are particularly well suited to interact with one another because each molecule has two polar O—H bonds and two lone pairs on the oxygen atom. This can lead to the association of four hydrogen atoms with one oxygen through a combination of covalent and hydrogen bonding as shown in Figure 1.3. Water molecules thus exist as trimers (H_6O_3) and boiling requires a high heat of vaporization to break the intermolecular hydrogen bonds among water molecules. Thus, water has the highest boiling point among the entire Group VIA hydrides as shown in Figure 1.4.

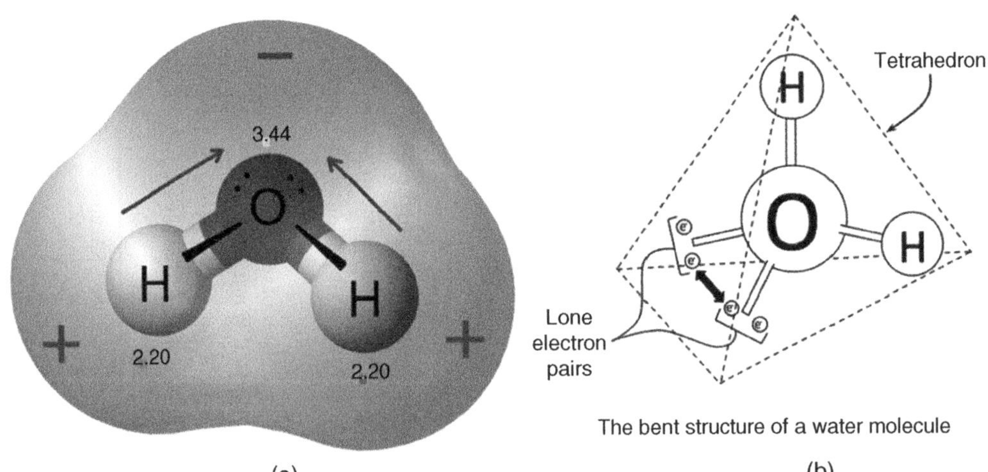

Figure 1.2 Shape of water molecules (a) Dipolar O—H bonds with electronegativity values; (b) Electronic structure with tetrahedral arrangement.

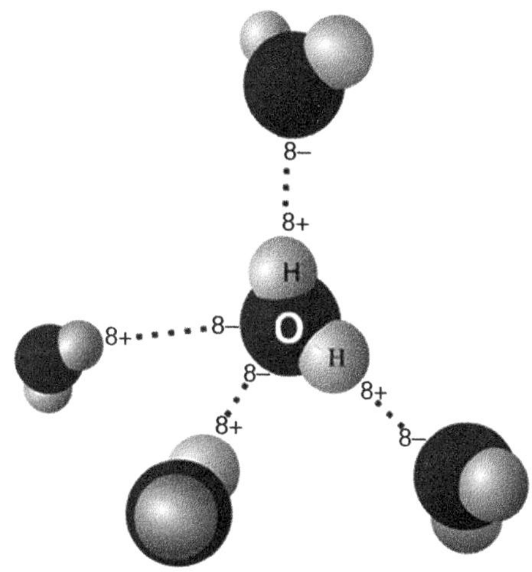

Figure 1.3 Interaction of water molecules through association of four hydrogen atoms with each oxygen atom.

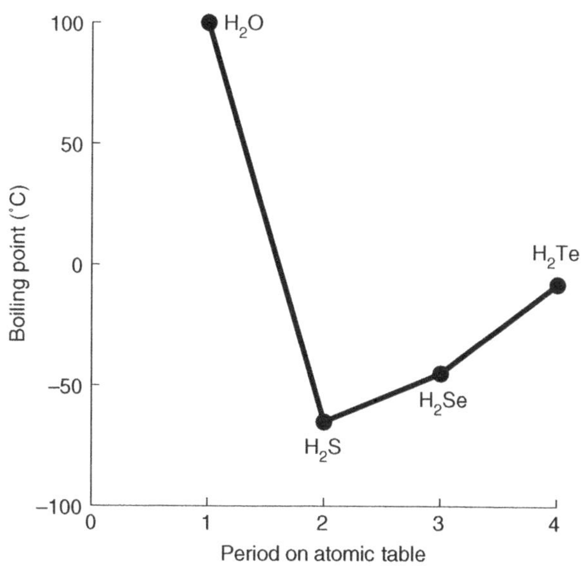

Figure 1.4 Anomalous boiling point behavior of H_2O in Group VIA hydrides.

Like dissolves like. Ionic compounds such as sodium chloride (NaCl) are highly soluble in water, which is an excellent polar solvent. When sodium chloride is added to water, the dipolar water molecules separate sodium from chloride ions forming a cluster of solvent molecules around them due to the ion–dipole interaction as presented in Figure 1.5. This interaction is known as hydration and the hydrated ionic radius of an ion is always greater than its ionic radius. The degree of hydration depends primarily on the charges and the atomic mass of the ions. Ions with higher charges, and similar masses, always are more hydrated, that is, divalent calcium ion (Ca^{2+}) is more hydrated than monovalent sodium ion (Na^+). For monatomic ions with identical charges, hydrated ionic radius increases with a decrease in atomic mass or crystal

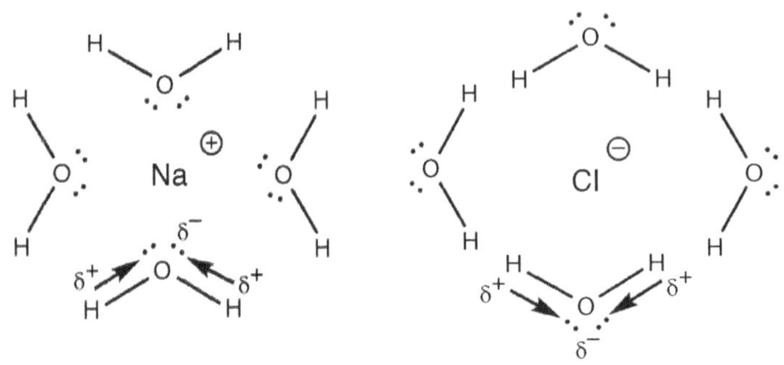

Figure 1.5 Illustration of ion–dipole interaction: Sodium chloride (ionic compound) solution in water (polar solvent).

Table 1.2 Hydrated ionic radius and atomic mass of typical monatomic ions of interest.

Ions	Atomic mass	Crystal ionic radii (pm)	Hydrated ionic radii (pm)
Li^+	6.94	59	382
Na^+	22.99	102	358
K^+	39.09	151	331
Rb^+	85.46	161	329
F^-	18.99	133	352
Cl^-	35.45	181	332
Br^-	79.9	196	330
Be^{2+}	9.01	27	459
Mg^{2+}	24.3	72	428
Ca^{2+}	40.07	100	412
Sr^{2+}	87.62	126	412
Ba^{2+}	137.33	142	404

Source: Conway 1981 [29]. Reproduced with permission of Elsevier.

ionic radius as illustrated in Table 1.2. Since the process of heterogeneous ion exchange inevitably involves hydrated ions, the following observations are universally true:

(i) Binding of an ion onto a rigid ion exchanger requires partial shedding of water of hydration and hence, all other conditions remaining identical, an ion with lower hydrated ionic radius shows higher affinity. For example, both K^+ and Na^+ are monovalent cations, but K^+ is preferred over Na^+ by cation exchange resins due to its lower hydrated ionic radius.

(ii) An ion with a larger hydrated ionic radius is less mobile, that is, it has a lower diffusion coefficient. The kinetics of ion exchange are often a diffusion-controlled process. Thus, binding of an ion with a higher hydrated ionic radius is always a kinetically slower process.

1.3 Constituents of an Ion Exchanger

An ion exchanger is ideally defined as a framework of fixed coions, which can be permeated and electrically neutralized by mobile counterions from the aqueous (liquid) phase. The underlined terms in the foregoing definition require further elaboration.

FRAMEWORK is much like a skeleton that constitutes a continuous phase, which is held together by covalent bonds or lattice energy. For polymeric ion exchangers, covalent bonds predominate and the framework is often referred to as the matrix. In inorganic ion exchangers, the lattice energy helps retain the ion exchange sites in the solid phase and the framework is constituted by amorphous or crystalline structures. FIXED COIONS are electric surplus charges (positive or negative) on the framework, or the matrix, unable to leave their phase. This surplus charge is due to covalent bonding for polymeric ion exchangers and isomorphous substitution for zeolites and clays. MOBILE COUNTERIONS are solutes with charges opposite to the fixed coions. They compensate the charges of fixed coions in the exchanger phase and can also be replaced by other ions of the same sign on an equivalent basis. Unlike fixed coions, the counterions can permeate in and out of the exchanger phase and by doing so, they maintain electroneutrality in both the liquid and the solid phase.

For synthetic ion exchangers, fixed coions are known as functional groups or ionogenic groups, while the exchanging ions are known as counterions. To readily grasp the underlying concept without loss of generality, let us consider a polymeric ion exchanger where the three-dimensional cross-linked polymer constitutes a separate insoluble phase or matrix. The covalently attached functional group is essentially the fixed coion that is permeated and electrically balanced by an exchangeable counterion. Figure 1.6 shows a simple schematic of a cation exchanger with sulfonic acid functional groups loaded with sodium counterions.

Thermodynamically, the activity or concentration of an ion exchanger is not a unique number, but it varies with the type and concentration of the counterion in the

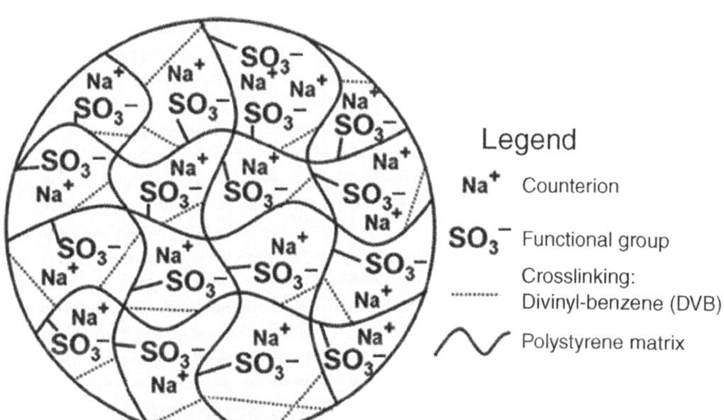

Commonly represented as: **$R\text{-}SO_3\text{–}Na^+$**

Figure 1.6 Schematic illustration of a strong acid cation exchange resin bead where matrix/framework is represented by R, fixed coions or functional groups by $—SO_3^-$ and counterions/exchanging ions by Na^+.

exchanger phase. However, the fixed coions in an ion exchanger are always balanced by permeating counterions, that is, the ion exchanger is always electrically neutral. Ideally, the ion exchange capacity is equal to the concentration of the fixed coions. We will later see that the capacity is not a constant and it depends, to some extent, on the external liquid phase concentration.

To be familiar with the basic premise and terminologies of ion exchange processes, let us consider the following cation exchange reaction between potassium and sodium ions:

$$\overline{(R - SO_3^-)Na^+} + K^+(aq) + Cl^-(aq) \leftrightarrow \overline{(R - SO_3^-)K^+} + Na^+(aq) + Cl^-(aq) \quad (1.4)$$

where the overbar denotes the exchanger phase; sulfonic acid functional group ($—SO_3^-$) is the fixed, non-diffusible coion and Na^+ and K^+ are the permeable or exchanging counterions. The chloride ion does not participate in the cation exchange reaction and is referred to as a mobile coion. Both the exchanger and aqueous-phase electroneutrality remain undisturbed at every stage of the cation exchange reaction. Likewise, the anion exchange process is fundamentally the same, but the exchanger phase has positively charged fixed coions (e.g., quaternary ammonium functional groups, R_4N^+) as shown for the nitrate-chloride exchange reaction below:

$$\overline{(R_4N^+)Cl^-} + NO_3^-(aq) + Na^+(aq) \leftrightarrow \overline{(R_4N^+)NO_3^-} + Cl^-(aq) + Na^+(aq) \quad (1.5)$$

While NO_3^- and Cl^- are the permeating counterions, R_4N^+ and Na^+ are the fixed and mobile coions, respectively.

1.4 What is Ion Exchange and What is it Not?

Prior to getting into the details of the various materials presented in this book, it is imperative that we present a scientifically coherent definition of what we call "ion exchange." A list of reactions, as shown below, are often mistakenly presented in the open literature as ion exchange simply because the process appears to involve an exchange of equivalent amounts of cations or anions:

Pseudo-cation exchange:

$$FeS(s) + Cu^{2+}(aq) \leftrightarrow CuS(s) + Fe^{2+}(aq) \quad (1.6)$$

$$Fe^{2+}(aq) + Zn^0(s) \leftrightarrow Fe^0(s) + Zn^{2+}(aq) \quad (1.7)$$

Pseudo-anion exchange:

$$BaCO_3(s) + SO_4^{2-}(aq) \leftrightarrow BaSO_4(s) + CO_3^{2-}(aq) \quad (1.8)$$

These are essentially precipitation–dissolution and redox reactions involving a pure solid phase denoted by "(s)." Since the activity of a pure independent solid phase (e.g., crystalline) is unity, the equilibrium constant of Reaction 1.6, considering ideality, is given by

$$K = \frac{[Fe^{2+}]}{[Cu^{2+}]} \quad (1.9)$$

All the foregoing reactions are identical in the sense that the equilibrium constants are influenced only by the dissolved species and are independent of the composition of the pure solid phases. Ion exchange phenomena are distinctly different from the above in this regard. An ion exchanger is a separate phase from the aqueous solution with a different dielectric constant and the exchanging counterions can be present at varying proportions to produce a continuous solid solution. The thermodynamic activity of an ion exchanger phase is not equal to unity, but is dependent on its composition. For the cation exchange reaction in (1.4), the idealized equilibrium constant is

$$K_{IX} = \frac{[\overline{RK^+}][Na^+]}{[\overline{RNa^+}][K^+]} \tag{1.10}$$

where overbar with a bracket represents the exchanger phase molar concentration while the bracket alone represents the aqueous phase concentration. Note that ion-exchanger phase activity is not unity and its relative proportion of Na^+ or K^+ will vary with the extent of each ion exchange reaction. Also, the mobile coion, Cl^-, does not influence the ion exchange equilibrium constant, K_{IX}. Mole or equivalent fraction (they are the same for monovalent ions) of Na^+ or K^+ in the exchanger phase is given by:

$$y_{Na} = \frac{[\overline{RNa^+}]}{[\overline{RNa^+}] + [\overline{RK^+}]} \tag{1.11}$$

$$y_K = \frac{[\overline{RK^+}]}{[\overline{RNa^+}] + [\overline{RK^+}]} \tag{1.12}$$

Since sodium and potassium are the only counterions present in the exchanger, the total capacity, Q, of the cation exchange is

$$Q = [\overline{RNa^+}] + [\overline{RK^+}] \tag{1.13}$$

Therefore,

$$y_{Na} = \frac{[\overline{RNa^+}]}{Q} \tag{1.14}$$

$$y_K = \frac{[\overline{RK^+}]}{Q} \tag{1.15}$$

$$y_{Na} + y_K = 1.0 \tag{1.16}$$

Thus, the equivalent fraction of Na^+ or K^+ (y_{Na} or y_K) in the exchanger phase is free to vary from zero to unity in accordance with Eq. (1.16). An ion exchanger, be it inorganic, polymeric or liquid, is essentially a separate phase or continuum, the composition of which can vary due to ion exchange reaction. An ion exchanger is thus distinctly different from a pure solid phase of single chemical composition. Instead, for an insightful understanding of diverse ion exchange phenomena, an ion exchanger may be viewed as a condensed and cross-linked polyelectrolyte where the anions (for a cation exchanger) or cations (for an anion exchanger) are immobilized and cannot permeate out of the condensed state.

1.5 Genesis of Ion Exchange Capacity

1.5.1 Inorganic

In accordance with the generalized definition of an ion exchanger, fixed coions are the true origins of ion exchange capacity. From a historical perspective, naturally occurring inorganic silicate minerals were the first materials to be studied for their ion exchange or, more specifically, cation exchange behavior. In such naturally occurring crystalline silicate materials with three-dimensional Si—O chains, a silicon atom, having an oxidation state of four, is often replaced by an aluminum atom having an oxidation state of three. Thus, there is a charge deficiency (excess negative charge) in the crystalline lattice at the defect location. To preserve electroneutrality, this deficiency must be balanced by the presence of a cation. It is this cation that becomes the exchangeable counterion. The above-mentioned defects are truly the seats of fixed coions. The higher the number of such defects per unit mass or volume in the silicate phase, the greater will be the cation exchange capacity. The process of such defect formation is often referred to as "Isomorphous Substitution." Since aluminum and silicon are the two most abundant elements in soil after oxygen, such substitutions are widespread in natural minerals, and these materials are often called zeolites.

Figure 1.7 provides a general schematic showing the formation of fixed coions through isomorphous substitutions in naturally occurring silicate phase or zeolites. It is to be noted that the substitution of Mg(II) for Al(III) gives rise to the same effect (i.e., generation of excess negative charges) as the substitution of Al(III) for Si(IV). The general stoichiometry of such silicate based inorganic ion exchangers or zeolites is given empirically as $M_{2/n}O \cdot Al_2O_3M_{2/n} \cdot xSiO_2 \cdot yH_2O$ where M is a cation of valence n (commonly $n = 1$ or 2) and x and y are integer values of coefficients.

The zeolites such as chabazite ($CaAl_2Si_6O_{16} \cdot 8H_2O$) and analcite ($Na_2O \cdot Al_2O_3 \cdot 4SiO_2 \cdot 2H_2O$) are essentially crystalline silicates with defects (fixed charges) to which sodium or calcium ions (counterions) are easily accessible through a three-dimensional network of pores. In the latter part of the nineteenth century, it was demonstrated that the zeolite mineral analcite could be converted stoichiometrically into leucite [$K(AlSi_2O_6)$] simply by leaching with an aqueous solution of potassium chloride, a synthesis step driven solely by ion exchange.

$$Na(AlSi_2O_6) \cdot H_2O(\text{Analcite}) + K^+ \leftrightarrow K(AlSi_2O_6)(\text{Leucite}) + H_2O + Na^+ \tag{1.17}$$

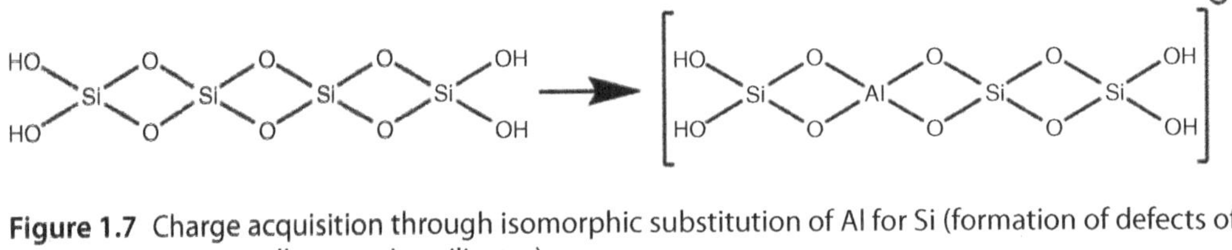

Figure 1.7 Charge acquisition through isomorphic substitution of Al for Si (formation of defects of fixed coions in naturally occurring silicates).

Table 1.3 List of some common inorganic ion exchangers.

Type of ion exchanger	Example
Smectite clays	Montmorillonite: $M^{n+}_{x/n}[Al_{4-x}Mg_x]Si_8O_{20}(OH)_4$
Zeolites	$Na_x(AlO_2)_x(SiO_2)_y \cdot zH_2O$
Substitute Aluminum Phosphates	Silica aluminum phosphates; metal aluminum phosphates $(M^{n+}_x Al_{1-x}O_2)(PO_2)(OH)_{2x/n}$
Phosphates of Gr IV elements	$Zr(HPO_4)_2 \cdot H_2O$; $Sn(HPO_4)_2 \cdot H_2O$
Metal oxides	$Fe_2O_3 \cdot xH_2O$; $ZrO_2 \cdot xH_2O$; $Al_2O_3 \cdot xH_2O$
Ferrocyanides	$M^{n+}_{4/n}Fe(CN)_6$, $M = Ag^+, Zn^{2+}, Cu^{2+}, Zr^{4+}$
Titanates	$Na_2Ti_nO_{2n+1}$; $n = 2–10$
Apatites	$Ca_{10-x}H_x(PO_4)_6(OH)_{2-x}$
Heteropolyacid salts	$M_nXY_{12}O_{40} \cdot xH_2O$; ($M = H, Na^+$; $X = P, As, Ge, Si, B$; $Y = Mo, W$)
Fast ion conductors	β-Aluminum oxides, $Na_{1+x}Al_{11}O_{17+x/2}$; NASICON, $Na_{1+x}Zr_2Si_xP_{3-x}O_{12}$
Anion exchanger	Hydrotalcite – $Mg_6Al_2(OH)_{16}CO_3 \cdot 4H_2O$

In recent years, zeolites with regular crystal structures have been synthesized and applied as ion-exchangers, catalysts and molecular sieves. Although chemical compositions of inorganic ion exchangers may be quite diverse, they are typically mixed metal oxides, insoluble salts of polyvalent metals and metal ferrocyanides [20,30,31]. Amorphous structures do exist, but inorganic ion exchangers are mostly crystalline polymers with a microporous framework. Table 1.3 provides a list of some common inorganic ion exchangers.

As ion exchangers, zeolites are of minor significance due to their chemical instability and poor regenerability. However, due to their narrow, rigid and strictly uniform pore structure, the zeolites act as "molecular sieves" and are capable of selectively sorbing molecules lower than specific sizes, while rejecting larger ones. Several types of molecular sieves are now commercially available both as microcrystalline powders and as pellets which consist of microcrystals in a clay binder [30,32,33]. Linde Sieve Type X and Type A have pore diameters of about 10 and 5 Å, respectively. Figure 1.8 shows structures of Zeolite A and Zeolite X and their cavities. Figure 1.9 illustrates how molecular sieves can effectively separate straight chain organic molecules from their branched-chain counterparts [33]. Since molecular sieves are essentially cation exchangers, the pore sizes can be adjusted to a certain degree by converting the materials into different ionic forms, resulting in other potential applications [31,35,36].

1.5.2 Organic/Polymeric Ion Exchanger

The advent of ion exchange technology began with the preparation and large-scale synthesis of polymeric ion exchangers, commonly referred to as ion exchange resins.

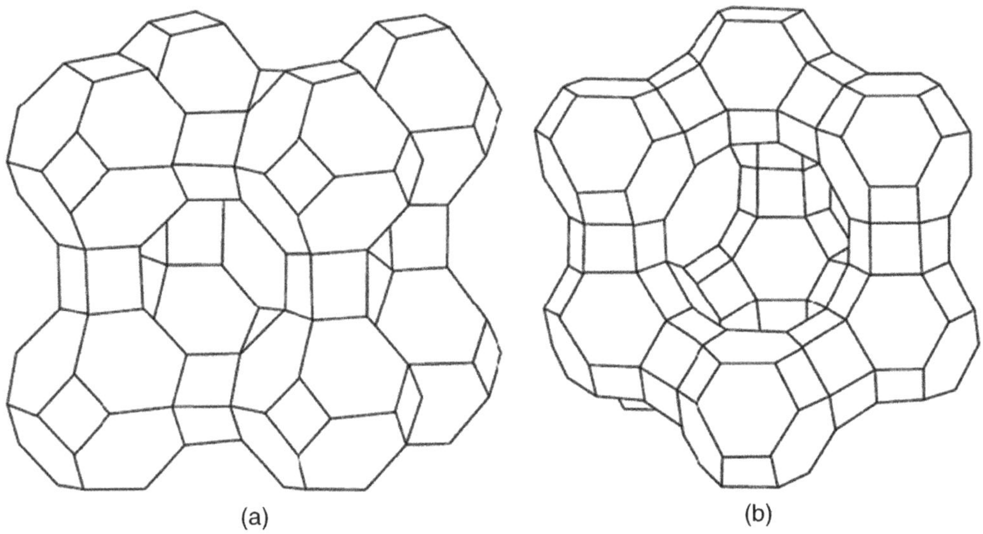

Figure 1.8 Zeolite A (a) and faujasite-type zeolites X and Y (b) formed by sodalite cages. *Source:* From Lutz 2014 [34].

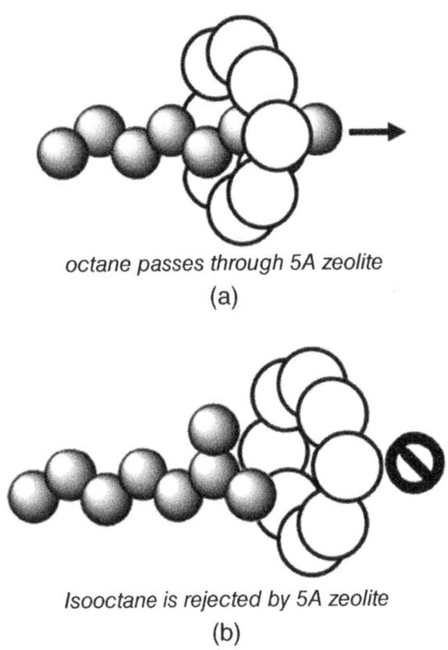

Figure 1.9 Illustration of molecular sieves for the separation of straight chain organic molecule. Straight chain normal octane molecule (a) passes through the eight-ring aperture of 5A zeolite; branched molecule of isooctane cannot (b). *Source:* From Bekkum *et al.* 1991 [33].

Ion exchange resins are cross-linked polyelectrolytes and their ion exchange capacities originate from functional groups covalently attached to the matrix or framework. Although details pertaining to chemical synthesis of polymeric ion exchange resins is beyond the scope of this book, Figures 1.10–1.13 provide the step-wise preparation of the four most widely used ion exchange resins, namely, weak-acid cation, strong-acid cation, weak-base anion and strong-base anion exchangers. The four ionogenic

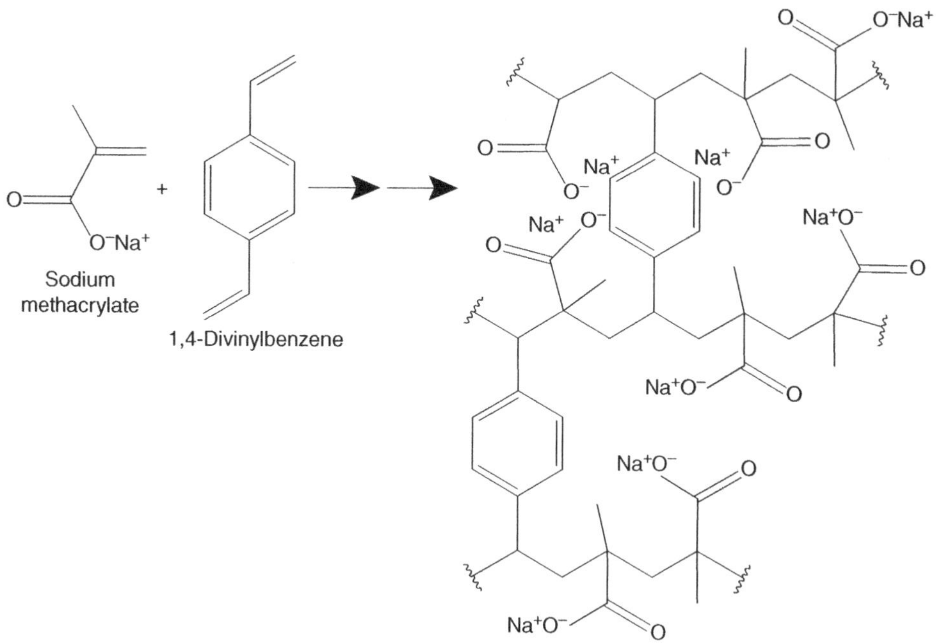

Figure 1.10 Synthesis of weak-acid cation exchanger by polymerization of sodium methacrylate with divinyl-benzene crosslinking.

groups or fixed coions are: carboxylic ($R - COO^-$), sulfonic ($R - SO_3^-$), tertiary amine ($R - N^+R_2H$) and quaternary ammonium ($R - N^+R_3$).

It is only appropriate to note some distinct differences in the synthesis process of these ion exchangers:

- For the weak-acid cation exchanger, the carboxylate ionogenic group is already present in the repeating methacrylic acid monomer prior to polymerization. Divinylbenzene or DVB introduces the cross-linking to attain the three-dimensional polymer network during the single-step synthesis process.
- For the strong-acid cation exchanger, the styrene–divinylbenzene copolymer is synthesized first and then sulfonated to introduce the ionogenic groups. The synthesis is a two-step process.
- Synthesis of both strong and weak-base anion exchange resins involve three consecutive steps; first synthesis of styrene–divinylbenzene copolymer; second, chloromethylation of the copolymer; and third, amination of the chloromethylated copolymer, culminating in positively charged functional groups.

The repeating unit for each ion exchanger without divinylbenzene cross-linking is provided in Table 1.4. Theoretical ion exchange capacity for an ion exchanger can be calculated from the molecular weight of the repeating unit. Example 1.1 shows theoretically calculated ion exchange capacity of each type of polymeric ion exchanger and discusses why the real ion exchange capacity of an anion exchanger is significantly less than its theoretical value. Since ion exchange takes place on an equivalent basis, the capacities should preferably be expressed in equivalent units such as equivalent/liter

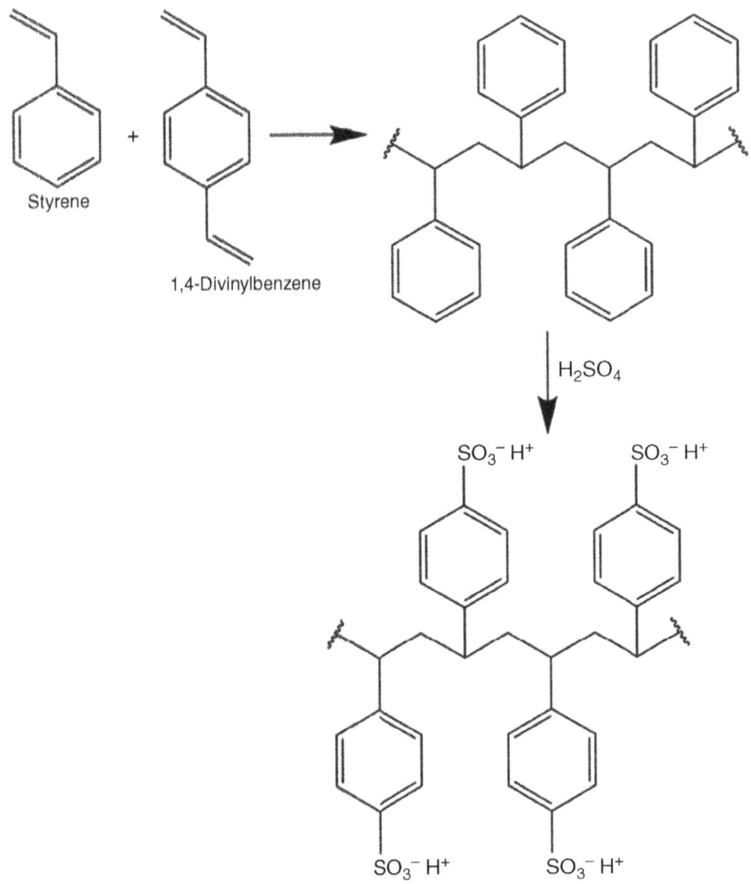

Figure 1.11 Synthesis of strong-acid cation exchange resin by polymerization of styrene with cross linking of divinyl benzene followed by sulfonation.

(eq/L) or milliequivalents/gram (meq/g). The commonly used abbreviated dimensions are eq/L, meq/mL, and meq/g. To avoid confusion, universally accepted conventions should be followed while expressing ion exchange capacity. The ionic form, that is, the specific type of counterion present will change the mass of the ion exchanger, all other conditions remaining identical. For example, the equivalent weight of sodium (Na^+) is 23 while that of lead (Pb^{2+}) is 103. Thus, the same ion exchanger (inorganic or organic) while loaded with lead will weigh much more than when it is in sodium form, that is, its specific gravity will increase. So, equivalent capacity per unit mass will vary. Capacity expressions for engineered applications thus need to be consistent with the ionic forms of the ion exchangers. In general, the capacity of strong-acid cation exchanger is expressed in Na-form, weak-acid cation in H-form, strong-base anion in chloride form and weak-base anion in OH or free base form.

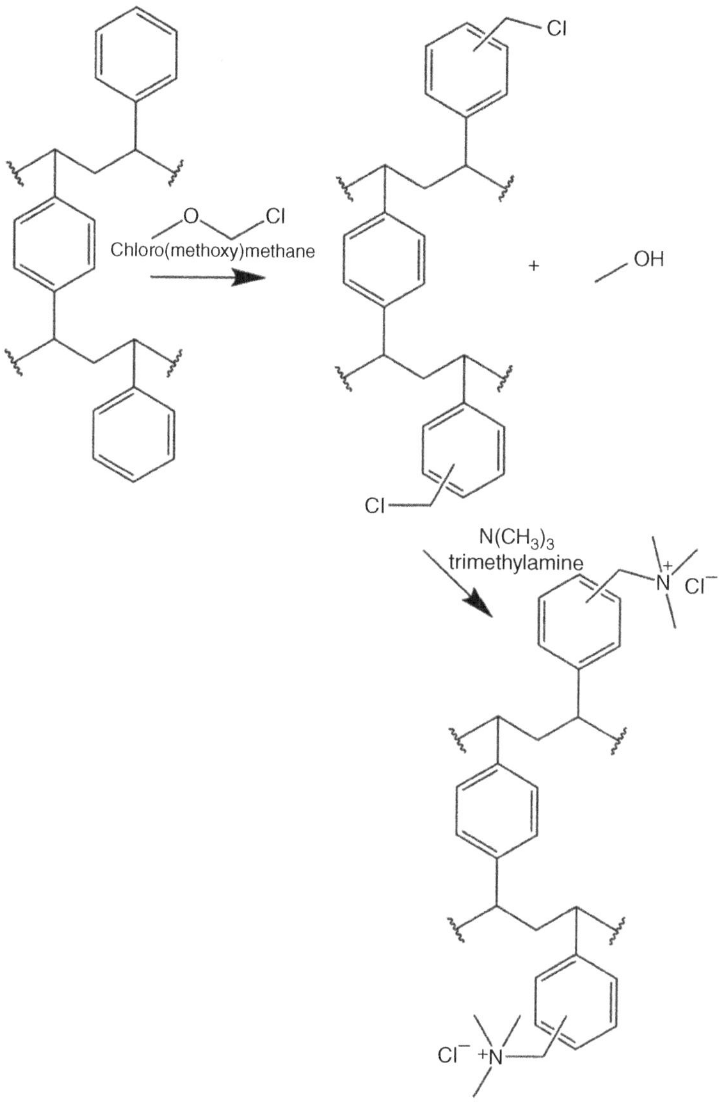

Figure 1.12 Synthesis of strong-base anion exchange resin through chloromethylation followed by amination with tertiary amine.

Example 1.1 *Compare the relative capacity of each ion exchanger using the information in Table 1.4 and explain any discrepancy with respect to anion exchangers. Comment on any anomaly.*

Weak-acid cation (WAC) exchanger with carboxylate functional group: For every ion exchange site, the repeating unit from Figure 1.10 contains – C: 4, O: 2, H: 5; Corresponding molecular mass $= 12 * 4 + 16 * 2 + 1 * 5 = 85$.

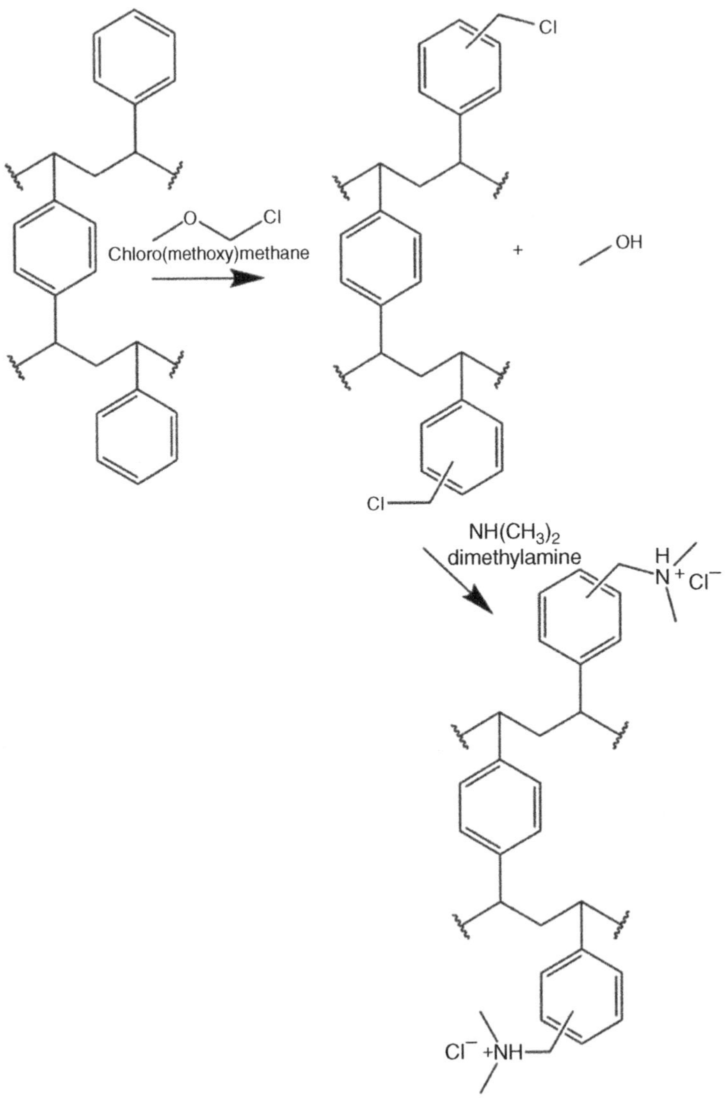

Figure 1.13 Synthesis of styrenic weak-base anion exchange resin through chloromethylation followed by amination with secondary amine.

Strong-acid cation (SAC) exchanger with sulfonic group: For every ion exchange site, the repeating unit contains – C: 8, O: 3, S: 1, H: 7; Corresponding molecular mass $= 12 * 8 + 16 * 3 + 32 * 1 + 1 * 7 = 183$.

Weak-base anion (WBA) exchanger with tertiary amine functionality: In a similar manner, the repeating unit contains – C: 11, N: 1, H: 15; Corresponding molecular mass $= 12 * 11 + 14 * 1 + 1 * 15 = 161$.

Strong-base anion (SBA) exchanger with quaternary amine: The repeating unit contains – C: 12, N: 1, H: 17; Corresponding molecular mass $= 12 * 12 + 14 * 1 + 1 * 17 = 175$.

Table 1.4 Repeating units of some common ion exchangers (without divinylbenzene crosslinking).

Type of ion exchanger	Functional group	Repeating units with functional group
Weak-acid cation	Carboxylate ($R - COO^-$)	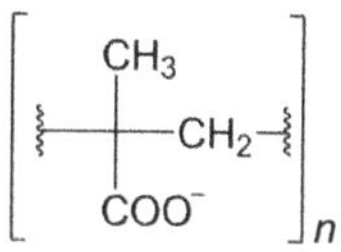
Strong-acid cation	Sulfonate ($R - SO_3^-$)	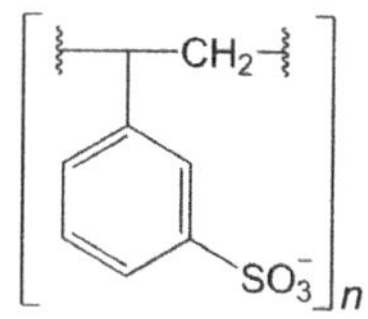
Weak-base anion	Tertiary amine ($R - N^+R_2H$)	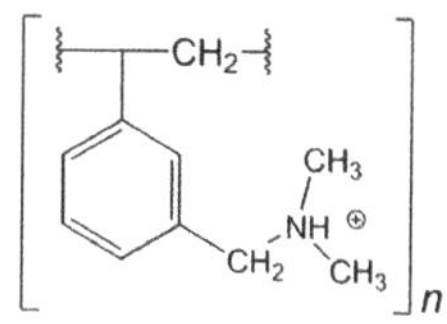
Strong-base anion	($R - N^+R_3$)	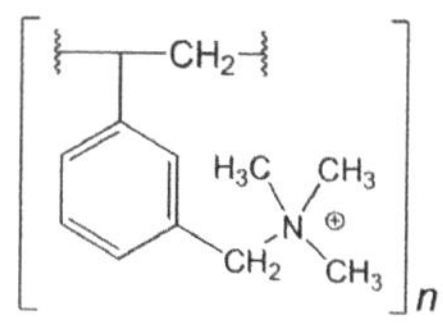

Ratio of the molecular mass of the four resins are WAC:SAC:WBA:SBA = 85:183:161:175.

Bulk densities of the fully ionized ion exchangers are nearly the same and equal to 1.0 g/mL. The lowest molecular mass of the WAC for each ion exchange site should render it the highest capacity, that is, ion exchange capacity is inversely proportional to the repeating molar mass. Thus, the ratio of the capacity when normalized with respect to the molecular weights of SBA, becomes

$$WAC:SAC:WBA:SBA = 2.05:0.96:1.1:1$$

Ion exchange capacities of the commercial gel resins are available in the open literature; the ratio of the volume-based capacities corresponds to the following trend:

$$WAC:SAC:WBA:SBA = 2.75:1.7:1:1$$

Agreement and Anomaly

The theoretical capacity thus calculated agrees reasonably well with the trend obtained from experimental/analytical determination for cation exchangers. However, for anion exchangers, actual ion exchange capacity is found to be significantly less than its theoretical value and this anomaly is due to the phenomenon of "methylene bridging." During the chloromethylation step, even with strictest control of process parameters, secondary cross-linking takes place between neighboring styrene molecules through methylene bridging as shown below:

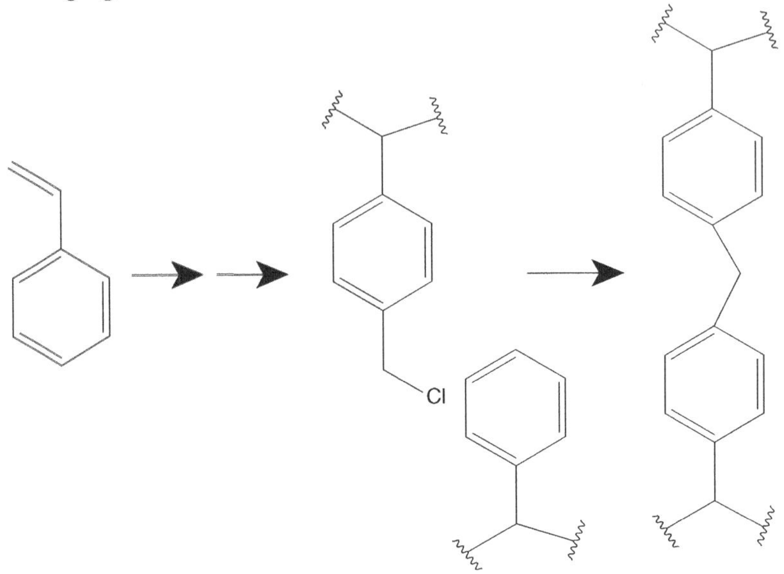

Figure 1. Secondary cross-linking during chloromethylation of polystyrene.

The styrene groups undergoing methylene bridging are difficult to be aminated (i.e., functionalized) during the next step due to enhanced steric hindrance. Hence, the total exchange capacity per unit mass or volume, all other parameters remaining identical, for anion exchange resins is relatively low.

1.5.3 Strong-Base Type I and Type II Anion Exchanger

Strong-base anion exchange resins possess quaternary amine functional groups corresponding to the general composition of R_4N^+. During the final stage of synthesis of an anion exchanger, chloromethylated copolymer is aminated with alkyl substituted amine. Use of trimethylamine, ($R - N^+(CH_3)_3$), yields the quaternary benzyltrimethyl ammonium functional group which is called Type I strongly basic anion exchange resin, as represented in Figure 1.14a in chloride form. If instead, the amination step

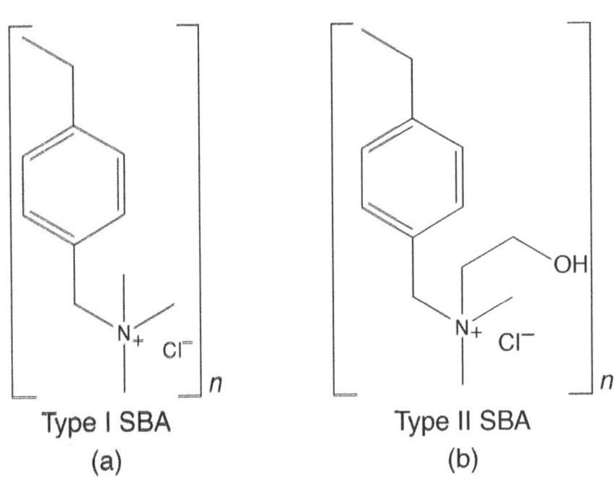

Figure 1.14 (a) Type I and (b) Type II functional groups of SBA resins.

is carried out by using dimethylethanolamine or $(R - N^+(CH_3)_2(C_2H_4OH))$, Type II strongly basic anion exchange resin is formed, as shown in Figure 1.14b.

Substitution of a methyl group with ethanol makes Type II anion exchange resins more hydrophilic. Hence, the efficiency of regeneration with NaOH is significantly greater for Type II SBA resins compared to its Type I counterpart. Conversely, relative chloride (Cl^-) affinity with respect to OH^- is greater for Type I resins compared to Type II resins.

Example 1.2 *A glass column containing 1.0 L of a strong-acid cation exchanger with sulfonic acid functional group is being tested in the laboratory for removal of trace lead (Pb^{2+}) from an industrial wastewater. At the start of the run the exchanger is in sodium form with bulk density 1.1 kg/L.*

(i) What is the approximate molecular weight of a bead with 1 mm diameter?
(ii) If the exchange capacity is 1.5 eq/L, how is the mass of the ion exchange bed expected to change upon saturation (i.e., specific gravity)?
(iii) For a resin bead of 1 mm diameter, how does the settling velocity change if the bead is fully converted into Pb^{2+} form?

State assumptions if any and briefly discuss the implications of ion exchanger density from an application viewpoint.

(i) Volume of the bead of $r = 0.05$ cm is

$$V = \frac{4}{3}\pi r^3 = 5.23 \times 10^{-4}\,\text{cm}^3 \cdot \frac{1\,\text{L}}{1000\,\text{cm}^3} = 5.23 \times 10^{-7}\,\text{L}$$

Capacity of the bead

$$q = 5.23 \times 10^{-7}\,\text{L} \cdot 1.5\frac{\text{eq}}{\text{L}} = 7.85 \times 10^{-7}\,\text{eq}$$

One equivalent consists of Avogadro's number of charges or in this case the same number of repeating units shown in Table 1.4 for strong-acid cation exchanger. Number of repeating units in one bead:

$$N = q \cdot 6.022 \times 10^{23}\frac{\text{repeating units}}{\text{eq}} = 4.73 \times 10^{17}\,\text{repeating units}$$

Molecular weight of each repeating unit from Example 1.1 is 183 Da per unit. Molecular weight of each spherical bead is

$$M = N \cdot 183.2 = 8.66 \times 10^{19}\,\text{Da}$$

(ii) Volume of resin = 1.0 L, initial bed mass = $1.1\frac{\text{kg}}{\text{L}} \times 1.0\,\text{L} = 1.1\,\text{kg}$ or 1100 g
During ion exchange 1.5 eq Na^+ (eq wt 23) is exchanged by 1.5 eq Pb^{2+} (eq wt 103.6)
Mass decrease from Na^+:

$$1.5\frac{\text{eq}}{\text{L}} \times 1.0\,\text{L} \times 23\,\text{g}\frac{\text{Na}^+}{\text{eq}} = 34.5\,\text{g}$$

Mass increase from Pb^{2+}:

$$1.5\frac{\text{eq}}{\text{L}} \times 1.0\,\text{L} \times 103.6\,\text{g}\frac{\text{Pb}^{2+}}{\text{eq}} = 155.4\,\text{g}$$

Final mass of

$$1\,\text{L resin} = (1100 - 34.5 + 155.4)\text{g} \approx 1221\,\text{g; density} = 1.221\,\text{kg/L}$$

Upon saturation mass of 1 L resin increases by = 1221 – 1100 = 121 g

(iii) The settling of the resin bead is assumed to follow Stoke's law, $V_s = \frac{D^2 g(\rho_s - \rho_f)}{18\mu}$, where, V_s = settling velocity, D = diameter of the bead (1 mm), g = acceleration due to gravity (9.81 m/s^2), ρ_s = density of solid bead, ρ_f = density of fluid (water ~ 0.997 kg/L), μ = dynamic viscosity (0.891×10^{-3} N s/m^2 at 298 K).
V_s in Na^+ form:

$$V_s = \frac{\left(1\,\text{mm} * \frac{1\,\text{m}}{1000\,\text{mm}}\right)^2 (1100\,^{\text{kg}}/_{\text{m}^3} - 997\,^{\text{kg}}/_{\text{m}^3})(9.81\,^{\text{m}}/_{\text{s}^2})}{18\left(0.891 \times 10^{-3}\,^{\text{N}\cdot\text{s}}/_{\text{m}^2}\right)}$$

$$V_s = 0.063\,\text{m/s}$$

V_s in Pb^{2+} form:

$$V_s = \frac{\left(1\,\text{mm} * \frac{1\,\text{m}}{1000\,\text{mm}}\right)^2 \left(1221\frac{\text{kg}}{\text{m}^3} - 997\frac{\text{kg}}{\text{m}^3}\right)\left(9.81\frac{\text{m}}{\text{s}^2}\right)}{18\left(0.891 \times 10^{-3}\frac{\text{N s}}{\text{m}^2}\right)}$$

$$V_s = 0.137\,\text{m/s}$$

Resin bead in Pb^{2+}-form settles 2.17× faster ($V_{s,Pb}/V_{s,Na}$) than Na^+-form. Differential settling can be a technique to separate resins in different ionic forms.

Some applications may require specially designed high specific gravity resins. High specific gravity resins require the presence of non-polymeric, immobilized materials, such as metal oxide nanoparticles. Raw and waste drilling fluids can have very high specific gravities, such that typical ion exchange resins float on top of the fluids, making flow-through operations challenging and ion exchange unsuccessful. By doping hydrated ferric oxide (HFO) nanoparticles, it is possible to increase the specific gravity of the resins to significantly greater than the drilling fluids. Figure 1(a) shows how a commercial anion exchange resin floats on Marcellus flowback wastewater collected from a gas well in Pennsylvania that has a total dissolved solids (TDS) of 150,000 mg/L; the resin is unusable in fixed bed operation. By doping iron oxide nanoparticles, the density of the anion exchanger can be appropriately increased to enable its use in fixed-bed columns, Figure 1(b), without any noticeable loss in exchange capacity.

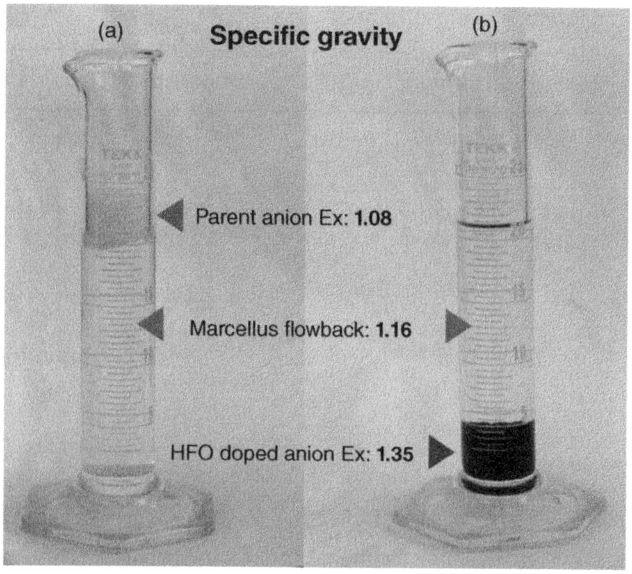

Figure 1. Specific gravity comparison of two different resins and Marcellus Shale flowback water. (a) Raw Purolite A850 resin; (b) Purolite A850 resin loaded with HFO nanoparticles to increase the specific gravity and to prevent floating during up-flow of high TDS and high specific gravity waste water.

1.6 Biosorbent, Liquid Ion Exchanger, and Solvent Impregnated Resin

Fundamentals of all organic and inorganic ion exchanging materials, including membranes, have a unifying commonality. Yet, it is imperative that we also become familiar with other types of ion exchangers and note their pros and cons for applications.

1.6.1 Biosorbent

Biosorption is defined as the *passive uptake* of cations and/or anions by dead microbial matter or other renewable biomass, including seaweed, chitosan and/or agricultural carbonaceous byproducts. "*Passive uptake*" is distinguished from bioaccumulation which is active and metabolically mediated by living cells. Many independent recent studies indicate that the biosorptive metal uptake mechanism heavily relies on ion exchange and is very sensitive to solution pH [37,38]. Functionally, biosorbents are quite similar to chelating ion exchangers and exhibit strong affinity toward transition metal ions through Lewis acid–base interactions. The ability of a biosorbent to sequester metal ions arises from the presence of lone electron pair donor atoms within its structure, namely, oxygen, nitrogen, phosphorus and sulfur. Table 1.5 provides the list of several commonly occurring metal-binding groups in biomass.

Note that the binding groups have weak-acid and weak-base functional groups and, so, metal sorption onto biosorbents is reversible through pH swings. A reducing environment persists inside the biomass due to anaerobic activity. Thus, metal sorption

Table 1.5 Metal-binding groups in biosorbents.

Binding group	Structural formula	pK_a	Donor atom	Occurrence in biomolecules
Hydroxyl (phenolic)	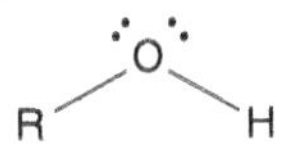	9–10	O	Polysaccharides
Carboxyl	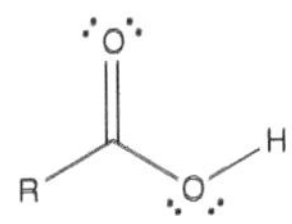	1.7–4.7	O	Humic acid, alginate
Sulfhydryl (thiol)	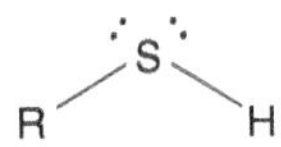	8–10	S	Amino acid, protein
Thioether	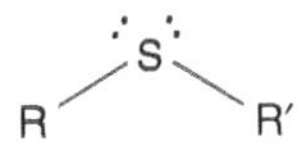	–	S	Amino acid, protein
Primary amine	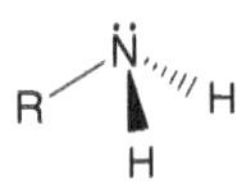	9–10	N	Amino acid, chitosan
Secondary amine	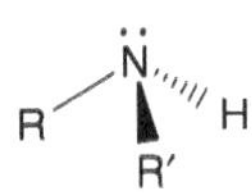	9–11	N	Amino acid, peptide bond
Amide	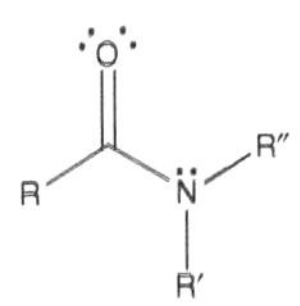	–	N	Amino acid
Imine	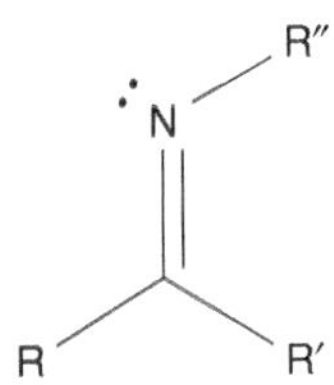	11.6–12.6	N	Amino acid
Imidazole	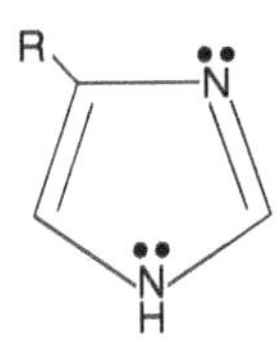	6.0	N	Amino acid

followed by reduction, for example, Cr(VI) to Cr(III), is also quite feasible. Besides metal sorption, chitin-based materials also have shown great promise as substrates for catalysis and nanotechnology [39,40].

The primary attribute of biosorbent materials is that they are biorenewable and hence attractive from a sustainability viewpoint. Poor chemical stability under extreme pH conditions and lack of mechanical strength are major impediments for large-scale application of biosorbent materials in flow-through or packed bed systems.

Through appropriate incorporation of cross-linking, significant progress is underway to improve the cost-effectiveness of biosorbents for specific applications [39,41].

1.6.2 Liquid Ion Exchange

The book is devoted primarily to solid-phase functionalized polymers and ion exchangers, both inorganic and organic. However, it is only appropriate that we briefly discuss the field of liquid ion exchange and its distinctive properties in comparison with solid-phase ion exchange. Liquid ion exchange is essentially a special case of solvent extraction where the principle of ion exchange is employed for transfer (or selective exchange) of solutes between two immiscible liquid phases, namely, an aqueous and an organic phase. The organic phase is the liquid exchanger and it contains highly hydrophobic ionogenic compounds dissolved in solvents like kerosene, trichloroethylene, chloroform and xylene that are immiscible with water. The most successful anion exchangers of this type are high molecular weight amine derivatives, while as cation exchangers, organophosphoric and carboxylic acids have proved particularly successful [42]. Over the years, liquid ion exchange has found major applications in extractive metallurgy for recovering metals from ore leachates.

The process of ion exchange is the same in both liquid and solid ion exchangers. Nevertheless, one major fundamental difference exists. For liquid ion exchangers, the exchange occurs at the phase boundaries formed between two immiscible liquids, such as kerosene, containing the liquid exchanger, and an aqueous phase, containing the counterions. On the contrary, the interior of a solid ion exchanger is permeable to the aqueous phase. Liquid ion exchangers have several advantages over their polymeric counterparts. Ion exchange rates are significantly higher due to efficient dispersion of the organic phase into the aqueous phase and rapid mobility of the functional groups within the organic phase. While for solid ion exchangers, polymeric or inorganic, functional groups are rigidly attached to the matrix through covalent or ionic bonds and hence, immobile. Intraparticle diffusion-limited ion exchange processes with solid ion exchangers are thus relatively slow.

Liquid ion exchangers can be easily prepared and their capacity can be varied by changing the relative volumes in the solvent organic phase. In principle, preparation and application of liquid ion exchange is in the public domain and requires no proprietary synthesis process. Figure 1.15 illustrates the three primary steps of liquid ion exchange for metal extraction from a dilute aqueous solution.

The primary disadvantages of liquid ion exchange are that the phase separation is difficult and not 100% efficient. Thus, the loss of ion exchange material escaping into the aqueous phase cannot be avoided. This shortcoming is particularly conspicuous if the components of the organic phase are partly soluble in water. With stricter environmental regulations, the presence of kerosene, xylene or chlorinated hydrocarbon in the treated aqueous phase is unacceptable and necessitates additional downstream treatment. In comparison, solid ion exchangers do not leach or impart any impurity into the aqueous phase and pose no environmental hazards. In a single-step process, the solid ion exchangers can achieve intended separation or removal of target ions without requiring any post-treatment. This attribute is the primary reason why synthesis

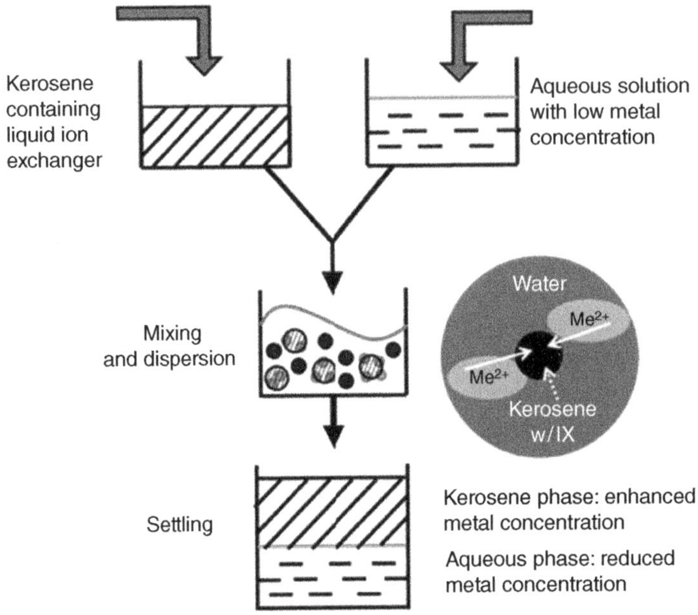

Figure 1.15 Illustration of a typical liquid ion exchange process followed by phase separation.

and application of functionalized polymers and solid ion exchangers have grown so steadily in every field during the last four decades.

More recently, membrane contactors have been used to improve the overall stability of liquid ion exchangers where the exchanger material dissolved in an organic solvent acts as a shuttle between the aqueous phases. Figures 1.16 and 1.17 show the general arrangement of a liquid ion exchange process to recover metals (M^{2+}) from a dilute stream using membrane contactors. In the first contactor on the left-hand side, metal

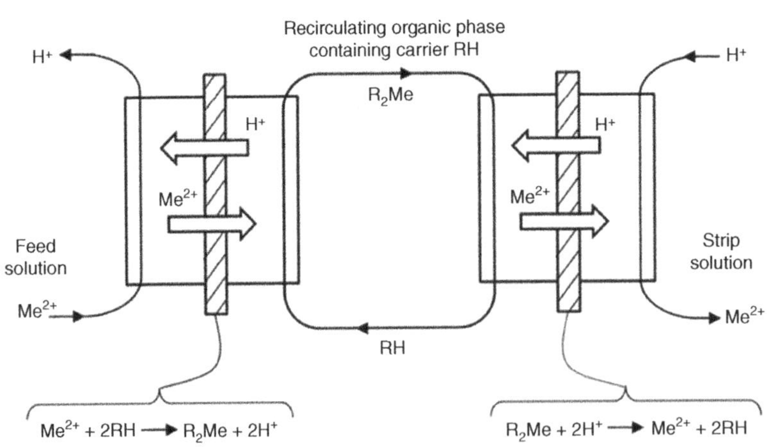

Figure 1.16 An illustration of facilitated transport in liquid ion exchange for recovering metals using two membrane contactors.

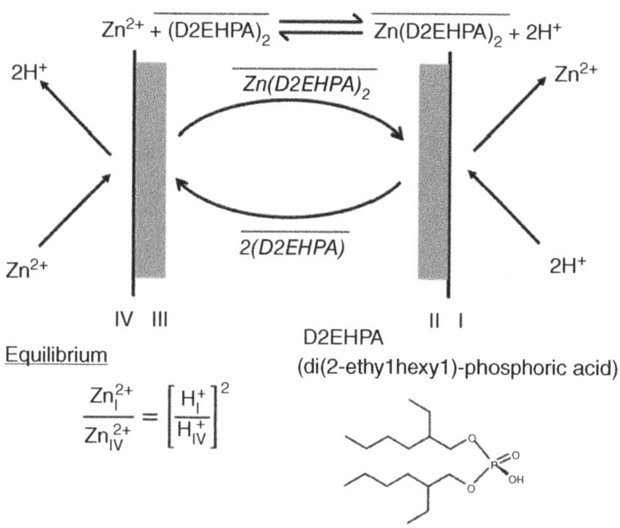

Figure 1.17 Zinc recovery with D2EHPA metal extractant.

ions diffuse across the microporous membrane into the liquid ion exchanger, exchanging hydrogen ions (H^+). The organic carrier solution containing the liquid exchanger is then brought in contact with the second membrane reactor on the right where the reaction is reversed and the metal ions are liberated into the strip solution. Similar systems, based on the principle of facilitated transport, have been used to separate metal ions, especially copper and zinc, using two steps, namely, sorption (i.e., uptake) and stripping (i.e., regeneration) [43,44].

Figure 1.17 illustrates recovery of zinc using liquid ion exchanger D2EHPA. While at the I/II interface, Zn^{2+} is finally recovered, separation of Zn^{2+} from the contaminated aqueous stream takes place at III/IV interface. Overall, hydrogen ions from the strip solution are transferred to the feed solution, exchanging equivalent amounts of metal ions. Loss of carrier solvent and instability of the membrane are two major hurdles for both facilitated and coupled transport processes in liquid ion exchange. Still, significant progress occurred during the last three decades to arrest solvent leakage primarily through the development of appropriate hydrophobic/hydrophilic contact materials.

1.6.3 Solvent-Impregnated Resins

In general, metal sorption onto polymeric chelating exchangers is quite selective, but kinetically very slow. Due to the rigid structure and tortuous pathways, the intraparticle diffusion rates for metals within solid ion exchangers are several orders of magnitude lower than they are in the solvent phase. Solvent impregnated resins (SIRs) can greatly overcome this shortcoming and SIRs are a compromise between solid ion exchange resins and liquid solvent extraction. Even more important, SIRs do not require covalent attachment of organic functional groups onto the parent polymer beads and, thus, conveniently avoid major steps in chemical synthesis for their preparation. SIRs can be easily deployed in packed-bed systems (i.e., plug flow reactor configuration) and are quite suitable for removal of trace concentrations of target metals. Liquid ion exchange

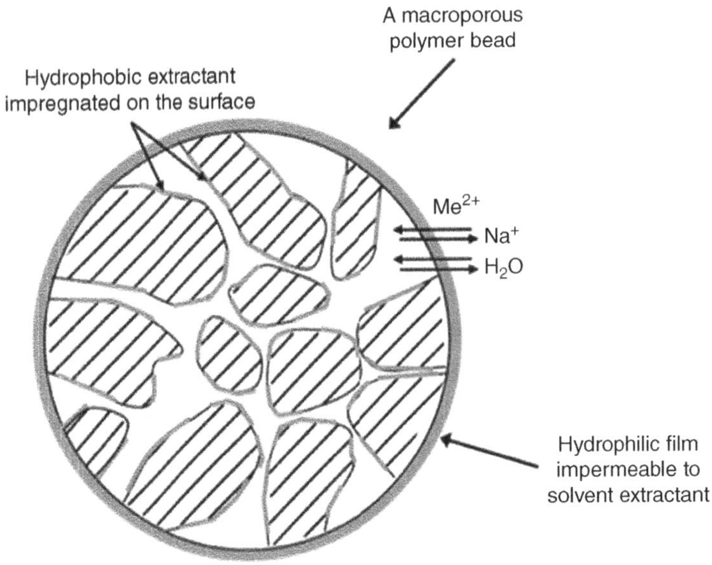

Figure 1.18 An illustration of the characteristic features of the modified solvent-impregnated resin (SIR).

processes, in comparison, use continuous stirred tank reactors (CSTRs) and require multiple reactor stages for low concentrations of target solutes.

In SIR, an organophilic complexant is sorbed within macroporous copolymer beads, and the combined material serves as the selective adsorbent [24,25]. The organic chelating agent is neither chemically bound nor physically entrapped within the matrix of the porous polymer. Instead, they remain attached to the hydrophobic surface of the parent adsorbent through weak van der Waals force. This phenomenon is confirmed by the ability for organic complexants to be removed completely from SIRs by washing with an appropriate organic solvent. One critical disadvantage of SIRs is the gradual loss of complexant through aqueous phase dissolution or physical separation from the parent polymer for flow-through conditions. This phenomenon is a significant problem and often precludes adaptations of SIRs in environmental applications. To eliminate the loss of complexant, a new SIR has been prepared wherein a thin coating is formed around each bead [45]. This coating is hydrophilic and prevents the transport of the hydrophobic complexant out of the bead, while permitting transport of hydrophilic cations and anions into the bead. Figure 1.18 illustrates the characteristic features of the modified SIR. Despite its potential ease and versatility in applications, use of SIRs has so far remained very limited.

1.7 Amphoteric Inorganic Ion Exchangers

Unlike zeolites, amphoteric inorganic ion exchangers are non-siliceous materials; they are hydrated oxides of polyvalent metals, namely, Al(III), Fe(III), Zr(IV) and Ti(IV). Traditional ion exchange literature does not include them as ion exchangers for their relatively low capacity. However, these materials, as will be shown here and in

Chapter 6, provide unique sorption behaviors for a host of different types of trace contaminants, mostly metals and ligands. These metal oxide surfaces exhibit concurrent Lewis acid–base (i.e., metal–ligand) interaction along with ion exchange. Depending on the pH at the solid–water interface, the surface metal hydroxyl groups, commonly referred to as $\overline{\text{MOH}}$, can undergo protonation or deprotonation accompanied by ion exchange as shown hereunder:

$$\overline{\text{MOH}_2^+} \leftrightarrow \overline{\text{MOH}} + \text{H}^+, K_{a1} \tag{1.18}$$

$$\overline{\text{MOH}} \leftrightarrow \overline{\text{MO}^-} + \text{H}^+, K_{a2} \tag{1.19}$$

where K_{a1} and K_{a2} are acid dissociation constants of the surface metal hydroxyl groups. One should be aware that the symbol $\overline{\text{MOH}}$ is strictly conceptual and does not have a rigorous chemical entity from a scientific viewpoint. Nevertheless, the symbol allows treatment of an amphoteric hydrated metal oxide as a diprotic weak acid with dissociation constants K_{a1} and K_{a2}. In most systems, there is no evidence of definite hydrates, but MOH groups have been identified by IR spectroscopy [46,47]. Water molecules are strongly bound to the MOH groups through hydrogen bonding and are given off only at elevated temperatures [46]. It is noteworthy that depending on pH, hydrated metal oxide surfaces may possess fixed positive charges, negative charges or be electrically neutral. Figure 1.19 shows how a protonated hydrated Fe(III) oxide selectively sorbs arsenate (an anionic ligand) and copper ion (transition metal cation) at different pH.

As ion exchangers, hydrated metal oxides possess some unique properties that are distinctive from zeolites and ion exchange resins:

1. Ion exchange or sorption properties of hydrated metal oxides essentially reside on the surface. Thus, the exchange capacity per unit mass increases with an increase in the surface area, that is, reduction of particle sizes. In contrast, zeolites and ion exchange resins have fixed ionic charges distributed through the entire mass. Thus, their ion exchange capacity per unit mass or volume is independent of size.
2. While in contact with water, every hydrated metal oxide surface has a characteristic pH value at which surface negative charges balance surface positive charges. This pH value is referred to as the pH at zero-point charge or pH_{ZPC}.

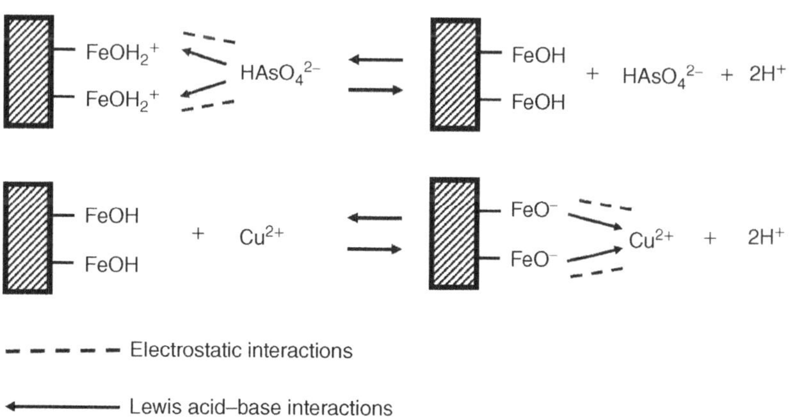

Figure 1.19 Illustration of amphoteric ion exchange behavior of polyvalent hydrated oxides.

Table 1.6 pH at zero-point charge (pH_{ZPC}) for different metal oxides of interest.

Oxide surface	pH_{ZPC}
MgO (periclase)	11.5–12.3
α-Al_2O_3 (corundum)	7.8–9.0
α-TiO_2 (rutile)	5.0–6.5
α-Fe_2O_3 (hematite)	7.0–8.5
HFO (hydrated ferric oxide)	7.3–8.3
Calcite (mainly $CaCO_3$)	8.0–9.0
Sepiolite ($Mg_4\ Si_5O_{15}(OH)_2{\cdot}6H_2O$)	6.5–7.5
Feldspar ($KAlSi_3O_8{\cdot}\ NaAlSi_3O_8{\cdot}\ CaAl_2Si_2O_8$)	3.0–4.0
α-SiO_2 (quartz)	2.0–3.0
β-MnO_2	6.5–7.5
$ZrSiO_4$	5.0–6.0
ZrO_2	6.5–7.5
Montmorillonite $(Na,Ca)_{0.33}(Al,Mg)_2(Si_4O_{10})(OH)_2{\cdot}nH_2O$	2.0–3.0
Kaolinite ($Al_2Si_2O_5(OH)_4$)	4.0–5.0

Table 1.6 includes pH_{ZPC} values for several metal oxides of interest [48]. There remain two common areas of misunderstanding or misinterpretation of pH_{ZPC} values. First, the value of pH_{ZPC} is not unique, it depends on the solution composition of the aqueous phase. The pH_{ZPC} values in Table 1.6 represent conditions where electrolytes (cations and anions) in the aqueous phase have negligible specific affinities toward ion exchange sites besides through electrostatic or coulombic interactions. Second, metal oxides, although electrically neutral at pH_{ZPC}, often exhibit high sorption capacities at $pH = pH_{ZPC}$, that is, both positive and negatively charged functional groups may act independently.

3. Metal oxides exhibit strong Lewis acid–base characteristics during ion exchange processes; the central metal atom, that is, Fe(III), Zr(IV), Ti(IV) may act as a Lewis acid (electron acceptor) and the oxygen atoms as an electron donor (Lewis base). Thus, they can selectively bind both transition metal cations (Lewis acids) and anionic ligands (Lewis bases), which, incidentally, comprise a large group of trace contaminants of environmental interest. In fact, concurrent sorption of both transition metal cations and anionic ligands is also possible.

To further explore selective ion exchange properties of hydrated metal oxides of interest, let us take the case of hydrated Fe(III) oxides or HFO particles that have been rigorously investigated by geochemists and environmental engineers [49–54]. HFO is a weak diprotic acid with the following dissociation constants:

$$\overline{FeOH_2^+} \leftrightarrow \overline{FeOH} + H^+ \quad pK_{a1} = 6.5 \tag{1.20}$$

$$\overline{FeOH} \leftrightarrow \overline{FeO^-} + H^+ \quad pK_{a2} = 8.8 \tag{1.21}$$

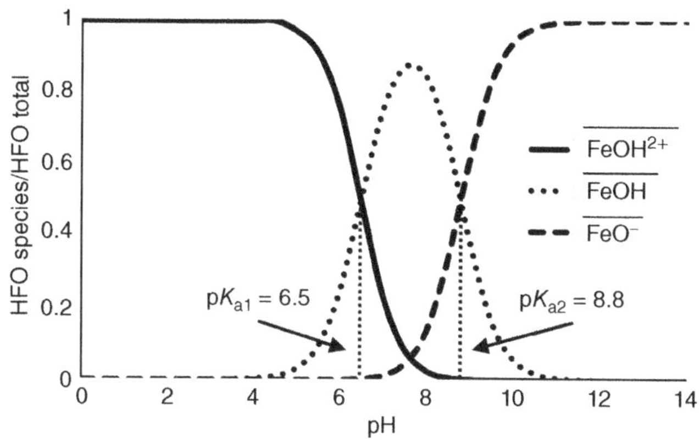

Figure 1.20 Relative distribution of HFO surface functional groups as a function of pH ($pK_{a1} = 6.5$, $pK_{a2} = 8.8$). *Source:* Cumbal and SenGupta 2005 [52]. Reproduced with permission of American Chemical Society.

Figure 1.20 illustrates the distribution of the three surface functional groups of HFO (e.g., $FeOH_2^+$, FeOH, and FeO^-) with respect to pH. At pH < 6.5, anionic ligands, such as arsenate, phosphate and citrate exhibit high sorption affinity toward positively charged surface hydroxyl groups. Such selective surface binding is often called formation of inner-sphere complexes because the solute-sorbent interaction involves an electron-pair donor (or Lewis base) and an electron-pair acceptor (or Lewis acid) as shown in Figure 1.21. Also, such Lewis acid–base interactions are often accompanied by electrostatic interactions. Note that commonly encountered anions, such as chloride, sulfate, and nitrate are poor ligands and, hence, unable to form inner-sphere

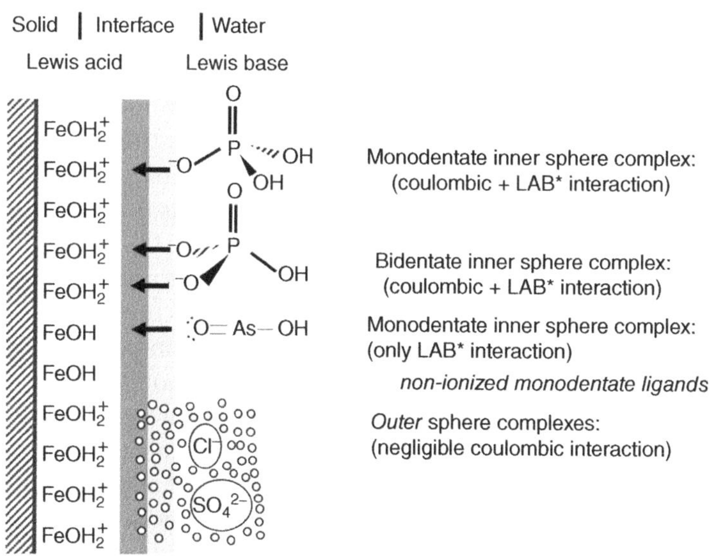

Figure 1.21 An illustration of interaction of ligands (e.g., phosphate) with HFO surface functional groups in the presence of chloride. *Note:* LAB* = Lewis acid–base. *Source:* Puttamaraju and SenGupta 2006 [53]. Reproduced with permission of American Chemical Society.

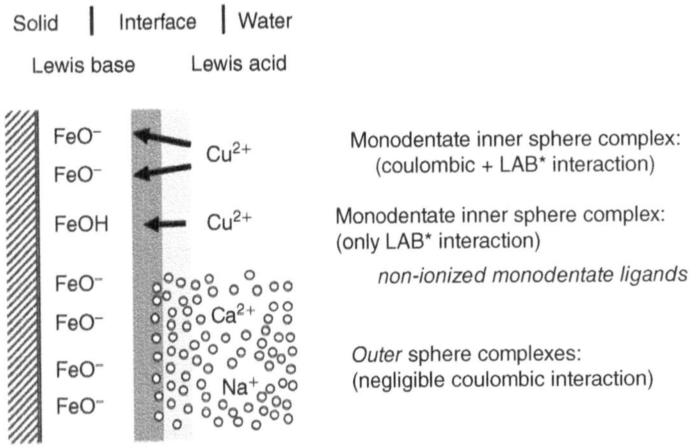

Figure 1.22 An illustration of interaction of transition metal cation (e.g., Cu^{2+}) with HFO surface in the presence of Ca^{2+}. *Source:* Puttamaraju and SenGupta 2006 [53]. Reproduced with permission of American Chemical Society.

complexes; they undergo only electrostatic interactions and form weak outer-sphere complexes.

Likewise, at pH ≥ 9.0, transition metal cations (e.g., Cu^{2+}, Pb^{2+}, Zn^{2+}) exhibit strong binding affinity toward negatively charged surface groups ($\overline{FeO^-}$) through formation of inner-sphere complexes accompanied by relatively weak electrostatic interaction. Surface $\overline{FeO^-}$ groups act as sites for Lewis bases due to the presence of oxygen donor atoms while transition metal cations act as Lewis acids. Commonly encountered alkaline and alkaline-earth metal cations (e.g., Na^+, Ca^{2+}, Mg^{2+}), on the contrary, form only outer-sphere complexes through electrostatic interactions. Figure 1.22 illustrates the underlying binding mechanism of Cu^{2+}, a transition metal cation of environmental significance.

In the open literature, the selective binding of transition metal ions or ligands onto hydrated metal oxides through Lewis acid–base interactions is sometimes called *chemisorption*. From both the thermodynamic and kinetic standpoints, that is an overstatement. Selective surface binding of transition metals and ligands is reversible and the target solutes are easily desorbed simply through alteration of pH [52–56].

Temperature dependence of metal or ligand exchange onto amphoteric metal oxides is rather minimal. Activation energy requirement is always less than 50 kJ/mol. Kinetically, these heterogeneous surface ion-exchange reactions are diffusion-controlled; chemical reactions are rarely the rate-limiting steps. Due to their amphoteric behaviors and preference for both trace ligands and metals, hydrated metal oxides have much potential as selective sorbent materials. However, their low mechanical strength and chemical instability are the primary obstacles against wide range of applications. Chapter 6 reports preparation and characterization of hybrid ion exchangers that combine attributes of polymeric ion exchange resins and hydrated metal oxides to overcome these deficiencies.

It is worth noting that hydrated metal oxides can selectively and simultaneously remove both a toxic metal cation (e.g., Cu^{2+}) and an anionic ligand (e.g., $HAsO_4^{2-}$). Figure 1.23 shows the effluent histories of Cu(II) and As(V) during a fixed-bed column run using granular ferric hydroxide (GFH) sorbent. Note that both copper

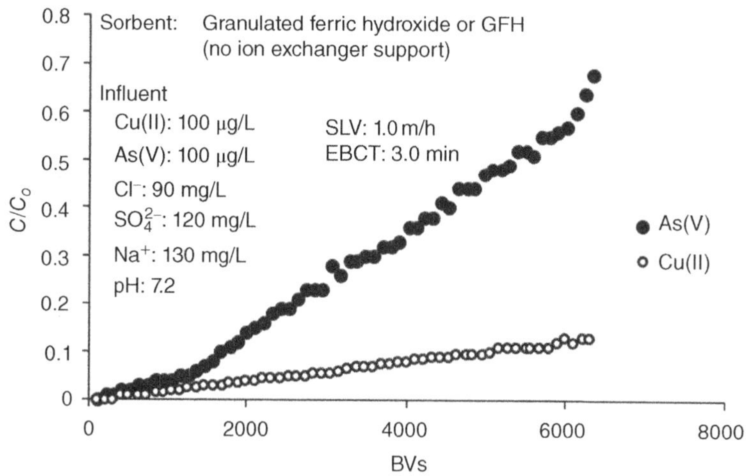

Figure 1.23 Effluent histories for As(V) and Cu^{2+} for a fixed bed column run with GFH. SLV = Superficial Liquid Velocity; EBCT = Empty Bed Contact Time. *Source:* Puttamaraju and SenGupta 2006 [53]. Reproduced with permission of American Chemical Society.

and arsenate are removed for several thousand bed volumes, while cations and anions (e.g., Na^+, Ca^{2+}, Cl^-, SO_4^{2-}) that form only outer-sphere complexes are poorly sorbed. In a succeeding chapter, we will discuss the preparation of tunable hydrated metal oxide materials that may remove either the ligands or the transition metals in accordance with the demand of the intended applications.

1.8 Ion Exchanger versus Activated Carbon: Commonalities and Contrasts

Activated carbon is probably the most widely used adsorbent and, expectedly, an enormous body of knowledge and scientific information is available about its properties and usage in the open literature. Activated carbon and ion exchangers have several properties in common: both are heterogeneous processes; fixed-bed column is the most common form of equipment configuration; both are diffusion-controlled processes; both have finite capacity and require reactivation/regeneration following exhaustion; and both exhibit different affinities to different solutes. Expectedly, there is often an inclination to interpret and understand ion exchange phenomena through the lens of activated carbon adsorption processes. As similar as they may seem to be, we need to be aware of the following distinct differences:

1. An ion exchanger, as already stated, is essentially a cross-linked polyelectrolyte with charged functional groups balanced by counterions, thus maintaining electroneutrality. Hence, it swells or shrinks very significantly due to the phenomenon of osmosis, which depends on the ionic strength and pH of the solution with which it is in contact. On the contrary, activated carbon barely exhibits any swelling/shrinking property in water.
2. Activated carbon adsorption is a surface phenomenon; so, adsorption capacity increases with an increase in surface area for a specific solute. The capacity of an ion exchanger is determined by the number of functional groups (i.e., fixed

coions) covalently attached per unit volume or mass of the ion exchanger. Thus, the capacity is not dependent on surface area. In fact, gel-type ion exchangers with practically no pores offer higher capacity than their macroporous counterparts with significantly higher surface area.

3. Activated carbon adsorption normally pertains to uncharged organic solutes and there is no electric potential gradient at the liquid/solid interface. So, solutes including ions of any sign and valence may permeate in and out of the porous solid phase. In contrast, there remains an electric potential gradient, often termed Donnan potential, at the ion exchanger/water interface inhibiting access of ions of specific charge inside the exchanger. The phenomenon is known as the Donnan exclusion effect. Several new and innovative processes have been developed through intelligent application of the Donnan exclusion effect and the subject has been adequately addressed in the succeeding chapters.

1.9 Ion Exchanger Morphologies

It is well recognized that the sorption behavior of an ion exchanger, inorganic or organic, is governed by its chemical makeup, that is, type of functional groups, matrix material and cross-linking. However, morphology or physical configuration of the ion exchanger also plays a significant role in deciding the success or appropriateness of an ion exchanger for an intended application. Type of reactor configuration, degree of pre-treatment required, environmental consideration, regenerability, durability and cost often decide the relative advantages of different types of ion exchanger morphologies that are available commercially. The following provides a broad classification of them [20,46,57–66]:

a. Granular
b. Spherical: Gel
c. Spherical: Macroporous
d. Pellicular
e. Ion exchange fibers (IXFs)
f. Composite ion exchanger (CIX) cloth
g. Hybrid ion exchanger (HIX)
h. Magnetic ion exchanger (MIEX)

Figure 1.24 provides a schematic and/or a photograph for each morphology depicting its physical configuration. Several succeeding chapters emphasize their distinctive properties and application opportunities.

1.10 Widely Used Ion Exchange Processes

Before leaving this chapter on *Introduction*, it is only appropriate that we briefly discuss the two most widely used ion exchange processes: Softening and Demineralization. These processes have been in use globally for over five decades, but the adverse

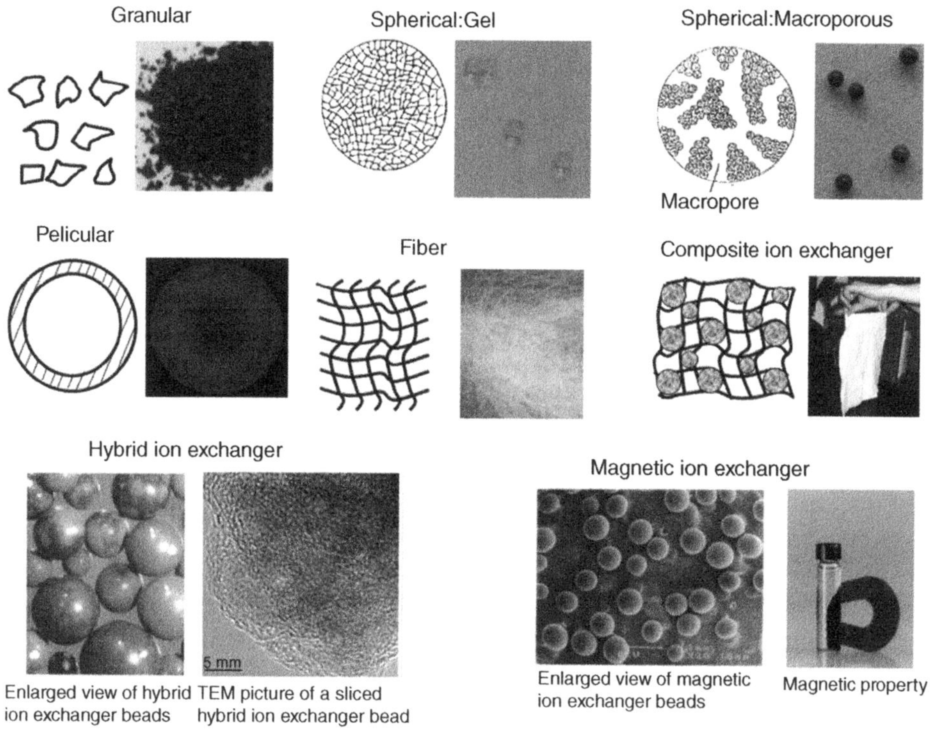

Figure 1.24 Illustration of morphologies of different ion exchange materials.

environmental impact of these processes from a sustainability viewpoint is currently under severe scrutiny. The primary intent of this section is to briefly discuss the key elements of these processes and highlight their major shortcomings and areas warranting improvement.

1.10.1 Softening

Softening is a process to remove hardness from water, that is, Ca^{2+} and Mg^{2+}. The presence of hardness in water increases the consumption of detergent and soap during cleaning and laundering. More importantly, hardness is highly undesirable in any heat exchanging equipment, including boilers, to avoid fouling and scaling. Membrane desalination of high hardness brackish water or sea water by reverse osmosis (RO) often requires the water to be pretreated by softening to avoid scaling of membrane surfaces. Lime and lime-soda softening processes can remove hardness, but the processes produce large volumes of sludge and is unable to remove hardness completely.

Complete softening may be achieved by sodium cycle cation exchange on a single fixed bed unit. As the feed or raw water containing hardness is passed through the fixed bed column, hardness causing cations (e.g., Ca^{2+}) displace sodium per the following ion exchange reaction:

$$\overline{2R^-Na^+} + Ca^{2+} \rightarrow \overline{(R^-)_2Ca^{2+}} + 2Na^+ \qquad (1.22)$$

Calcium is removed from the feed water and the treated water is soft and in compliance with the specifications for the intended applications. Once the capacity is exhausted and calcium in the treated water starts increasing, the cation exchanger in the fixed bed warrants regeneration to desorb Ca^{2+} and return the cation resin to Na^+-form for the next cycle as shown below:

$$\overline{(R^-)_2Ca^{2+}} + 2Na^+ \rightarrow 2\overline{R^-Na^+} + Ca^{2+} \tag{1.23}$$

The regeneration of the exhausted resin is often, if not always, carried out by brine or NaCl solutions. The regeneration step is often accomplished in <10 BVs (bed volumes). *One bed volume or BV corresponds to the volume of the cation exchange resin in the fixed bed.* Equivalents of NaCl solution used for regeneration are always greater than the equivalents of Ca^{2+} removed by the bed during the service or sorption cycle. Thus, the spent regenerant contains a significant excess of brine in addition to the calcium removed. From a sustainability or efficiency viewpoint, the softening process is poor-performing, because it invariably delivers more electrolytes to the surrounding waterways than it removes. To appropriately assess the overall sustainability of the softening process, let us consider the following example.

Figure 1.25 shows the schematic of the cation exchange softening process in a fixed-bed system where the solid line shows the service/sorption cycle and the

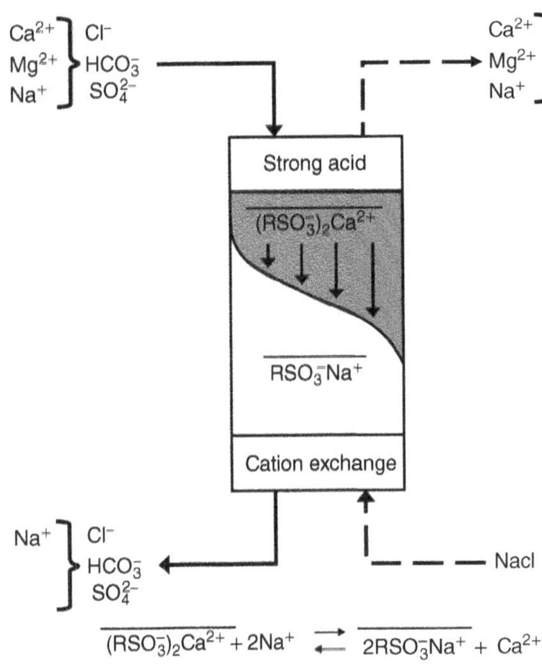

Figure 1.25 A schematic of the cation exchange softening process in a fixed-bed system. The solid line shows the service/sorption cycle and the dashed line represents regeneration with brine.

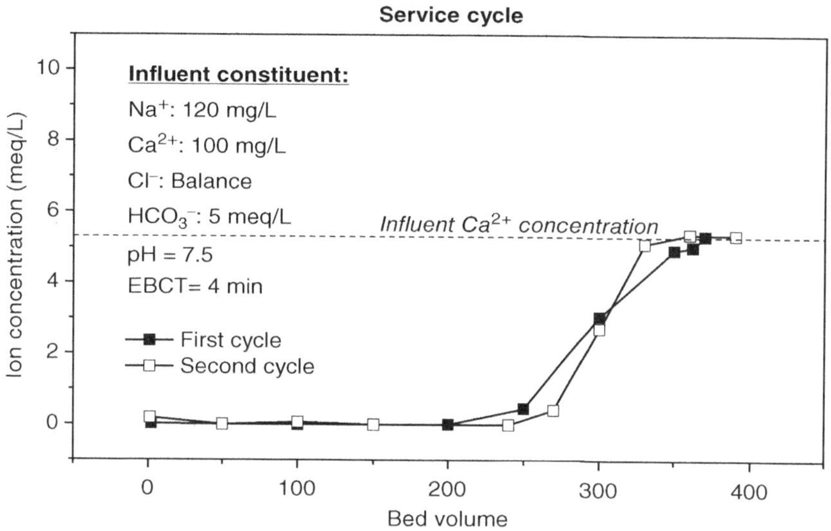

Figure 1.26 Calcium history of two consecutive service cycles by a strong-acid cation exchange softener.

dashed line represents regeneration with brine. Figure 1.26 presents the effluent or breakthrough history of calcium from the cation exchange column in sodium form for two consecutive runs. The feed water composition is also included in the figure [67].

Between the two successive runs, the column was regenerated with 5% NaCl. Figure 1.27a shows elution or desorption profiles of calcium during regeneration: over 90% calcium is recovered after 15 BVs of regeneration.

It is apparent that from an operational viewpoint, the cyclic process – sorption followed by regeneration and rinsing – is sustainable and can be carried out for tens of cycles with desired hardness removal. From an environmental sustainability viewpoint, however, a critical question is now being asked: How many equivalents of Na^+ are consumed and/or discharged to the environment per equivalent of Ca^{2+} removed from the hard water? This sustainability parameter is of great consequence, particularly in arid areas with low rainfall or precipitation. High volumes of spent regenerant could cause adverse effects by elevating the salinity level in natural water bodies in the region. Ideally, based on the stoichiometry of the exchange reactions in Eqs (1.22) and (1.23), one equivalent of Na^+ is needed for removal of one equivalent of hardness.

Figure 1.27b presents the ratio of equivalents of Na^+ used per equivalent of Ca^{2+} removed during the regeneration cycle. Note that the ratio after 12 BVs is 8.5 implying that the process consumes and discharges several times more equivalents of NaCl to the environment than is stoichiometrically required. This inefficiency is embedded in an ion exchange softening process and is the sole reason why brine regeneration is being banned in many places, including the state of California [68].

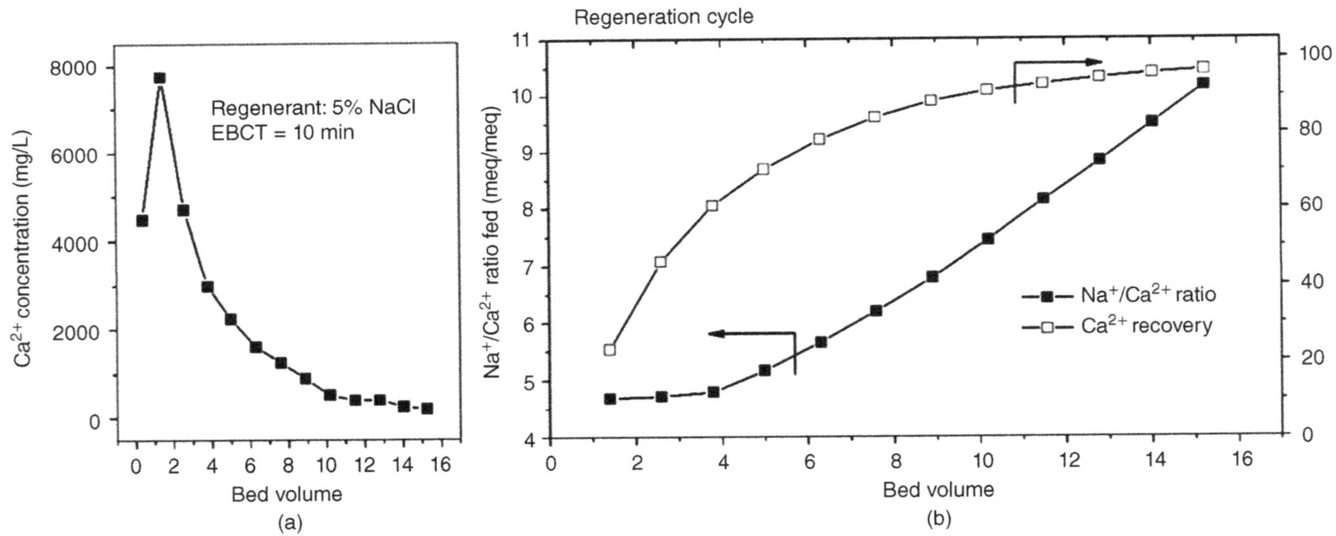

Figure 1.27 (a) Regeneration curve for exhausted strong-acid cation exchange resin with 5% NaCl; (b) A comparison of regeneration efficiency (Na^+: Ca^{2+}) versus Ca^{2+} recovery. *Source:* Reprinted with permission from Li *et al.* 2016 [67].

Example 1.3 *Consider the effluent history of calcium in Figure 1 during the softening cycle. Superimpose the Na^+ effluent history in the same figure. What is the chloride concentration at the exit of the column after 100 BVs?*

Solution:

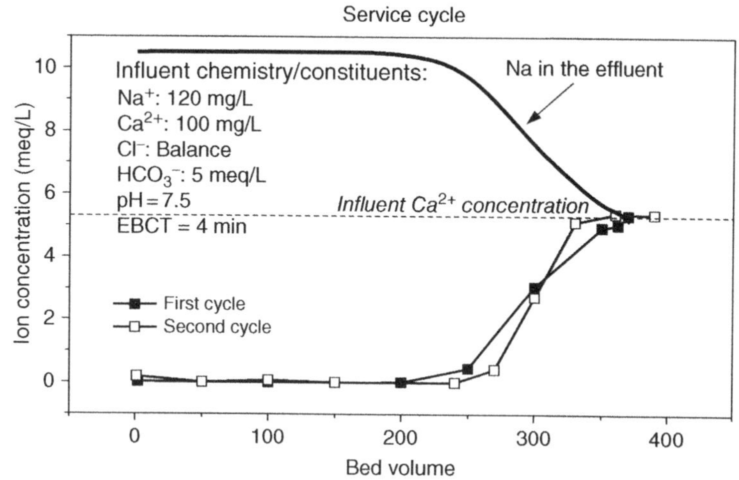

Figure 1. Calcium and sodium history of two consecutive service cycles by a strong-acid cation exchange softener. *Source:* Li *et al.* 2016 [67]. Reproduced with permission of American Chemical Society.

The sodium concentration in the treated water can be calculated by applying the principle of electroneutrality. Same equivalent amount of Na^+ was released to the effluent as Ca^{2+} was removed. $Ca^{2+}_{effluent}$ was read from the figure provided.

$$Na^+{}_{effluent} = Na^+_{initial} + Ca^{2+}_{initial} - Ca^{2+}_{effluent}$$

$$Na^+_{effluent} = \left(\frac{120\,mg}{L} \times \frac{1\,mmol}{23\,mg} \times \frac{1\,meq}{mmol}\right) + \left(\frac{100\,mg}{L} \times \frac{mmol}{40\,mg} \times \frac{2\,meq}{mmol}\right) - Ca^{2+}_{effluent}$$

$$Na^+_{effluent} = 5.22\frac{meq}{L} + 5\frac{meq}{L} - Ca^{2+}_{effluent}$$

Thus, for each data point corresponding to $Ca^{2+}_{effluent}$ in Figure 1, $Na^+_{effluent}$ can be computed. Figure 1 includes the effluent history of Na^+ during the column run.

Chloride concentration may also be computed using the principle of electroneutrality. Concentration of H^+ and OH^- can be ignored at pH = 7.5.

Cations:

$$Na^+ = \frac{120\,mg/L}{23\,mg/mmol} \times 1\,meq/mmol = 5.22\,meq\,L$$

$$Ca^{2+} = \frac{100\,mg/L}{40\,mg/mmol} \times 2\,meq/mmol = 5\,meq/L$$

Anion:

$$HCO_3^- = 5\,meq/L$$

Thus,

$$Cl^- = (5.22 + 5 - 5)\frac{\text{meq}}{\text{L}} \times 1\frac{\text{mmol}}{\text{meq}} \times 35.5\frac{\text{mg}}{\text{mmol}} = 185.3\,\text{mg/L}$$

There is no anion exchange during the process. Thus, the chloride concentration remains the same throughout and is equal to 185.3 mg/L.

1.10.2 Deionization or Demineralization

Deionization or demineralization is a process of removing electrolytes by ion exchange to produce pure water, that is, removal of all cations and anions. In the simplest form of this process, two ion exchange units in series constitute the heart of the process; a cation exchange bed in hydrogen form is followed by an anion exchange bed in hydroxyl form. As the water to be treated is passed through the two-bed system, all the cations (say monovalent M^+) are exchanged releasing equivalent amounts of hydrogen ions:

$$\overline{RH} + M^+ + N^- \rightarrow \overline{R^-M^+} + H^+ + N^- \tag{1.24}$$

Thus, at the exit of the cation exchange column, the water is essentially a dilute acid containing all the anions (N^-) that were originally present in the feed water balanced by equivalent H^+ cations. Upon passing through the anion exchange column, all anions are exchanged for OH^-, which then associates with H^+ to produce pure water or H_2O.

$$\overline{ROH} + H^+ + N^- \rightarrow \overline{R^+N^-} + H^+ + OH^- \tag{1.25}$$

$$H^+ + OH^- \rightarrow H_2O \tag{1.26}$$

Thus, the overall reaction is

$$\overline{RH} + \overline{ROH} + M^+ + N^- \rightarrow \overline{R^-M^+} + \overline{R^+N^-} + H_2O \tag{1.27}$$

Dissolved solutes that are not ionized cannot be removed by the deionization process. Also, the product water is not free of electrolytes due to early "leakage" or "breakthrough" of ions from the ion exchange columns. Once the capacities are exhausted, the columns need to be regenerated: cation exchangers with an acid (HCl or H_2SO_4) and the anion exchanger with a base (NaOH). The regeneration process is often not efficient, that is, the consumption of acid and base is significantly more than their stoichiometric requirement.

As the electrolyte concentration in the feed water increases, regeneration is needed more frequently, as capacity is a function of equivalents of electrolytes removed. The general guideline is that once the TDS of the feed water increase beyond 500 mg/L, conventional deionization becomes economically less viable. From an environmental sustainability viewpoint, spent regenerant from the demineralization process is a challenge that requires innovation and improvement both in the efficiency of the regeneration process and choice of the regenerant chemicals. The following example attempts to illustrate several pertinent issues.

Example 1.4 Design of a Two-Bed Deionizer

Given: Raw water of the following quality:

Hardness = 3.0 meq/L
Bicarbonate = 2.0 meq/L

pH = 7.8
Chloride = 1.0 meq/L
Sulfate = 2.0 meq/L
Sodium = to balance anions

Find: The design parameters for a two-bed, 400 L/min deionizer system which must run for 8 hours before breakthrough.

(a) The volume of cation resin required if the two-bed deionizer has a capacity of 1.0 eq/L of resin after regeneration with H_2SO_4 at 162 g/L (100% basis) and a concentration of 2.0 N.
(b) The volume of anion resin required if it has a capacity of 0.7 eq/L of resin after regeneration with NaOH at 120 g/L of resin (100% basis) and a concentration of 1.0 N.
(c) The analysis in mg/L for all the ions in the mixed waste regenerant solution after neutralization. Assume that the slow rinses are collected with the regenerants and that they comprise 2 BV for each bed. Deionized water is used to make up the regenerant solutions. Neutralization is done with the same acid or base as used for regeneration.
(d) In the above two-bed deionization system, do you anticipate any problem if the anion-exchanging unit precedes the cation exchanging unit?
(e) Draw a sketch of the process and indicate approximate pH at different locations of the systems during exhaustion run.
(f) Find the electrolyte concentration in the spent regenerant and compare this to the amount of removed contaminants to find a measure of the ion exchange efficiency.

Solution:

(a) *Concentration of cations:*

Hardness = $[Ca^{2+}] + [Mg^{2+}] = 3.0\,\text{meq/L}$

Total anions = Bicarbonate + Chloride + Sulfate = 5.0 meq/L $[Na^+] = 2.0\,\text{meq/L}$

Total cations = 5.0 meq/L

Total mass of exchanged cation:

$$5.0\frac{\text{meq}}{\text{L}} \cdot 400\frac{\text{L}}{\text{min}} \cdot 60\frac{\text{min}}{\text{h}} \cdot 8\,\text{h} = 9.60 \times 10^5\,\text{meq} = 960\,\text{equivalents}$$

The volume of cation exchange resin needed:

$$\frac{960\,\text{equivalents}}{1.0\frac{\text{eq}}{\text{L}}} = 960\,\text{L} = 0.96\,\text{m}^3$$

(b) *Concentration of anions:*

$$2.0\frac{\text{meq}}{\text{L}}(HCO_3^-) + 2.0\frac{\text{meq}}{\text{L}}(SO_4^{2-}) + 1.0\frac{\text{meq}}{\text{L}}(Cl^-) = 5.0\frac{\text{meq}}{\text{L}}$$

Total equivalents of exchanged anion same as cations: 960 eq
The volume of anion exchange resin needed:

$$\frac{960\,\text{eq}}{0.7\frac{\text{eq}}{\text{L}}} = 1370\,\text{L} = 1.37\,\text{m}^3$$

(c) First, let us identify and record the volumes of separate streams in the mixed waste regenerant. Note that the mixed waste will contain: (i) all the electrolytes removed during the 8-h period of the service cycle, (ii) excess H_2SO_4 and NaOH used during the regeneration of the cation exchanger and anion exchanger, (iii) excess acid and/or base added to neutralize the pH from the previous steps.

Volume of rinse water (V_1):

$$V_1 = 2\,\mathrm{BVs} \cdot 0.96\,\frac{\mathrm{m}^3}{\mathrm{BV}} + 2\,\mathrm{BVs} \cdot 1.37\,\frac{\mathrm{m}^3}{\mathrm{BV}} = 4.66\,\mathrm{m}^3$$

Volume of H_2SO_4 (V_2):

$$V_2 = \frac{162\,\frac{\mathrm{g}}{\mathrm{L_{resin}}} \cdot 960\,\mathrm{L_{resin}}}{49\,\frac{\mathrm{g}}{\mathrm{eq}} \cdot 2\,\frac{\mathrm{eq}}{\mathrm{L}}} = 1587\,\mathrm{L} = 1.59\,\mathrm{m}^3$$

Volume of NaOH (V_3):

$$V_3 = \frac{120\,\frac{\mathrm{g}}{\mathrm{L_{resin}}} \cdot 1370\,\mathrm{L_{resin}}}{40\,\frac{\mathrm{g}}{\mathrm{eq}} \cdot 1\,\frac{\mathrm{eq}}{\mathrm{L}}} = 4110\,\mathrm{L} = 4.11\,\mathrm{m}^3$$

Volume of acid for neutralizing base (V_4):
Excess of acid

$$\frac{162\,\frac{\mathrm{g}}{\mathrm{L_{resin}}} \cdot 960\,\mathrm{L_{resin}}}{49\,\frac{\mathrm{g}}{\mathrm{eq}}} - 960\,\mathrm{eq} = 2214\,\mathrm{eq}$$

Excess of base

$$\frac{120\,\frac{\mathrm{g}}{\mathrm{L_{resin}}} \cdot 1370\,\mathrm{L_{resin}}}{40\,\frac{\mathrm{g}}{\mathrm{eq}} \cdot 1\,\frac{\mathrm{eq}}{\mathrm{L}}} - 960\,\mathrm{eq} = 3150\,\mathrm{eq}$$

Acid needed to neutralize base

$$3150 - 2214\,\mathrm{eq} = 936\,\mathrm{eq}$$

$$V_4 = \frac{936\,\mathrm{eq}}{2\,\frac{\mathrm{eq}}{\mathrm{L}}} = 468\,\mathrm{L} = 0.47\,\mathrm{m}^3$$

Total volume:

$$V_\mathrm{T} = V_1 + V_2 + V_3 + V_4 = 10.8\,\mathrm{m}^3$$

Concentration in mixed solution:

$$[\mathrm{Ca}^{2+}] + [\mathrm{Mg}^{2+}] = \frac{3\,\frac{\mathrm{meq}}{\mathrm{L}} \cdot 400\,\frac{\mathrm{L}}{\mathrm{min}} \cdot 60\,\frac{\mathrm{min}}{\mathrm{h}} \cdot 8\,\mathrm{h}}{10.8\,\mathrm{m}^3 \cdot 10^3\,\frac{\mathrm{L}}{\mathrm{m}^3}} = 52.8\,\frac{\mathrm{meq}}{\mathrm{L}}$$

$$= 1057\,\frac{\mathrm{mg}}{\mathrm{L}}\ \mathrm{as\ Ca}^{2+}$$

$$[Na^+] = \frac{2\frac{meq}{L} \cdot 400\frac{L}{min} \cdot 60\frac{min}{h} \cdot 8\,h + 4110\,L \times 10^3\frac{meq}{L}}{10.8\,m^3 \times 10^3\frac{L}{m^3}}$$

$$= 412\frac{meq}{L} = 9483\frac{mg}{L}$$

$$[HCO_3^-] = \frac{2\frac{meq}{L} \cdot 400\frac{L}{min} \cdot 60\frac{min}{h} \cdot 8\,h}{10.8\,m^3 \times 10^3\frac{L}{m^3}} = 35.2\frac{meq}{L} = 2149\frac{mg}{L}$$

$$[Cl^-] = \frac{1\frac{meq}{L} \cdot 400\frac{L}{min} \cdot 60\frac{min}{h} \cdot 8\,h}{10.8\,m^3 \times 10^3\frac{L}{m^3}} = 17.6\frac{meq}{L} = 625\frac{mg}{L}$$

$$[SO_4^{2-}] = \frac{2\frac{meq}{L} \cdot 400\frac{L}{min} \cdot 60\frac{min}{h} \cdot 8\,h}{10.8\,m^3 \times 10^3\frac{L}{m^3}} = 35.2\frac{meq}{L} = 1690\frac{mg}{L}$$

(d) If an anion exchange bed precedes the cation exchange bed, Ca^{2+} and Mg^{2+} will probably precipitate in the anion exchange bed due to the low solubility of their hydroxide and carbonate salts at high pH.

$$M^{2+} + 2Cl^- + 2\overline{R^+OH^-} \leftrightarrow 2\overline{R^+Cl^-} + 2OH^- + M^{2+} \leftrightarrow M(OH)_2(s)\downarrow;$$
$$M^{2+} = Ca^{2+}, Mg^{2+}$$

(e)

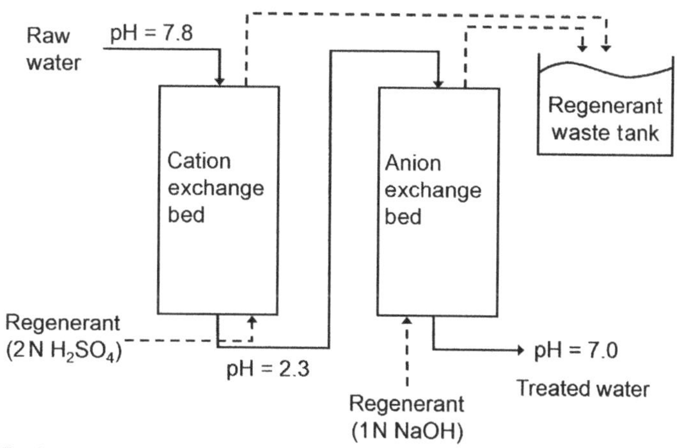

During an exhaustion run, H^+ in the effluent from the cation exchange bed is 5.0×10^{-3} M (pH = 2.3).

(f) Part c of the solution provides various concentrations and total masses of different cations and anions present in the spent regenerant. Ideally, the most efficient deionization system would be when the equivalents removed from the feed water during the service cycle is equal to the equivalents present in the spent regenerant. Equivalents in the spent regenerant are always greater than what have been removed. Thus, a dimensionless sustainability indicator for ion exchange, SI_{IX}, can be presented as the ratio of the equivalents in the spent regenerant over equivalents removed.

$$\text{Cations and anions removed per L of water treated}$$

$$= [Ca^{2+}] + [Mg^{2+}] + [Na^{+}] + [HCO_3^{-}] + [Cl^{-}] + [SO_4^{2-}]$$

$$= 5.0\frac{\text{meq}}{\text{L}}\text{cations} + 5.0\frac{\text{meq}}{\text{L}}\text{anions} = 10\frac{\text{meq}}{\text{L}}$$

$$\text{Regenerant added} = [H^{+}]_{\text{added}} + [OH^{-}]_{\text{added}}$$

$$= 2214\,\text{eq}\,H^{+} + 3150\,\text{eq}\,OH^{-} = 5364\,\text{eq}$$

$$\text{pH neutralization requirement} = [OH^{-}]_{\text{added}} - [H^{+}]_{\text{added}}$$

$$= 3150\,\text{eq}\,OH^{-} - 2214\,\text{eq}\,H^{+} = 936\,\text{eq}$$

$$= \frac{\text{Regenerant Added} + \text{pH Neutralization Requirement}}{\text{Cations and Anions Removed}}$$

$$= \frac{([H^{+}]_{\text{added}} + [OH^{-}]_{\text{added}}) + ([OH^{-}]_{\text{added}} - [H^{+}]_{\text{added}})}{[Ca^{2+}] + [Mg^{2+}] + [Na^{+}] + [HCO_3^{-}] + [Cl^{-}] + [SO_4^{2-}]}$$

$$= \frac{2214\,\text{eq}\,H^{+} + 3150\,\text{eq}\,OH^{-} + 936\,\text{eq}\,H^{+}}{10\frac{\text{meq}}{\text{L}} \cdot \frac{1\,\text{eq}}{1000\,\text{meq}} \cdot 400\frac{\text{L}}{\text{min}} \cdot 60\frac{\text{min}}{\text{h}} \cdot 8\,\text{h}} = 3.28$$

SI_{IX} values greater than unity represent the deviation of the process from its theoretical limit – the process in this example is very inefficient. Later in Chapter 2, we will see how the regeneration efficiency can be significantly improved and SI_{IX} value lowered by modifying the deionization process.

Summary

- The phenomenon of cation exchange was first recognized and reported in 1850 by two British soil scientists, J.T. Way and H.S. Thompson, while investigating properties of natural soils. They provided experimental evidence that the process of ion exchange is reversible and takes place through equivalent exchange of ions. Also, different cations have different selectivity toward the soil. In general, natural materials were found to have very poor anion exchange behavior.
- In 1935, Adams and Holmes synthesized the first organic-based (polymeric) cation exchanger. In less than 10 years, D'Alelio prepared the first synthetic anion exchange resin. Since then, synthesis of new ion exchange resins has never slowed down. The Second World War and, more specifically, the race for nuclear technology helped catalyze the growth and maturity of the field of ion exchange.
- Ion exchange resins have charged functional groups (positive or negative) that are covalently attached to a cross-linked polymer, often called matrix. Inorganic ion

exchangers, namely clays or zeolites, have cation exchange capacity that is obtained through isomorphous substitutions of silicon with aluminum.
- Many redox and precipitation–dissolution reactions have the appearance of ion exchange reactions, but they are not. An ion exchanger is not a pure solid phase. It contains counterions at varying distributions, like a continuous solid solution.
- The large-scale synthesis of polymeric ion exchangers has been standardized and is practiced globally. The four most widely used ion exchange resins are: weak-acid cation exchange resins, strong-acid cation exchange resins, weak-base anion exchange resins and strong-base anion exchange resins. Due to uncontrolled methylene bridging, anion exchange resins possess lower capacity than their cation counterparts.
- Biosorbent materials often exhibit high sorption affinity toward heavy metals and other trace contaminants of environmental concern. They are biorenewable and attractive from a sustainability viewpoint. Poor chemical stability under extreme pH and redox conditions and lack of mechanical strength are major impediments for large-scale application of biosorbent materials.
- Liquid ion exchangers have exchange at the phase boundaries of the immiscible liquids and have much faster kinetics versus solid phase ion exchangers that are controlled by intraparticle diffusion. But effluent solutions are often contaminated with organic solvent. SIRs use an organophilic extractant and are easy to synthesize. However, the gradual loss of complexant during lengthy fixed-bed processes is a major shortcoming of the SIRs.
- Oxides of polyvalent metals, namely, Fe(III), Zr(IV) and Ti(IV) are amphoteric. They exhibit high sorption affinity toward both transition metal cations and anionic ligands.
- Activated carbon and ion exchange resins are the two most commonly used adsorbent materials for packed-bed water treatment processes and have similar operating procedures. But the genesis of their sorption capacity is distinctly different. While carbon adsorption is a surface phenomenon, the ion exchange capacity is derived from the functional group density, that is, the number of covalently attached functional groups per unit volume of the ion exchanger.
- Softening and demineralization (DM) are the two most common applications of ion exchange. These industrial processes improve system efficiency through hardness scaling prevention and TDS elimination, respectively. Both have sustainability concerns due to poor regeneration efficiency on an equivalent basis and consequent production of spent regenerant with large amounts of very high TDS.

References

1 Thompson, H.S. (1850) On the absorbent power of soils. *Journal of the Royal Agricultural Society of England*, **11**, 68–74.
2 Way, J.T. (1850) On the power of soils to absorb manure. *Journal of the Royal Agricultural Society of England*, **11**, 313–379.
3 Adams, B.A. and Holmes, E.L. (1935) Adsorptive properties of synthetic resins. *Journal of the Society of Chemical Industry, London*, **54** (1), 1–6.

4 D'Alelio, G.F., inventor. General Electric, assignee (1944). Process for removing cations from liquid media. US Patent 2340110A. 1944 1944-01-25.
5 D'Alelio, G.F., inventor. D'Alelio GF, assignee (1944). Production of synthetic polymeric compositions comprising sulphonated polymerizates of poly-vinyl aryl compounds and treatment of liquid media therewith. US Patent 2,366,007A.
6 D'Alelio GF, inventor. General Electric, assignee (1944) Production of synthetic polymeric compositions comprising aminated polymerizates of poly-vinyl aryl compounds and treatment of liquid media therewith. USA Patent US2,366,008A. 1944-12-26.
7 Google Patents Search "Anion Exchange". (2016) https://patents.google.com/?q=%22anion+exchange%22 (accessed 30 November 2016).
8 Google Patents Search "Cation Exchange". (2016) https://patents.google.com/?q=%22cation+exchange%22 (accessed 30 November 2016).
9 Boyd, G., Adamson, A., and Myers, L. Jr., (1947) The exchange adsorption of ions from aqueous solutions by organic zeolites. II. Kinetics. *Journal of the American Chemical Society*, **69** (11), 2836–2848.
10 Boyd, G., Myers, L. Jr., and Adamson, A. (1947) The exchange adsorption of ions from aqueous solutions by organic zeolites. III. Performance of deep adsorbent beds under non-equilibrium conditions. *Journal of American Chemical Society*, **69** (11), 2849–2859.
11 Boyd, G., Schubert, J., and Adamson, A. (1947) The exchange adsorption of ions from aqueous solutions by organic zeolites. I. Ion-exchange equilibria. *Journal of American Chemical Society*, **69** (11), 2818–2829.
12 Lemberg, J. (1876) Uber Silicatumwandlungen. *Zeitschr. Deutsch. Geol. Gesell.*, **28**, 519–621.
13 Lemberg, J. (1887) Zur Kenntnis der Bildung und Umbildung von Silicaten. [S. 583: Verhalten d. Natrolith v. Hohentwiel.], *Zeitschrift der Deutschen Geologischen Gesellschaft* **39**. S. 559–600. [Erschienen 188.] – Ausz.:Neues Jahrb f. Min. u. s. w., 1889. II Ref. S. 34–36.
14 Gans, R. (1905) Zeolites and similar compounds, their constitution and significance for technology and agriculture. *Jahrb. Preuss Geol. Landesanst (Berlin)*, **26**, 179.
15 Adams, B.A. and Holmes, E.L. (1935b) Base exchange resins. *Chemical Age of India*, **38**, 117.
16 Marinsky, J.A., Glendenin, L.E., and Coryell, C.D. (1947) The chemical identification of radioisotopes of neodymium and of element 61. *Journal of the American Chemical Society*, **69** (11), 2781–2785.
17 Swinton, E. and Weiss, D. (1953) Counter current adsorption separation processes. 1. Equipment. *Australian Journal of Applied Science*, **4** (2), 316–328.
18 Higgins, I. and Chopra, R. (1970) CHEM-SEPS continuous ion exchange contactor and its applications to de-mineralization processes. *Ion Exchange in the Process Industries: Society of Chemical Industry London*, 121–126.
19 Arden, T.V., Davis, J.B., Herwig, G.L. *et al.* (September 1958) *Extraction of Uranium from Acid Leach Pulps by Jigged Bed Ion Exchange*, Second UN International Conference on The Peaceful Uses of Atomic Energy, Geneva.
20 Helfferich, F.G. (1962) *Ion Exchange*, McGraw Hill Book Co., Inc., New York.
21 Cloete, F.L.D. and Streat, M. (1963) A new continuous solid–fluid contacting technique. *Nature*, **200**, 1199–1200.

22 Kressman, T.R.E. (1969) Properties of some modified polymer networks and derived ion exchangers. *Ion Exchange in the Process Industries, Society of Chemical Industry, London*, 3–9.
23 Kunin, R. (1969) Pore structure of macroreticular ion exchange resins. *Ion Exchange in the Process Industries; Society of Chemical Industry: London*, 10–15.
24 Warshawsky, A., Strikovsky, A., Jerabek, K., and Cortina, J. (1997) Solvent-impregnated resins via acid–base interaction of poly(4-vinylpyridine) resin and di(2-ethylhexyl)dithiophosphoric acid. *Solvent Extraction and Ion Exchange*, **15** (2), 259–283.
25 Cortina, J., Warshawsky, A., Miralles, N. *et al.* (1994) Removal of heavy metal ions from liquid effluents by solvent-impregnated resins, in *Hydrometallurgy'94*, Institution of Mining and Metallurgy, the Society of Chemical Industry, Springer, pp. 725–739.
26 Muraviev, D.N. (1979) *Zhurnal Fizicheskoi Khimii*, **53** (2), 438–444.
27 Muraviev, D.N. and Khamizov, R. (2004) Ion-exchange isothermal supersaturation: concept, problems, and applications, in *Ion Exchange and Solvent Extraction*, A Series of Advances, vol. **16** (eds A. SenGupta and Y. Marcus), Marcel Dekker, New York, pp. 119–210.
28 Schwellnus, A.H. and Green, B.R. (1990) The chemical stability, under alkaline conditions, of substituted imidazoline resins and their model compounds. *Reactive Polymers*, **12** (2), 167–176.
29 Conway, B.E. (1981) *Ionic Hydration in Chemistry and Biophysics (Studies in Physical and Theoretical Chemistry)*, Elsevier Science Ltd, Amsterdam.
30 Clearfield, A. (1982) *Inorganic Ion Exchange Materials*, CRC Press, Inc., Boca Raton.
31 Qureshi, M. and Varshney, K.G. (1991) *Inorganic Ion Exchangers in Chemical Analysis*, CRC Press Inc., Boca Raton.
32 Szostak, R. (1992) *Hand Book of Molecular Sieves: Structures*, Van Nostrand Reinhold, New York, NY.
33 Flanigen EM. Chapter 2. Zeolites and molecular sieves – an historical perspective. In: van Bekkum H, Flanigen EM, Jansen J, editors. *Introduction to Zeolite Science and Practice* Amsterdam: Elsevier; 1991. p. 13–34.
34 Lutz, W. (2014) Zeolite Y: synthesis, modification, and properties – a case revisited. *Advances in Materials Science and Engineering*, **22**, 2014.
35 Stöcker, M., Karge, H., Jansen, J., and Weitkamp, J. (1994) *Advanced Zeolite Science and Applications (Volume 85 of Studies in Surface Science and Catalysis)*, Elsevier, Amsterdam.
36 Cejka, J. (2007) *Introduction to Zeolite Science and Practice (Volume 168 of Studies in Surface Science and Catalysis)*, Elsevier, Amsterdam.
37 Volesky, B. (1990) *Biosorption of Heavy Metals*, CRC Press, Inc., Boca Raton.
38 Streat, M., Mailk, D.J., and Saha, B. (2004) Adsorption and ion-exchange properties of engineered activated carbons and carbonaceous materials, in *Ion Exchange and Solvent Extraction: A Series of Advances*, vol. **16** (eds A.K. SenGupta and Y. Marcus), Marcel Dekker, Inc., New York, pp. 1–84.
39 Inoue, K., Baba, Y., and Yoshizuka, K. (1993) Adsorption of metal ions on chitosan and crosslinked copper(II)-complexed chitosan. *Bulletin of the Chemical Society of Japan*, **66** (10), 2915–2921.

40 Pomogailo, A.D., Dzhardimalieva, G.I., Rozenberg, A.S., and Muraviev, D.N. (2003) Kinetics and mechanism of in situ simultaneous formation of metal nanoparticles in stabilizing polymer matrix. *Journal of Nanoparticle Research*, **5** (5–6), 497–519.

41 Guibal, E. (2004) Interactions of metal ions with chitosan-based sorbents: a review. *Separation and Purification Technology*, **38** (1), 43–74.

42 Bart H, Stevens GW. Reactive solvent extraction, in *Ion Exchange and Solvent Extraction: A Series of Advances* (vol. **17**, edited by Marcus Y., SenGupta A. K.), Marcel Dekker Inc. New York, 2004.

43 Baker, R.W. (2000) *Membrane Technology and Applications*, 3rd edn, Wiley Online Library, Chichester, United Kingdom.

44 Winston, W.S.W. and Sirkar, K.K. (1992) *Membrane Handbook Volume I*, Springer Science + Business Media, LLC, New York.

45 Alexandratos, S.D. and Ripperger, K.P. (1998) Synthesis and characterization of high-stability solvent-impregnated resins. *Industrial and Engineering Chemistry Research*, **37** (12), 4756–4760.

46 Dorfner, K. (1991) *Ion Exchangers*, Walter de Gruyter, New York.

47 Brown, G.E., Henrich, V.E., Casey, W.H. *et al.* (1999) Metal oxide surfaces and their interactions with aqueous solutions and microbial organisms. *Chemical Reviews*, **99** (1), 77–174.

48 Stumm, W. and Morgan, J.J. (1996) *Aquatic Chemistry: Chemical Equilibria and Rates in Natural Waters*, 4th edn, John Wiley & Sons, New York.

49 Dzombak, D.A. and Morel, F.M.M. (1990) *Surface Complexation Modeling: Hydrous Ferric Oxides*, Wiley Interscience, John Wiley and Sons Inc., New York.

50 Cowan, C.E., Zachara, J.M., and Resch, C.T. (1991) Cadmium adsorption on iron oxides in the presence of alkaline-earth elements. *Environmental Science & Technology*, **25** (3), 437–446.

51 Manning, B.A., Fendorf, S.E., and Goldberg, S. (1998) Surface structures and stability of arsenic(III) on goethite: spectroscopic evidence for inner-sphere complexes. *Environmental Science & Technology*, **32** (16), 2383–2388.

52 Cumbal, L. and SenGupta, A.K. (2005) Arsenic removal using polymer-supported hydrated iron(III) oxide nanoparticles: role of Donnan membrane effect. *Environmental Science & Technology*, **39** (17), 6508–6515.

53 Puttamaraju, P. and SenGupta, A.K. (2006) Evidence of tunable on-off sorption behaviors of metal oxide nanoparticles. *Industrial and Engineering Chemistry Research*, **45**, 7737–7742.

54 Kney, A. (1999) *Synthesis and characterization of a new heavy-metal-selective inorganic ion exchanger*. Ph.D. Dissertation: Lehigh University, Bethlehem, PA.

55 Blaney, L.M., Cinar, S., and SenGupta, A.K. (2007) Hybrid anion exchanger for trace phosphate removal from water and wastewater. *Water Research*, **41** (7), 1603–1613.

56 Sarkar, S., Blaney, L.M., Gupta, A. *et al.* (2008) Arsenic removal from groundwater and its safe containment in a rural environment: validation of a sustainable approach. *Environmental Science & Technology*, **42** (12), 4268–4273.

57 Kunin, R. and Meyers, R.J. (1951) *Ion Exchange Resins*, John Wiley and Sons Inc., New York.

58 SenGupta, A.K. (1995) *Ion Exchange Technology: Advances in Pollution Control*, Technomic Publishing Co. Inc., Lancaster, PA.

59 Sengupta, S.K. and SenGupta, A.K. (1993) Characterizing a new class of sorptive/desorptive ion exchange membranes for decontamination of heavy-metal-laden sludges. *Environmental Science & Technology*, **27** (10), 2133–2140.
60 Soldatov VS, Pawlowski L, Wasag H, Schunkievich A. *New Materials and Technologies for Environmental Engineering. Part I. Structure and Syntheses of Ion Exchange Fibers*. Vol. 21. Monographs of the Polish Academy of Sciences Lublin, Poland: Komitet Inzynierii Srodowiska PAN; 2004. p. 1–127.
61 Zagorodni, A.A. (2006) *Ion Exchange Materials: Properties and Applications: Properties and Applications*, Elsevier, Amsterdam, Netherlands.
62 SenGupta, A.K. (2001) *Environmental Separation of Heavy Metals: Engineering Processes*, CRC Press, Boca Raton, FL.
63 Hashida, I. and Nishimura, M. (1976) Adsorption and desorption of sulfur dioxide by macroreticular strong-base anion exchanger. *Journal of the Chemical Society of Japan, Chemistry and Industrial Chemistry*, **4**, 131–135.
64 Leun, D. and SenGupta, A.K. (2000) Preparation and characterization of magnetically active polymeric particles (MAPPs) for complex environmental separations. *Environmental Science & Technology*, **34** (15), 3276–3282.
65 Greenleaf, J.E. and SenGupta, A.K. (2006) Environmentally benign hardness removal using ion-exchange fibers and snowmelt. *Environmental Science & Technology*, **40** (1), 370–376.
66 SenGupta AK, Cumbal LH, inventors. (2007) SenGupta AK, assignee. Hybrid anion exchanger for selective removal of contaminating ligands from fluids and method of manufacture thereof. USA patent US7291578B2. 2007-11-06.
67 Li, J., Koner, S., German, M., and SenGupta, A.K. (2016) Aluminum-cycle ion exchange process for hardness removal: a new approach for sustainable softening. *Environmental Science & Technology*, **50** (21), 11943–11950.
68 Santa Clarita Sanitation District of Los Angeles County. SANTA CLARA RIVER CHLORIDE REDUCTION ORDINANCE OF 2008. 2008;DOC# 1035050:1–6.

2

Ion Exchange Fundamentals

Throughout this chapter, the goal is to first present easy-to-comprehend physical-chemical phenomena that accompany ion exchange processes. Later, principles and theories governing these phenomena are gradually introduced.

2.1 Physical Realities

When an ion exchange resin, for example, a spherical gel type cation exchanger in sodium form, is introduced into a dilute solution containing an electrolyte, for example, KCl, the following three phenomena occur simultaneously until equilibrium is attained:

(a) The ion exchange resin beads swell; they may also shrink in a very concentrated solution.
(b) Sodium ions in the cation exchanger are partly exchanged with potassium ions from the solution. Consequently, the external solution now contains sodium ions along with potassium. Both sodium and potassium ions are also present at the ion exchange sites inside the cation exchanger.
(c) Trace amounts of coions (Cl^- in this case) also enter the exchanger phase.

Figure 2.1 depicts a representative cartoon and it is important to recognize that "ion exchange" is more than the mere exchange of ions. Swelling or shrinking of the exchanger results from osmosis; redistribution of Na^+ and K^+ in both the exchanger and the aqueous phase is a sequel to their relative selectivity toward the ion exchanger; and the presence of Cl^- in the exchanger phase is caused by electrolyte penetration or Donnan coion invasion. The three abovementioned phenomena essentially cover the entire gamut of ion exchange equilibrium. Although they occur simultaneously, these phenomena will be treated individually in the succeeding sections in an attempt to visualize the physical realities with associated scientific insight.

It is worth noting that ion exchange fundamentals have emerged and matured primarily through the works of physical chemists who were, interestingly enough, never in the mainstream of synthesizing wide ranges of organic and inorganic ion exchangers available today. Understandably, the current body of knowledge in ion exchange fundamentals and, more importantly, the relative importance of embedded assumptions and theoretical complexities have still not been well assimilated by others engaged

Ion Exchange in Environmental Processes: Fundamentals, Applications and Sustainable Technology,
First Edition. Arup K. SenGupta.

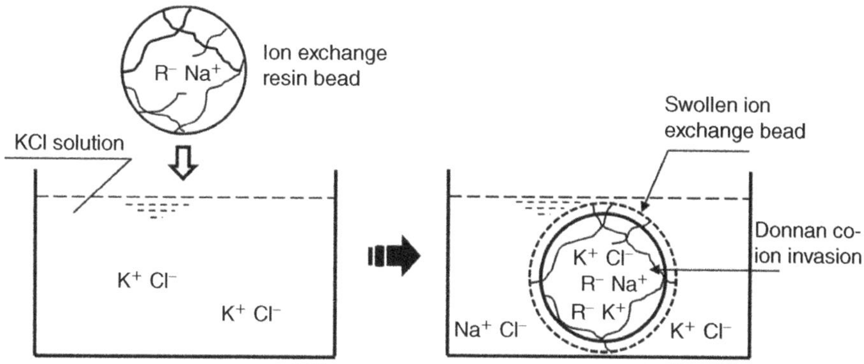

Figure 2.1 Illustration of an ion exchange process showing ion exchange, swelling and coion invasion.

in synthesizing resins or in modifying and applying ion exchange processes for the multitude of applications. The study of fundamentals of the "ion exchange process" as a physical-chemical phenomenon has led to many quantitative models. Rigorous attempts have, however, been conspicuously absent to bridge the gap among many of these models. Many approaches, although theoretically elegant, become cumbersome in real scenarios, and are not amenable to application beyond simple systems. An overly empirical model, on the contrary, fails to address physical realities under varying conditions. The primary intent in succeeding sections of this chapter is to connect the key physical-chemical phenomena with the basic scientific principles, while embracing empiricism and avoiding undue complexity.

2.2 Swelling/Shrinking: Ion Exchange Osmosis

An ion exchange resin can be viewed as a cross-linked polyelectrolyte gel or a pseudo-liquid, within which a high concentration of fixed charges is present. So, the internal osmotic pressure of an ion exchanger is very high. Once introduced in water, the covalently attached fixed charges are unable to diffuse into water due to their immobility. Despite the significant concentration gradient, the mobile counterions cannot diffuse into the aqueous phase because that will result in an imbalance in electroneutrality. The ion exchanger–water interface thus acts as a semi-permeable membrane and water molecules move inside the ion exchanger to decrease the difference in osmotic pressure, that is, the process of osmosis is initiated. The exchanger phase thus tends to dilute itself, that is, to swell. The resin is stretched and it finally comes to equilibrium when the "swelling pressure" in the resin balances the osmotic pressure gradient. The mechanistic model to describe the behavior of an ion exchange resin primarily from a swelling-shrinking viewpoint was first presented by Gregor [1–3]. The matrix of the resin, per this model, is a network of elastic springs on which the ionogenic groups are attached. An illustration of Gregor's concept of elastic force vis-à-vis osmotic force on swelling-shrinking is exhibited in Figure 2.2 [4].

As the water or solvent molecules enter the ion exchanger due to the osmotic pressure difference, the resin swells and the springs are stretched. Hence, the pore liquid

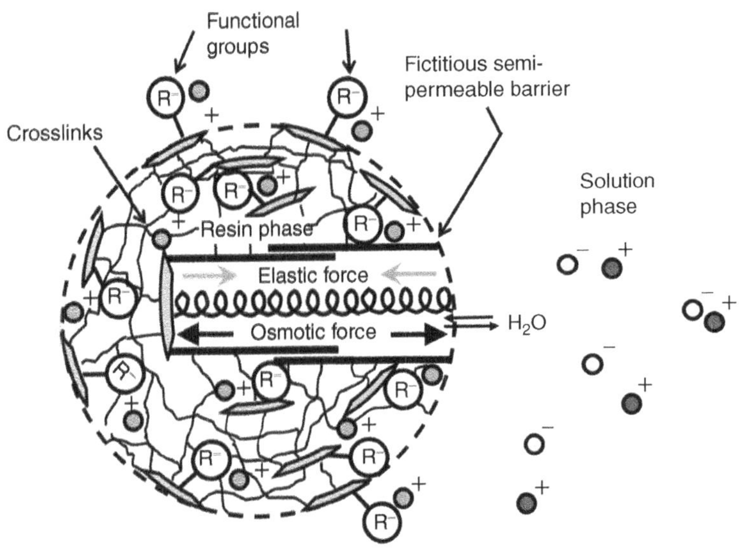

Figure 2.2 Schematic illustration of elastic force vis-à-vis osmotic force in swelling and shrinking of ion exchange resin. *Source:* Sarkar *et al.* 2010 [4]. Reproduced with permission of American Chemical Society.

within the exchanger experiences a pressure, known as "swelling pressure" caused by the elastic force of the cross-linked resin. Note that the elastic force and the osmotic force act in opposite directions. It is significant that inorganic ion exchangers, specifically zeolites, can be viewed as solid solutions of aluminosilicates and are amenable to interpretation through similar spring models. Table 2.1 provides a summary delineating the effects of different process variables on swelling and the underlying scientific rationale. Interestingly, however, the swelling-shrinking behavior also allows a methodology to determine the relative selectivity of ions with the same valence. For the most common univalent counterions of a strong-acid cation exchange resin, the sequence of swelling is

$$K^+ < Na^+ < Li^+$$

Selectivity or sorption affinity is inversely related to the sequence of swelling and stands as follows:

$$K^+ > Na^+ > Li^+$$

Later in Section 2.5.1, we will further revalidate the genesis of selectivity based on Coulombic interaction. In this series, swelling decreases with a decrease in hydrated ionic radius, i.e., the hydrated ionic radius of K^+ is smaller than Na^+, and that of Na^+ is smaller than Li^+.

Although the swelling-hydrated ionic radius sequence conforms inversely to the order of selectivity of the counterions, the use of swelling equilibria to determine the relative selectivity of target ions has not progressed much. The degree of swelling of an ion exchanger is directly related to its water content and the swelling data for a styrene-DVB cation exchange resin in different ionic forms is presented in Figure 2.3 [5,6].

Table 2.1 Effect of variables on swelling.

Variable	Ion exchanger swelling	Remarks
Exchanger capacity increases	Increases	Osmotic pressure of the ion exchanger increases
Solution ionic strength increases	Decreases	Osmotic pressure difference between the solution and exchanger decreases
Degree of cross-linking increases	Decreases	Stretching of the ion exchanger is opposed
Counterion hydrated ionic radius increases	Increases	More water is imbibed into the exchanger for identical equivalent exchange

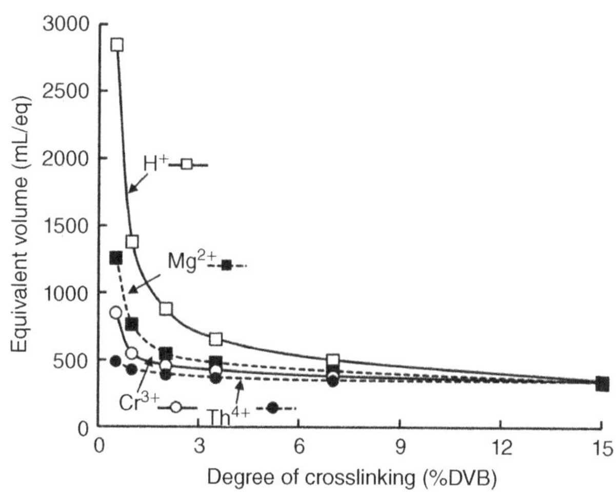

Figure 2.3 Plot showing dependence of swelling on the degree of cross-linking and the counterion valence for a strong-acid cation exchanger. *Source:* Adapted from Calmon 1952 [5] and Calmon 1953 [6][5,6].

The degree of DVB cross-linking has a very significant effect on the water content of an ion exchanger. Thus, the "pore water" experiences greater swelling pressure in a more highly cross-linked ion exchanger. For an analogy, one may consider two balloons with different skin thicknesses. To inflate, the balloon with a greater skin thickness will require a higher pressure to arrive at the same inflated volume, that is, air molecules inside will be subjected to greater pressure. The following points also are noteworthy:

- Swelling or shrinking is a reversible process and is indeed accompanied by an uptake or expulsion of water molecules.
- When a counterion inside an ion exchanger is replaced by a counterion with higher valence, the osmotic pressure in the exchanger phase decreases and hence, the resin shrinks.

- Beyond a certain percentage of DVB cross-linking, the type of counterion has minimal effect on the swelling observed. So, ion exchangers with lower cross-linking undergo greater swelling-shrinking effects as the external solution concentration changes from the sorption cycle (dilute) to the regeneration cycle (concentrated) resulting in mechanical fracturing and the formation of fines. The operational life of ion exchangers with lower cross-linking is, therefore, relatively short. However, the higher water content of ion exchangers with lower cross-linking tends to offer faster ion exchange kinetics, as will be demonstrated in a later chapter.

Example 2.1 *A swelling experiment is conducted with 1 g polystyrene-divinylbenzene (PS-DVB) matrix cation exchanger* ($\overline{R - SO_3^- H^+}$) *in dilute 1 mM NaCl solution. Note: Overbar indicates resin phase. The volume of water uptake by the resin is found to be 1200 mL/eq. The number of bound water molecules for each functional site* ($\overline{R - SO_3^- H^+}$) *is 6. The capacity of the cation exchanger is 5.3 meq/g. Calculate the quantity of bound and unbound water in the cation exchanger. Comments welcome.*

Basis: 1 g of cation exchange resin

Ion exchange capacity:

$$5.3\,\text{meq} = 5.3 \times 10^{-3}\,\text{eq}$$

Number of functional sites (based on Avogadro's number):

$$5.3 \times 10^{-3} \cdot 6.022 \times 10^{23} = 3.19 \times 10^{21}$$

Number of bound water molecules:

$$6 \times 3.19 \times 10^{21} = 1.91 \times 10^{22}$$

Mass of water = 18 mg/mmol, that is, 18 mg for 6.02×10^{20} molecules

Mass of water bound to the functional sites:

$$6 \frac{\text{mmol}\,H_2O}{\text{meq resin}} \cdot 5.3\,\text{meq resin} \cdot 18 \frac{\text{mg}\,H_2O}{\text{mmol}\,H_2O} = 572\ \text{mg}\,H_2O = 0.572\ \text{g}\,H_2O$$

Volume of bound water (assumed density 1 g/mL):

$$0.572\ \text{g} \cdot \left(1 \frac{\text{g}\,H_2O}{\text{mL}\,H_2O}\right)^{-1} = 0.572\,\text{mL}\,H_2O$$

Water uptake = 1200 mL/eq = 1.2 mL/meq

Total water uptake by 1 g ion exchanger:

$$1.2 \frac{\text{mL}\,H_2O}{\text{meq resin}} \cdot 5.3 \frac{\text{meq resin}}{1\,\text{g resin}} \cdot 1\,\text{g resin} = 6.36\,\text{mL}\,H_2O$$

Quantity of free water (not bound with the hydration shell of the functional groups):

$$6.36\,\text{mL} - 0.572\,\text{mL} = 5.79\,\text{mL}\,H_2O$$

So, the amount of free water (>90%) is much more than water bound to functional sites.

Comment: During the process of osmosis, water flows from a lower solute concentration to a higher solute concentration through a semi-permeable membrane that allows water to diffuse, but does not allow ions to diffuse. In this swelling example, water permeated into the ion exchanger and caused swelling, although there was no semi-permeable membrane. This was possible due to the presence of fixed non-diffusible functional groups in the exchanger imparting semi-permeability. Such a process of osmosis is known as Donnan osmosis and is due to the Donnan membrane principle, which will be detailed in a later section of this chapter.

2.3 Ion Exchange Equilibrium

Of the three ion exchange phenomena illustrated in Figure 2.1, the redistribution of counterions between the exchanger and the solvent (water) phase is the most significant for nearly every application pertaining to separation processes. Understandably, there are different approaches toward defining ion exchange equilibrium. Of them, the Law of Mass Action is universally used and will be used throughout this book, as it can be readily extended to comprehend physical realities of complex systems. The Law of Mass Action, as applied to ion exchange systems, is essentially an extension of the second law of thermodynamics under less rigorous conditions. A derivation of the ion exchange equilibrium constant, K_{IX}, substantiates this claim in the Supplementary Reading S2.1. However, a reader can proceed straight into the next section without any loss of continuity.

Supplementary Reading S2.1 Ion Exchange Equilibrium Constant: Convergence of Thermodynamics and the Law of Mass Action

At equilibrium, the electrochemical potential of a freely permeating component, an ion, solute or water, is equal to that in the exchanger and in the aqueous phase. The electrochemical potential of a component "i" in each phase is

$$\eta_i = \mu_i^0 + RT\ \ln(a_i) + (P-1)\overline{V}_i \pm Z_i F \Psi_i \qquad \text{(S2.1)}$$

Where μ_i^0 is its chemical potential at the standard state of unit activity and one atmosphere pressure, a_i its activity, P is the pressure, $\overline{V_i}$ is its partial molar volume, $\pm Z_i$ is its electrovalency, F is the Faraday constant and Ψ is the electric potential. Note that the total electrochemical potential of a species is made up from its activity (a), pressure (P) and electric potential (Ψ), all compared with the reference state of unit activity (a = 1), infinite dilution (activity coefficient = 1) and one atmosphere pressure.

Let us consider the familiar ion exchange reactions with counterions A^+ and B^+

$$\overline{R^-B^+} + A^+(aq) \leftrightarrow \overline{R^-A^+} + B^+(aq) \qquad \text{(S2.2)}$$

Accordingly, the stoichiometry of an ion exchange reaction between two counterions A and B with charges z_A and z_B can be presented as follows:

$$\frac{1}{z_B}\overline{B^{z_B}} + \frac{1}{z_A}A^{z_A} \leftrightarrow \frac{1}{z_A}\overline{A^{z_A}} + \frac{1}{z_B}B^{z_B} \qquad \text{(S2.3)}$$

Although universally accepted, Eq. (S2.3) fails to include a few accompanying phenomena for ion exchange equilibrium. First, the effects of swelling pressure and the resulting volume

(Continued)

Supplementary Reading S2.1 (Continued)

change have been ignored; second, the activity of water is considered equal in both phases; and, third, the coions associated with A and B only maintain electroneutrality in the aqueous phase with no effects on equilibrium whatsoever. It is noteworthy that the foregoing assumptions are well in order for ion exchange processes of practical relevance. Ignoring the effect of swelling pressure and because no external electrical potential gradient is being applied, electrochemical potential (η_i) and chemical potential (μ_i) are essentially the same

$$\eta_i = \mu_i = \mu_i^0 + RT \ln a_i \tag{S2.4}$$

The free energy change associated with the ion exchange reaction of counterions A and B in Eq. (S2.2) at equilibrium is

$$\Delta G_{IX} = \overline{\mu_A} d\overline{n_A} + \overline{\mu_B} d\overline{n_B} + \mu_A dn_A + \mu_B dn_B = 0 \tag{S2.5}$$

where n_i is the number of moles for the corresponding ion. For convenience in nomenclature, we have designated the exchanger phase with an overbar throughout this book, unless otherwise specified. The mass balance provides

$$d\overline{n_A} = -dn_A \tag{S2.6}$$

$$d\overline{n_B} = -dn_B \tag{S2.7}$$

From electroneutrality consideration,

$$z_A dn_A = -z_B dn_B \tag{S2.8}$$

$$z_A d\overline{n_A} = -z_B d\overline{n_B} \tag{S2.9}$$

Substitution of Eqs ((S2.6))–((S2.9)) into (S2.5) yields,

$$\frac{1}{z_A}\overline{\mu_A} - \frac{1}{z_B}\overline{\mu_B} + \frac{1}{z_B}\mu_B - \frac{1}{z_A}\mu_A = 0 \tag{S2.10}$$

Again, the free energy change corresponding to ion exchange reaction in Eq. (S2.3) at equilibrium is

$$\Delta G_{IX} = 0 = \Delta G_{IX}^0 + RT \ln K_{IX} \tag{S2.11}$$

where K_{IX} is the thermodynamic equilibrium constant and ΔG_{IX}^0 represents the free energy change at the standard state. R is the universal gas constant and T is the temperature in degree Kelvin.

Considering the equality of (S2.10), (S2.11), the relationship in (S2.4) and through minor algebraic manipulations, the thermodynamic equilibrium constant for reaction (S2.3) is computed as

$$K_{IX} = \frac{\overline{a}_A^{\,1/z_A} \cdot a_B^{\,1/z_B}}{\overline{a}_B^{\,1/z_B} \cdot a_A^{\,1/z_A}} \tag{S2.12}$$

Note that Eq. (S2.12) is identical to the expression of the thermodynamic equilibrium constant one would easily derive from the Law of Mass Action. It, however, needs to be recognized that the convergence of the classical thermodynamic approach and the Law of Mass Action into Eq. (S2.12) results from two simplifying assumptions: swelling-shrinking *of the exchanger and the difference in water activities between the two phases have been ignored. As already stated, in the realm of ion exchange, these assumptions have insignificant impacts on the relative distribution of ions of interest. Further, contrary to Eq. (S2.3), the stoichiometry of an ion exchange reaction between two counterions A and B with charges z_A and z_B is more universally presented as*

$$z_A\overline{B^{z_B}} + z_B A^{z_A} \leftrightarrow z_B\overline{A^{z_A}} + z_A B^{z_B} \qquad (S2.13)$$

From the Law of Mass Action, the thermodynamic equilibrium constant for reaction (S2.13) is

$$K_{IX} = \frac{\overline{a}_A^{\,z_B} \cdot a_B^{\,z_A}}{\overline{a}_B^{\,z_A} \cdot a_A^{\,z_B}} \qquad (S2.14)$$

Due to the difference in stoichiometry, the two equilibrium constants are different but related to each other based on stoichiometry:

$$K_{IX}(\text{Reaction S2.13}) = (K_{IX})^{Z_A Z_B}(\text{Reaction S2.3}) \qquad (S2.15)$$

For compatibility with the current body of literature, we will resort to the thermodynamic equilibrium constant, K_{IX}, as stated in Eq. (S2.14) corresponding to the ion exchange reaction in Eq. (S2.13).

2.3.1 Genesis of Non-Ideality

For meaningful application of K_{IX} in real systems, measurement and/or deduction of activities of ions in both phases are required. Let us define ionic activity as follows:

$$\text{In water (or solvent)}, a_i = \gamma_i c_i \qquad (2.1)$$

$$\text{In the exchanger phase}, \overline{a_i} = \overline{\gamma_i c_i} \qquad (2.2)$$

where γ_i is the activity coefficient and c_i is the molar concentration and the overbar denotes the exchanger phase. Let us consider the general ion exchange reaction in (S2.13)

$$z_A\overline{B^{z_B}} + z_B A^{z_A} \leftrightarrow z_B\overline{A^{z_A}} + z_A B^{z_B} \qquad (S2.13)$$

Using the equalities in Eqs (2.1) and (2.2), the thermodynamic equilibrium constant, K_{IX}, in Eq. (S2.14) becomes

$$K_{IX} = \frac{\overline{C_A}^{Z_B} C_B^{\,Z_A}}{\overline{C_B}^{Z_A} C_A^{\,Z_B}} * \frac{\gamma_B^{\,Z_A}}{\gamma_A^{\,Z_B}} * \frac{\overline{\gamma_A}^{Z_B}}{\overline{\gamma_B}^{Z_A}} \qquad (2.3)$$

Precise calculation of the value of the thermodynamic equilibrium constant, K_{IX}, requires determination of activity coefficients in the exchanger phase, that is, $\overline{\gamma_i}$ values. Sadly, a direct activity measurement technique in the exchanger phase is currently absent and the theoretical approaches are based on simplifying assumptions. Despite the absence of theoretical or experimental tools to quantify the exchanger phase activity coefficient, it is imperative to take an insightful look into its physical realities. An ion exchanger is essentially a continuous phase with immobile ion exchange

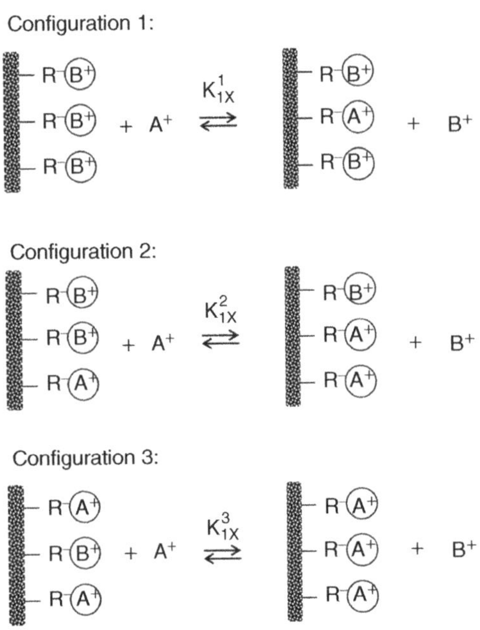

Figure 2.4 Illustration of an ion exchange process with three different lateral configurations.

sites in close proximity to each other. Thus, the thermodynamic activity of an ion exchange site is not dependent solely on its own physicochemical properties, but is also influenced by its nearest neighbors. The genesis of non-ideality in the exchanger phase stems from effects exerted by neighboring ion exchange sites, occasionally referred to as lateral effects. To elucidate such a non-ideality effect, let us consider the following simple case of cation exchange for a specific site:

$$\overline{R^-B^+} + A^+ \leftrightarrow \overline{R^-A^+} + B^+ \tag{2.4}$$

Now, due to different degrees of counterion loading onto the exchanger and associated heterogeneities, the ion exchange in Eq. (2.4) can be presented with at least three different types of configurations for the neighboring sites as shown in Figure 2.4.

Note that the overall ion exchange for each configuration represents Eq. (2.4). However, due to the differences in the energy states for each configuration, the thermodynamic equilibrium constant, K_{IX}, is different from each other

$$K_{IX}^1 \neq K_{IX}^2 \neq K_{IX}^3 \tag{2.5}$$

Using these three different configurations for neighboring sites, Hogfeldt [7,8] developed a three-parameter algebraic model to account for non-ideal behavior in ion exchangers. In a similar vein, Soldatov offered a statistical approach to quantify the dependence of the equilibrium constant on ionic loading of the exchanger [9,10]. Any further discussion is deliberately being avoided here due to the empiricism embedded in such models and the insurmountable difficulty in applying them in real systems with more than two counterions. However, the following guidelines are noteworthy when one tries to bridge the gap between fundamental concepts and physical realities:

- Of all the variables influencing the exchanger phase activity coefficient, $\overline{\gamma_i}$, of a counterion ("i"), the ionic composition of the exchanger is by far the most important one.

While comparing K_{IX} values for two different ion exchangers for exchange of ions A and B, experimental data should be obtained close to near-identical relative loadings of the exchanger, that is, similar values of $\overline{C_A}$ and $\overline{C_B}$ or y_A and y_B.

- The effect of the activity coefficient in the ion exchanger phase is more pronounced with heterovalent ion exchange than with homovalent ion exchange.
- Configuration 1 in Figure 2.4 represents a situation where the counter ion A is a trace species. For trace species, exchanger phase activity coefficients are essentially constant and Henry's law prevails under such conditions.

2.4 Other Equilibrium Constants and Equilibrium Parameters

Due to obvious complexities in the determination of the thermodynamic equilibrium constant, K_{IX}, other less rigorous but relatively easy-to-measure equilibrium constants and equilibrium parameters are widely used. Such "pseudo" constants will be defined and their underlying embedded assumptions will be highlighted, based on the terminologies in the literature.

2.4.1 Corrected Selectivity Coefficient

The corrected selectivity coefficient, K_{IX}^{c}, assumes ideality in the exchanger phase but not in the aqueous phase and becomes equal to

$$K_{IX}^{c} = \frac{\overline{C_A}^{Z_B} C_B^{Z_A}}{\overline{C_B}^{Z_A} C_A^{Z_B}} \cdot \frac{\gamma_B^{Z_A}}{\gamma_A^{Z_B}} \tag{2.6}$$

where C_i and $\overline{C_i}$ values of counterions can be determined experimentally and aqueous-phase activity coefficient values can be computed independently. Considering ion exchange reaction (S2.13) and using Eq. (2.3), the corrected selectivity coefficient, K_{IX}^{c}, is related to the thermodynamic equilibrium constant, K_{IX}, in the following way:

$$K_{IX}^{c} = K_{IX} \cdot \frac{\overline{\gamma_B}^{Z_A}}{\overline{\gamma_A}^{Z_B}} \tag{2.7}$$

Ionic strength, I, is the single most important variable for non-ideality in the aqueous phase and is equal to:

$$I = \frac{1}{2}\sum_{i=1}^{n} c_i z_i^2 \tag{2.8}$$

where c_i is the molar concentration of the ion "i" with a charge of z_i and "n" is the number of ionic species present in the aqueous phase.

The activity coefficient of an ion is related to the ionic strength (I) per the Debye–Hückel equation as follows:

$$\log(\gamma_i) = -Az_i^2\sqrt{I} \tag{2.9}$$

$$A = 1.82 \times 10^4 \cdot (\varepsilon \cdot T)^{-3/2} \tag{2.10}$$

where ε = dielectric constant; T = temperature (K); $A = 0.5$ for water at 25 °C.

The empirical correlation provided by Davies equation [11] is more widely used for ionic strength (I) values up to 0.5 mol/L

$$\log(\gamma_i) = -Az_i^2\left(\frac{\sqrt{I}}{1+\sqrt{I}} - 0.2I\right) \tag{2.11}$$

If the charge remains the same, the aqueous phase activity coefficient values of ions tend to remain the same

$$\gamma_{Na^+} = \gamma_{Cl^-} = \gamma_{K^+} = \gamma_{NO_3^-} = \gamma_1 \tag{2.12}$$

$$\gamma_{Ca^{2+}} = \gamma_{Mg^{2+}} = \gamma_{SO_4^{2-}} = \gamma_{CrO_4^{2-}} = \gamma_2 \tag{2.13}$$

where γ_1 and γ_2 represent activity coefficients of monovalent and divalent ions, respectively. Using Debye–Hückel relationships it can be easily shown that

$$\gamma_2 = (\gamma_1)^4$$

Since aqueous-phase activity coefficients can be computed for a given solution composition (i.e., from its ionic strength), the corrected selectivity coefficient, K_{IX}^{c}, can be obtained from the experimental data for exchanging ions A and B using Eq. (2.6).

2.4.2 Selectivity Coefficient, K_{IX}^{se}

Selectivity coefficient is essentially equal to the thermodynamic equilibrium constant assuming ideality in both aqueous and exchanger phase. Thus,

$$K_{IX}^{se} = \frac{\overline{C_A}^{Z_B} C_B^{Z_A}}{\overline{C_B}^{Z_A} C_A^{Z_B}} = K_{IX} * \frac{\overline{\gamma_B}^{Z_A}}{\overline{\gamma_A}^{Z_B}} * \frac{\gamma_A^{Z_B}}{\gamma_B^{Z_A}} = K_{IX}^{c} * \frac{\gamma_A^{Z_B}}{\gamma_B^{Z_A}} \tag{2.14}$$

Selectivity coefficient is by far the most widely used pseudo-equilibrium constant in the practicing world.

2.4.3 Separation Factor (α_B^A)

Separation factor is essentially an expression of relative affinity of counterions A and B toward the exchanger and identical to the expression of relative volatility in the distillation process. In principle, separation factor is the ratio of the distribution coefficient values (λ_i) of counterions A and B:

$$\alpha_B^A = \frac{\lambda_A}{\lambda_B} = \left(\frac{\overline{C_A}}{C_A}\right) * \left(\frac{C_B}{\overline{C_B}}\right) = \frac{\left(\frac{\overline{C_A}}{Q}\right)}{\left(\frac{C_A}{C_T}\right)} * \frac{\left(\frac{C_B}{C_T}\right)}{\left(\frac{\overline{C_B}}{Q}\right)} = \frac{y_A}{x_A} * \frac{x_B}{y_B} \tag{2.15}$$

where Q and C_T denote total exchanger capacity and total aqueous phase counterion concentration, that is,

$$Q = \overline{C}_A + \overline{C}_B \tag{2.16}$$

$$C_T = C_A + C_B \tag{2.17}$$

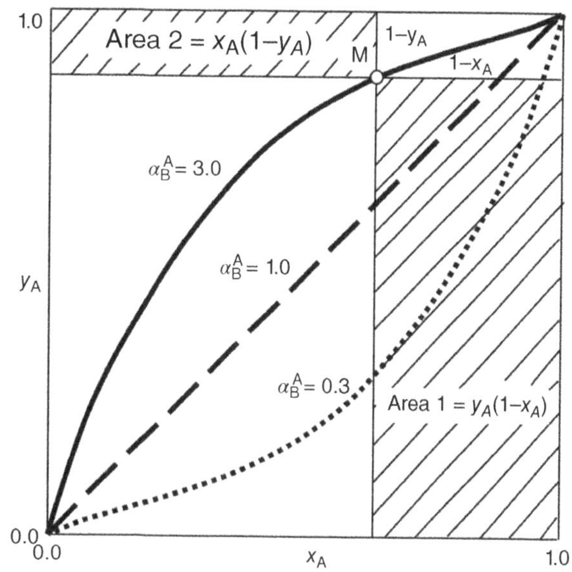

Figure 2.5 Illustration of binary equilibrium isotherm plots (y_A vs x_A) and their relationship with separation factors (Eq. 2.18).

y_i and x_i are equivalent fractions of counterion "i" in the exchanger phase and the aqueous phase, respectively.

Separation factor can be directly computed from the binary equilibrium isotherm plot, y_A versus x_A, as illustrated in Figure 2.5. The diagonal represents an isotherm corresponding to an α_B^A value of unity, that is, both A and B are equally preferred by the exchanger. Figure 2.5 also includes the plots of two isotherms at constant separation factor values of $\alpha_B^A = 3.0$ and $\alpha_B^A = 0.3$. Note that for $\alpha_B^A > 1.0$ (i.e., A is preferred over B), the isotherm always resides above the diagonal while for $\alpha_B^A < 1.0$, that is, when B is preferred over A, the isotherm is consistently below the diagonal. Also, for a given point (say M) in the isotherm, α_B^A is essentially the ratio of the two shaded areas in Figure 2.5 as follows:

$$\alpha_B^A = \frac{y_A x_B}{y_B x_A} = \frac{y_A(1 - x_A)}{x_A(1 - y_A)} = \frac{\text{Area 1}}{\text{Area 2}} \qquad (2.18)$$

In a fixed-bed column, separation factor, and not the equilibrium constant or selectivity coefficient, is the true determinant of the chromatographic behavior of the counterions. However, separation factor may not be a true constant even for a specific ion exchange process as discussed hereunder.

2.4.4 Separation Factor: Homovalent Ion Exchange

In homovalent ion exchange, the exchanging counterions have identical charges. Without loss of generality, let us consider nitrate–chloride homovalent ion exchange due to its environmental significance:

$$\overline{R^+Cl^-} + NO_3^- \leftrightarrow \overline{R^+NO_3^-} + Cl^- \qquad (2.19)$$

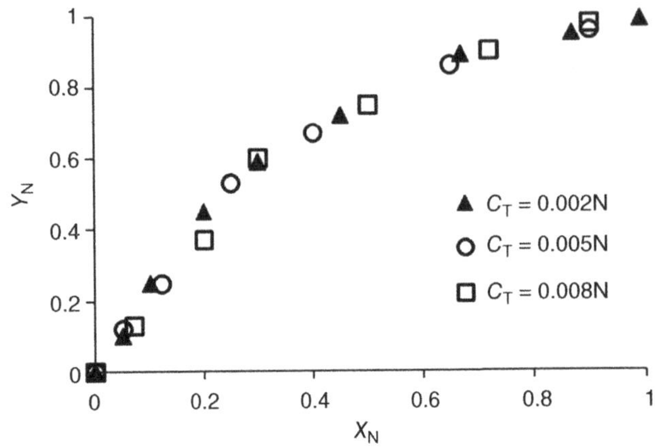

Figure 2.6 Nitrate–chloride equilibrium isotherms at three different electrolyte concentrations. *Source:* Data taken with permission from Clifford 1978 [12].

Nitrate–chloride separation factor value is

$$\alpha_{N/Cl} = \frac{y_N x_{Cl}}{x_N y_{Cl}} = \frac{\overline{C}_N C_{Cl}}{\overline{C}_{Cl} C_N} = K_{se} \tag{2.20}$$

Thus, for homovalent exchange, separation factor is essentially equal to the selectivity coefficient. Since activity coefficient values in the exchanger and aqueous phase tend to remain identical for ions of equal valence, separation factor is not influenced significantly by aqueous phase concentration (C_T), ion exchanger capacity (Q) and fractional loading (y_i) for homovalent ion exchange. Isotherm plots (i.e., y_i vs x_i) should, therefore, remain the same for homovalent exchange at different electrolyte concentrations. Figure 2.6 shows nitrate–chloride isotherms at three different electrolyte concentrations [12]. The plot essentially validates the premise that separation factor values tend to be constant for homovalent ion exchange.

Similar rules apply also to counterions with charge greater than one.

$$\overline{(R^-)_2Ca^{2+}} + Ba^{2+} \leftrightarrow \overline{(R^-)_2Ba^{2+}} + Ca^{2+} \tag{2.21}$$

Thus, the barium/calcium separation factor is,

$$\alpha_{Ba/Ca} = \frac{y_{Ba} x_{Ca}}{x_{Ba} y_{Ca}} = \frac{\overline{C}_{Ba} C_{Ca}}{\overline{C}_{Ca} C_{Ba}} = K_{se} \tag{2.22}$$

2.4.5 Separation Factor: Heterovalent Exchange

In heterovalent ion exchange, exchanging counterions have dissimilar charges. The following monovalent–divalent exchange reactions are of great significance in both environmental separation processes and water treatment:

$$2\overline{R^+Cl^-} + CrO_4^{2-} \leftrightarrow \overline{(R^+)_2CrO_4^{2-}} + 2Cl^- \tag{2.23}$$

$$\overline{2R^+Cl^-} + SO_4^{2-} \leftrightarrow \overline{(R^+)_2SO_4^{2-}} + 2Cl^- \tag{2.24}$$

$$\overline{2R^-Na^+} + Ca^{2+} \leftrightarrow \overline{(R^-)_2Ca^{2+}} + 2Na^+ \tag{2.25}$$

To illustrate the variability of separation factor in heterovalent ion exchange, let us consider the case of chloride-sulfate exchange in reaction (2.24).

The selectivity coefficient is given by,

$$K_{se}^{S/Cl} = \frac{\overline{C}_S (C_{Cl})^2}{(\overline{C}_{Cl})^2 C_S} \tag{2.26}$$

where subscript S and Cl refer to sulfate and chloride species, respectively.

Now,

$$\overline{C}_S + \overline{C}_{Cl} = Q \tag{2.27}$$

$$C_S + C_{Cl} = C_T \tag{2.28}$$

Also,

$$y_S = \frac{\overline{C}_S}{Q} \tag{2.29}$$

$$y_{Cl} = \frac{\overline{C}_{Cl}}{Q} \tag{2.30}$$

$$x_S = \frac{C_S}{C_T} \tag{2.31}$$

$$x_{Cl} = \frac{C_{Cl}}{C_T} \tag{2.32}$$

Dividing the numerator and denominator of the right-hand side of Eq. (2.26) by ${C_T}^2$ and Q^2, and applying equalities of Eqs (2.29)–(2.32), we get

$$K_{se}^{S/Cl} = \frac{y_S (x_{Cl})^2}{(y_{Cl})^2 x_S} * \frac{C_T}{Q} \tag{2.33}$$

For the binary isotherm,

$$x_{Cl} = 1 - x_S$$

$$y_{Cl} = 1 - y_S$$

Thus,

$$K_{se}^{S/Cl} = \frac{\frac{y_S}{(1-y_S)^2}}{\frac{x_S}{(1-x_S)^2}} * \left(\frac{C_T}{Q}\right) \tag{2.34}$$

Considering the selectivity coefficient remaining constant, two observations can be readily made.

I. Effect of C_T: As C_T increases at constant Q, for the selectivity coefficient to remain constant, y_S must decrease at a given x_S. Therefore, y_{Cl} increases at a given x_{Cl}. Sulfate–chloride separation factor, $\alpha_{S/Cl}$, thus decreases with an increase in C_T, all other conditions remaining identical. In general, divalent-monovalent separation factor, $\alpha_{2/1}$, always decreases with an increase in C_T and vice versa. Unlike homovalent ion exchange, separation factor is thus not a constant with variation in electrolyte concentration in the solution phase. The phenomenon that the aqueous-phase electrolyte concentration significantly influences the separation factor of counterions in heterovalent ion exchange, is often referred to as electroselectivity effect.

II. Effect of *Q*: In a similar vein, when C_T remains constant, the sulfate–chloride separation factor ($\alpha_{S/Cl}$) increases with an increase in Q per Eq. (2.34). Thus, an increase in ion exchanger's capacity (i.e., number of functional groups per unit volume) helps improve the preference of counterion of higher valence (say sulfate) over a counterion of lower valence (say chloride). In general, the total capacity, Q, of an ion exchanger remains constant and is not a very tunable process variable. Yet, Q influences the relative selectivity of ions in heterovalent ion exchange. All other factors being identical, an increase in Q in a binary system will enhance the preference of the exchanger toward the ion of higher valence.

Figure 2.7 shows the plots of sulfate–chloride isotherms at three different electrolyte concentrations, namely 10, 170, and 400 meq/L. Values of respective separation factors also are mentioned in the figure. In accordance with the electroselectivity effect in heterovalent exchange, the preference for divalent sulfate diminishes with an increase in electrolyte concentration. The $\alpha_{S/Cl}$ value decreases from 10 at 10 meq/L to 1 at 170 meq/L and then to 0.5 at 400 meq/L. Thus, at electrolyte concentrations greater than 170 meq/L, the preference of the anion exchanger shifts from divalent sulfate anion to monovalent chloride anion, that is, $\alpha_{S/Cl}$ becomes less than unity. This phenomenon is known as "electroselectivity reversal" effect. Such an effect of

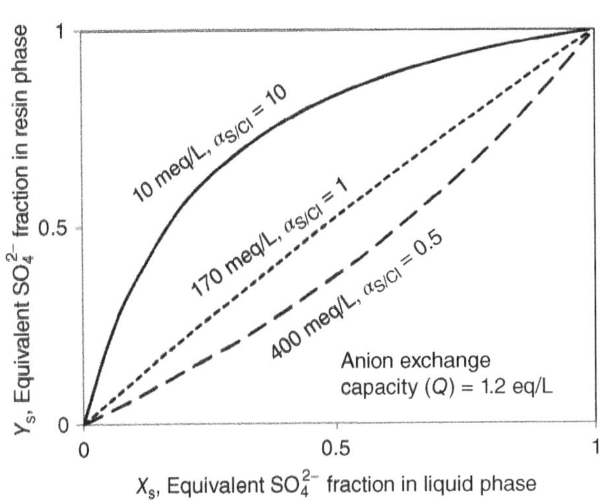

Figure 2.7 Plots of sulfate–chloride isotherms at three different electrolyte concentrations showing impact of electrolyte concentration on separation factors.

aqueous-phase electrolyte concentration is often used in many real-life processes to achieve high efficiency of regeneration with monovalent Na^+ for cation and Cl^- in anion exchange processes. The effect of electrolyte concentration in heterovalent exchange is characteristically analogous to that of temperature in "parametric pumping" in adsorption processes [13–16]. For dilute solutions, the ions of higher valence exhibit greater affinity over competing ions during the sorption step. For efficient regeneration, use of higher concentration leads to thermodynamically favorable desorption processes. Electroselectivity reversal is routinely used in many ion exchange applications.

2.4.6 Physical Reality of Selectivity Reversal: Role of Le Châtelier's Principle

The question has been raised umpteen times, and rightly so, that Eq. (2.35) mathematically explains why the sulfate/chloride separation factor drops with an increase in C_T, but still fails to provide any scientific insight into the observed phenomenon. The best way to comprehend this somewhat counterintuitive phenomenon is to use the two-century old Le Châtelier's principle. Let us again consider the exchange of chloride and sulfate:

$$2\overline{R^+Cl^-} + SO_4^{2-} \leftrightarrow \overline{(R^+)_2SO_4^{2-}} + 2Cl^- \tag{2.35}$$

Ion exchange is by and large a constant volume reaction, that is, the combined volume of ion exchange resin and water remains unaltered by ion exchange reaction. Note that the forward reaction causes an increase in the molar concentration in the aqueous phase, that is, two moles of chloride are released in the aqueous phase with the removal of one mole of sulfate. Thus, any increase in the total aqueous phase molar concentration, according to Le Châtelier's principle, will favor the backward reaction to diminish the aqueous-phase concentration, C_T. The exchanger will exhibit preference of chloride over sulfate under this condition. Conversely, the molar concentration in the exchanger phase is reduced by the forward reaction. In a similar vein, an increase in the exchanger-phase capacity, Q, will favor the forward reaction, that is, a greater preference for divalent sulfate to monovalent chloride.

The above phenomenon is analogous to the widely-used gas-phase ammonia synthesis reaction (Haber's process) as shown below:

$$N_2(g) + 3H_2(g) \leftrightarrow 2NH_3(g) \tag{2.36}$$

Note that the number of moles increases with the reverse reaction, that is, two moles of products produce four moles of reactants. At constant volume, the pressure is directly proportional to the number of moles in the gas phase and thus, the forward reaction causes a decrease in the system pressure. The pressure in reaction (2.36) has a similar effect as the total aqueous-phase molar concentration for the sulfate–chloride ion exchange in reaction (2.35). An increase in pressure will always favor the forward reaction with a greater yield of ammonia even at a constant temperature, that is, at the same value of the equilibrium constant. In practice, ammonia synthesis processes are carried out around the globe at high pressure to attain high yield. Characteristically, the system pressure in reaction (2.36) and the electrolyte concentration in ion exchange reaction (2.35) are equivalent.

2.4.7 Equilibrium Constant: Inconsistencies and Potential Pitfalls

Prior to leaving this section, it is imperative that we define the stoichiometry of ion exchange consistent with physical realities. Almost without any exception, the single ion exchange site has a charge of ±1 (negative for cation and positive for anion) for every inorganic or organic exchanger known to date. For that reason, the single site of an ion exchanger should serve as a single unit (i.e., molecule) in ion exchange stoichiometry.

Attempts have been made in some places to present ion-exchange stoichiometry as follows [17]:

$$\overline{NaKZ} + Ca^{2+} \leftrightarrow \overline{CaZ} + Na^+ + K^+ \tag{2.37}$$

While theoretically correct, Z denotes a site charge of −2 in reaction (2.37); this situation is practically non-existent. Thus, an equilibrium constant based on reaction (2.37) is not representative of the real situation and, therefore, depicts incorrect stoichiometry. The appropriate stoichiometry consistent with the physical reality, is as follows:

$$\overline{NaZ} + \overline{KZ} + Ca^{2+} \leftrightarrow \overline{CaZ_2} + Na^+ + K^+ \tag{2.38}$$

Sadly, no international standard currently exists regarding uniformity in nomenclature and definitions of equilibrium parameters in ion exchange. The outcome of a workshop on the subject may be valuable for professionals working in the area [18].

Example 2.2 Selectivity Reversal

Below is an isotherm between Na^+ and Ca^{2+} for a strong-acid cation (SAC) exchange resin with a matrix of polystyrene and divinyl benzene cross-linking (PS-DVB) and sulfonate functional groups. The capacity of the exchanger is 2.0 eq/L.

$$\overline{2R^-Na^+} + Ca^{2+} \rightarrow \overline{(R^-)_2Ca^{2+}} + 2Na^+$$

Chloride is the only coion in the inlet aqueous phase and the total electrolyte concentration, $C_T = 0.1\,N$,

(i) Find the corrected selectivity coefficient at $x_{Ca} = 0.3$;
(ii) Find the selectivity coefficient at $x_{Ca} = 0.3$;
(iii) Find the separation factor ($\alpha_{Ca/Na}$) at $x_{Ca} = 0.3$;
(iv) Assuming the selectivity coefficient to be remaining constant, compute and plot $\alpha_{Ca/Na}$ for $C_T = 0.1N$ to $C_T = 5.0N$;
(v) Determine the ionic strength at which the Ca/Na separation factor becomes equal to unity. Comments welcome.

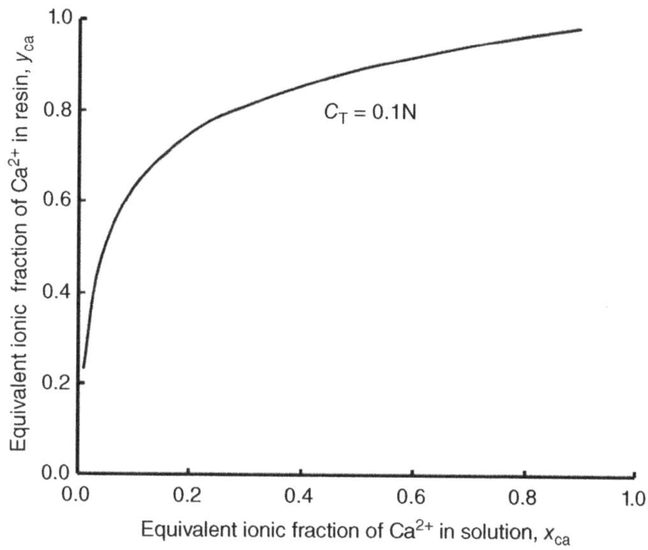

Figure 1. Ca^{2+}/Na^{+} isotherm for a PS-DVB SAC resin with sulfonate functionality, $Q = 2.0$ eq/L.

Solution:

(i) The corrected selectivity coefficient,

$$K_{SE}^{C} = \frac{x_{Na}^{2} y_{Ca}}{y_{Na}^{2} x_{Ca}} \frac{\gamma_{Na}^{2}}{\gamma_{Ca}} \frac{C_T}{Q}$$

At $C_T = 0.1\,N$, $x_{Ca} = 0.3$, $C_{Ca} = 0.03\,N = 0.015\,M$, $x_{Na} = 0.7$, $C_{Na} = 0.07\,N = 0.07\,M$, $C_{Cl} = 0.1\,N = 0.1\,M$.

From the isotherm, at $x_{Ca} = 0.3$, $y_{Ca} = 0.81$

Since ionic strength

$$I = \frac{1}{2}\sum C_i Z_i^2 = \frac{1}{2}(0.015 \times (2)^2 + 0.07 \times (1)^2 + 0.1 \times (-1)^2) = 0.115 < 0.5$$

Thus, Davies approximation is appropriate to calculate the activity coefficient,

$$\log(\gamma) = -Az^2\left(\frac{\sqrt{I}}{1+\sqrt{I}} - 0.2I\right)$$

At 25 °C, $A \approx 0.5$, thus,

$$\log(\gamma_{Ca}) = -0.46,\ \gamma_{Ca} = 0.346$$

Similarly,

$$\log(\gamma_{Na}) = -0.115,\ \gamma_{Na} = 0.767$$

Thus, the corrected selectivity coefficient

$$K_{SE}^{C} = \frac{x_{Na}^{2} y_{Ca}}{y_{Na}^{2} x_{Ca}} \frac{\gamma_{Na}^{2}}{\gamma_{Ca}} \frac{C_T}{Q} = \frac{0.7^2 \cdot 0.81}{0.19^2 \cdot 0.3} \cdot \frac{0.77^2}{0.35} \cdot \frac{0.1N}{2\frac{eq}{L}} = 3.1$$

(ii) Normal" selectivity coefficient does not take into consideration non-ideality in any phase, thus,

$$x_{Ca} = 0.3,\ x_{Na} = 0.7,\ y_{Ca} = 0.81,\ y_{Na} = 0.19,\ C_T = 0.1\,N$$

The "normal" selectivity coefficient,

$$K_{SE} = \frac{x_{Na}^2 y_{Ca}}{y_{Na}^2 x_{Ca}} \frac{C_T}{Q} = \frac{0.7^2 \cdot 0.81}{0.19^2 \cdot 0.3} \cdot \frac{0.1N}{2\frac{eq}{L}} = \mathbf{1.83}$$

(iii) The separation factor is

$$\alpha_{Ca/Na} = \frac{y_{Ca} x_{Na}}{y_{Na} x_{Ca}} = \mathbf{9.95}$$

(iv) First the separation factor $\alpha_{Ca/Na}$ must be solved for in terms of C_T. Starting with the relationship:

$$K_{SE} = \frac{x_{Na}^2 y_{Ca}}{y_{Na}^2 x_{Ca}} \frac{C_T}{Q}$$

Since this is a binary system,

$$x_{Na} = 1 - x_{Ca}, \text{ and } y_{Na} = 1 - y_{Ca}$$

Substituting values,

$$K_{SE} = \frac{(1 - x_{Ca})^2 y_{Ca}}{(1 - y_{Ca})^2 x_{Ca}} \frac{C_T}{Q} = \frac{(1 - 0.3)^2 y_{Ca}}{(1 - y_{Ca})^2 \times 0.3} \times \frac{C_T}{2\,eq/L}$$

Now K_{SE} is equal to 1.83, $Q = 2.0$ eq/L and $x_{Ca} = 0.3$

For every chosen value of C_T, we may now determine y_{Ca} and the corresponding $\alpha_{Ca/Na}$ as tabulated below:
Calculated values are shown *in the following table,*

C_T	x_{Ca}	x_{Na}	y_{Ca}	y_{Na}	$\alpha_{Ca/Na}$
0.1	0.3	0.7	0.81	0.19	9.95
1	0.3	0.7	0.52	0.48	2.52
2	0.3	0.7	0.40	0.60	1.56
3	0.3	0.7	0.33	0.67	1.16
4	0.3	0.7	0.28	0.72	0.93
5	0.3	0.7	0.25	0.75	0.78

$\alpha_{Ca/Na}$ is plotted versus C_T.

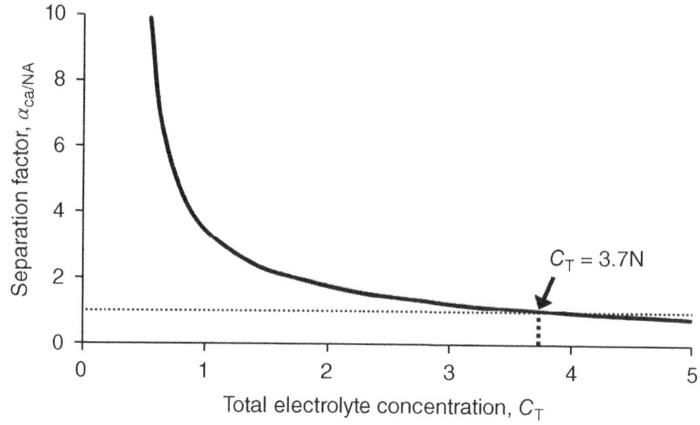

Figure 2. Separation factor ($\alpha_{Ca/Na}$) as a function of total electrolyte concentration ($C_T = 0.1N - 5.0N$). Note: selectivity reversal occurs at $C_T = 3.7N$.

(v) Comments:
The plot above shows the selectivity reversal, that is, $\alpha_{Ca/Na} = 1$ occurs at $C_T = 3.66$ N For lower value of Q, the C_T corresponding to selectivity reversal will decrease.

2.5 Electrostatic Interaction: Genesis of Counterion Selectivity

Electrostatic or Coulombic interaction between counterions in the aqueous phase and the fixed coions (i.e., functional group) on the exchanger forms the heart of the ion exchange process. Every ion exchange reaction of interest essentially involves replacing one ion with another while maintaining electroneutrality in both exchanger and aqueous phase. Thus, the relative selectivity or affinity of one counterion with another is an important prerequisite to the viability of an ion exchange process. Energy of Coulombic interaction is directly proportional to the charge of the counterions. Hence, between two counterions of unequal valence, one with higher valence always exhibits greater affinity toward the fixed sites of opposite charge in an ion exchanger and that can be readily recognized. However, for homovalent ion exchange, the counterions have identical charges. Specifically, for the example described in Section 2.1, both Na^+ and K^+ are monovalent cations located in the same group of the periodic table. The pertinent question is: what is the genesis of relative selectivity between two counterions of identical charges? Let us consider it for monovalent counterions.

2.5.1 Monovalent–Monovalent Coulombic Interaction

We consider a simple homovalent cation exchange reaction involving counterions A^+ and B^+

$$\overline{R^-B^+} + A^+(aq) \leftrightarrow \overline{R^-A^+} + B^+(aq) \tag{2.39}$$

To get a fundamental insight into the individual steps of the reaction, we divide the above reaction into two halves:

$$\overline{R^-} + A^+(aq) \leftrightarrow \overline{R^-A^+} \tag{2.39a}$$

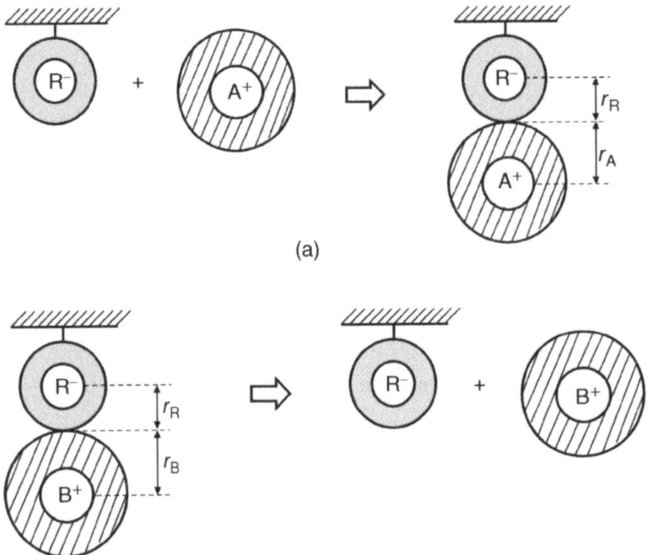

Figure 2.8 An illustration of the ion exchange half reactions through formation and splitting of the solvated ion pair. (a) One half of reaction where A^+ gets attached with R^-; (b) The other half of reaction, that is, splitting of R^-B^+.

$$\overline{R^-B^+} \leftrightarrow \overline{R^-} + B^+(aq) \tag{2.39b}$$

Let us first consider the half-reaction (2.39a). Since an ion exchanger may be viewed as a condensed polyelectrolyte, both the fixed coions and the exchanging counterions will remain solvated (hydrated) in the exchanger phase through ion-solvent interactions. As a result, solvated ion pairs (SIP) are formed and the half-reaction in Eq. (2.39a) can be schematically presented as shown in Figure 2.8a. In a similar vein, Figure 2.8b illustrates splitting of SIP in accordance with Reaction (2.39b).

The free energy change associated with the first half reaction is essentially equal to the negative amount of electrical work for bringing the counterion A^+ from the bulk aqueous-phase next to the fixed charge R^-. Using Coulomb's law and considering the hydrated ionic radius of R^- and A^+ to be equal to r_R and r_A in the exchanger phase, free energy change at the standard state is:

$$\Delta G_1^0 = \int_{r_R+r_A}^{\infty} \frac{(-e)(+e)}{\varepsilon_D}\frac{dr}{r^2} = \frac{-e^2}{(r_R + r_A)\varepsilon_D} \tag{2.40}$$

where, e = charge of an electron

ε_D = dielectric constant of the ion exchanger

r = charge radius (subscripts A and R refer to the counterion and fixed coion, respectively).

In a similar way, for the other half reaction as shown in Figure 2.39b, the counterion B^+ must be moved away from the fixed charge.

$$\overline{R^-B^+} \leftrightarrow \overline{R^-} + B^+(aq) \tag{2.39b}$$

The energy required is positive and equal to

$$\Delta G_2^0 = \frac{\mathrm{e}^2}{(r_\mathrm{R} + r_\mathrm{B})\varepsilon_\mathrm{D}} \tag{2.41}$$

Thus, the overall free energy change per mole or equivalent is

$$\Delta G^0 = (\Delta G_1^0 + \Delta G_2^0)N = -\frac{Ne^2}{\varepsilon_\mathrm{D}}\left(\frac{1}{r_\mathrm{R} + r_\mathrm{A}} - \frac{1}{r_\mathrm{R} + r_\mathrm{B}}\right) \tag{2.42}$$

where N is Avogadro's number.

Ignoring shrinking/swelling effect, the thermodynamic equilibrium constant for reaction (2.39) is related to ΔG^0 as follows:

$$-\mathrm{RT}\ \ln K = \Delta G^0 = -\frac{Ne^2}{\varepsilon_\mathrm{D}}\left(\frac{1}{r_\mathrm{R} + r_\mathrm{A}} - \frac{1}{r_\mathrm{R} + r_\mathrm{B}}\right) \tag{2.43}$$

$$\log K = \frac{Ne^2}{2.303\varepsilon_\mathrm{D} RT}\left(\frac{1}{r_\mathrm{R} + r_\mathrm{A}} - \frac{1}{r_\mathrm{R} + r_\mathrm{B}}\right) \tag{2.44}$$

The following three scenarios can be readily noted:

Case 1. For $r_\mathrm{A} = r_\mathrm{B}$, $\log K = 0$, that is, $K=1$.
Two ions of identical hydrated ionic radius are equally preferred by the ion exchanger.

Case 2. For $r_\mathrm{B} < r_\mathrm{A}$, $\log K =$ negative, that is, $K < 1$.
When the hydrated ionic radius of A^+ is bigger than B^+, A^+ is less preferred by the ion exchanger.

Case 3. For $r_\mathrm{B} > r_\mathrm{A}$, $\log K =$ positive, that is, $K > 1$.
When the hydrated ionic radius of A^+ is smaller than B^+, A^+ is preferred by the ion exchanger.

Equation (2.44) provides a quantitative relationship to compute the equilibrium constant for the exchange of monovalent ions and highlights the fact that, for an ion-exchange process involving only electrostatic interaction, an ion with lower hydrated ionic radius exhibits a greater selectivity. The dielectric constant within the exchanger phase is significantly lower than pure water but the selectivity sequence of various ions of identical valence always follows the order of their hydrated ionic radii in water. Figures 2.9 and 2.10 show the chromatograms of the various cations and anions during the elution ion chromatography with pellicular exchangers [19] (i.e., functional groups present at the surface of the spherical beads).

Considering $r_\mathrm{R} \ll r_\mathrm{B}$ or r_A, Eq. (2.6) becomes

$$\log K = \frac{Ne^2}{2.303\varepsilon_\mathrm{D} RT}\left(\frac{1}{r_\mathrm{A}} - \frac{1}{r_\mathrm{B}}\right) \tag{2.45}$$

Since earlier elution presents lower selectivity, the selectivity sequence of cations and anions of identical charges stands as follows in descending order in accordance with the elution chromatographs in Figures 2.9 and 2.10:

$Br^- > Cl^- > F^-$
$K^+ > Na^+ > Li^+$

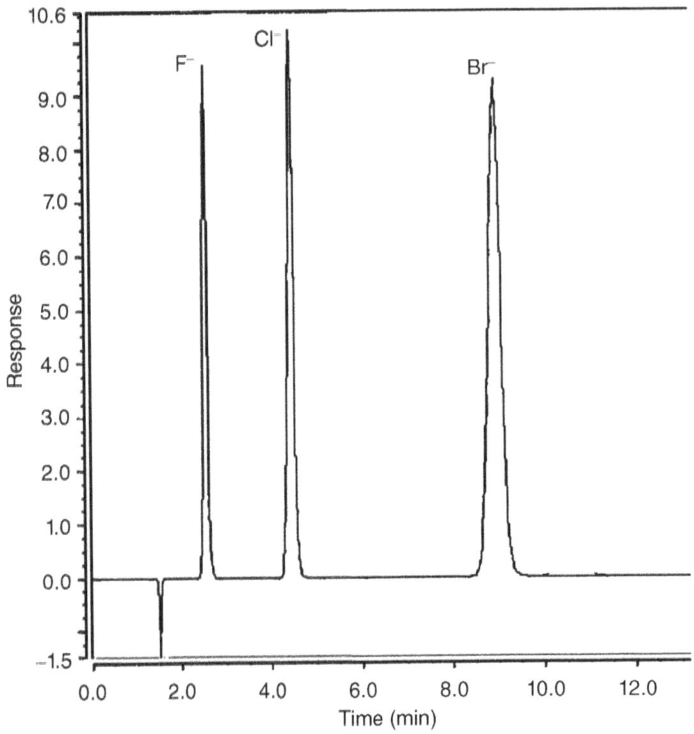

Figure 2.9 Chromatograms of different anions in ion chromatography elution with anion exchange resin. *Source:* Mukherjee and SenGupta [19]. Reproduced with permission of American Chemical Society.

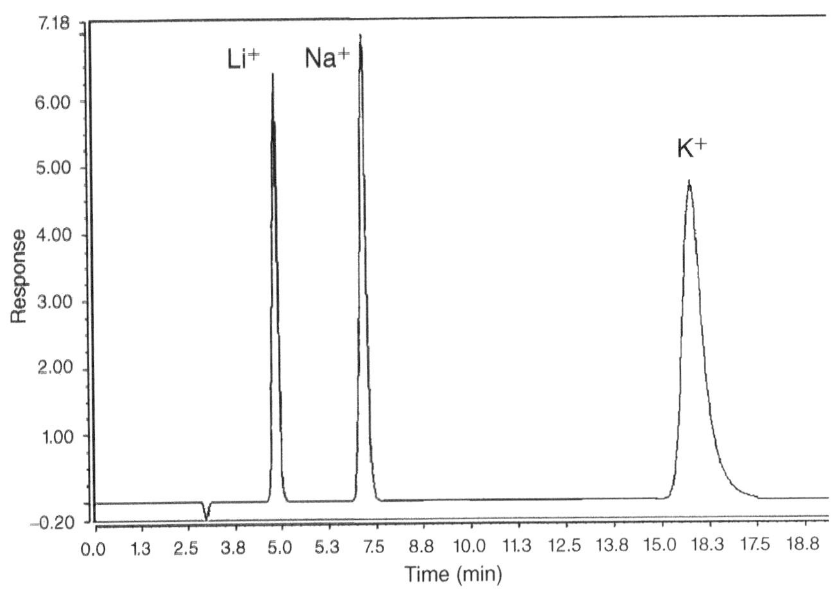

Figure 2.10 Chromatograms of different cations in ion chromatography elution with cation exchange resin. *Source:* Mukherjee and SenGupta [19]. Reproduced with permission of American Chemical Society.

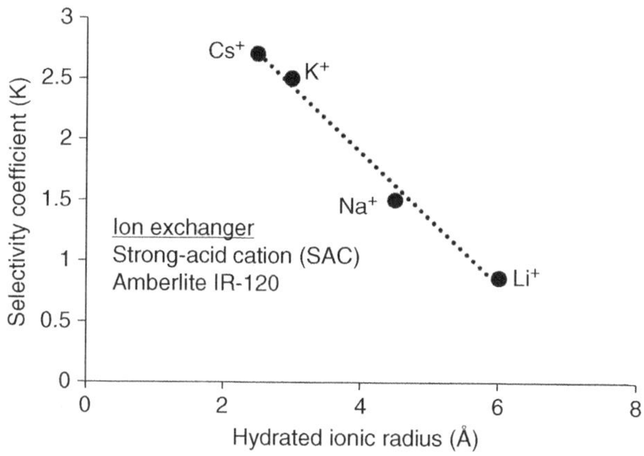

Figure 2.11 Plot showing dependence of ion exchange selectivity on hydrated ionic radii for monovalent cations. *Source:* Data taken with permission from Helfferich [20], Harned *et al.* [21], and Dilts [22].

The available data on hydrated ionic radii for various cations and anions tend to strongly support such an approach to identify selectivity sequence. Figure 2.11 shows the plot of ion-exchange selectivity and hydrated ionic radii for various monovalent cations [20–22]; an inverse relationship between the hydrated ionic radius and ion-exchange selectivity can be readily noted. Polyatomic anions are often non-spherical and their hydration is also governed by polarizability. However, the selectivity for an anion also follows the inverse relationship with equivalent hydrated ionic radii.

Readers should take note that the ion-exchange selectivity discussed in this section is governed solely by electrostatic or Coulombic interactions. Such ion-exchange selectivity, as will be shown in later chapters, can be altered, both enhanced or diminished and in some cases even reversed, by deliberately incorporating other interactions in conjunction with electrostatic one. Tables 2.2A *and* 2.2B *provide estimated separation factor values for cations and anions in dilute solutions for strong-acid and strong-base polymeric ion exchangers with reference to* H^+ *and* OH^- *ions, respectively. Particularly noteworthy is the high separation factor values of organic monovalent anions at the bottom of the table. The genesis of their high selectivity has been discussed in Chapter 3.*

2.6 Ion Exchange Capacity: Isotherms

A sorption/adsorption isotherm represents the distribution of a solute or solutes between the solution and the sorbent/adsorbent phase at a given temperature at equilibrium. The isotherm for an ion exchange process is essentially identical, but it involves the distribution of ions between the ion exchanger and solution phase. Thermodynamically, an isotherm is analogous to the equilibrium constant for a chemical reaction. Both are constants at a given temperature in accordance with the second law of thermodynamics and independent of concentrations under ideal conditions. However, solutes or ions do not undergo any chemical transformation

Table 2.2A Estimated separation factor values (compared with the hydrogen ion) in dilute solutions for sulfonated polystyrene cation exchange resins of different cross-linking amounts.

Counterion	4% DVB	8% DVB	10% DVB	16% DVB
Li^+	0.76	0.79	0.77	0.68
H^+	1.00	1.00	1.00	1.00
Na^+	1.20	1.56	1.61	1.62
NH_4^+	1.44	2.01	2.15	2.27
K^+	1.72	2.28	2.54	3.06
Rb^+	1.86	2.49	2.69	3.14
Cs^+	2.02	2.56	2.77	3.17
Ag^+	3.58	6.70	8.15	15.6
Tl^+	5.08	9.76	12.6	19.4
UO_2^{2+}	1.79	1.93	2.00	2.27
Mg^{2+}	2.23	2.59	2.62	2.39
Zn^{2+}	2.37	2.73	2.77	2.57
Co^{2+}	2.45	2.94	2.92	2.59
Cu^{2+}	2.49	3.03	3.15	3.03
Cd^{2+}	2.55	3.06	3.23	3.37
Ni^{2+}	2.61	3.09	3.08	2.76
Ca^{2+}	3.14	4.06	4.42	4.95
Sr^{2+}	3.56	5.13	5.85	6.87
Pb^{2+}	4.97	7.80	8.92	12.2
Ba^{2+}	5.66	9.06	9.42	14.2

during ion exchange processes. Theoretical exchange capacity of an ion exchanger solely depends on the concentrations of the ionogenic or functional groups in the exchanger phase. However, precise information in this regard is rarely available for commercially produced or naturally occurring inorganic and polymeric exchangers. The majority of environmental separation processes with ion exchange pertain to sorption/desorption of target ions of interest in the presence of others. Batch equilibrium tests, although most commonly used for capacity determination, are often inaccurate for target ions in a multi-component system. Two primary reasons contributing toward such inaccuracies are:

- Sorption of a target ion is often an extremely slow process due to intraparticle diffusion-limited kinetics. It is rather difficult to predict a realistic equilibration time which may vary from hours to months.
- Target ion sorption is often very sensitive towards pH; obtaining equilibrium isotherm data at a pre-determined pH poses obvious complexity with the traditional batch equilibrium technique.

Isotherm data invariably provide the foundation for designing and/or evaluating real-life sorption processes. It is thus imperative that one becomes familiar with

Table 2.2B Estimated separation factor values of various anions (compared with the hydroxyl ion) on polystyrene- divinylbenzene strong-base anion exchange resins with Type I and Type II strong-base functional groups.

Counterion	Type I	Type II
OH^-	1.0	1.0
I^-	175	17
HSO_4^-	85	15
ClO_3^-	74	12
NO_3^-	65	8
Br^-	50	6
CN^-	28	3
HSO_3^-	27	3
$BrSO_3^-$	27	3
NO_2^-	24	3
Cl^-	22	2.3
HCO_3^-	6.0	1.2
IO_3^-	5.5	0.5
Formate	4.6	0.5
Acetate	3.2	0.5
Propionate	2.6	0.3
F^-	1.6	0.3
Benzene sulphonate	500	75
Salicylate	450	65
Citrate	220	23
Phenate	110	27

different approaches and their relative pros and cons to generate the equilibrium or isotherm data. In addition to the batch technique, the procedures for regenerable mini-column method and step-feed frontal column-run method are presented for generating isotherm data.

2.6.1 Batch Technique

In a given volume of feed solution containing the target species (A) along with background solutes, a fixed amount of ion exchanger in a specific form is added. Upon attaining equilibrium, the final concentration of the target species is analyzed. The solution volume is often quite large compared to the amount of exchanger added. While the target ion concentration changes measurably, the concentrations of other counterions in the solution remain fairly unchanged. As shown in Figure 2.12, it is advisable that the batch isotherm technique should include a control run to confirm that sorption onto ion exchanger is the sole mechanism for the dissipation of the target ion, A, from the solution phase.

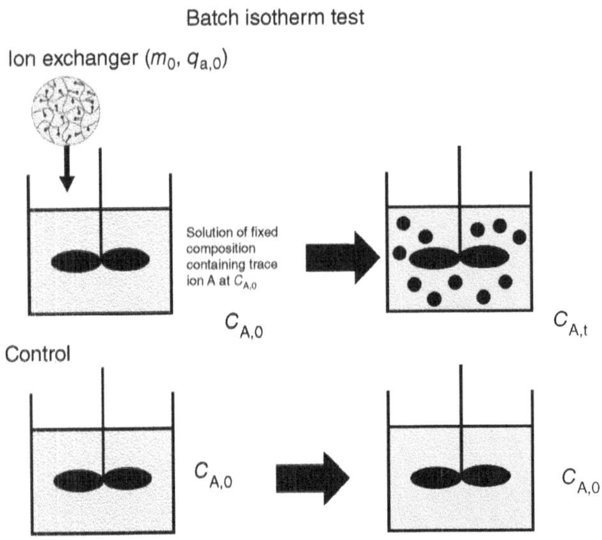

Figure 2.12 Diagram of the batch technique for testing isotherms.

The mass balance for the ion, A, can be written as follows:

$$m_{IX}q_{A,o} + V_L C_{A,o} = m_{IX}q_{A,f} + V_L C_{A,f} \quad (2.46)$$

$q_{A,0}$ and $q_{A,f}$ are initial and final loadings of the exchanger phase with A in meq/g, respectively. $C_{A,0}$ and $C_{A,f}$ are the initial and final concentrations of A in the solution phase in meq/L. From Eq. (2.46),

$$q_{A,f} = \frac{V_L(C_{A,o} - C_{A,f})}{m_{IX}} + q_{A,o} \quad (2.47)$$

In general, experiments are conducted in a way that

$$q_{A,o} = 0 \quad (2.48)$$

Thus,

$$q_{A,f} = \frac{V_L(C_{A,o} - C_{A,f})}{m_{IX}} \quad (2.49)$$

By varying m_{IX} at a constant temperature, isotherms, that is, a plot of $q_{A,f}$ versus $c_{A,f}$ can be easily constructed for a specific ion exchanger and a solution composition. Despite its operational simplicity, the batch technique, as already stated, lacks precision to determine isotherms of metals and ligands because their sorption affinity is very sensitive to pH, and pH is difficult to control in a batch system.

Example 2.3 *Sulfate–nitrate binary isotherm was carried out with a weak-base anion exchange resin (Amberlite IRA-45) at pH = 3.0 using the batch technique.*

Volume of the solution = 100 mL
Sulfate concentration = 5.0 meq/L
Nitrate concentration = 0.0 meq/L

Ion exchanger was originally in nitrate form
Resin capacity = 3.0 meq/g

Table 1. Data for Sulfate/ Nitrate Isotherm on Amberlite IR- 45; sample volume = 100 mL. [12].

Mass of resin (g)	$[SO_4^{2-}]$ (meq/L)	$[NO_3^-]$ (meq/L)
0.03	4.02	0.900
0.10	2.15	2.60
0.20	0.496	3.97
0.40	0.065	4.56
1.200	0.030	5.30

Plot the sulfate–nitrate isotherm for Amberlite IRA-45, that is, plot y_s versus x_s and y_N versus x_N. Describe the isotherms.
Find sulfate–nitrate average separation factor.
Show calculations

Solution:

(1) Calculate meq. NO_3^- on resin at equilibrium:
At 0.10 g resin/100 mL

$$[\text{Nitrate on resin}]_{\text{initial}} - [\text{Nitrate in solution}]_{\text{equilibrium}} = [\text{Nitrate}]_{\text{resin}}$$

$$[m_{\text{resin}} \times Q_{\text{resin}}] - [NO_3^-] \times V_{\text{soln}} = \text{meq of } NO_3^- \text{ in resin}$$

$$\left[0.10\,\text{g resin} \times 3\frac{\text{meq } NO_3^-}{\text{g resin}}\right] - \left[2.60\frac{\text{meq } NO_3^-}{\text{L}} \times 0.1\,\text{L}\right] = 0.040\,\text{meq } NO_3^-$$

(2) Calculate meq. SO_4^{2-} on resin at equilibrium:
At 0.10 g resin/100 mL

$$[\text{Sulfate in solution}]_{\text{initial}} - [\text{Sulfate in solution}]_{\text{equilibrium}} = [\text{Sulfate}]_{\text{resin}}$$

$$[SO_4^{2-}]_{\text{initial}} \times V_{\text{soln}} - [SO_4^{2-}]_{\text{equilibrium}} \times V_{\text{soln}} = m_{SO_4^{2-}\ \text{resin}}$$

$$\left[5.00\frac{\text{meq } SO_4^{2-}}{\text{L}} \times 0.1\,\text{L}\right] - \left[2.15\frac{\text{meq } SO_4^{2-}}{\text{L}} \times 0.1\,\text{L}\right] = 0.285\,\text{meq } SO_4^{2-}$$

(3) Calculate equivalent fraction of SO_4^{2-} on resin at equilibrium:

$$y_S = \frac{\overline{[SO_4^{2-}]}}{\overline{[SO_4^{2-}]} + \overline{[NO_3^-]}}$$

$$y_S = \frac{0.285\,\frac{\text{meq SO}_4^{2-}}{0.1\,\text{g resin}}}{0.285\,\frac{\text{meq SO}_4^{2-}}{0.1\,\text{g resin}} + 0.040\,\frac{\text{meq NO}_3^-}{0.1\,\text{g resin}}} = 0.877$$

(4) Calculate equivalent fraction of SO_4^{2-} in liquid at equilibrium:

$$x_S = \frac{[SO_4^{2-}]}{[SO_4^{2-}] + [NO_3^-]}$$

$$x_S = \frac{2.15\,\frac{\text{meq}}{\text{L}}}{2.15\,\frac{\text{meq}}{\text{L}} + 2.60\,\frac{\text{meq}}{\text{L}}} = 0.453$$

(5) Calculate separation factor $\alpha_{S/N}$:

$$\alpha_{S/N} = \frac{y_S(1 - x_S)}{x_S(1 - y_S)}$$

$$\alpha_{S/N} = \frac{(0.877)\times(0.547)}{(0.453)\times(0.123)} = 8.61$$

This separation factor value is not constant and will change at different y_s values. Plot curve y_S versus $x_{S,}$ that is, the sulfate–nitrate isotherm. See Figure 1a.
Again, $y_N = 1 - y_s$ and $x_N = 1 - x_s$
Plot y_N versus x_N for the same isotherm. See Figure 1b.

(6) Plot isotherms:

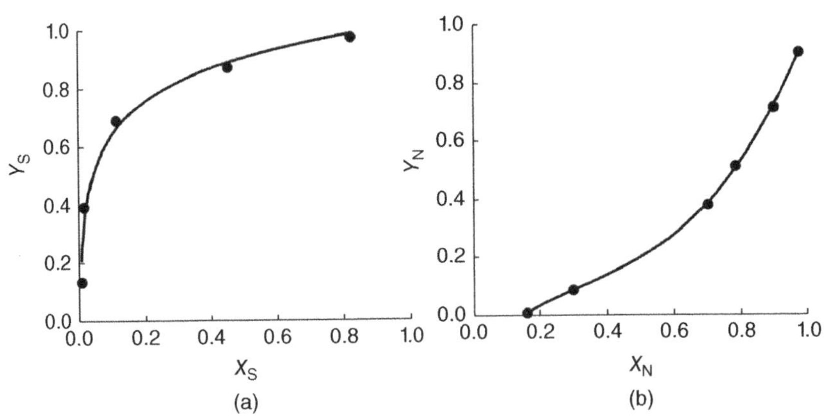

Figure 1. Sulfate–nitrate binary isotherm with a weak-base anion exchange resin (Amberlite IRA-45) at pH = 3.0 using the batch technique. (a) Plot of y_s versus x_s. (b) Plot of y_N versus x_N. Data taken with permission from Clifford [12].

Table 2. Average separation factors for sulfate–nitrate at pH = 3.0 with the batch technique.

X_S	X_N	Y_s	Y_N	$\alpha_{S/N}$
0.0141	0.9859	0.400	0.600	46.53
0.1110	0.8890	0.689	0.311	17.76
0.4526	0.5474	0.876	0.124	8.61

Note that although the aqueous-phase concentration remains the same ($C_T = 5.0$ meq/L), sulfate–nitrate separation factor values vary widely with y_S. For heterovalent ion exchange, separation factor values are very sensitive to resin composition: with an increase in exchanger phase equivalent fraction, y_i, the separation factor of "i" with respect to other counterions decreases.

2.6.2 Regenerable Mini-Column Method

Solutions containing known concentrations of ions, A, along with other electrolytes or competing ions are passed through mini-columns containing a small packed bed of an ion exchanger as shown in Figure 2.13. The solutions are always passed in far excess to attain equilibrium. The feed solution conditions, namely, pH, target ion concentrations and background electrolytes, represent equilibrium conditions. Subsequently, after a short rinse with DI, the mini-column is regenerated with an appropriate solution of known volume and the concentration of the target species is determined in the spent regenerant. Equilibrium capacity corresponding to the solution concentration of $C_{A,1}$ is given as

$$q_{A,1} = \frac{V_{R1} C_{A,R1}}{m_1} \tag{2.50}$$

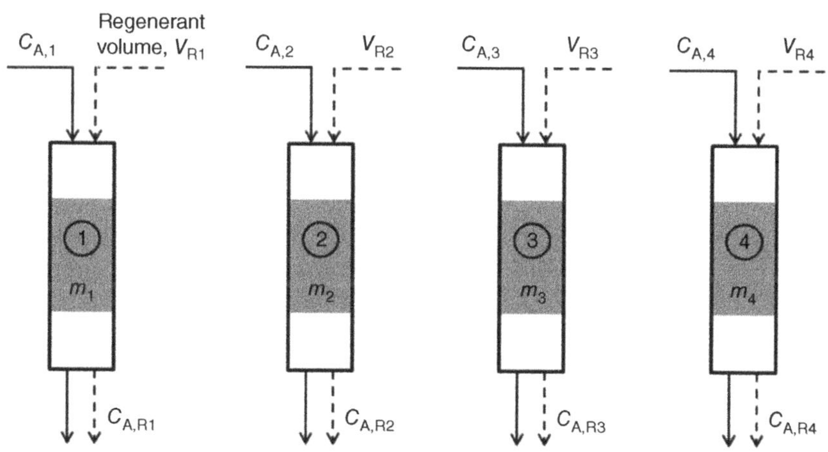

Figure 2.13 Illustration of determination of ion exchange capacity using regenerable mini-column method, where service runs and regeneration runs are represented by solid and dashed lines, respectively and $q_{Ai} = \frac{V_{Ri} C_{A,Ri}}{m_i}$.

While

V_{R1}	Regenerant volume collected from mini-column 1
m_1	Amount of ion exchanger in mini-column 1
$C_{A,R1}$	Concentration of A in the regenerant.

Mini-column equilibration techniques are operationally simple and appropriate when sorption is very sensitive to pH fluctuation. It is, however, more time-consuming than batch technique and not suitable for target ions that are not amenable to efficient regeneration.

Example 2.4 Mini-column problem
In a mini-column sorption experiment of phosphate sorption onto hybrid anion exchanger (HAIX), influent phosphate concentration is varied, whereas background anion concentrations of Cl^- and SO_4^{2-} are constant at 100 and 120 mg/L, respectively. The experiment is carried out at constant room temperature. The inlet solutions are fed far in excess to make sure equilibrium is achieved. The quantity of HAIX is maintained at 1.2 g in each run and other experimental conditions, namely empty bed contact time, liquid velocity is kept unchanged. HAIX is amenable to efficient regeneration with 2% NaOH and 1% NaCl. With the help of the following experimental results, draw the sorption isotherm.

Table 1. Experimental results of HAIX.

Inlet phosphate concentration (C_A) (mg/L)	Regenerant volume (V_R) (mL)	Phosphate concentration in regenerant ($C_{A,R}$) (mg/L)
0.04	30	38
0.06	35	46.3
0.08	40	57
0.1	45	61.3
0.12	50	60

Source: Blaney *et al.* 2007 [23]. Reproduced with permission of Elsevier.

The equilibrium capacity (q_A) of phosphate uptake by HAIX:

$$q_A = \frac{V_R C_{A,R}}{m} \tag{1}$$

For the condition $C_A = 0.04$ mg/L, $V_R = 30$ mL, and $C_{A,R} = 38$ mg/L.
Hence $q_A = (0.03\,\text{L} \times 38\,\text{mg/L})/1.2\,\text{g} = 0.95$ mg phosphate/g HAIX.

Similarly, q_A at different inlet phosphate concentrations can be calculated.

Since inlet solution is fed in far excess, the sorption capacity in each case would be the equilibrium capacity corresponding to the inlet phosphate concentration.

Table 2. Sorption isotherm from experimental results of HAIX.

Equilibrium liquid phase concentration, C_A (mg/L)	Equilibrium sorption capacity, q_A (mg/g)
0.04	0.95
0.06	1.35
0.08	1.90
0.1	2.30
0.12	2.50

Source: Blaney *et al.* 2007 [23]. Reproduced with permission of Elsevier.

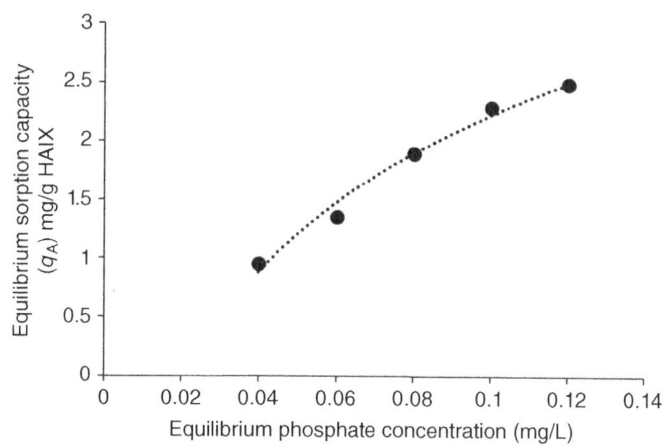

Figure 1. Sorption isotherm from experimental results of HAIX. *Source:* Blaney *et al.* 2007 [23]. Reproduced with permission of Elsevier.

2.6.3 Step-Feed Frontal Column Run

A solution containing the target species along with other background electrolytes is passed through a fixed-bed column containing the ion exchanger. The effluent from the exit of the column is sampled regularly and analyzed. Once the effluent concentration becomes equal to the influent concentration (i.e., the exchanger is in equilibrium with the influent), the target species concentration in the influent is increased to a predetermined value, all other conditions remaining identical. The column run is continued again until the effluent concentration becomes equal to the new influent concentration. The process is repeated for several gradually increasing influent target ion concentrations. From the detailed breakthrough histories, ion sorption capacities are then

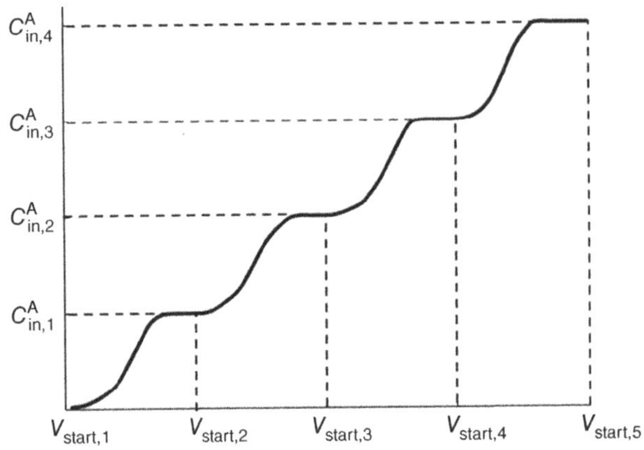

Figure 2.14 Illustration of effluent histories for step-feed frontal column run.

determined corresponding to every feed concentration as illustrated in Figure 2.14. For a given inlet target ion concentration, the sorption capacity is given by,

$$q_j^A = q_{j-1}^A + \int_{V_{START,j}}^{V_{STOP,j}} (C^A_{in,j} - C^A_{out,j})dV \tag{2.51}$$

where $j \geq 2$ integer number for each step-feed, and q_j^A is the equilibrium sorption capacity for the jth step corresponding to $C^A_{in,j}$. $V_{STOP,j}$ is the bed volume (or normalized volume of feed with respect to the mass or volume of the exchanger) where the exit concentration $C^A_{out,j}$ equals $C^A_{in,j}$. Under experimental conditions as stated,

$$V_{STOP,j} = V_{START,j+1} \tag{2.52}$$

and

$$V_{START,j} = V_{STOP,j-1} \tag{2.53}$$

Each boxed area in Figure 2.14 corresponds to the mass of the target solute sorbed onto the exchanger for the incremental increase in the concentration of the target ion in the feed. This technique, although precise and representative of real situations, involves careful analyses of a multitude of samples. Batch or mini-column techniques are unreliable when the sorption process is pH sensitive or efficient regeneration is not viable . The step-feed columnar technique is particularly appropriate under such circumstances.

Example 2.5 Frontal column problem

The following is the effluent history for arsenic removal by a hybrid ion exchanger (HIX) in a step-feed frontal column run (Figure 1). The inlet arsenic concentration is increased stepwise as marked in the figure once the effluent arsenic concentration from the previous run equals to its influent concentration. The rest of the background concentrations and other experimental conditions remain identical between the runs. Develop the isotherm for arsenic sorption onto HIX (i.e., equilibrium arsenic sorption capacity vs. equilibrium arsenic concentration).

The equilibrium sorption capacity (q_1) corresponding to the equilibrium concentration 10 µg/L (step 1) is calculated from the area over the curve, that is, shaded area marked (1)

Sorption capacity q_1 = Triangular area = $(C_{out} - C_{in})$*(Liter fed/g HIX)/2

So, $q_1 = (10 - 0)$ µg/L * (8.5 L/g)/2 = 42.5 µg/g

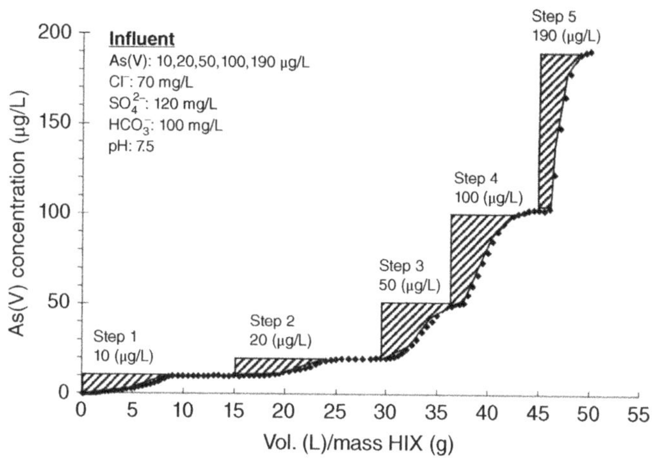

Figure 1. Effluent history for a step feed frontal column run of HIX for arsenic removal. *Source:* Greenleaf *et al.* 2003 [24]. Reproduced with permission of Elsevier.

Equilibrium sorption capacity (q_2) corresponding equilibrium concentration 20 µg/L (Step 2) is given by:

$q_2 = q_1$ + shaded area marked (2) = 97.5 µg/g

$q_3 = q_2$ + shaded area marked (3) = 194.5 µg/g (for eq. concentration 50 µg/L)

$q_4 = q_3$ + shaded area marked (4) = 344.5 µg/g (for eq. concentration 100 µg/L)

$q_5 = q_4$ + Shaded area marked (5) = 536 µg/g (for eq. concentration 190 µg/L)

Table 1. Equilibrium sorption capacity as a function of liquid phase concentration.

Equilibrium sorption capacity, *q* (µg/g)	Equilibrium liquid phase concentration, *C* (µg/L)
42.5	10
97.5	20
194.5	50
344.5	100
536	190

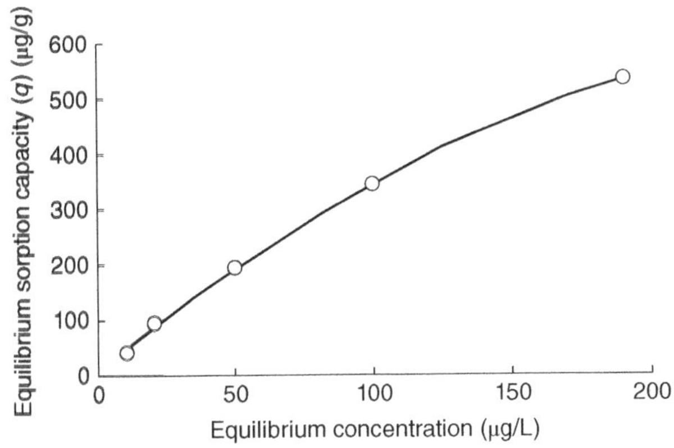

Figure 2. Sorption isotherm for arsenic on HIX during a step-feed frontal column run.

Example 2.6 *Consider the arsenic isotherm data from Example 2.5, Figure 2. Now you ran a mini-column regenerable isotherm with 125 ppb influent arsenic with 2 g of resin. If you used 100 mL of regenerant, what will be the* $C_{A,R}$.

$$\text{At}\, C_{\text{equilib}} = 125\,\text{ppb},\ Q_{\text{equilib}} = 410\frac{\mu\text{g As}}{\text{g resin}}$$

$$\text{For}\, m_{\text{resin}} = 2\,\text{g},\ m_{\text{As}} = m_{\text{resin}} * Q_{\text{equilib}} = 2\,\text{g resin} * 410\frac{\mu\text{g As}}{\text{g resin}} = 820\,\mu\text{g As}$$

$$C_{\text{regen}} = \frac{820\,\mu\text{g As}}{100\,\text{mL}} = 8200\,\mu\text{g}\frac{\text{As}}{\text{L}} = 8.2\,\text{mg As/L}$$

2.7 The Donnan Membrane Effect in Ion Exchanger

It is pertinent to mention at the outset that the Donnan effect or the Donnan membrane equilibrium is essentially a specific domain of the second law of thermodynamics dealing with completely ionized electrolytes. It was Frederick G. Donnan, an English Physical Chemist, who propelled the quantitative description and various implications of this effect to the forefront in the early twentieth century [25]. In ion exchange processes, the conditions leading to the Donnan membrane equilibrium arise from the inability of fixed coions (i.e., charged functional groups) to diffuse out from the polymer phase to water or polar solvents. For the example provided in the beginning of the chapter (Figure 2.1), some chloride ions are also present inside the cation exchanger; this phenomenon is referred to as coion invasion.

2.7.1 Coion Invasion or Electrolyte Penetration

The Donnan membrane equilibrium and coion invasion, or exclusion, are interwined and in this section, we will provide a theoretical basis to both understand and quantify this phenomenon.

To get an insightful understanding of various ramifications of the Donnan membrane principle, let us consider a cation exchanger with capacity $\overline{C}_R$ (equivalents per liter) in sodium form in contact with a solution of NaCl. From electroneutrality conditions:

Aqueous phase:

$$C_{Na^+} = C_{Cl^-} \tag{2.54}$$

And in the exchanger phase:

$$\overline{C}_{R^-} + \overline{C}_{Cl^-} = \overline{C}_{Na^+} \tag{2.55}$$

At equilibrium:

$$(\overline{a}_{Na^+} * \overline{a}_{Cl^-})_{Resin} = (a_{Na^+} * a_{Cl^-})_{Water} \tag{2.56}$$

where, "*a*" denotes the activity of the species, *C* is the molar concentration and overbar denotes the exchanger phase.

Considering ideality in both the exchanger phase and the aqueous phase (i.e., activities and molar concentrations are equal), and applying equalities from Eqs (2.54) and (2.55) into Eq. (2.56),

$$(\overline{C}_{R^-} + \overline{C}_{Cl^-})\overline{C}_{Cl^-} = C^2_{Cl^-} \tag{2.57}$$

$$\overline{C}^2_{Cl^-} + \overline{C}_{R^-}\overline{C}_{Cl^-} - C^2_{Cl^-} = 0 \tag{2.58}$$

Solving for $\overline{C}_{Cl^-}$,

$$\overline{C}_{Cl^-} = \frac{1}{2}\left[\left(\sqrt{\overline{C}^2_{R^-} + 4C^2_{Cl^-}}\right) - \overline{C}_{R^-}\right] \tag{2.59}$$

Note that the true sodium loading or sodium exchange capacity obtained from Eq. (2.55) is

$$\overline{C}_{Na^+} = \overline{C}_{R^-} + \frac{1}{2}\left[\left(\sqrt{\overline{C}^2_{R^-} + 4Cl^2_{Cl^-}}\right) - \overline{C}_{R^-}\right] \tag{2.60}$$

For conditions $\overline{C}_{R^-} \gg C_{Cl^-}$

From Eq. (2.57)

$$\overline{C}_{Cl^-} \approx 0$$

Thus, the presence of non-diffusible R^- functional group in the cation exchange resin imparts semi-permeability, that is, exchanger phase is very permeable to Na^+ but practically impermeable to Cl^- in dilute solutions. Conversely, an anion exchange resin is permeable to Cl^- but nearly impermeable to Na^+. This phenomenon is also commonly referred to as coion exclusion or the Donnan exclusion effect. However, coion invasion (i.e., Cl^- permeability for a cation exchanger and Na^+ for an anion exchanger) increases with an increase in the aqueous phase concentration. All other conditions remaining identical, the coion invasion diminishes with an increase in its charge.

Scientifically, the Donnan membrane effect is distinctly different from the effect of surface charges often present at the solid/liquid surface. It is appropriate to note that there are frequent occurrences in research literature to incorrectly interpret coion exclusion as a surface charge phenomenon while the Donnan membrane effect is the underlying reason. Nearly a century ago, the concept, as well as the detailed thermodynamic explanation of the underlying principle, was introduced by Donnan [25–27]. Very recently, the English translation of the original paper has been published [28]. One unique and somewhat counterintuitive feature of the processes and materials that use the principle is that the physical existence of a semi-permeable membrane is not essential. It is the inability of the charged functional groups to diffuse out from the solid to the solvent phase that leads to the phenomenon of semi-permeability, that is, the existence of a fictitious semi-permeable membrane. Thus, a cation exchange resin with fixed negative charges (R^-) or an anion exchange resin with fixed positive charges (R^+) exhibit semi-permeable behaviors that are customarily represented as illustrated in Figure 2.15a and b. Note that both A^+ and B^- are present in the aqueous phase and balance each other from an electroneutrality consideration. Yet, while A^+ enjoys an easy access inside a cation exchange resin in sodium form, B^- is highly impeded. For an anion exchange resin, the reverse is true, that is, B^- can move back and forth between the exchanger and the aqueous phase but A^+ is restricted from entry inside the anion exchanger.

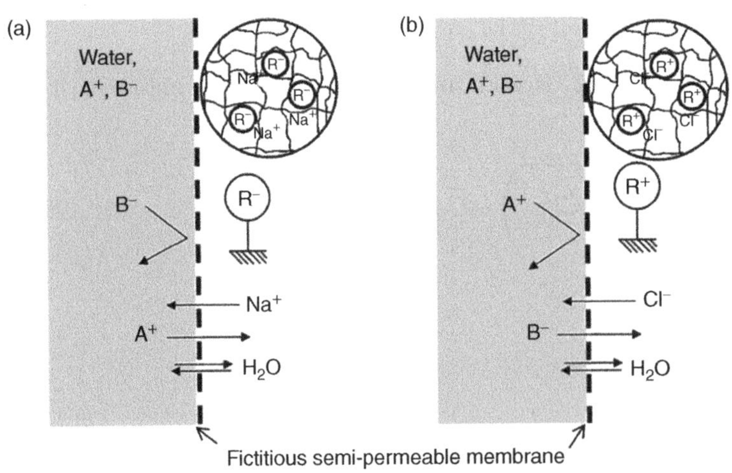

Figure 2.15 Illustration of semi-permeable behavior of ion exchange resins due to the presence of fixed charges in the exchanger phase: (a) cation exchanger with fixed negative charges; (b) anion exchanger with fixed positive charges. *Source:* Sarkar *et al.* 2010 [4]. Reproduced with permission of American Chemical Society.

Example 2.7 illustrates the dependence of coion invasion on the exchanger capacity, electrolyte concentration and charge of the coion.

Example 2.7 *An illustration of the effects of solution concentration, exchanger capacity and coion charge on coion invasion:*

I. Effect of solution concentration (C_T) on coion exclusion

An anion exchange resin with a capacity (fixed positive charge) of 2.0 eq/L (i.e., $R^+ = 2.0\,M$) in chloride form is added separately to two different NaCl solutions: (i) 0.1 M and (ii) 0.05 M. Compute Na^+ concentration inside the anion exchanger.

In solution:

$$[Na^+]_{aq} = [Cl^-]_{aq} = 0.1\,M$$

For electroneutrality in the exchanger phase:

$$\overline{[Na^+]} + \overline{[R^+]} = \overline{[Cl^-]}$$

Principle of equilibrium under ideal conditions:

$$[Na^+]_{aq} \cdot [Cl^-]_{aq} = \overline{[Na^+]} \cdot \overline{[Cl^-]}$$

Substituting and rearranging,

$$[Na^+]_{aq}^2 = \overline{[Na^+]} \cdot (\overline{[Na^+]} + \overline{[R^+]})$$

Here, aq subscript and overbar denote solution and exchanger phase, respectively.
The monovalent resin functional group $\overline{[R^+]} = 2\frac{eq}{L} = 2\,M$
Solving,

$$(0.1\,M)^2 = \overline{[Na^+]} \cdot (\overline{[Na^+]} + 2\,M)$$

$$\overline{[Na^+]} \approx 0.005\,M$$

$$20 \cdot \overline{[Na^+]} \approx [Na^+]_{aq}$$

The exchanger concentration of Na^+ is 20× lower than the solution concentration of Na^+.

When $[Na^+]_{aq} = [Cl^-]_{aq} = 0.05\,M$, $\overline{[Na^+]} \approx 0.00125\,M$ and $40 \cdot \overline{[Na^+]} \approx [Na^+]_{aq}$.

Note: While coion exclusion means the ability of a cation exchanger to reject an anion or an anion exchanger to reject a cation, coion invasion (or electrolyte penetration) refers to the coion concentration within the exchanger. Thus, they are negatively correlated, that is, one increases at the expense of the other.

II. Effect of resin capacity

The anion exchanger capacity is now increased to 4.0 eq/L. and the NaCl concentration is 0.1 M.

Again,

$$[Na^+]_{aq} = [Cl^-]_{aq} = 0.1\,M$$

Solving

$$[Na^+]_{aq}^2 = \overline{[Na^+]} \cdot (\overline{[Na^+]} + \overline{[R^+]})$$
$$\overline{[Na^+]} \approx 0.0025\,M$$

It shows that coion (Na^+) invasion into the anion exchanger decreases with an increase in resin capacity, that is, coion exclusion is more intense.

III. Effect of coion charge

For this scenario, ion exchanger capacity is kept unchanged at $R^+ = 2\,eq/L\ (2\,M)$
Coion charge is varied using 0.1 M NaCl or 0.1 M $CaCl_2$ solution. Coion Na^+ is monovalent whereas Ca^{2+} is divalent
For 0.1 M NaCl from the previous example, $\overline{[Na^+]} \approx 0.005\,M$.
For 0.1 M $CaCl_2$ solution,

$$2\,[Ca^{2+}]_{aq} = [Cl^-]_{aq}$$

Or,

$$[Ca^{2+}] = 0.1\,M$$

Exchanger phase electroneutrality states that,

$$2\overline{[Ca^{2+}]} + \overline{[R^+]} = \overline{[Cl^-]}$$

Principle of equilibrium;

$$[Ca^{2+}]_{aq} \cdot [Cl^-]_{aq}^2 = \overline{[Ca^{2+}]} \cdot \overline{[Cl^-]}^2$$

Rearranging,

$$4[Ca^{2+}]_{aq}^3 = \overline{[Ca^{2+}]} \cdot (2\overline{[Ca^{2+}]} + \overline{[R^+]})^2$$

Other parameters being known,

$$\overline{[Ca^{2+}]} = 0.001\,M$$

Or

$$[Ca^{2+}]_{aq} = 100\overline{[Ca^{2+}]}$$

Thus, Ca^{2+} is rejected more by the anion exchanger than Na^+ under otherwise identical situations. The following figure shows comparison of coion concentration inside an anion exchanger phase for monovalent (Na^+) and divalent (Ca^{2+}), all other conditions remaining the same. Higher rejection (i.e., lower coion invasion) of divalent Ca^{2+} is readily noted.

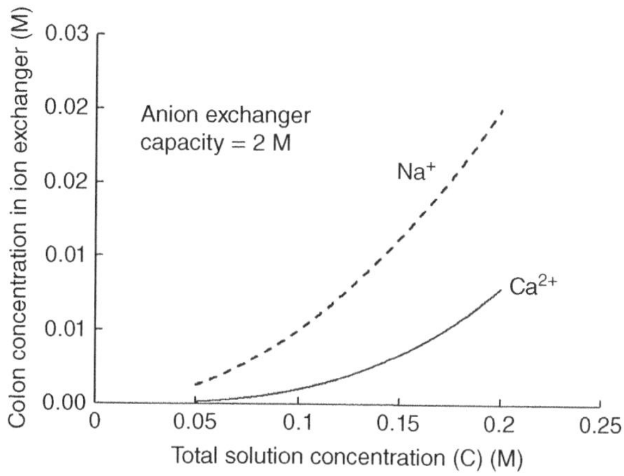

Figure 1. Effect of coion charge on coion invasion at different C_T values.

IV. Effect of counterion charge

One cation-exchange resin with a capacity of 2.0 M is separately in contact with two following solutions:

(i) 0.05 M NaCl
(ii) 0.05 M $AlCl_3$

Now the goal is to find Cl^- concentration inside the strong-acid cation exchanger.

(i) Following the same equality as before,

$$[Na^+]_{eq}[Cl^-]_{eq} = [Na^+]_R[Cl^-]_R$$
$$0.05\,M \times 0.05M = (2 + X)(X), \text{ where } X = [Cl^-]_R$$

Upon solution,

$$X = 0.00125\,M$$
$$[Al^{3+}]_{eq}[Cl^-]^3_{eq} = [Al^{3+}]_R[Cl^-]^3_R$$

From electroneutrality,

$$3[Al^{3+}]_{eq} = [Cl^-]_{eq}$$

Thus,

$$0.05\,M \times (0.15\,M)^3 = \left(2 + \frac{X}{3}\right)(X)^3$$

Upon solution,

$$X = [Cl^-]_R = 0.0438\,M$$

Note that the concentration of chloride, a coion, is significantly greater inside the cation exchanger for a trivalent counterion (Al^{3+}) than that of monovalent (Na^+).

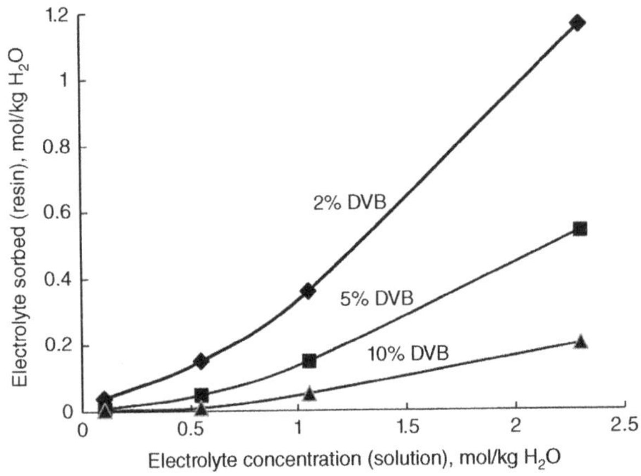

Figure 2.16 Effect of crosslinking percentage on coion invasion. *Source:* Pepper *et al.* 1952 [29]. Reproduced with permission of Royal Society of Chemistry.

2.7.2 Role of Cross-linking

When immersed in a dilute NaCl solution, a cation exchanger swells due to the difference in osmotic pressure between the two phases. Swelling lowers volumetric capacity of the ion exchanger (eq/L), which in turn allows more electrolyte (NaCl) sorption or coion invasion. Cross-linking (expressed % DVB for polystyrene-DVB matrix) imparts mechanical strength to resist swelling. Thus, a higher degree of cross-linking results in less swelling and hence, less electrolyte sorption or coion invasion. Mechanistically, electrolyte sorption is a measure of the incompleteness of the Donnan exclusion effect. Figure 2.16 shows how DVB cross-linking of 2%, 5%, and 10% influence electrolyte (NaCl) sorption onto a cation exchanger [29].

2.7.3 Genesis of the Donnan Potential

Although no electric potential gradient can physically be measured at the ion exchanger-water interface, its existence can be easily conceived and recognized. For example, a cation exchanger has a negatively charged potential that prevents anions (say chloride) from entering inside the exchanger. For an anion exchanger, this potential is positive and it rejects cations. This electric potential at the interface, which is experimentally non-detectable, but operative, is referred to as the Donnan potential and results from the charged but non-diffusible functional groups covalently bonded to the exchanger.

Physically, when a cation exchanger in sodium form is placed in a dilute solution of sodium chloride, there are considerable concentration differences between the two phases. The concentration of the cation (Na^+) is larger in the ion exchanger, while the anion (Cl^-) is larger in the solution. If the ions carried no electric charges, these concentration differences would be leveled out by diffusion. However, for ions, such a process would disturb electroneutrality and cannot proceed spontaneously. Migration of cations into the solution and of anions into the ion exchanger will result in an accumulation of negative charges in the exchanger and positive charges in the solution.

The first few ions which diffuse thus build up an electric potential differences between the two phases. This so called Donnan potential pulls cations back into the negatively charged cation exchanger and anions back into the positively charged solution. The situation with anion exchanger is analogous, but the Donnan potential has the opposite (positive) sign.

Although the Donnan Potential does not alter electroneutrality and is not measurable by physical means, it can be computed using the condition for equilibrium as applied for ions. In equilibrium, the electrochemical potential, η_i, of the ionic species "*i*" is the same in both phases

$$(\eta_i)_R = (\eta_i)_L \tag{2.61}$$

where subscripts R and L denote the resin phase and liquid phase, respectively.

Ignoring the effect of the swelling pressure in the resin phase,

$$(\eta_i)_R = \mu_i^0 + RT\ \ln \overline{a}_i + Z_i F\overline{\Phi} \tag{2.62}$$

$$(\eta_i)_L = \mu_i^0 + RT\ \ln a_i + Z_i F\Phi \tag{2.63}$$

where Z_i = valence of species *i*, F = Faraday's constant, Φ = electric potential and overbar denotes the exchanger phase.

Thus,

$$E_{\text{Don}} = \overline{\Phi} - \Phi = \frac{RT}{Z_i F}\ \ln \frac{a_i}{\overline{a}_i} \tag{2.64}$$

For a cation exchanger, E_{Don} is negative while it is positive for an anion exchanger. Note that as the concentration of the solution phase increases, Donnan potential, E_{Don}, decreases. It is important to recognize that the Donnan potential, coion invasion and electrolyte penetration are all intertwined and influenced by the same process variables. Example 2.7 helps illustrate such phenomena. The Donnan membrane principle or the Donnan potential is often viewed only as a theoretical concept in many quarters, but the development of several relatively new processes and materials in environmental separation is rooted into the core of this principle introduced over a century ago [4].

Example 2.8 *A cation exchanger of 2 M capacity is in equilibrium with 0.05 M NaCl and 0.05 M Na_2SO_4 solution respectively. Find out Donnan potential in each type of solution. Compute Cl^- and SO_4^{2-} concentration in exchanger phase. Assume ideality.*

$$E_{\text{Don}} = \frac{RT}{z_i F}\ \ln \frac{a_i}{\overline{a}_i}$$

where, $R = 8.314\,\text{J/K mol}$, $T = 298\,\text{K}$ (25 °C), $F = 96485$ Coulomb/mol, a_i activity of species i in solution, $\overline{a}_i$ activity of species "*i*" in exchanger, z_i charge of "*i*"

$$\frac{RT}{F} = \frac{8.314\frac{\text{J}}{\text{K}\cdot\text{mol}} \cdot 298\,\text{K}}{96485\frac{\text{C}}{\text{mol}}} = 0.0257\,\text{V} = 25.7\,\text{mV}$$

Assuming ideality, a is replaced by concentration [] in the respective phases.

0.05 M NaCl

$$a_i = [\mathrm{Na}^+]_{\mathrm{aq}} = 0.05\,\mathrm{M},\ \overline{a_i} = \overline{[\mathrm{Na}^+]} = 2\mathrm{M},\ z_i = +1$$

$$E_{\mathrm{Don}} = \frac{25.7\,\mathrm{mV}}{+1}\ln\frac{0.05}{2} = -94.8\,\mathrm{mV}$$

0.05 M Na_2SO_4

$$a_i = [\mathrm{Na}^+]_{\mathrm{aq}} = 2[\mathrm{SO}_4^{2-}]_{\mathrm{aq}} = 0.1\,\mathrm{M},\ z_i = +1$$

$$E_{\mathrm{Don}} = \frac{25.7\,\mathrm{mV}}{+1}\ln\frac{0.1}{2} = -77.0\,\mathrm{mV}$$

Cl^- in exchanger phase

$$E_{\mathrm{Don}} = -94.8\,\mathrm{mV},\ [\mathrm{Cl}^-]_{\mathrm{aq}} = 0.05\,\mathrm{M},\ z_i = -1$$

$$-94.8\,\mathrm{mV} = \frac{25.7\,\mathrm{mV}}{-1}\ln\frac{0.05}{\overline{[\mathrm{Cl}^-]}}$$

$$\overline{[\mathrm{Cl}^-]} = 1.25\cdot 10^{-3}\,\mathrm{M}$$

Concentration of chloride in the resin is 40× lower than the equilibrium aqueous concentration.

SO_4^{2-} in exchanger phase

$$E_{\mathrm{Don}} = -77.0\,\mathrm{mV},\ [\mathrm{SO}_4^{2-}]_{\mathrm{aq}} = 0.05\,\mathrm{M},\ z_i = -1$$

$$-77.0 = \frac{25.7\,\mathrm{mV}}{-2}\ln\frac{0.05}{\overline{[\mathrm{SO}_4^{2-}]}}$$

$$\overline{[\mathrm{SO}_4^{2-}]} = 1.25\times 10^{-4}\,\mathrm{M}$$

Sulfate concentration in the exchanger phase is 400× lower than its concentration in the solution phase.

Note: Similar results were reached based on the equilibrium principle explained in Example 2.7 part III, while explaining coion invasion for divalent Ca^{2+} in the exchanger phase for 0.05 M $CaCl_2$ solution.

2.8 Weak-Acid and Weak-Base Ion Exchange Resins

In aqueous solutions, the dissolved weak-acid (or weak-base) molecules are free to move around in water without interference from others. Thus, every monoprotic weak acid has a unique acid dissociation constant (K_a) or pK_a (i.e., $-\log K_a$) value. In contrast, for weak-acid (or weak-base) ion exchange resin, the functional groups are fixed, that is, they are covalently attached and reside near each other, often less than 1 nm to each other. So, there is a lateral interaction or interference among neighboring sites, often through hydrogen bonding as illustrated in Figure 2.17 for an ion exchanger with carboxylic acid functional groups.

Also, heterogeneous distribution of functional sites and the Donnan membrane effect caused by neighboring sites influence dissociation of weakly ionized functional

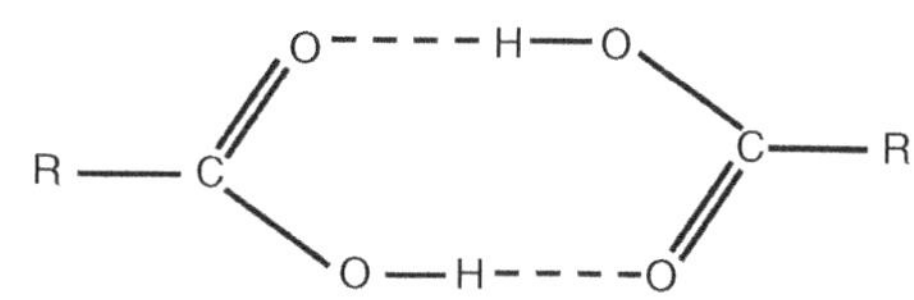

Figure 2.17 Formation of cyclic structure of two neighboring carboxylate functional groups through hydrogen bonding.

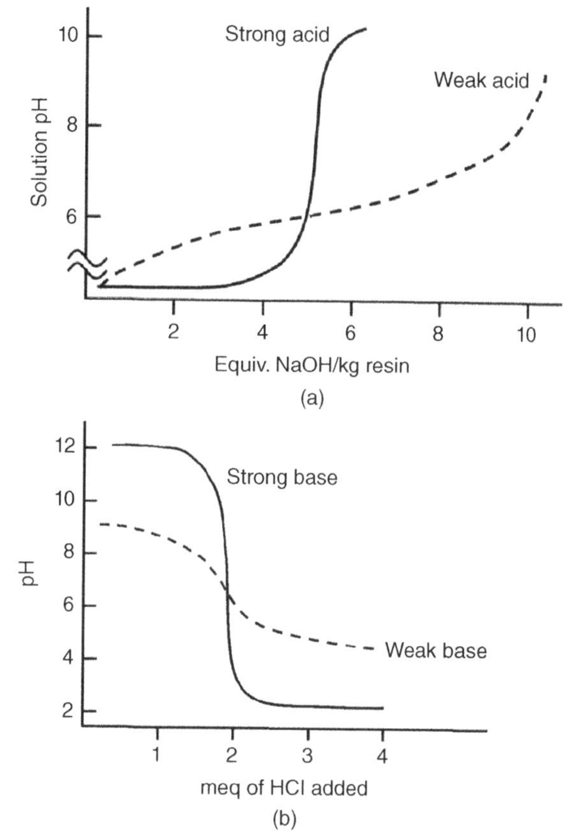

Figure 2.18 (a) Illustration of pH titration curves of strong-acid and weak-acid cation exchange resins. (b) Illustration of pH titration curves of strong-base and weak-base anion exchange resins.

groups. Weak-acid or weak-base ion exchange resins, therefore, do not have a single pK_a value even in dilute solution. Figure 2.18a represents typical titration curves of strong-acid and weak-acid ion exchange resins with gradual addition of base where those of strong-base and weak-base anion exchange resins are presented in (b). For reasons explained above, end or equivalence points for weak-acid and weak-base resin titration curves are difficult to identify during the progress of titration.

Customarily, dissociation of weak-acid and weak-base ion exchange resins is expressed as proton-release reactions as follows:

$$\overline{R - COOH} \leftrightarrow \overline{R - COO^-} + \overline{H^+} \tag{2.65}$$

$$\overline{R_3 - NH^+} \leftrightarrow \overline{R_3 - N} + \overline{H^+} \tag{2.66}$$

In addition, as the ionic strength of the aqueous solution in contact with the weak-acid ion exchange resin is gradually increased (say by adding NaCl), the weak-acid functional groups get increasingly dissociated through partial displacement of H^+ with Na^+, thus increasing the K_a value (i.e., lower pK_a). The foregoing phenomena clearly distinguish weak-acid or weak-base ion exchange resins from their counterparts in the aqueous phase. The following section provides a methodology to provide a theoretical framework to determine Ka values of weakly functionalized resins.

2.8.1 p*K*a Values of Weak Ion Exchange Resins

The dissociation of a cation exchanger can be written as

$$\overline{\mathrm{RH}} \leftrightarrow \overline{\mathrm{R}^-} + \overline{\mathrm{H}^+} \tag{2.67}$$

From the law of mass action, acid dissociation constant for cation exchanger (CIX) is given as

$$K_a^{\mathrm{CIX}} = \frac{[\overline{\mathrm{R}^-}][\overline{\mathrm{H}^+}]}{[\overline{\mathrm{RH}}]} \tag{2.68}$$

The dissociation of an anion exchanger (AIX) with amine functional group can be presented as

$$\overline{\mathrm{RNH}^+} \leftrightarrow \overline{\mathrm{RN}} + \overline{\mathrm{H}^+} \tag{2.69}$$

and

$$K_a^{\mathrm{AIX}} = \frac{[\overline{\mathrm{RN}}][\overline{\mathrm{H}^+}]}{[\overline{\mathrm{RNH}^+}]} \tag{2.70}$$

As for normal acids and bases, the acid dissociation constants are expressed as their negative logarithm values

$$\mathrm{p}K_a^{\mathrm{CIX}} = -\log K_a^{\mathrm{CIX}} \tag{2.71}$$

and

$$\mathrm{p}K_a^{\mathrm{AIX}} = -\log K_a^{\mathrm{AIX}} \tag{2.72}$$

For strong-acid cation exchangers (e.g., with sulfonic acid group), $\mathrm{p}K_a^{\mathrm{CIX}} \leq 1$ and for strong-base anion exchangers (e.g., with quaternary ammonium group), $\mathrm{p}K_a^{\mathrm{AIX}} \geq 13$. Thus, acid dissociation constants are of minor consequence for strong resins and their capacity is available for the entire pH range. Ion exchange capacities of weak-acid and weak-base resins, in contrast, are pH-dependent and they can be titrated with standard bases and acids. The neutralization of the resins can be observed by recording the pH of the supernatant solution while the titration is in progress. Such a titration curve, however, does not provide the pH inside the ion exchange resin and, therefore, needs to be appropriately evaluated to determine the capacity and the pK_a value.

The degree of dissociation, $\propto$, and the pH in the weak-acid resin are defined as follows:

$$\propto \equiv \frac{[\overline{R^-}]}{[\overline{R^-}] + [\overline{RH}]} \tag{2.73}$$

$$\overline{pH} = -\log[\overline{H^+}] \tag{2.74}$$

Combining Eqs (2.68), (2.73), and (2.74), one obtains

$$\overline{pH} = pK_a^{CIX} - \log\frac{1-\propto}{\propto} \tag{2.75}$$

Note that this equation involves pH in the resin and that is different from the pH in external solution. Also, when the resin is loaded 50% with Na^+, that is, $\alpha = 0.5$, pKa of the cation resin is equal to its resin-phase pH. To relate pH in the solution to the pH in resin during titration, let us assume that the concentration ratio $[Na^+] : [H^+]$ is the same in the ion exchanger as in the aqueous phase, that is,

$$[\overline{H^+}] = \frac{[H^+][\overline{Na^+}]}{[Na^+]} \tag{2.76}$$

At 50% conversion, (i.e., $\propto = 0.5$) the Na^+ concentration in the resin is

$$[\overline{Na^+}] = \frac{[\overline{X}]}{2} \tag{2.77}$$

where $[\overline{X}]$ is the total concentration of dissociated and undissociated ionogenic groups and equal to: $[\overline{X}] = [\overline{RH}] + [\overline{R^-}]$.

Thus, using Eqs (2.75)–(2.77) one obtains

$$pK_a^{CIX} = pH_{0.5} + \log[Na^+] - \log\frac{[\overline{X}]}{2} \tag{2.78}$$

$pH_{0.5}$ denotes aqueous-phase pH at $\propto = 0.5$. In a similar vein, the corresponding relation for weak-base anion exchangers when titrated with HCl is

$$pK_a^{AIX} = pH_{0.5} - \log[Cl^-] + \log\frac{[\overline{X}]}{2} \tag{2.79}$$

Equations (2.78) and (2.79) can be used for computing pK_a values of weak-acid cation and weak-base anion exchangers from pH titrations.

pH titration experiments are kinetically slow. For determination of apparent pK_a values, a series of samples of exchanger materials in the H^+ form (or OH^- for weak-base anion) must be contacted with basic (or acidic) solutions of different initial composition. These solutions contain NaCl to guarantee an approximately constant ionic strength. Figures 2.19 and 2.20 show pH titration curves for both weak-acid and weak-base resins with and without NaCl. Note that in the presence of NaCl, both weak-acid and weak-base resins are more ionized, that is, the apparent pK_a values are influenced by electrolyte concentration in the aqueous phase.

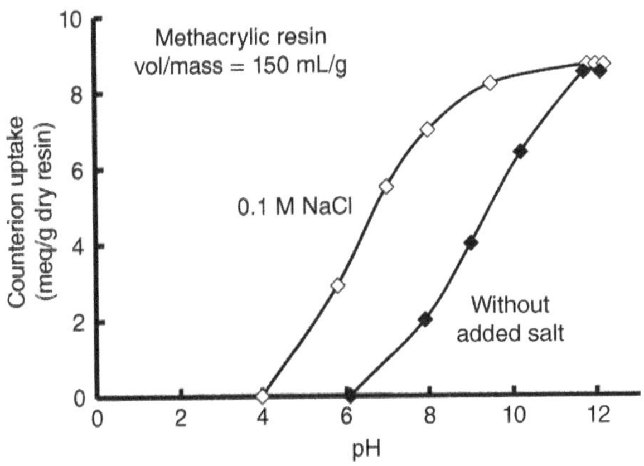

Figure 2.19 Experimental pH titration curves of weak-acid cation exchange resins with and without NaCl. The counterion uptake is plotted versus the pH of the aqueous phase. The ratio of solution volume to resin dry-weight is 150 mL : 1 g. *Source:* Adapted from Pepper *et al.* 1952 [29] and Topp and Pepper 1949 [30].

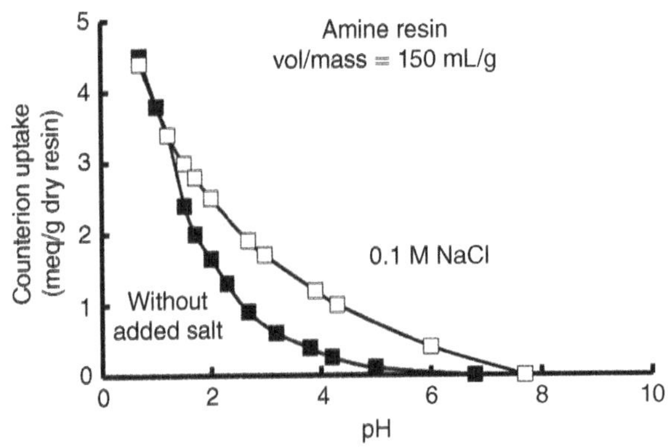

Figure 2.20 Experimental pH titration curves of weak-base anion exchange resins with and without NaCl. The counterion uptake is plotted versus the pH of the aqueous phase. The ratio of solution volume to resin dry-weight is 150 mL : 1 g. *Source:* Adapted from Pepper *et al.* 1952 [29] and Topp and Pepper 1949 [30].

2.8.2 Weak-Acid and Weak-Base Functional Groups

Weak-acid and weak-base ion exchange resins, although solid and insoluble in most solvents, exhibit high H^+ and OH^- affinities like their water-soluble analogs. Consequently, weak-acid cation resins and weak-base anion resins can be regenerated very efficiently with dilute acid and base, respectively. High regeneration efficiency of the weak resins compared to their strong counterparts, is the primary attribute for their diverse applications wherever appropriate. However, ion exchange of neutral salts by the weak-acid resin in H-form (or weak-base resin in free base or OH^- form) must yield free hydrogen ions (or OH^- ions), which would promptly displace the exchange equilibrium in the direction of functional group association. Thus, ion exchange reactions with neutral salts cannot be sustained and it is commonly said that weak-acid and weak-base resins possess only a limited salt-splitting capacity. However, they are efficient for ion exchange reactions involving salts of weak acids and weak bases .

The following examples demonstrate ion exchange behaviors of weak-acid and weak-base resins for different salts over a wide range of pH.

Weak-Acid Ion Exchange Resin

$$2\overline{R-COOH} + CaCl_2 \rightleftarrows \overline{(R-COO^-)_2Ca^{2+}} + 2HCl \tag{2.80}$$

The forward reaction is unfavorable due to the production of a strong acid, for example, HCl. Thus, the weak-acid resin does not have any salt-splitting capacity but the reverse reaction or regeneration with mineral acid is very efficient.

$$2\overline{R-COOH} + Ca(HCO_3)_2 \rightleftarrows \overline{(R-COO^-)_2Ca^{2+}} + 2H_2CO_3 \qquad (2.81)$$

Here, the forward reaction is favorable because H_2CO_3 is a weak acid. Weak-acid resins are, thus, quite suitable for removing temporary hardness.

Typical ion exchange reactions are possible at pH ≥ 5.0

$$2\overline{(RCOO^-)Na^+} + CaCl_2 \xleftrightarrow{pH>5.0} \overline{(RCOO^-)_2Ca^{2+}} + 2Na^+ \qquad (2.82)$$

It is important to note that weak-acid carboxylate resins offer significantly higher Ca^{2+} selectivity over Na^+ than strong-acid sulfonic acid resins. Thus, upon exhaustion, weak-acid resins are first regenerated with acid, followed by neutralization with $NaOH/NaHCO_3$.

Weak-Base Ion Exchange Resin

At alkaline pH, weak-base anion resins remain deprotonated (i.e., in free base form) and are thus unable to split neutral salts. However, reactions with acid solutions are quite favorable and proceed to completion because the produced water pH remains neutral. A tertiary weak-base resin, $RCH_2(CH_3)_2N$, is used as an example to depict the reactions.

$$\overline{RCH_2(CH_3)_2N} + NaCl + H_2O \rightleftarrows \overline{RCH_2(CH_3)_2NH^+Cl^-} + Na^+ + OH^- \qquad (2.83)$$

The forward reaction is highly unfavorable due to the formation of OH^- as a potential product.

$$\overline{RCH_2(CH_3)_2N} + HCl \rightarrow \overline{RCH_2(CH_3)_2NH^+Cl^-} \qquad (2.84)$$

Uptake of anions at acidic pH may be viewed as a neutralization step and is highly favorable.

Anions of very weakly dissociated acids with $pK_a \geq 7$ exhibit poor uptake onto the weak-base ion exchange resins. Thus, silica (or silicate anion) and sulfide (HS^-) are poorly sorbed onto the weak-base resins. At acidic pH, weak-base resins participate in anion exchange reactions like their strong-base counterparts as follows:

$$\overline{RCH_2(CH_3)_2NH^+Cl^-} + NO_3^- \xleftrightarrow{pH<5} \overline{RCH_2(CH_3)_2NH^+NO_3^-} + Cl^- \qquad (2.85)$$

Obviously, weak-base resins are amenable to efficient regeneration with weakly basic solutions, namely, ammonium hydroxide.

$$\overline{RCH_2(CH_3)_2NH^+NO_3^-} + NH_4OH \rightarrow \overline{RCH_2(CH_3)_2N} + NH_4^+ + NO_3^- + H_2O \qquad (2.86)$$

Methyl groups (—CH_3) are electron-donating groups and, therefore, by substituting hydrogen with a methyl group, the dissociation of the weak-acid groups is weakened. So, the pK_a values are increased. Such an approach is confirmed by observing the pK_a values of the repeating functional groups with and without substitution of hydrogen by methyl group:

Acidic Group—

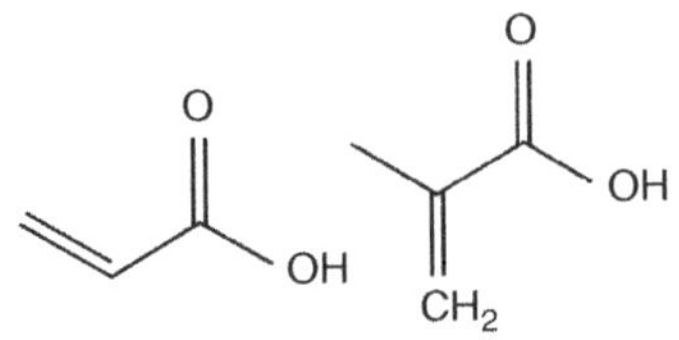

Acrylic acid
$pK_a = 4.2$

Methacrylic acid
$pK_a = 4.7$

Basic Group--
Ammonium:
NH_4^+, $pK_a = 9.3$
Secondary amine:

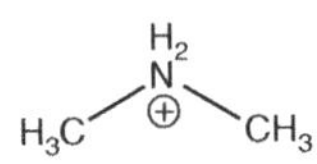

Dimethylammonium
$pK_a = 10.8$

Ion exchangers with multiple weak-acid and/or weak-base functional groups exhibit dissociation/association over a wider range of pH. Dissociation of a popular chelating exchanger with weak iminodiacetate functional group with an increase in pH is illustrated hereunder in Figure 2.21.

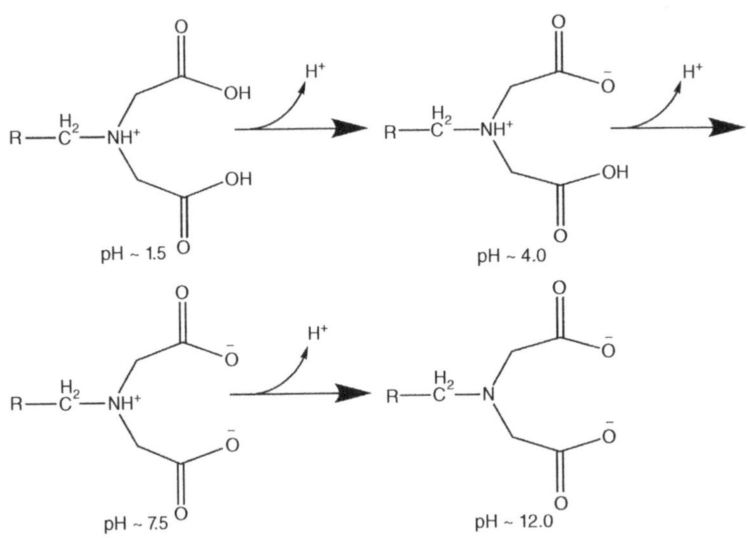

Figure 2.21 Gradual deprotonation of a cation exchanger with weak iminodiacetate functional group with an increase in pH (R represents repeating styrene matrix).

2.9 Regeneration

In principle, ion exchange is a sorption process and its viability for any specific application is influenced by the two following considerations:

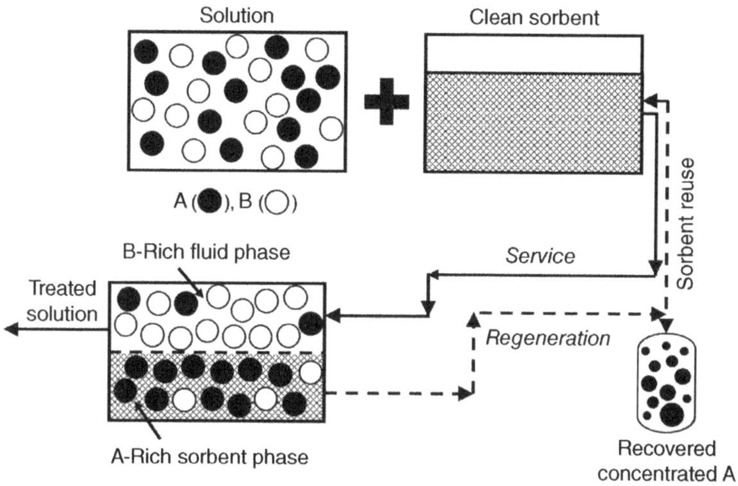

Figure 2.22 A schematic depicting the most commonly used ion exchange system configuration in environmental separation to remove target solute "A" from the solution and reuse of the ion exchanger.

– Capacity and sorption affinity of the ion exchangers for target solutes.
– Regeneration or desorption efficiency of the ion exchangers.

Figure 2.22 provides a schematic of a generic ion exchange process where the goal is to separate and concentrate solute A from a mixture of A and B in the fluid (solution) and then reuse the sorbent after regeneration. Ion exchange is, thus, a cyclic process with two major steps: sorption (or separation) and desorption (or regeneration).

For the ion exchange process to be viable, the ion exchangers should be amenable to regeneration or desorption so that they may be used for hundreds of cycles. In fact, the overall economy of an ion exchange process is often dictated by the operating costs of regeneration as opposed to the fixed cost of the ion exchangers. During the last two decades, environmental sustainability of the ion exchange process has been a major focus of many applications. In general, the spent regenerant resulting from regeneration, that is, its volume, type of chemicals it contains, long-term fates and ecological impact, tend to deeply influence the overall acceptability of the process. Ideally, an ion exchange process should be reversible so that the target solutes can be desorbed efficiently, thus leading to energy-efficient separations. However, efficiency of desorption (or regeneration) tends to diminish for highly selective sorbents. To strike a balance between selectivity and regenerability, the intensity of solute-sorbent interaction must lie within an envelope where ion exchange-type sorption is selective, yet reversible. Figure 2.23 helps quantify such a working regime for various types of interactions.

Efficient regeneration can be attained by appropriately exploiting fundamentals of ion exchange for many specific processes. In principle, every regeneration process has three goals in common: (i) reducing the sorption affinity of target ions loaded onto the ion exchanger; (ii) lowering the volume of spent regenerant; (iii) using inexpensive and environmentally benign chemicals or avoiding chemicals altogether. Specific examples emphasizing underlying fundamentals to achieve this goal are discussed hereunder.

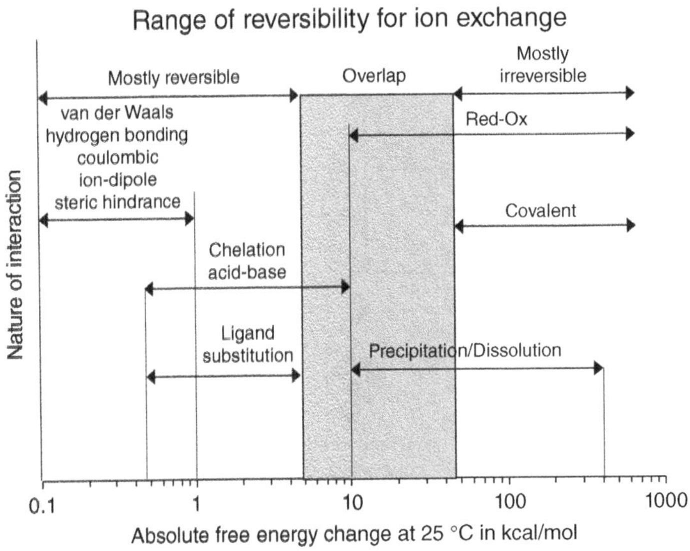

Figure 2.23 A quantitative measure of various interactions in ion exchange-type sorption processes.

2.9.1 Selectivity Reversal in Heterovalent Ion Exchange

Removal of hardness (e.g., Ca^{2+}) from surface and groundwater is the most widely used heterovalent ion exchange process.

$$2\overline{R^-Na^+} + Ca^{2+} \xrightarrow{\text{Favorable}} \overline{(R^-)_2Ca^{2+}} + 2\,Na^+ \tag{2.87}$$

Calcium–sodium separation factor, $\alpha_{Ca/Na}$, is significantly greater than unity for the water to be treated and, thus, calcium removal is favorable. For efficient regeneration with NaCl, $\alpha_{Ca/Na}$ should preferably be less than unity. Earlier, Section 2.4.5 is devoted to heterovalent ion exchange and discusses how divalent–monovalent separation factor drops with an increase in electrolyte concentration. So, at high sodium chloride concentration, $\alpha_{Ca/Na}$ can be lowered less than unity, that is, calcium can be efficiently and favorably desorbed from the ion exchanger using sodium. Example 2.2, solved earlier in the chapter, shows how theoretically computed $\alpha_{Ca/Na}$ varies with concentration of regenerant NaCl for a cation exchange resin with sulfonic acid functional groups.

High sodium chloride concentration (10% mass/volume) is routinely used for regeneration in the hardness removal process.

$$\overline{(R^-)_2Ca^{2+}} + 2\,Na^+(aq) \xrightarrow{\text{Favorable}} 2\overline{R^-Na^+} + Ca^{2+}(aq) \tag{2.88}$$

For chromate ion exchange, monovalent chromate ($HCrO_4^-$) is used for sorption at acidic pH to attain higher removal capacity.

Sorption:

$$\overline{R^+Cl^-} + HCrO_4^-(aq) \xrightarrow{\text{Favorable}} \overline{R^+HCrO_4^-} + Cl^-(aq) \tag{2.89}$$

For regeneration, high chloride concentration is used at alkaline pH to take advantage of selectivity reversal by first transforming monovalent $HCrO_4^-$ into divalent CrO_4^{2-}.

$$2\overline{R^+HCrO_4^-} + 2NaOH \xrightarrow{\text{Favorable}} \overline{(R^+)_2CrO_4^{2-}} + 2Na^+(aq) + 2H_2O \tag{2.90}$$

$$\overline{(R^+)_2CrO_4^{2-}} + 2Cl^-(aq) \xrightarrow{\text{Favorable}} 2\overline{R^+Cl^-} + CrO_4^{2-}(aq) \tag{2.91}$$

Note that the pK_a value of $HCrO_4^-$ is $pK_a = 6.5$, that is, at a pH value greater than pH = 6.5, CrO_4^{2-} predominates in the aqueous phase over $HCrO_4^-$.

2.9.2 pH Swings

Protonation and deprotonation of weak-acid and weak-base ion exchange resins are thermodynamically very favorable and hence, by using near-stoichiometric amount of acid or base, the ion exchangers can be regenerated efficiently. Essentially, these exchangers sorb and desorb at different pH values distinctive of their functional groups. Typically, a weak-acid cation exchanger sorbs calcium present as temporary hardness at pH > 6.0, while regeneration can be carried out at pH ≤ 3.0.

Sorption at pH > 6.0:

$$2\overline{R - COOH} + Ca(HCO_3)_2 \xrightarrow{\text{Favorable}} \overline{(R - COO^-)_2Ca} + 2H_2O + 2CO_2 \tag{2.92}$$

Regeneration at pH ≤ 3.0:

$$\overline{(R - COO^-)_2Ca} + 2H^+ \xrightarrow{\text{Favorable}} 2\overline{R - COOH} + Ca^{2+}(aq) \tag{2.93}$$

For weak-base anion exchange resins, sorption cycle is carried out at pH < 5.0 and desorption at pH ≥ 10.

Sorption: pH < 5

$$\overline{R_3NH^+Cl^-} + NO_3^-(aq) \rightleftarrows \overline{R_3NH^+NO_3^-} + Cl^-(aq) \tag{2.94}$$

Regeneration: pH ≥ 10

$$\overline{R_3NH^+NO_3^-} + OH^-(aq) \rightarrow \overline{R_3N} + NO_3^-(aq) + H_2O \tag{2.95}$$

Pre-conditioning:

$$\overline{R_3N} + H^+(aq) + Cl^-(aq) \rightarrow \overline{R_3NH^+Cl^-} \tag{2.96}$$

Example 2.9 Design of a Three-Bed Deionizer with Decarbonation (WAC-SAC-Decarbonator-WBA)

Revisit Example 1.4 from Chapter 1, but now the system is changed to a three-bed deionizer with a decarbonator tank between cation and anion exchangers. Also, now the systems include a weak-acid cation (WAC) exchanger at the head of the train and a weak-base anion (WBA) exchanger at the end.

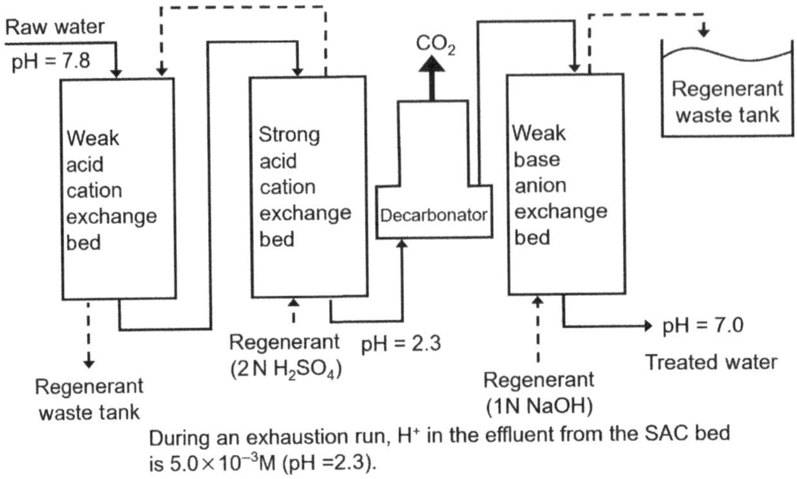

Given: Raw water of the following quality:

Hardness = 3.0 meq/L
Bicarbonate = 2.0 meq/L
pH = 7.8
Chloride = 1.0 meq/L
Sulfate = 2.0 meq/L
Sodium = to balance anions
The decarbonator is 90% effective at removing dissolved gases after the SAC resin.

The objective is to design a three-bed, 400 L/min deionizer system with decarbonation which must run for 8 hours before breakthrough.

Find:

(a) The volume of weak-acid cation (WAC)-strong-acid cation (SAC) exchange resin required if the three-bed deionizer has a capacity of 3.0 eq/L for WAC resin and 1.0 eq/L for SAC resin after thoroughfare regeneration with H_2SO_4 at 130% of the capacity of the cation exchange resins at a concentration of 2.0 N. In thoroughfare regeneration, the regenerant is reused in two resin columns before being wasted, for example, passed through the SAC before being passed through the WAC. Assume that the WAC column is used solely to protonate all alkalinity and the SAC column is used for removal of other cations.

(b) The volume of weak-base anion (WBA) exchange resin required if it has a capacity of 0.7 eq/L of resin after regeneration with NaOH at 120% of the capacity of the WBA and a concentration of 1.0 N.

(c) The analysis in mg/L for all the ions in the mixed waste regenerant solution after neutralization. Assume that the slow rinses are collected with the regenerants and that they comprise 2 BV for each bed. Deionized water is used to make up the regenerant solutions. Neutralization is done with the same acid or base as used for regeneration.

(d) What are the advantages or disadvantages of having a deaerator in the deionization system?
(e) Find the electrolyte concentration in the spent regenerant and compare this to the amount of removed contaminants to find a measure of the ion exchange efficiency or sustainability index. Commentswelcome.

Solutions:

(a) *WAC*

Concentration of alkalinity: $[HCO_3^-] = 2.0\frac{meq}{L}$

Total mass of protons needed:

$$2.0\frac{meq}{L} \cdot 400\frac{L}{min} \cdot 60\frac{min}{hr} \cdot 8hr = 3.84 \cdot 10^5\,meq = 384\,equivalents$$

The volume of resin needed:

$$\frac{384\,equivalents}{3.0\frac{eq}{L}} = 128\,L = 0.13\,m^3$$

Total mass of hardness removed:

$$2.0\frac{meq}{L} \cdot 400\frac{L}{min} \cdot 60\frac{min}{hr} \cdot 8hr = 3.84 \cdot 10^5\,meq = 384\,equivalents$$

SAC

Concentration of cations:

$$\text{Hardness} = [Ca^{2+}] + [Mg^{2+}] = 3.0\,meq/L$$

$$[Na^+] = 2.0\,meq/L$$

$$\text{Total} = 5.0\,meq/L$$

Total mass of exchanged cation

$$\left(5.0\frac{meq}{L} - 2.0\frac{meq}{L}\right) \cdot 400\frac{L}{min} \cdot 60\frac{min}{hr} \cdot 8\,h = 5.76 \times 10^5\,meq = 576\,equivalents$$

The volume of SAC resin needed:

$$\frac{576\,equivalents}{1.0\frac{eq}{L}} = 576\,L = 0.58\,m^3$$

(b)

Concentration of anions:

$$(1 - 0.9) \cdot 2.0\frac{meq}{L}(HCO_3^-) + 2.0\frac{meq}{L}(SO_4^{2-}) + 1.0\frac{meq}{L}(Cl^-) = 3.2\frac{meq}{L}$$

Total mass of exchanged anions

$$3.2\frac{meq}{L} \cdot 400\frac{L}{min} \cdot 60\frac{min}{hr} \cdot 8h = 6.14 \times 10^5\,meq = 614\,equivalents$$

The volume of WBA resin needed:

$$\frac{614\,eq}{0.7\frac{eq}{L}} = 877\,L = 0.88\,m^3$$

(c) *Volume of rinse water (V_1)*

$$V_1 = 2\,\text{BVs} \cdot 0.13\,\frac{\text{m}^3}{\text{BV}} + 2\text{BVs} \cdot 0.58\,\frac{\text{m}^3}{\text{BV}} + 2\text{BVs} \cdot 0.88\frac{\text{m}^3}{\text{BV}} = 3.18\,\text{m}^3$$

Volume of H_2SO_4 (V_2)

$$V_2 = \frac{1.3 \cdot \left(3\,\frac{\text{eq}}{\text{L}_{\text{WAC}}} \cdot 128\,\text{L}_{\text{WAC}} + 1\,\frac{\text{eq}}{\text{L}_{\text{SAC}}} \cdot 576\,\text{L}_{\text{SAC}}\right)}{2\frac{\text{eq}}{\text{L}_{\text{H}_2\text{SO}_4}}} = 624\text{L}_{\text{H}_2\text{SO}_4} = 0.62\,\text{m}^3\text{H}_2\text{SO}_4$$

Volume of NaOH (V_3)

$$V_3 = \frac{1.2 \cdot \left(0.7\,\frac{\text{eq}}{\text{L}_{\text{WBA}}} \cdot 877\,\text{L}_{\text{WBA}}\right)}{1\frac{\text{eq}}{\text{L}_{\text{NaOH}}}} = 737\text{L}_{\text{NaOH}} = 0.74\,\text{m}^3\,\text{NaOH}$$

Volume of base for neutralizing acid (V_4)

$$624\,\text{L}\,\text{H}_2\text{SO}_4 \cdot 2\frac{\text{eq}\,\text{H}^+}{\text{L}_{\text{H}_2\text{SO}_4}} - 737\,\text{L}\,\text{NaOH} \cdot 1\frac{\text{eq}\,\text{OH}^-}{\text{L}_{\text{NaOH}}} = 511\,\text{eq}\,\text{H}^+$$

$$511\,\text{eq}\,\text{H}^+ \cdot 1\frac{\text{eq}\,\text{OH}^-}{\text{eq}\,\text{H}^+} \cdot 1\frac{\text{L}_{\text{NaOH}}}{\text{eq}\,\text{OH}^-} = 511\,\text{L}_{\text{NaOH}} = 0.51\,\text{m}^3\,\text{NaOH}$$

Total Volume

$$V_T = V_1 + V_2 + V_3 = (3.18 + 0.63 + 0.74 + 0.51)\,\text{m}^3 = 5.06\,\text{m}^3$$

Concentration in mixed solution:

$$[\text{Ca}^{2+}] + [\text{Mg}^{2+}] = \frac{3\frac{\text{meq}}{\text{L}} \cdot 400\frac{\text{L}}{\text{min}} \cdot 60\frac{\text{min}}{\text{h}} \cdot 8\text{h}}{5.06\,\text{m}^3 \times 10^3\,\frac{\text{L}}{\text{m}^3}} = 113.8\frac{\text{meq}}{\text{L}} = 2277\,\frac{\text{mg}}{\text{L}}\,\text{as}\,\text{Ca}^{2+}$$

$$[\text{Na}^+] = \frac{2\frac{\text{meq}}{\text{L}} \cdot 400\frac{\text{L}}{\text{min}} \cdot 60\frac{\text{min}}{\text{hr}} \cdot 8\,\text{h} + (737\text{L} + 511\text{L}) \cdot 1 \times 10^3\frac{\text{meq}}{\text{L}}}{5.06\,\text{m}^3 \times 10^3\,\frac{\text{L}}{\text{m}^3}}$$

$$= 322.5\,\frac{\text{meq}}{\text{L}} = 7418\,\frac{\text{mg}}{\text{L}}$$

$$[\text{HCO}_3^-] = \frac{0.1 \cdot 2\frac{\text{meq}}{\text{L}} \cdot 400\frac{\text{L}}{\text{min}} \cdot 60\frac{\text{min}}{\text{h}} \cdot 8\,\text{h}}{5.06\,\text{m}^3 \times 10^3\,\frac{\text{L}}{\text{m}^3}} = 7.6\,\frac{\text{meq}}{\text{L}} = 463\,\frac{\text{mg}}{\text{L}}$$

$$[\text{Cl}^-] = \frac{1\frac{\text{meq}}{\text{L}} \cdot 400\frac{\text{L}}{\text{min}} \cdot 60\frac{\text{min}}{\text{h}} \cdot 8\,\text{h}}{5.06\,\text{m}^3 \times 10^3\,\frac{\text{L}}{\text{m}^3}} = 38\,\frac{\text{meq}}{\text{L}} = 1347\,\frac{\text{mg}}{\text{L}}$$

$$[\text{SO}_4^{2-}] = \frac{2\frac{\text{meq}}{\text{L}} \cdot 400\frac{\text{L}}{\text{min}} \cdot 60\frac{\text{min}}{\text{h}} \cdot 8\,\text{h} + 624\text{L}\,\text{H}_2\text{SO}_4 \cdot 2 \times 10^3\,\frac{\text{meq}}{\text{L}_{\text{H}_2\text{SO}_4}}}{5.06\,\text{m}^3 \times 10^3\,\frac{\text{L}}{\text{m}^3}}$$

$$= 322.5\,\frac{\text{meq}}{\text{L}} = 15{,}481\,\frac{\text{mg}}{\text{L}}$$

(d)

Ion exchange sustainability index

$$\begin{aligned}\text{Cations and anions removed} &= [Ca^{2+}] + [Mg^{2+}] + [Na^{+}] \\ &\quad + [HCO_3^-] + [Cl^-] + [SO_4^{2-}] \\ &= 5.0\frac{\text{meq}}{\text{L}}\text{cations} + 5.0\frac{\text{meq}}{\text{L}}\text{anions} = 10\frac{\text{meq}}{\text{L}}\end{aligned}$$

$$\text{Regenerant added} = [H^+]_{\text{added}} + [OH^-]_{\text{added}} = \left(2\frac{\text{eq}}{L_{H_2SO_4}} \cdot 624L_{H_2SO_4}\right) + \left(1\frac{\text{eq}}{L_{NaOH}} \cdot 1248\,L_{NaOH}\right) = 2496\,\text{eq}$$

$$\begin{aligned}\text{pH neutralization requirement} &= \frac{\text{regenerant added} + \text{pH neutralization eequirement}}{\text{cations and anions removed}} \\ &= \frac{([H^+]_{\text{added}} + [OH^-]_{\text{added}})}{[Ca^{2+}] + [Mg^{2+}] + [Na^+] + [HCO_3^-] + [Cl^-] + [SO_4^{2-}]} \\ &= \frac{\left(2\frac{\text{eq}}{L_{H_2SO_4}} \cdot 624L_{H_2SO_4}\right) + \left(1\frac{\text{eq}}{L_{NaOH}} \cdot 1248\,L_{NaOH}\right)}{10\frac{\text{meq}}{\text{L}} \cdot \frac{1\,\text{eq}}{1000\,\text{meq}} \cdot 400\frac{\text{L}}{\text{min}} \cdot 60\frac{\text{min}}{\text{h}} \cdot 8\,\text{h}} \\ &= 1.30\end{aligned}$$

Remarks:

Note that the sustainability index for the three-bed deionization system with a degasser/decarbonator is closer to 1.0 than the two-bed system described in Example 1.4 from Chapter 1, that is, the amount of regenerant used and spent regenerant discharged to the environment is significantly lower.

Introducing the degasser after the cation exchangers reduced the load onto the anion exchanger and offered regenerant-free removal of alkalinity.

Weak-acid and weak-base ion exchange resins greatly improved the efficiency of regeneration.

From an environmental sustainability viewpoint, the disposal of spent regenerant continues to be the single most area of concern.

2.9.3 Ligand Exchange with Metal Oxides

Polyvalent metal oxides, namely, oxides of iron, aluminum, titanium and zirconium exhibit high sorption affinities for anionic ligands at neutral to slightly acidic pH. These oxides are also amphoteric and they can be regenerated through pH swings as illustrated in Figure 2.24 for arsenate, fluoride and arsenite sorption and desorption where "M" represents a polyvalent metal, namely, Fe(III) or Zr(IV).

Note that at alkaline pH, the surface hydroxyl groups get deprotonated and negatively charged, thus causing desorption of negatively charged arsenic species very efficiently

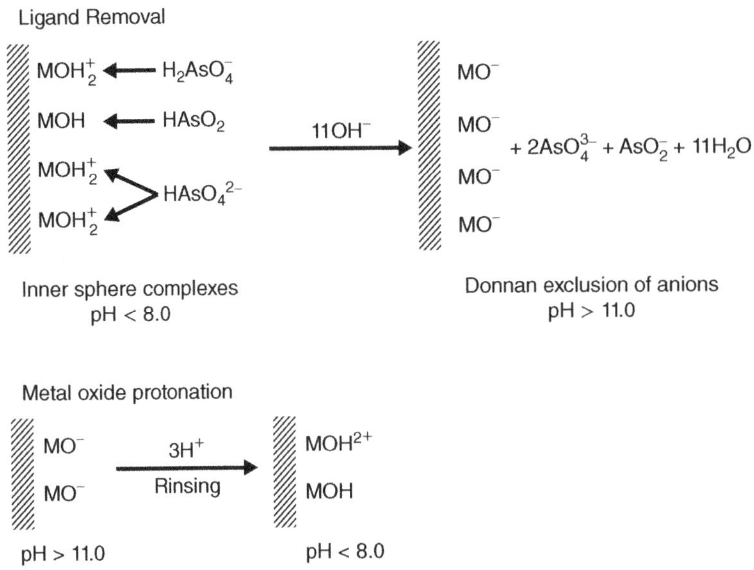

Figure 2.24 Regeneration of metal oxides exhausted by anionic ligands through pH swings.

through the Donnan coion exclusion effect. Subsequent rinsing with dilute acid allows formation of protonated surface functional groups with arsenic sorption affinity.

2.9.4 Use of Co-Solvent

Desorption of hydrophobic ionizable organic compound (HIOC), such as pentachlorophenate (PCP^-) or aromatic anions, from an ion exchanger can be facilitated through use of a co-solvent along with water. The co-solvent has a lower dielectric constant than water (i.e., less polar) and helps diminish the interaction between the resin matrix and the non-polar moiety (NPM) of the HIOC. Using PCP^- as a model HIOC, Figure 2.25 shows the plot of PCP^-/Cl^- separation factor values for a strong-base anion exchanger (e.g., IRA-900 from Rohm and Hass Co., PA) versus the dielectric constant (ε) of the solvent medium. A meaningful corelation is observed and noted that the separation factor value drops from 145 with pure water ($\varepsilon = 78$) to less than unity with pure methanol ($\varepsilon = 32$).

To investigate the regenerability of the PCP^- loaded anion exchanger, the exhausted IRA-900 from the column run was divided into three portions. They were separately regenerated using first, 50/50 methanol–water solution with 5% NaCl; second, 5% NaCl in water; and third, 100% methanol only. Figure 2.26 shows concentration profiles of desorbed PCP^- during the three regeneration processes [31]. Note that while the combination of methanol and sodium chloride in water provides very efficient regeneration (82% recovery of PCP^- in 15 bed volumes), aqueous solution of sodium chloride or methanol alone is practically unable to desorb PCP^-. From a mechanistic viewpoint, the experimental observations clearly suggest that both ion exchange (caused by the presence of chloride anions) and an enhanced NPM-methanol interaction (due to reduced solvent dielectric constant) are simultaneously operative toward achieving efficient regeneration.

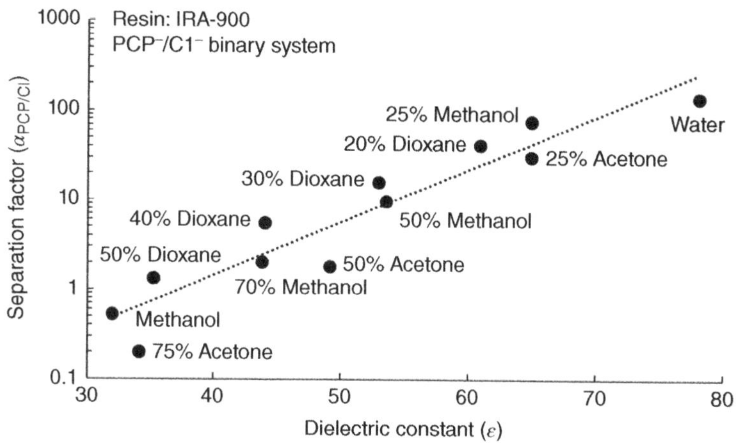

Figure 2.25 A plot of experimentally determined PCP^-/Cl^- separation factor values against dielectric constants of the solution phase illustrating the effect of co-solvent on PCP^- desorption. *Source:* Li and SenGupta 1998 [31]. Reproduced with permission of American Chemical Society.

Figure 2.26 Concentration profile of desorbed PCP^- during separate regenerations with (a) 5% NaCl in methanol/water; (b) 5% NaCl in water; and (c) 100% methanol only. *Source:* Li and SenGupta 1998 [31]. Reproduced with permission of American Chemical Society.

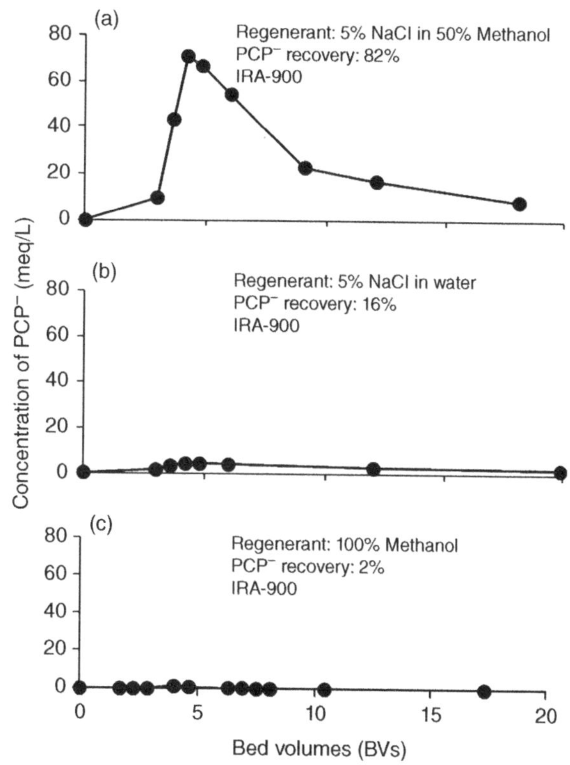

The results of the three regeneration processes in Figure 2.26 can be explained as follows:

Regenerant: Cl^- in water
Remark: unfavorable equilibrium

$$\overline{R^+PCP^-} + Cl^-(aq) \rightarrow \text{poor } PCP^- \text{ desorption} \tag{2.97}$$

Regenerant: Methanol alone, no counterion
Remark: Absence of ion exchange

$$\overline{R^+PCP^-} + \text{methanol} \rightarrow \text{poor } PCP^- \text{ desorption} \tag{2.98}$$

Regenerant: Cl^- in methanol/water solvent with reduced dielectric constant
Remark: Enhanced NPM-solvent interaction coupled with ion exchange

$$\overline{R^+PCP^-} + Cl^-(\text{co} - \text{solvent}) \rightarrow \overline{R^+Cl^-} + PCP^-(\text{co} - \text{solvent}) \tag{2.99}$$

In principle, any selective ion exchange process that results from the hydrophobic interaction between the counterion and the resin matrix can be efficiently reversed using a co-solvent with a lower dielectric constant, all other conditions remaining identical.

2.9.5 Dual-Temperature Regeneration

The main drawback of conventional ion exchange processes arises from the use of chemicals (most often salts, acids and alkalis) during the regeneration step. The spent regenerants, often in excess of their stoichiometric requirements, warrant downstream treatment to comply with environmental regulations. A great deal of research and development work are underway to improve efficiency of regeneration and, thus, reduce the volume of waste. If specific contaminants are being removed during the sorption step, additional treatment or containment of the contaminant poses a major challenge. One solution to this problem is complete elimination of chemicals during the regeneration step and its replacement with parametric separation through dual-temperature processes.

The dual-temperature technique exploits the temperature dependence of ion exchange processes, that is, exothermicity or endothermicity. Since ion exchange processes are often, if not always, carried out at ambient temperature, the dual-temperature approach is viable only when the enthalpy of ion exchange reaction (ΔH) for the reaction is significant, that is, greater than 10 kJ/eq. Between polymeric and inorganic ion exchangers, the latter remains chemically stable at temperature higher than 70 °C and are thus more appropriate for thermal treatment. At the same time, polymeric ion exchangers may have greater varieties of functional groups with relatively large ΔH values. During the last three decades, a significant amount of research and development in temperature-driven ion exchange processes have been carried out in Russia, former Soviet Union, and Australia [32]. In the following section, we will discuss the underlying fundamentals and some noticeable success of the dual-temperature processes for regeneration of exhausted ion exchangers absent any regenerant chemicals.

Table 2.4 Temperature dependence of separation factor ($\alpha_{A/B}$) values in homovalent ion exchange.

Ion exchanger	**Exchanging ions ($\overline{RB} \rightarrow \overline{RA}$)**	***T* (°C)**	$\alpha_{A/B}$
Weak-acid cation (polymethacrylic cation resin)	$Mg^{2+} \rightarrow Ca^{2+}$	15 80	4.9 1.4
Clinoptilolite	$Na^{+} \rightarrow K^{+}$	13 70	26.4 12.5
Strong base (e.g., Dowex 118)	$Cl^{-} \rightarrow Br^{-}$	25 90	4.2 2.8
	$Cl^{-} \rightarrow I^{-}$	12 80	45 16

Source: Adapted from Khamizov *et al.* 2011 [16].

Let us consider a typical ion exchange (cation or anion) as follows:

$$\overline{RB} + A \rightleftarrows \overline{RA} + B \tag{2.100}$$

Considering ideality, the dependence of the equilibrium constant, K_{AB} on standard enthalpy change, ΔH^o, is given by Van't Hoff equation as follows:

$$\frac{d \ln K}{dT} = \frac{\Delta H^o}{RT^2} \tag{2.101}$$

The temperature dependence of enthalpy in the temperature range of interest (e.g., 5–75 °C) is rather small and may be ignored. Upon integration between two temperatures of interest, T_1 and T_2, Eq. (2.99) becomes

$$\ln \frac{K_{T_2}}{K_{T_1}} = \frac{T_2 - T_1}{RT_1T_2} \cdot \Delta H^o \tag{2.102}$$

Table 2.4 includes separation factor or selectivity coefficient values for homovalent ion exchange reactions for different exchangers. Note that the separation factor values decrease with an increase in temperature for the counterion with higher affinity, confirming exothermicity of the ion exchange reactions. On the contrary, divalent-monovalent ion exchange reactions tend to be endothermic and so are the sorption of hydrophobic ionizable organic compounds (HIOCs) [33]. Thus, theoretically, by bringing the ion exchange resin in contact with solutions at two different temperatures, T_1 and T_2, sorption and regeneration can be carried out as a cyclic process without requiring any chemicals as shown in Figure 2.27. Note that the concentration of A, C_A, increased beyond the feed concentration, $C_{A,\text{feed}}$, during the regeneration step at T_2.

Separation of divalent calcium ions onto an ion exchanger from a predominantly monovalent sodium ion solution is an endothermic process, that is, an increase in temperature enhances calcium sorption while a decrease causes desorption. Figure 2.28 shows how calcium was enriched (or desorbed) during the purification of a mixture of 2.5 M Na^+ and 0.01 M Ca^{2+} with a gel-type polymethacrylic resin for a temperature change from 76 °C (purification) to 10 °C (regeneration) [16].

The simple dual-temperature cyclic process can also be applied successfully for reagent-free concentration of iodides from underground aquifers, especially

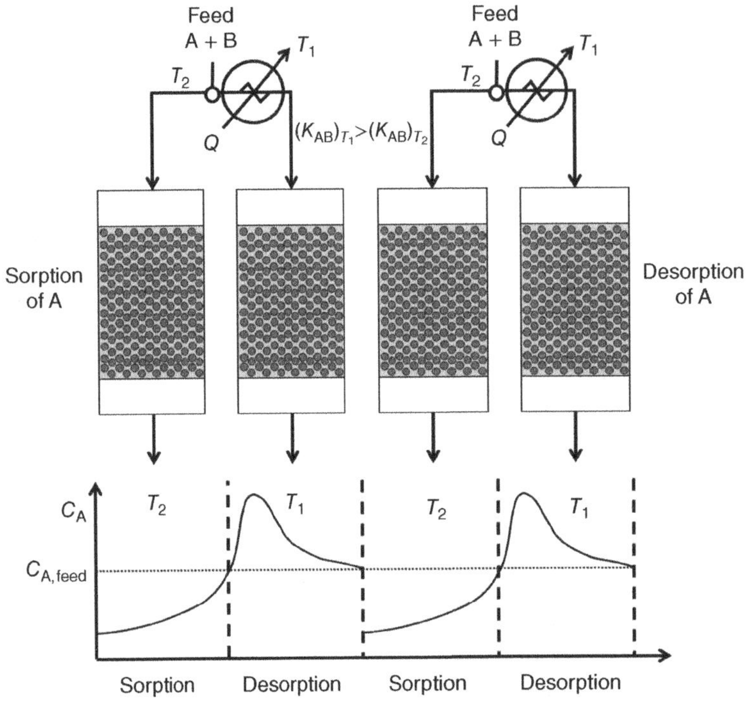

Figure 2.27 A simple schematic of a dual-temperature process illustrating both sorption and desorption of A from the feed through alteration of temperature from T_2 to T_1. *Source:* Adapted from Khamizov *et al.* 2011 [16].

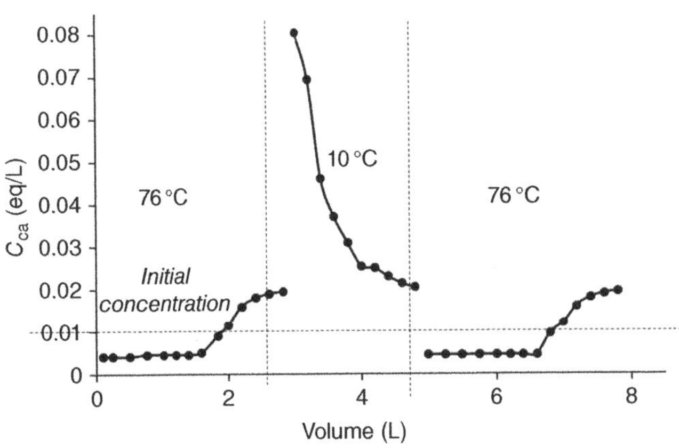

Figure 2.28 Desorption of Ca^{2+} from a mixture of 2.5 M NaCl + 0.01 M $CaCl_2$ using a polymethacrylic resin (KB-4). Bed volume = 180 mL. *Source:* Adapted from Khamizov *et al.* 2011 [16].

geothermal water. Unlike Na^+-Ca^{2+} exchange, Cl^--I^- exchange is exothermic and iodide selectivity compared to chloride decreases significantly with an increase in temperature for strong-base anion exchangers.

Figure 2.29 shows dual-temperature breakthrough curves of three successive cycles for the model feed solution of geothermal water (60 g/L NaCl and 30 mg/L NaI) [16].

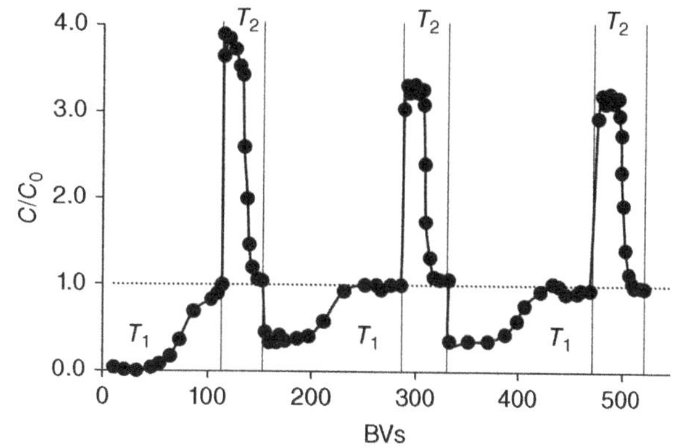

Figure 2.29 Iodide (I^-) enrichment during a cyclic process with a strong-base anion exchange resin at $T_1 = 15\,^\circ C$ and $T_2 = 75\,^\circ C$. *Source:* Adapted from Khamizov *et al.* 2011 [16].

The feed solution was continuously passed through a column with a strong-base anion exchange resin and the temperature was periodically changed from $T_1 = 15\,^\circ C$ (sorption) to $T_2 = 75\,^\circ C$ (desorption). A solution enriched with iodide was produced during each hot half-cycle of the dual-temperature cyclic process.

The much-anticipated follow-up question is: Can we desalinate or demineralize water solely through temperature swing in an ion exchange process? Although yet to make any significant mark toward real-life application, the Sirotherm Process, originally developed in Australia, offered some promise for partial desalination of brackish water through temperature swings. Specialty weak-acid and weak-base anion exchange resins form the heart of the process and the reversible desalination can be presented as follows where temperature T_2 is greater than temperature T_1:

$$\overline{R_3N} + \overline{RCOOH} + Na^+(aq) + Cl^-(aq) \underset{T_2}{\overset{T_1}{\rightleftarrows}} \overline{R_3NH^+Cl^-} + \overline{RCOO^-Na^+} \tag{2.103}$$

At higher temperature, water dissociates more leading to higher concentrations of H^+ and OH^-, which, in turn, improve the efficiency of regeneration of the weak-acid and weak-base resins. For example, as the temperature changes from 20 to 80 °C, water dissociation increases nearly 30×, that is, significant enhancement of H^+ and OH^- concentrations. Sorption and regeneration curves for dual-temperature desalination are analogous to those in Figure 2.27. Development of appropriate weak ion exchangers amenable to regeneration at elevated temperature has been the primary obstacle toward viable application of the Sirotherm Process. Interested readers can consult the discussion of the underlying polymer chemistry of the exchangers presented by Bolto and his colleagues for additional insight into the process [34].

2.9.6 Carbon Dioxide Regeneration

Carbon dioxide or CO_2, an acidic gas, has the potential of regenerating both cation and anion exchange resins in an environmentally benign way. Upon dissolution in water, CO_2 produces carbonic acid dissociating into H^+ and HCO_3^-:

$$CO_2 + H_2O \leftrightarrow H_2CO_3(aq) \tag{2.104}$$

$$H_2CO_3(aq) \leftrightarrow H^+ + HCO_3^- \tag{2.105}$$

While H^+ may be used to regenerate weak-acid cation exchange resins, HCO_3^- may regenerate strong-base anion exchanger. The subject has been presented later in Chapter 5.

2.9.7 Regeneration with Water

In some special cases, water may be used as an effective regenerant. For example, concentrated hydrochloric acid used in the steel galvanizing process complexes with Fe(III) and Zn(II) to form chloro-complex anions, for example, $FeCl_4^-$, $ZnCl_4^{2-}$. Strong-base anion exchange resins in chloride form readily sorb these anionic complexes and can purify the acid as follows:

$$\overline{R^+Cl^-} + FeCl_4^- \rightarrow \overline{R^+FeCl_4^-} + Cl^- \quad (2.106)$$

$$2\overline{R^+Cl^-} + ZnCl_4^{2-} \rightarrow \overline{(R^+)_2ZnCl_4^{2-}} + 2Cl^- \quad (2.107)$$

Regeneration is carried out with water, which breaks down the anionic complex in dilute solution into metal cations that are rapidly desorbed due to the Donnan exclusion effect:

$$\overline{R^+FeCl_4^-} + H_2O \rightarrow \overline{R^+Cl^-} + [Fe(OH)]^{2+} + H^+ + 3Cl^- \quad (2.108)$$

$$\overline{(R^+)_2ZnCl_4^{2-}} + H_2O \rightarrow 2\overline{R^+Cl^-} + [Zn(OH)]^+ + H^+ + 2Cl^- \quad (2.109)$$

Fe(III) is eluted before Zn(II); thus, the metals may also be separated and recovered.

2.10 Resin Degradation and Trace Toxin Formation

Intrinsically, every complex organic molecule or polymeric substance is chemically unstable and, from a thermodynamic viewpoint, amenable to oxidation, even under atmospheric conditions. Ion exchange resins are no exception, although their shelf life in the absence of sunlight is very long and often exceeds well beyond a decade. However, degradation or deterioration of ion exchange resins during the treatment process, and consequent loss in exchange capacity, is of concern in many applications. In general, the quality of polymer-based ion exchange resins has improved globally regarding their durability, chemical stability and resistance to mechanical attrition. Still, resin deterioration occurs and the following are the primary causes for deterioration:

- Osmotic shocks resulting in swelling/shrinking
- Thermal and/or chemical degradation

The beads are gradually fragmented due to periodic volume changes in cyclic operations where ionic forms and concentration of the solution in surrounding media change routinely. Understandably, the same ion exchanger beads may undergo different degrees of fragmentation with different applications. Weak-acid resins undergo large cyclic volume changes between the free-acid form and monovalent-salt form and are subject to osmotic-shock breakage in such cycles. Resin fines resulting from fragmentation caused by osmotic shocks do not lose ion exchange capacity but create increased pressure drops in fixed-bed columns and other related operational problems. Compared to gel-type resins, macroporous resins are less susceptible to osmotic shock losses due to their higher degree of cross-linking accompanied by their macroporosity. Loss of resin or the annual replacement requirement for typical softening and demineralizing plants is less than 10% of the total volume.

Regarding both thermal and chemical deterioration, breakage of chemical bonds and loss of functional groups are of greater significance. Such adverse impacts are pronounced at elevated temperatures (mostly above 50 °C) and in the presence of oxidizing agents, namely, chlorine, chromate, peroxide and permanganate, over a prolonged period of use. Most of the modern strong-acid cation exchange resins with polystyrene matrix, divinylbenzene cross-linking and sulfonic acid functional groups are extremely stable, even at elevated temperatures, and breaking of the covalent bonds in $R - SO_3^-$ functionality and the matrix are unlikely. Osmotic shock accompanied by swelling/shrinking is the primary reason for continued loss of gel-type cation exchange resins at a slow rate.

Anion exchange resins are more susceptible to both thermal and chemical degradation and the situation is the worst when strong-base anion exchange resins are in OH^- form. All quaternary ammonium compounds tend to undergo a Hoffman degradation when in hydroxide form, resulting in a cleavage of one of the carbon—nitrogen bonds. Figure 2.30 shows the cleavage of the strong-base Type 1 structure in two possible pathways and both are equally likely.

Note that when a methyl group is lost, a strong-base functional group is converted into a weak-base functional group. When the amine group is split off, there is a loss in total exchange capacity. Anion exchange resins with tertiary amine functional

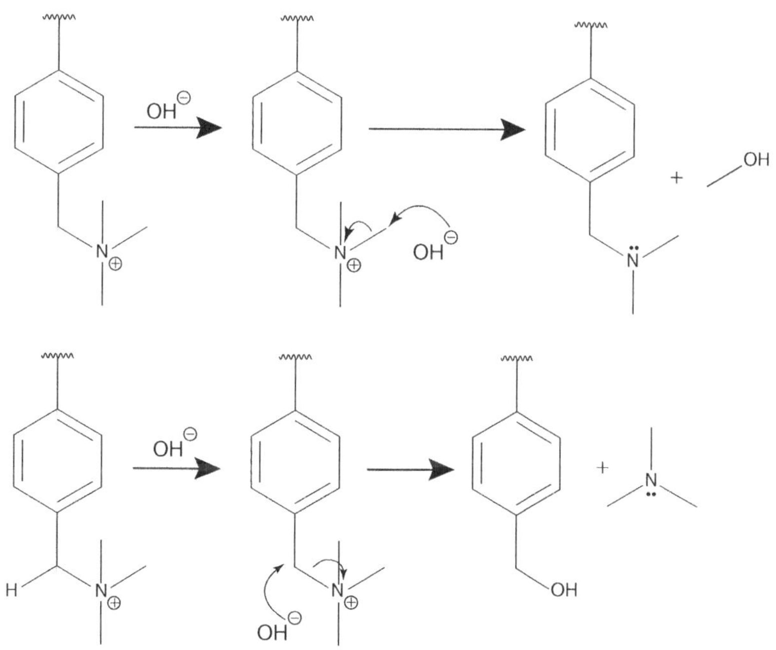

Figure 2.30 Degradation of Type 1 quaternary ammonium functional group in two possible pathways: (a) S_N2 hydroxyl addition, (b) Hoffman degradation.

groups may also undergo degradation releasing dimethylamine. It is recommended that strong-base resins, both Type I and Type II, be not exposed to temperatures much above 60 °C when in hydroxide form. The rate of loss of structure decreases as a resin ages. Trimethylamine accounts for the fishy odor common to Type I anion resins. Under no conditions should anion exchange resins be brought in contact with nitric acid, for it may lead to run-away explosive reactions. In the presence of oxidizing agents, namely, chlorine and chloramine, the breakaway dimethylamine or dialkylamine tends to form nitrosamine compounds, including NDMA (nitrosodimethylamine), that are classified as carcinogenic and toxic.

2.10.1 Formation of Trace Nitrosodimethylamine (NDMA) from Resin Degradation

Nitrosodimethylamine (NDMA) is a member of the general group of nitrosamine compounds that were first classified as toxic and carcinogenic by Magee and Barnes in the mid-1950s [35,36]. Nitrosamines have been studied for their high occurrence in various foodstuffs (especially nitrite-preserved items) and tobacco products. High concentrations (>100 ng/L) in drinking water is uncommon in unaffected, surface-water-fed water utilities [37], but a concern because of low-dose chronic exposures. Potable water reuse systems should be wary of the concentration of nitrosamines, including NDMA, because of wastewater chlorination and high nitrosamine risk (>100 ng/L) [38,39].

Although the situation is completely different compared to wastewater reuse, strong- and weak-base anion exchange resins can be a source of NDMA or nitrosamine precursors, namely, dimethylamine (DMA) or trimethylamine (TMA), in the two following ways: first, incomplete washing, i.e., leftover DMA and TMA from the synthesis of anion exchange resins; and, second, gradual degradation of anion exchange resins at high pH, elevated temperatures or in contact with oxidizing agents, as discussed in the previous section. Obviously, leftover or residual DMA or TMA can be easily removed through extended washing. We will focus primarily on the DMA and TMA formed during the degradation of anion exchange resins and their conversion to potentially carcinogenic NDMA or other nitrosamine products.

A strong-base anion exchange resin with quaternary ammonium functional groups, both Type 1 and Type 2, as stated in the previous section, can undergo demethylation (or dealkylation) resulting in the production of weak-base resins with tertiary amine functionality. Deamination of a tertiary amine functional group is favored by an oxidizing chlorine species, especially chloramines, to form NDMA, the most common form of nitrosamine. Figure 2.31 illustrates the postulated mechanism [39].

Flowers and Singer investigated the presence of nitrosamine precursors and nitrosamines including NDMA during anion exchange resin initial washing, early resin regeneration and exposure to oxidizing agents (e.g., chloramine, free chlorine). Regeneration (10% NaCl) occurred after 100 BVs of throughput and a 12 h flow interruption occurred after 150 BVs. Monochloramine was seen to produce higher amounts of nitrosamines than free chlorine. Of the 15 resins tested in continuous flow, eight produced quantifiable amounts of nitrosamines during the first 10 BVs of washing.

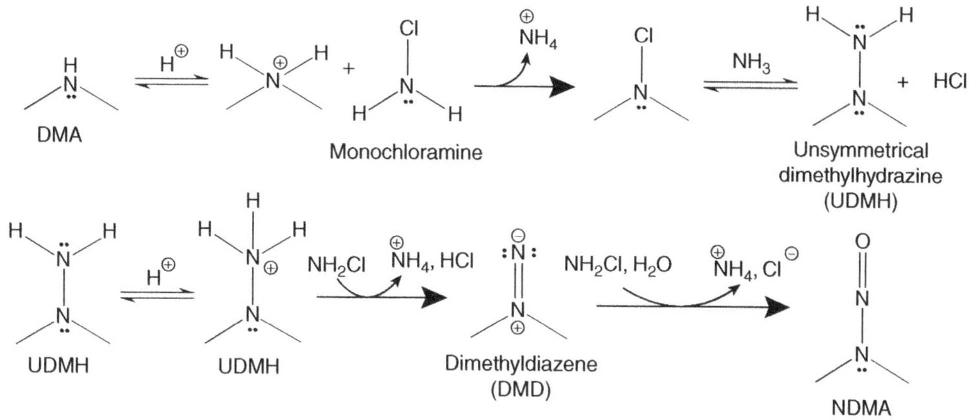

Figure 2.31 NDMA formation by chlorination of DMA suggested by Mitch and Sedlak. *Source:* Mitch and Sedlak 2002 [39]. Reproduced with permission of American Chemical Society.

After the recommended manufacturer start-up procedure of pre-washing and regeneration, or the first 10 BVs of column flow, the anion exchange resins produced significantly lower concentrations of nitrosamines than before. However, many resins had nitrosamine precursors significantly above 10 ng/L – a potential concern for later nitrosamine production. Exposure of anion exchange resins to oxidizing agents or UV light is never recommended because of enhanced degradation, which lead to much higher nitrosamine release than during normal continuous flow; after chlorination, low concentrations of nitrosamine precursors were released because precursors were transformed to nitrosamines [40]. Thus, the importance of dechlorination, for example, activated carbon sorption, prior to anion exchange treatment is important for the longevity and performance of the anion exchange resins and, also, to avoid nitrosamine production and public health concerns.

2.11 Ion Exclusion and Ion Retardation

Both ion exclusion and ion retardation are essentially ion exchanger mediated processes to separate strong electrolytes from weak electrolytes and non-electrolytes. Most importantly, water (the solvent) is used as an eluent in both cases. While the Donnan exclusion principle constitutes the foundation for *Ion Exclusion*, an amphoteric exchanger with cationic and anionic functional groups in close proximity to one another, often referred to as a "snake-cage" polyelectrolyte, serves as the source of separation for *Ion Retardation*. For ion exclusion, the strong electrolyte is eluted out of the column ahead of weak electrolytes and non-electrolytes. The order of elution is the opposite for ion retardation.

2.11.1 Ion Exclusion

In ion exclusion, the ion exchanger merely acts as a sorbent and no actual ion exchange occurs. A strong electrolyte, say XY, can be separated from a nonelectrolyte in a column with a cation exchange resin in X form (or an anion exchange resin in Y form).

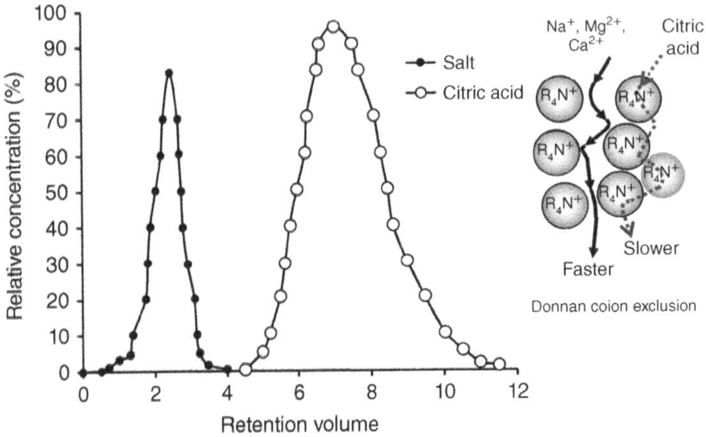

Figure 2.32 Separation of citric acid and its salts from a fermentation broth in a chromatographic column with water as the eluent. *Source:* Sarkar *et al.* 2010 [4]. Reproduced with permission of American Chemical Society.

After introduction in the upper part of the ion exchange column, both solutes are eluted with water. The strong electrolyte XY is rejected by the ion exchanger due to the Donnan exclusion effect and, hence, appears at the outlet before the weak- or non-electrolyte. The weak electrolyte is sorbed onto the ion exchanger and thereby retained longer in the absence of the Donnan effect. Figure 2.32 illustrates the separation of citric acid (undissociated) from its salts through use of an anion exchange resin in citrate form. The citric acid in relatively pure form appears later at the outlet of the column.

Separation efficiency is improved by decreasing the particle size of the resin bead and the flow rate. Lower electrolyte concentrations in the solution and higher capacity of the exchanger promote efficient ion exclusion. Donnan exclusion is stronger with higher valency of the coion. Also, internal mixing or non-ideality in the fixed-bed column interferes with the separation resolution.

Very recently, a major enhancement in separation during ion exclusion processes has been achieved by Khamizov and his coworkers by introducing an immiscible solvent (e.g., decanol) in the column in place of water [41]. Non-polar decanol has a lower specific gravity than water and greatly reduces non-ideality from column dynamics. Figure 2.33 shows the separation of nitric acid from aluminum nitrate from metallurgical plant wastewater.

Ions or electrolytes do not pass through non-polar decanol. Instead, they pass from one bead to the next through the thin water film layers surrounding the beads. This innovative approach of using an appropriate non-polar solvent for enhancing separation efficiency of ion exclusion holds promise for many similar applications.

2.11.2 Ion Retardation

Snake-cage amphoteric ion exchangers form the heart of the ion retardation process. A snake-cage bed resembles a mixed bed of cation and anion exchange resins, except

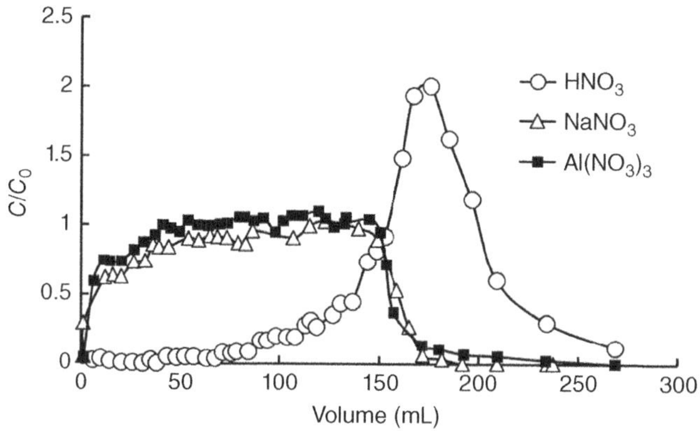

Figure 2.33 Breakthrough concentration curves for separation of nitric acid and nitrates. Column loadings: 110 mL strong-acid cation exchange resin AV-17 (Russia). Flow rate: 2 BV/h. *Source:* Adapted from Khamizov *et al.* 2012 [41].

Figure 2.34 Experimental effluent concentration history in separation of NaCl from glycerol and polyglycerols by ion retardation. Feed: 12.5% (w/w) glycerol, 6.2% (w/w) polyglycerols, 6% (w/w) NaCl; temperature 70 °C; resin: Retardion 11A8. *Source:* Hatch *et al.* 1957 [42]. Reproduced with permission of American Chemical Society.

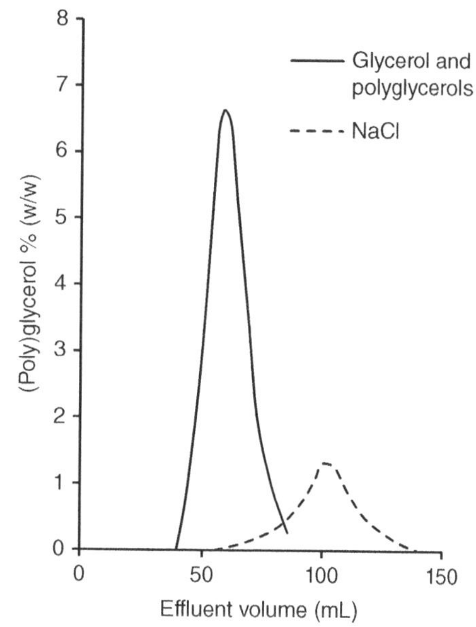

that the sorbed electrolytes can be eluted from the snake-cage resin by passing water through the bed. In contrast, the mixed bed requires regeneration with acid and base. A snake-cage or *snake in cage* material consists of a cross-linked polymeric network with fixed charges (cages) and trapped linear polyelectrolytes of the opposite charge (snakes). Two oppositely charged functional groups in close proximity can sorb especially small cations and anions (e.g., Na^+ and Cl^-) that can enter the cage. These counterions are also amenable to desorption by water.

Ion retardation is carried out in the same way as ion exclusion. No regenerant is needed. The only difference is that during ion retardation, the electrolyte is sorbed and, hence, appears later in the effluent; non-electrolyte sorption should be small. Thus, the separation of macromolecular non-electrolytes becomes feasible. Figure 2.34 shows the separation of glycerol and polyglycerol from salt impurities.

Despite its early promise, the application of ion retardation processes has been very limited to date. Very low ion exchange capacity of the snake-cage polymer is the primary reason for its lack of market growth versus ion exclusion processes.

2.12 Zwitterion and Amino Acid Sorption

Amino acids are zwitterions that can exist as both cations and as anions, and are the building blocks of proteins and life. A typical amino acid consists of a primary amine connected to a carboxylic group by a methyl bridge that supports a wide variety of side chains. Figure 2.35 shows the general structure of an amino acid with two acid dissociation constants.

There are 21 amino acids in eukaryotes used in protein formation and they vary based on their side chain. By having both a primary amine and carboxylic group, amino acids can have a net positive charge (i.e., a cation) at low pH, a net neutral charge at a pH between the pK_a values of the amino/carboxyl groups, or a net negative charge (i.e., an anion) at a pH above the pK_a of the amino group. Although an amino acid might have a net neutral charge, both a positive and a negative charge exist on the amino acid on the amino and carboxyl group, respectively. Having both charges has significant impacts on the chemistry of amino acids versus having no charges, especially regarding their interactions with ion exchangers.

The stepwise dissociation of a typical amino acid with two pK_a values can be presented as follows:

$$\mathrm{NH_3^+CHRCOOH} \xleftrightarrow{K_{a1}} \mathrm{NH_3^+CHRCOO^-} + \mathrm{H^+} \xleftrightarrow{K_{a2}} \mathrm{NH_2CHRCOO^-} + \mathrm{H^+} \quad (2.110)$$

The above dissociation is often presented as

$$\mathrm{R^+COOH} \xleftrightarrow{K_{a1}} \mathrm{R^+COO^-} + \mathrm{H^+} \quad (2.111)$$

$$\mathrm{R^+COO^-} \xleftrightarrow{K_{a2}} \mathrm{RCOO^-} + \mathrm{H^+} \quad (2.112)$$

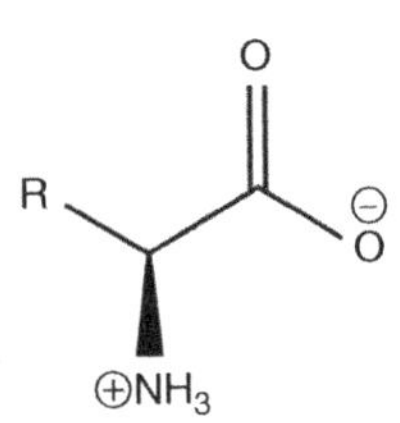

Figure 2.35 A typical amino acid with two acid dissociation constants or pK_a values.

Example 2.10 *Alanine is an amino acid for which the two acid dissociation constants are as follows:*

Find out the pH at which alanine becomes electrically neutral, that is, total positive charges are same as total negative charges. (Such a pH is often called isoelectric point and referred to as pI.)

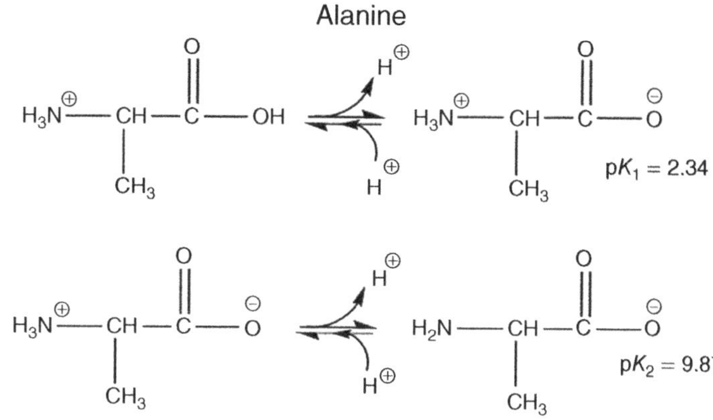

Answer:
pI is the pH at which [R^+COOH] = [$RCOO^-$]
Using the above equality, it can be easily shown that

pI = isoelectric point = the pH at which the aminio acid is a zwitterion

$$pI = \frac{1}{2}(pK_{a1} + pK_{a2})$$

$$pI_{alanine} = \frac{1}{2}(2.34 + 9.69) = \mathbf{6.0}$$

At **pH 6.0**, alanine is a **zwitterion** with neutral net charge

It is noteworthy that although electrically neutral, alanine may be sorbed onto a cation exchanger or an anion exchanger due to the independent presence of a negatively charged carboxylate group and a positively charged amino group. Additional complications in amino acid-ion exchanger behavior come from the variety of side chains, especially the side chains that can be protonated/deprotonated. The side groups of amino acids can significantly change the pK_a values of the amino and carboxylate groups. Table 2.5 includes the pK_a and pI values of amino acids including those with dissociating side chains, that is, with pK_3 values.

2.12.1 Interaction with a Cation Exchanger: Role of pH

At the outset, we will assume that $NH_2CHRCOO^-$ (or $RCOO^-$) has no affinity for cation exchange resins. Only R^+COOH and R^+COO^- are likely to be sorbed onto a cation exchanger.

Considering that the total concentration of the amino acid (C_T) comprises R^+COOH, R^+COO^- forms at the prevailing pH, and $RCOO^-$ is negligible:

$$C_T = C_{R^+COOH} + C_{R^+COO^-} \tag{2.113}$$

Again,

$$K_{a_1} = \frac{C_{R^+COO^-} \cdot C_H}{C_{R^+COOH}} \tag{2.114}$$

Table 2.5 The pK_a and pI values of amino acids.

Amino acid	Three letter code	Single letter	p*K*a (α-carboxylic acid)	p*K*a (α-amino)	p*K*a (side chain)	pI
Alanine	ALA	A	2.34	9.69	—	6.00
Arginine	ARG	R	2.17	9.04	12.48	10.76
Asparagine	ASN	N	2.02	8.80	—	5.41
Aspartic acid	ASP	D	1.88	3.65	9.60	2.77
Cysteine	CYS	C	1.96	8.18	10.28	5.07
Cystine	—	—	<1.00	1.70	7.48, 9.02	4.60
Glutamic Acid	GLU	E	2.19	4.25	9.67	3.22
Glutamine	GLN	Q	2.17	9.13	—	5.65
Glycine	GLY	G	2.34	9.60	—	5.97
Histidine	HIS	H	1.82	6.00	9.17	7.59
Hydroxyproline	HYP	—	1.92	9.73	—	5.83
Isoleucine	ILE	I	2.36	9.68	—	6.02
Leucine	LEU	L	2.36	9.60	—	5.98
Lysine	LYS	K	2.18	8.95	10.53	9.74
Methionine	MET	M	2.28	9.21	—	5.74
Phenylalanine	PHE	F	1.83	9.13	—	5.48
Proline	PRO	P	1.99	10.96	—	5.48
Serine	SER	S	2.21	9.15	—	5.68
Threonine	THR	T	2.71	9.62	—	6.16
Tryptophan	TRP	W	2.38	9.39	—	5.89
Tyrosine	TYR	Y	2.20	9.11	10.07	5.66
Valine	VAL	V	2.32	9.62	—	5.96

Combining eqs (2.113) and (2.114)

$$C_{\mathrm{R^+COO^-}} = \frac{C_{\mathrm{T}} \cdot K_{\mathrm{a_1}}}{K_{\mathrm{a_1}} + C_{\mathrm{H}}} \quad (2.115)$$

$$C_{\mathrm{R^+COOH}} = \frac{C_{\mathrm{T}} \cdot C_{\mathrm{H}}}{K_{\mathrm{a_1}} + C_{\mathrm{H}}} \quad (2.116)$$

The separation factor of the two forms of the amino acid with respect to sodium are:

$$\alpha_1 = \frac{\overline{C_{\mathrm{R^+COOH}}} \cdot C_{\mathrm{Na}}}{C_{\mathrm{R^+COOH}} \cdot \overline{C_{\mathrm{Na}}}} \quad (2.117)$$

$$\alpha_2 = \frac{\overline{C_{\mathrm{R^+COO^-}}} \cdot C_{\mathrm{Na}}}{C_{\mathrm{R^+COO^-}} \cdot \overline{C_{\mathrm{Na}}}} \quad (2.118)$$

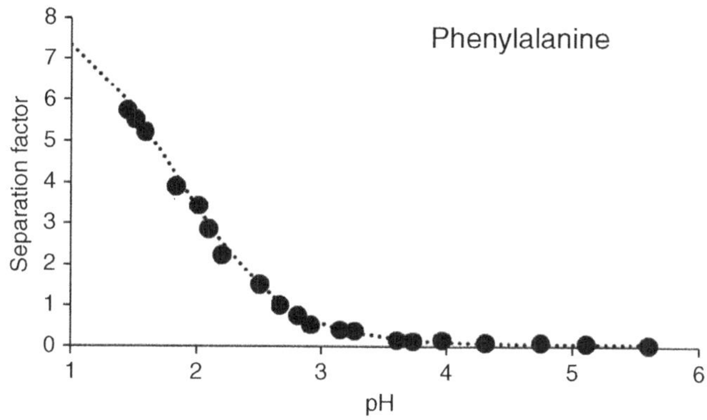

Figure 2.36 Overall separation factor values of phenylalanine relative to sodium as a function of pH for a cation exchange resin. *Source:* Yu *et al.* 1987 [43]. Reproduced with permission of Elsevier.

Defining the overall separation factor, α_T, for the total concentration of amino acid:

$$\alpha_T = \frac{\overline{(C_{R^+COOH} + C_{R^+COO^-})} \cdot C_{Na}}{(C_{R^+COOH} + C_{R^+COO^-}) \cdot \overline{C_{Na}}} \tag{2.119}$$

Thus, eliminating all concentration terms in both the aqueous and solid phases, the overall separation factor α_T, is simplified in terms of pH and the first pKa:

$$\alpha_T = \frac{\alpha_1 + \alpha_2 10^{(\text{pH}-\text{p}K_a)}}{1 + 10^{(\text{pH}-\text{p}K_a)}} \tag{2.120}$$

The apparent separation factor is highly dependent on the pH and pK_a of the carboxylic group and is very sensitive when pH ~ p$K_{a.}$ Figure 2.36 presents the overall separation factor α_T for phenylalanine with changes in pH [43]; similar separations are feasible with anion exchange resins at pH ~ pK_a amino group [44].

Test your understanding by solving Problem 2.1.

Problem 2.1

i. Using the data in Figure 2.36, compute α_1 and α_2 values and state assumptions, if any.
ii. Deduce Eq. (2.120).

2.13 Solution Osmotic Pressure and Ion Exchange

For a solution, osmotic pressure is referred to as one of the four colligative properties along with freezing point depression, boiling point elevation and vapor pressure depression caused by a nonvolatile compound dissolved in solution. The Latin root of the word "colligative" means "*binding together.*" The criterion that binds these four properties together is that, for ideal solutions, they all depend on the number

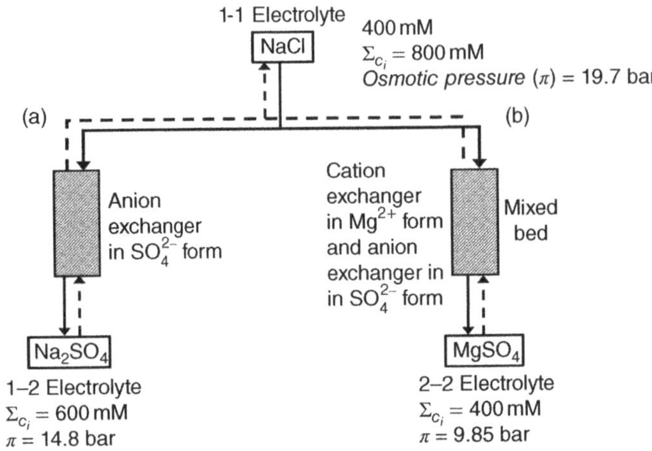

Figure 2.37 Reduction in osmotic pressure of sodium chloride following passage through ion exchangers pre-saturated with divalent ions. *Source:* Sarkar and SenGupta 2008 [45]. Reproduced with permission of Elsevier.

of dissolved particles, that is, molarity of solutes. Ideally, the osmotic pressure of a solution is given by:

$$\Pi = \sum_{i=1}^{n} C_i RT \tag{2.121}$$

where C_i is the molar concentration of dissolved solute "*i*."

Ion exchange processes, on the contrary, work on the exchange of equivalents of ions. Thus, osmotic pressure of an aqueous solution containing sodium chloride (i.e., Na^+ and Cl^- or 1 : 1 monovalent cation: monovalent anion) can be significantly reduced by using an ion exchanger in pre-saturated forms of divalent or polyvalent ions, for example, magnesium or sulfate, as illustrated in Figure 2.37 (Situations A and B). In general, transformation of monovalent cations: monovalent anions into monovalent cations: divalent anions or divalent cations: divalent anions will always be accompanied by a reduction in osmotic pressure. The ion exchange process is reversible and sodium chloride can be produced from Na_2SO_4 or $MgSO_4$ when the reaction occurs in reverse.

Figure 2.38a–c shows the schematic and the results of a column run where 560 mM NaCl solution is passed through a strong-base anion exchange resin (Purolite A-850) in sulfate form; 560 mM NaCl corresponds to approximately 32,000 mg/L as NaCl, a representative sea water. The effluent histories of sulfate and chloride are presented in Figure 2.38b. Note that chloride is exchanged for sulfate strictly on an equivalent basis; for every two moles of chloride in the feed, one mole of sulfate or Na_2SO_4 is produced at the exit of the anion-exchange column. Osmotic pressure of the solution dropped from 24 bar in the feed (NaCl) to 15 bar in the effluent following anion exchange, that is, conversion of NaCl to Na_2SO_4 (Figure 2.38c).

This phenomenon- reduction of osmotic pressure of monovalent cation: monovalent anion solution through ion exchange – has been the focal point of a hybrid

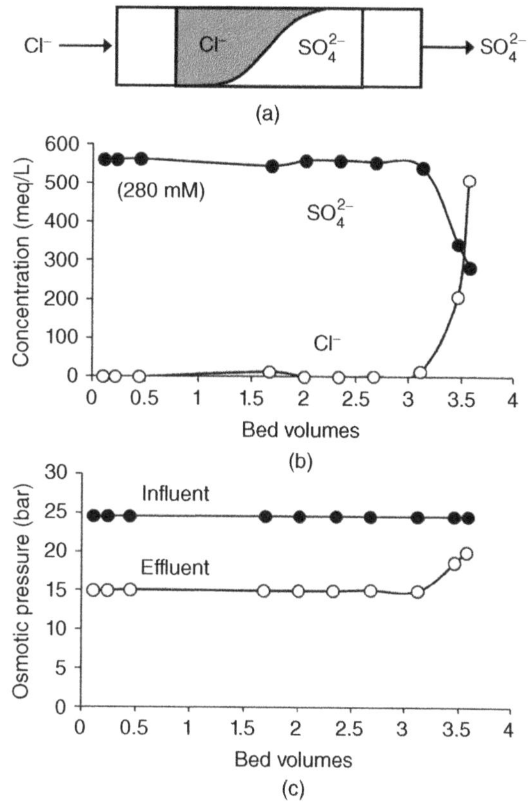

Figure 2.38 (a) Schematic showing exchange of chloride with sulfate in an anion exchanger. (b) Sulfate elution profile with a feed chloride solution of 560 meq/L. (c) Evidence of osmotic pressure reduction through chloride–sulfate exchange. *Source:* Sarkar and SenGupta 2008 [45]. Reproduced with permission of Elsevier.

ion exchange-reverse osmosis (HIX-RO) process for energy efficient desalination [45].

Test your understanding by solving Problem 2.2.

Problem 2.2 The molar concentration of the major ions in a brackish ground water supply are as follows:

Cation	Concentration (M)	Anion	Concentration (M)
Na^+	0.02	Cl^-	0.025
Mg^{2+}	0.005	HCO_3^-	0.001
Ca^{2+}	0.01	NO_3^-	0.002
K^+	0.001	SO_4^{2-}	0.012

1. The brackish water is to be subject to reverse osmosis for 75% recovery of permeate or product water. What will be the osmotic pressure of the reject water or concentrate?
2. What will be the osmotic pressure of the reject if the brackish water is first passed through a mixed-bed ion exchange column containing cation exchange resin in Mg^{2+} and anion exchange resin in SO_4^{2-} form?

2.14 Ion Exchanger as a Catalyst

Many important organic and inorganic processes are often catalyzed in acidic or basic medium. Introducing acid (i.e., H^+) or base (i.e., OH^-) as a solution is always accompanied by addition of equivalent amounts of an anion (i.e., Cl^-) or cation (i.e., Na^+), respectively, to maintain electroneutrality. Such additional solutes (i.e., Cl^-, Na^+) thus impart impurities and total dissolved solids, which may need to be separated through downstream treatment. Esterification, ester hydrolysis, alcohol dehydration, and condensation reactions are all acid/base catalyzed reactions. Strongly acidic cation exchange resins or strongly basic anion exchange resins are essentially solid acids and solid bases that can catalyze the foregoing reactions without introducing any impurity to the product stream, that is, H^+ alone can be introduced through ion exchange without any accompanying anion. Also, ion exchange resins can be retained in a packed-bed column, separated from the solution phase, reloaded with H^+ and reused.

Inversion of sucrose in solution is frequently practiced to increase the sweetness, increase moisture retention, and decrease crystallization of a sweetener, as seen in Figure 2.39.

The proton form of strongly acidic cation exchange resins is used to catalyze the inversion of sucrose. The heterogeneous hydrolysis of sucrose to the respective monosaccharides, glucose and fructose, using cation exchangers is a first-order reaction. As expected, the reaction rate increases with an increase in temperature and is inversely related to resin cross-linking and particle size. Macroporous resins are normally preferred for their superior resistance to high temperature and osmotic shock. Use of cation exchangers in H^+ form, as an insoluble solid acid, avoids the addition of unwanted impurities associated with the conventional practice of homogeneous hydrolysis through acid addition, followed by neutralization with caustic soda.

Many environmentally significant synthetic organic compounds, including pesticides, degrade through chemical hydrolysis that are strongly dependent on pH. The hydrolysis reactions consist of the cleavage of one bond and the formation of another

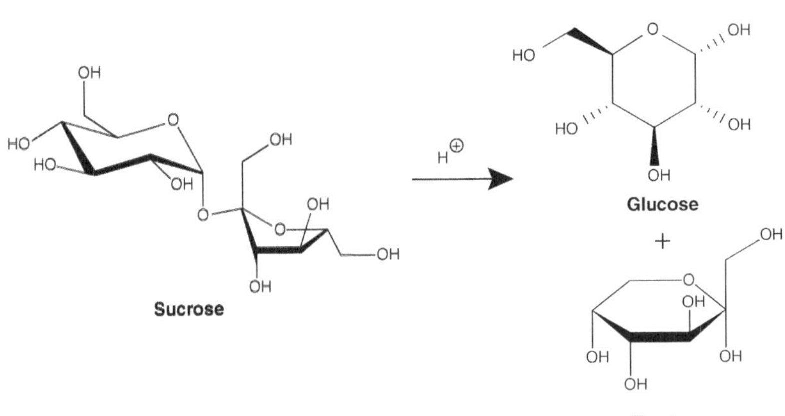

Figure 2.39 The acid-catalyzed inversion of sucrose to glucose and fructose.

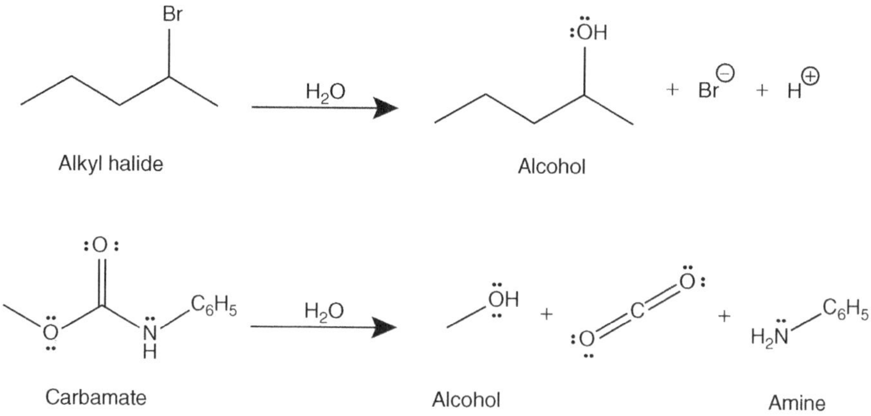

Figure 2.40 Hydrolysis of alkyl halide and carbamate.

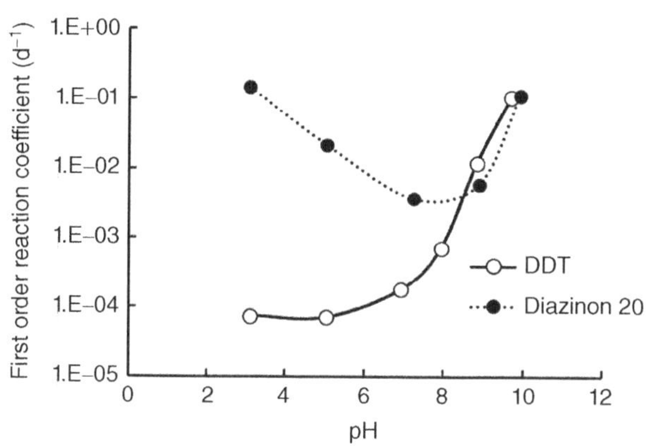

Figure 2.41 Effect of pH on hydrolysis rate constants of synthetic organic pesticides. *Source:* Data taken with permission from Sato *et al.* 1987 [46].

bond with components of water molecules, for example, H^+ or OH^-, as depicted in Figure 2.40.

Such hydrolysis cleavage reactions, by and large, follow first-order kinetics. Figure 2.41 shows the first-order kinetic coefficients of two pesticides of global significance. The coefficients are strongly dependent on the aqueous phase pH.

Understandably, an intelligent use of solid acid (i.e., cation exchanger in H^+-form) or a solid base (i.e., anion exchanger in free base/hydroxide form) may greatly enhance the degradation of persistent synthetic organic compounds into environmentally benign products.

It is also appropriate to emphasize that many ligand exchange processes are pH-dependent. Thus, sorption and desorption of environmentally significant ligands, namely, arsenic, fluoride, phosphate, etc., can be mediated through cation and anion exchangers for pH control. This subject will be addressed in the next chapter pertaining to trace fluoride sorption from contaminated groundwater.

Summary

- The process of ion exchange is always accompanied by swelling or shrinking of ion exchange resins caused by the difference in osmotic pressure between the aqueous and the exchanger phase. During these steps, coions also invade inside the ion exchanger.
- Separation factor $\left(\alpha_{\mathrm{B}}^{\mathrm{A}} = \frac{y_{\mathrm{A}}}{x_{\mathrm{A}}} * \frac{x_{\mathrm{B}}}{y_{\mathrm{B}}}\right)$ is a dimensionless expression of the relative affinity of counterions A and B toward the exchanger and it is identical to the expression of relative volatility in the distillation process. For homovalent exchange, the separation factor is equal to the equilibrium constant under ideal conditions and does not depend on the aqueous-phase electrolyte concentration, C_{T}. However, for heterovalent ion exchange, separation factor depends on C_{T} and the phenomenon is known as the electroselectivity effect. For example, at high values of C_{T}, monovalent Na^+ is preferred by a cation exchange resin to divalent Ca^{2+}, that is, $\alpha_{\mathrm{Na}}^{\mathrm{Ca}} < 1$.
- For ion exchange involving only Coulombic interactions, counterion selectivity is governed by valence and the hydrated ionic radius. For ions of identical valence, lower hydrated ionic radius offers higher selectivity.
- The conditions leading to the Donnan membrane equilibrium arise from the inability of fixed coions (i.e., charged functional groups) to diffuse out from the polymer phase into water or other polar solvents, thus creating a fictitious (i.e., non-physical) semi-permeable membrane at the interface of the ion exchanger and the aqueous phase. The Donnan membrane effect provides a quantitative description of the rejection of coions from within the ion exchanger and its dependence on C_{T}, cross-linking, Q and the valence of both coions and counterions.
- Weak-acid cation exchange resins function at neutral to alkaline pH range, while weak-base anion exchange resins are efficient in the acidic pH range. Integrating weak-acid and weak-base ion exchange resins with their strong counterparts greatly improves the regeneration efficiency of the demineralization process.
- With increasingly stringent environmental regulations and sustainability considerations, overall economy of an ion exchange process is often dictated by the operating costs of regeneration as opposed to the fixed cost of the ion exchangers. More emphasis is now underway to develop efficient and/or chemical-free regeneration processes.
- Resin deterioration occurs via physical mechanisms, such as osmotic shocks, resulting in swelling-shrinking cycles (especially pronounced in weak-acid/-base resins) and thermal stress (>50 °C). Deterioration also occurs via chemical mechanism, such as exposure to oxidizing agents and UV light. Macroporous resins are more mechanically robust than gel-type resins. Anion exchange resins are most susceptible to thermal/chemical degradation, especially in OH^- form. Exposure of anion exchange resins to oxidizing agents and UV light is not recommended due to the enhanced likelihood of nitrosamine release from the normal continuous-flow packed-bed systems.
- Ion exclusion and ion retardation are essentially ion exchanger-mediated separations of strong electrolytes from weak electrolytes and non-electrolytes. During ion

exclusion, the strong electrolyte is eluted out of the column ahead of weak electrolytes and non-electrolytes. The order of elution is the opposite of ion retardation. Most importantly, water (the solvent) is used as an eluent in both cases.

- Amino acids are zwitterions that can exist as both cations and anions. Their affinities for cation and anion exchangers are pH-dependent and they can be separated from each other by appropriately altering pH of the eluent solution.
- Osmotic pressure of a solution consisting of electrolytes can be altered – increased or decreased – through ion exchange.
- Strong-acid cation exchangers in H-form are essentially solid acids while strong-base anion exchangers in OH-form are solid bases. They are often used as solid catalysts in place of liquid acids and bases to improve downstream separation efficiency.

References

1 Gregor, H.P. (1948) A general thermodynamic theory of ion exchange processes. *Journal of the American Chemical Society,* **70** (3), 1293–1293.

2 Gregor, H.P. (1951) Gibbs-Donnan equilibria in ion exchange resin systems. *Journal of the American Chemical Society,* **73** (2), 642–650.

3 Lazare, L., Sundheim, B.R., and Gregor, H.P. (1956) A model for cross-linked polyelectrolytes. *The Journal of Physical Chemistry,* **60** (5), 641–648.

4 Sarkar, S., SenGupta, A.K., and Prakash, P. (2010) The Donnan membrane principle: opportunities for sustainable engineered processes and materials. *Environmental Science & Technology,* **44** (4), 1161–1166.

5 Calmon, C. (1952) Application of volume change characteristics of sulfonated low cross-linked styrene resin. *Analytical Chemistry,* **24** (9), 1456–1458.

6 Calmon, C. (1953) Application of volume characteristics of sulfonated polystyrene resins as a tool in analytical chemistry. *Analytical Chemistry,* **25** (3), 490–492.

7 Högfeldt, E. (1952) On ion exchange equilibria. II. Activities of the components in ion exchangers. *Akriv Kemi,* **5**, 147–171.

8 Hogfeldt, E. (1955) On ion exchange equilibria III. An investigation of some empirical equations. *Acta Chemica Scandinavica,* **9**, 151–165.

9 Soldatov, V. (1995) Application of basic concepts of chemical thermodynamics to ion exchange equilibria. *Reactive and Functional Polymers,* **27** (2), 95–106.

10 Soldatov VS. *Ion exchangers.* : Walter de Gruyter; 1991.

11 Davies CW. *Ion Association.* : Butterworths; 1962. p. 37-53.

12 Clifford, D.A. (1978) Nitrate Removal from Water Supplies by Ion Exchange. EPA-600/2-78-052.

13 Pigford, R., Baker, B. III,, and Blum, D. (1969) Equilibrium theory of parametric pump. *Industrial & Engineering Chemistry Fundamentals,* **8** (1), 144–149.

14 Wilhelm, R.H. and Sweed, N.H. (1968) Parametric pumping: separation of mixture of toluene and *n*-heptane. *Science,* **159** (3814), 522–524.

15 Sweed, N.H. and Wilhelm, R.H. (1969) Parametric pumping. Separations via direct thermal mode. *Industrial & Engineering Chemistry Fundamentals,* **8** (2), 221–231.

16 Khamizov, R.K., Ivanov, V.A., and Tikhonov, N.A. (2011) Dual temperature methods of separation and concentration of elements in ion exchange columns, in

Ion Exchange and Solvent Extraction: A Series of Advances Boca Raton (ed. A.K. SenGupta), CRC Press, FL, pp. 171–231.

17 Vulava, V., Kretzschmar, R., Rusch, U. *et al.* (2000) Cation competition in a natural subsurface material: modeling of sorption equilibria. *Environmental Science & Technology,* **34** (11), 2149–2155.

18 Lehto, J. and Harjula, R. (1995) Proceedings of the Workshop on Uniform and Reliable Formulations, Nomenclature and Experimentation for Ion Exchange-Preface.

19 Mukherjee, P. and SenGupta, A.K. (2003) Ion exchange selectivity as a surrogate indicator of relative permeability of ions in reverse osmosis processes. *Environmental Science & Technology,* **37** (7), 1432–1440.

20 Helfferich, F.G. (1962) *Ion Exchange,* Courier Corporation.

21 Harned, H.S., Owen, B.B., and King, C. (1959) The physical chemistry of electrolytic solutions. *Journal of the Electrochemical Society,* **106** (1), 15C–15C.

22 Dilts, R.V. (1974) *Analytical Chemistry; Methods of Separation,* Van Nostrand Reinhold Company.

23 Blaney, L.M., Cinar, S., and SenGupta, A.K. (2007) Hybrid anion exchanger for trace phosphate removal from water and wastewater. *Water Research,* **41** (7), 1603–1613.

24 Greenleaf, J., Cumbal, L., Staina, I., and SenGupta, A.K. (2003) Abiotic As (III) oxidation by hydrated Fe (III) oxide (HFO) microparticles in a plug flow columnar configuration. *Process Safety and Environment Protection,* **81** (2), 87–98.

25 Donnan, F.G. (1911) Theorie der Membrangleichgewichte und Membranpotentiale bei Vorhandensein von nicht dialysierenden Elektrolyten. Ein Beitrag zur physikalisch-chemischen Physiologie. *Zeitschrift für Elektrochemie und Angewandte Physikalische Chemie,* **17** (14), 572–581.

26 Donnan, F. and Guggenheim, E. (1932) Exact thermodynamics of membrane equilibrium. *Zeitschrift für Physikalische Chemie,* **162**, 346–360.

27 Donnan, F.G. (1934) The thermodynamics of membrane equilibria. *Zeitschrift für Physikalische Chemie A.,* **A168**, 369–380.

28 Donnan, F.G. (1995) Theory of membrane equilibria and membrane potentials in the presence of non-dialysing electrolytes. A contribution to physical-chemical physiology. *Journal of Membrane Science,* **100** (1), 45–55.

29 599Pepper, K., Reichenberg, D., and Hale, D. (1952) Properties of ion-exchange resins in relation to their structure. Part IV. Swelling and shrinkage of sulphonated polystyrenes of different cross-linking. *Journal of the Chemical Society (Resumed),* **Part IV**, 3129–3136, DOI: 10.1039/JR9520003129.

30 690Topp, N. and Pepper, K. (1949) Properties of ion-exchange resins in relation to their structure. Part I. Titration curves. *Journal of the Chemical Society (Resumed),* **Part I**, 3299–3303, DOI: 10.1039/JR9490003299.

31 Li, P. and SenGupta, A.K. (1998) Genesis of selectivity and reversibility for sorption of synthetic aromatic anions onto polymeric sorbents. *Environmental Science & Technology,* **32** (23), 3756–3766.

32 Khamizov, R.K., Ivanov, V.A., and Madani, A.A. (2010) Dual-temperature ion exchange: a review. *Reactive and Functional Polymers,* **70** (8), 521–530.

33 Li, P. and SenGupta, A.K. (2004) Sorption of hydrophobic ionizable organic compounds (HIOCs) onto polymeric ion exchangers. *Reactive and Functional Polymers,* **60**, 27–39.

34 Bolto, B.A. and Weiss, D.A. (1977) in *Ion Exchange and Solvent Extraction* (eds J.A. Marinsky and Y. Marcus), CRC Press, Boca Raton, FL, p. 222.
35 Magee, P. (1971) Toxicity of nitrosamines: Their possible human health hazards. *Food and Cosmetics Toxicology*, **9** (2), 207–218.
36 Magee, P. (1996) Nitrosamines and human cancer: introduction and overview. *European Journal of Cancer Prevention*, **5**, 7–10.
37 Woods, G.C., Trenholm, R.A., Hale, B. *et al.* (2015) Seasonal and spatial variability of nitrosamines and their precursor sources at a large-scale urban drinking water system. *Science of the Total Environment*, **520**, 120–126.
38 Mitch, W. and Sedlak, D. (2002) Factors controlling nitrosamine formation during wastewater chlorination. *Water Science and Technology: Water Supply*, **2** (3), 191–198.
39 Mitch, W.A. and Sedlak, D.L. (2002) Formation of N-nitrosodimethylamine (NDMA) from dimethylamine during chlorination. *Environmental Science & Technology*, **36** (4), 588–595.
40 Flowers, R.C. and Singer, P.C. (2013) Anion exchange resins as a source of nitrosamines and nitrosamine precursors. *Environmental Science & Technology*, **47** (13), 7365–7372.
41 Khamizov, K.R., Krachak, A., and Khamizov, K.S. (2012) Separation of ionic mixtures in sorption columns with two liquid phases. *Proceedings of the IEX*, **71-72**, 14–23.
42 Hatch, M.J., Dillon, J.A., and Smith, H.B. (1957) Preparation and use of snake-cage polyelectrolytes. *Industrial & Engineering Chemistry*, **49** (11), 1812–1819.
43 Yu, Q., Yang, J., and Wang, N. (1987) Multicomponent ion-exchange chromatography for separating amino acid mixtures. *Reactive Polymers, Ion Exchangers, Sorbents*, **6** (1), 33–44.
44 Jandik, P., Cheng, J., and Avdalovic, N. (2004) Analysis of amino acid–carbohydrate mixtures by anion exchange chromatography and integrated pulsed amperometric detection. *Journal of Biochemical and Biophysical Methods*, **60** (3), 191–203.
45 Sarkar, S. and SenGupta, A.K. (2008) A new hybrid ion exchange-nanofiltration (HIX-NF) separation process for energy-efficient desalination: process concept and laboratory evaluation. *Journal of Membrane Science*, **324** (1), 76–84.
46 Sato, C. McKechnie, D. Schnoor, J.L. and Sahoo, D. (1987) Processes, Coefficients, and Models for Simulating Toxic Organics and Heavy Metals in Surface Waters. Processes, coefficients, and models for simulating toxic organics and heavy metals in surface waters;EPA/600/3-87/015.

3

Trace Ion Exchange

Trace ion exchange represents a scenario where the ion of interest or the target ion in the liquid phase is present in concentrations significantly lower than others. Separation or removal of the target trace ion from the background of others in the aqueous (or solvent) phase thus demands high selectivity, that is, favorable partitioning of the trace ion in the exchanger phase. The equilibrium properties that influence the separation and recovery of the target ion in such ion exchange processes constitute the primary focus of this chapter.

These properties stem from the following three areas:

(a) Intrinsic properties of the target trace ion and other bulk competing species;
(b) Properties of the ion exchanger; and
(c) Properties of the solution

3.1 Genesis of Selectivity

The discussion in this regard will treat "ion exchanger" or sorbents as an insoluble solid phase, both organic and inorganic. The scientific treatment presented can, however, be easily extended to liquid ion exchanger without any loss of generality. Depending on the nature of separation and intended applications, the target ions often fall into a wide spectrum of compounds, such as toxic metals, radionuclides, metalloids, hydrophobic ionizable organic compounds or HIOCs, inorganic and organic ligands, surfactants and others. Although the process of ion exchange always involves electrostatic or coulombic interaction, high (or low) selectivity of a target ion stems from the nature and intensity of solute–sorbent (i.e., ion–ion exchanger) interaction beyond electrostatic phenomenon. Table 3.1 lists a group of ions of general interest and highlights their intrinsic physical–chemical properties often used for enhanced solute–sorbent interaction [1–19]. The commonly encountered interactions including the electrostatic one in selective ion exchange are:

- Electrostatic (coulombic type)
- Hydrophobic (van der Waals type)
- Brønsted Lowry acid–base (proton donor–acceptor)
- Lewis acid–base (lone electron pair donor–acceptor)

Ion Exchange in Environmental Processes: Fundamentals, Applications and Sustainable Technology, First Edition. Arup K. SenGupta.

Table 3.1 Ions of interest, physic-chemical properties and types of interactions [1–19].

Inorganic and organic ions	Properties	Remarks
Na^+, K^+, Ca^{2+}, Mg^{2+}	Hard cations; Electronic configuration similar to inert gases	Only electrostatic or coulombic interaction possible
Cs^+, Ra^{2+}	Electronic configuration similar to inert gases	Electrostatic or coulombic interaction; Higher selectivity than other Group IA or IIA cations, for example, $Ra^{2+} > Ca^{2+}$, Mg^{2+}, Ba^{2+}, and $CS^+ > Na^+$, K^+, Li^+
Cu^{2+}, Ni^{2+}, Hg^{2+}, Co^{2+}	Transition metal cations; incomplete electronic orbitals; good electron pair acceptors or Lewis acids; toxic to all living organism	Popularly known as heavy metals; concurrent electrostatic and Lewis acid–base interactions
NO_3^-, ClO_4^-	Poorly hydrated anions; no ability to form complexes with metal ions	Preferred by ion exchangers with hydrophobic functional groups and matrices
HPO_4^{2-}, $HAsO_4^{2-}$	Anions with electron pair donating oxygen atoms, that is, fairly strong ligands	Concurrent electrostatic and metal–ligand (i.e., Lewis acid–base) interaction
$HAsO_2$, NH_3	No ionic charge; able to donate electron pair from oxygen/nitrogen atom	Only ligand exchange; no electrostatic interaction
U(VI)	Often found as uranyl carbonate complexes with multiple negative charges, for example, $UO_2(CO_3)_2^{2-}$, $UO_2(CO_3)_3^{4-}$, and so on	Electrostatic interaction; very high selectivity in dilute solution over other anions
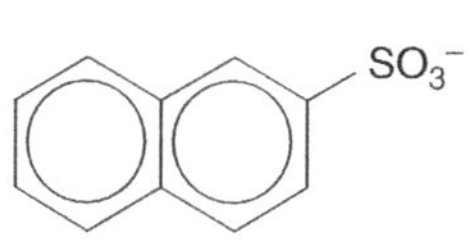 Naphthalene sulfonate	Presence of both hydrophobic non-polar moiety and hydrophilic sulfonic acid group	Concurrent hydrophobic and electrostatic interaction
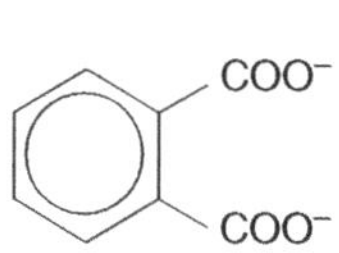	Non-polar aromatic group containing charged carboxylate groups with oxygen donor atoms	Concurrent hydrophobic, electrostatic and Lewis acid–base interaction

Table 3.1 (Continued)

Inorganic and organic ions	Properties	Remarks
$CH_3{-}(CH_2)_n{-}COO^-$	Amphiphilic surface active anions capable of forming micelles	Formation of micelles on anion exchange surface
$-N(C_3H_7)_4^+$	Organic quaternary ammonium cations	Transforms hydrophilic surface to hydrophobic upon sorption onto a cation exchanger

- Donnan effect/ion exclusion
- Ion dipole (non-aqueous or mixed solvent)
- Steric effect/ion sieving
- Charge density (distance among neighboring sites).

Table 3.2 provides schematic illustrations emphasizing each of the foregoing interactions with specific examples.

It is worth noting that more than one type of interaction is often operative for the sorption (ion exchange) of a single solute (or ion). In such cases, individual free energy changes are additive and may enhance the overall sorption affinity. To receive a broader perspective on the subject from a thermodynamic viewpoint, consider a simple homovalent anion exchange process where the counterion, B^-, is removed from the ion exchanger in the presence of competing ion A^-.

$$\overline{R^+B^-} + A^-(aq) \leftrightarrow \overline{R^+A^-} + B^-(aq) \tag{3.1}$$

The reaction (3.1) may also be written for a cation exchange reaction without any loss of generality. Considering "n" other type of interactions to be simultaneously operative along with the electrostatic one, the overall free energy change for the reaction in Eq. (3.1) at the standard state may be written as

$$\Delta G^0_{\text{overall}} = \Delta G^0_{\text{el}} + \sum_{i=1}^{i=n} G^{\text{o}}_i \tag{3.2}$$

where subscript "el" is an abbreviation for electrostatic or coulombic interaction and subscript "i"–"n" are other types of interactions. Ignoring any volume change of the ion exchanger due to swelling or shrinking, Eq. (3.2) can be broken down into respective equilibrium constants as follows:

$$-RT \ln K_{\text{overall}} = -RT \ln K_{\text{el}} - RT \ln \prod_{i=1}^{i=n} K_i \tag{3.3}$$

or

$$K_{\text{overall}} = \prod_{i=1}^{i=n} K_i K_{\text{el}} \tag{3.4}$$

where R is the universal gas constant and T is the temperature in kelvin.

Table 3.2 Schematic illustrations of different solute (ions) sorbent interactions.

Electrostatic or coulombic 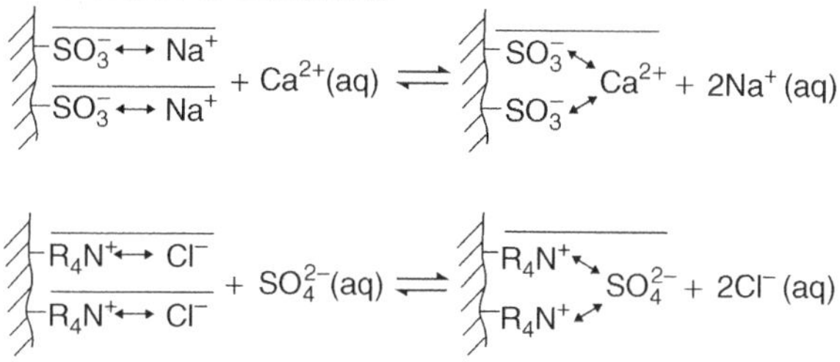	Two-sided arrows represent electrostatic interactions. With only electrostatic interactions, both fixed- and mobile-counterions remain in the hydrated forms leading to the highest possible swelling
Hydrophobic 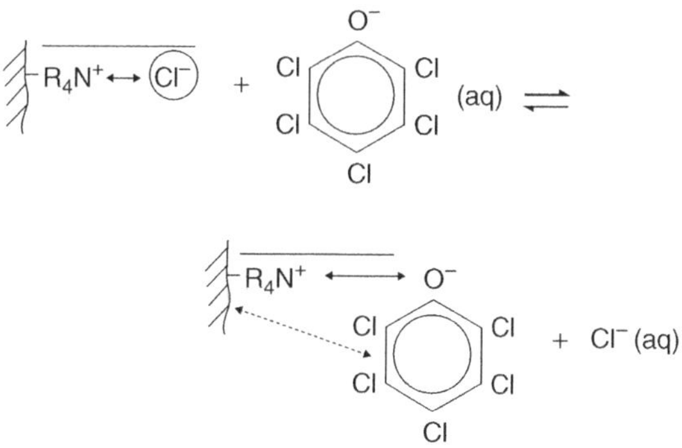	Two-sided arrows on the dashed line represent hydrophobic interactions between the polymer matrix and the aromatic moiety of the counterion. Note that electrostatic interactions are present concurrently
Brønsted Lowry acid–base 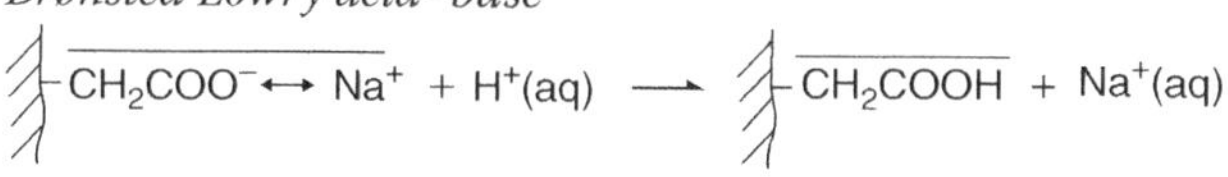	In the presence of hydrogen ion, the electrostatic interaction between the carboxylate functional groups and sodium ions are replaced by more favorable acid–base reactions; the carboxylate groups act as proton acceptors and, hence, a Brønsted–Lowry base
Lewis acid–base 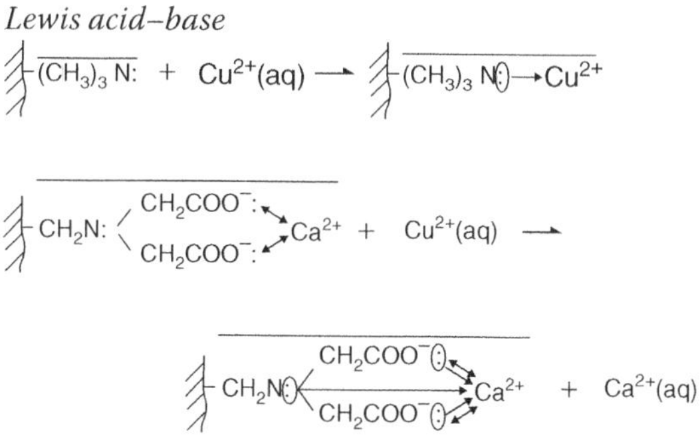	A single-sided arrow denotes transfer of a lone pair of electrons to the coordination sphere of copper ions, that is, the tertiary amine or the iminodiacetate functional groups are the electron-pair donating Lewis base, while copper is the Lewis acid. Lewis acid–base interactions are also widely known as inner sphere complexes. Note that for the tertiary amine functional groups, electrostatic interactions are absent, while they are concurrently present for iminodiacetate functional groups

(Continued)

Table 3.2 (Continued)

Ion sieving 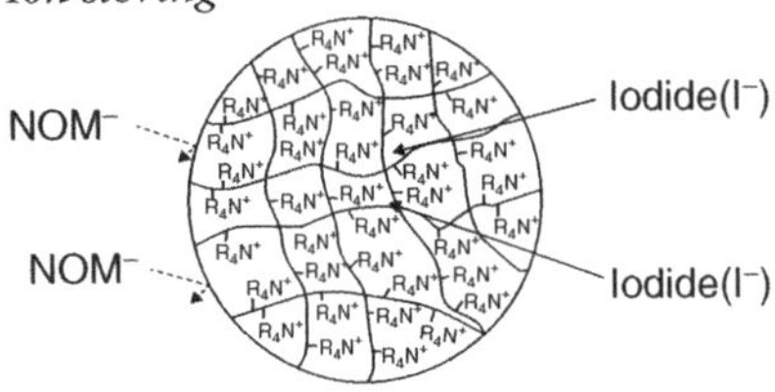	The gel phase of anion exchange resins are often not accessible to large anions comprising natural organic matter (NOM^-), for example, iodide (I^-) can be selectively removed by a gel anion exchange resin in the presence of competing NOM^-.
Donnan effect/coion exclusion 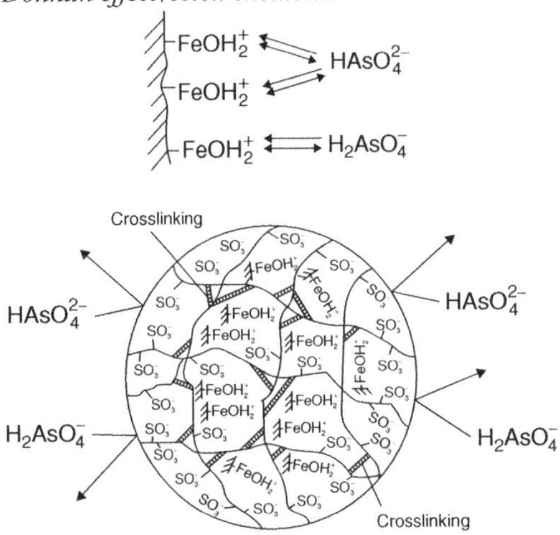	The surface functional groups of hydrated Fe(III) oxide particles offer strong sorption affinity toward arsenate. When dispersed within a cation exchanger, HFO particles are, however, unable to remove arsenate effectively, which is excluded by the fixed negatively-charged functional groups of the cation exchanger due to the Donnan coion exclusion effect
Ion dipole or ion solvent 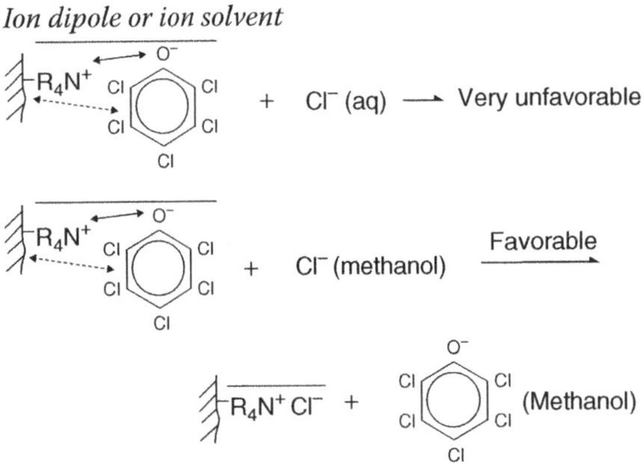	Pentachlorophenate or PCP^- has very high affinity toward strongly basic anion exchangers due to concurrent hydrophobic interactions between the aromatic moiety of the solute and the polystyrene matrix. In the presence of chloride ions (Cl^-) in the aqueous phase, sorption of PCP^- is essentially irreversible, that is, PCP^- cannot be desorbed. However, in methanol, which is much less polar than water, chloride can effectively exchange with PCP^-. Hydrophobic PCP^-–matrix interaction is considerably reduced in the presence of a relatively non-polar solvent, such as methanol

Table 3.2 (Continued)

Charge density (distance between neighboring sites)	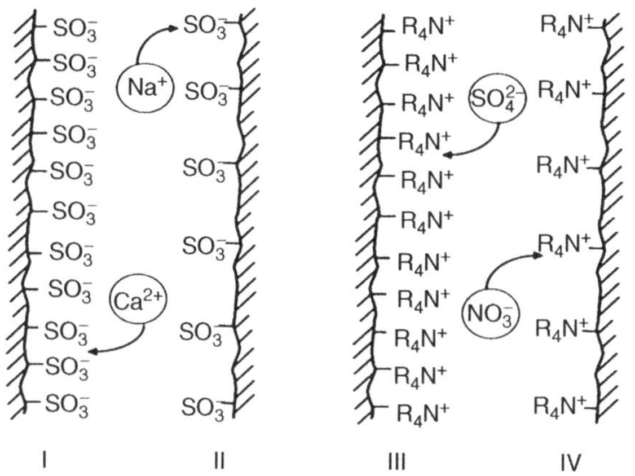	Shorter distances among neighboring sites (i.e., high charge density) in columns I and III exhibit higher preference for divalent ions (e.g., Ca^{2+} or $SO_4{}^{2-}$) over monovalent ions (e.g., Na^+ and $NO_3{}^-$) Ion exchange sites in columns II and IV with lower charge density show higher monovalent ion selectivity.

Regardless of the nature of interaction, Eq. (3.4) provides the following equalities:

1. In the absence of any other interaction,

$$\sum_{i=1}^{i=n} \Delta G_i^{\circ} = 0 \tag{3.5}$$

thus,

$$\prod_{i=1}^{i=n} K_i = 1 \tag{3.6}$$

hence,

$$K_{\text{overall}} = K_{\text{el}} \tag{3.7}$$

2. If individual free energy changes are present and favorable (i.e., ΔG_i° values are all negative), every K_i will then be greater than unity. So, there will be a synergistic effect and, K_{overall}, which is a direct measure of selectivity of anion A^- over anion B^- will be greatly enhanced. Similarly, if the free energy changes are unfavorable (i.e., K_i values are less than unity), the selectivity will diminish.
3. Considering ideality for the ion exchange reaction (3.1),

$$K_{\text{overall}} = \frac{q_A C_B}{q_B C_A} = \frac{y_A X_B}{y_B X_A} = \alpha_{AB} \tag{3.8}$$

where α_{AB} is the separation factor, q_i and C_i are the concentrations of species 'i' in exchanger phase and solution phase, y_i and x_i are the equivalent fractions of 'i' in exchanger and solution phases, respectively. Thus, every solute–sorbent interaction, explicit or not, is embedded in the separation factor value. Enhancement or alteration of separation factor values through engineered solute–sorbent interaction is the goal of selective ion exchange.

3.2 Trace Isotherms

Let us again consider the ion exchange reaction (3.1)

$$\overline{R^+B^-} + A^-(aq) \leftrightarrow \overline{R^+A^-} + B^-(aq) \tag{3.9}$$

Assuming ideality in both aqueous and ion exchanger phases, the selectivity coefficient (K_{AB}) or separation factor for the above reaction is the same and given by,

$$K_{AB} = \frac{q_A C_B}{q_B C_A} = \alpha_{AB} \tag{3.10}$$

Considering all concentrations to be in equivalent units (i.e., meq/L or meq/g) and the fact that

$$q_A + q_B = Q \tag{3.11}$$

and

$$C_A + C_B = C^0 \tag{3.12}$$

where Q and C^0 are the total exchanger capacity and aqueous phase anion concentrations, respectively, Eq. (3.8) now takes the following form:

$$K_{AB} = \frac{q_A}{Q - q_A} \frac{C^0 - C_A}{C_A} \tag{3.13}$$

After rearrangement,

$$q_A = \frac{K_{AB} C_A Q}{C^0 + (K_{AB} - 1)C_A} \tag{3.14}$$

or,

$$q_A = \frac{Q K_{AB}(C_A/C^0)}{1 + (K_{AB} - 1)(C_A/C^0)} \tag{3.15}$$

Let us now consider different possibilities,

Case I. Trace species A

Under such a condition, $C_A \ll C^0$. Thus, for real values of K_{AB}, Eq. (3.15) becomes

$$q_A = \left(\frac{K_{AB} Q}{C^0}\right) C_A \tag{3.16}$$

Since K_{AB}, Q, and C^0 are constant, q_A and C_A are linearly dependent on each other, that is, it is a linear isotherm,

$$q_A = \lambda C_A \tag{3.17}$$

Such linearity is independent of relative selectivity between A and B and analogous to Henry's law; Figure 3.1a demonstrates the same.

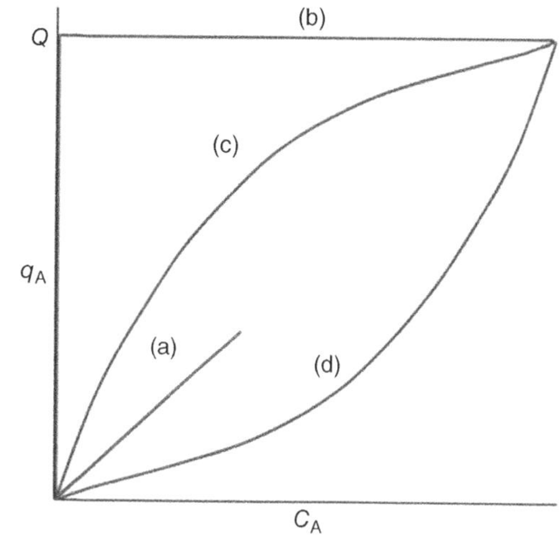

Figure 3.1 Plot of different types of isotherms (a) linear, (b) rectangular, (c) favorable or convex upward, and (d) unfavorable or concave upward.

Case II. Infinite selectivity

If the anion A^- is infinitely preferred by the exchanger over B^-, then, $K_{AB} \gg 1$. From Eq. (3.14) or (3.15),

$$q_A = Q \tag{3.18}$$

Equation (3.18) is analogous to what is known as "rectangular isotherm" and is represented by Figure 3.1b.

Case III. Non-trace conditions

Here the condition $C_A \ll C^0$ is no longer valid. An isotherm resulting from $K_{AB} > 1$ is referred to as a favorable isotherm while the one corresponding to $K_{AB} < 1$ is an unfavorable isotherm. Figure 3.1c and d represent the scenario for some realistic values of K_{AB}, Q, and C^0.

Some distinctive characteristics with respect to the first derivative (slope) and the second derivative (curvature) of these isotherms are noteworthy; they influence both intraparticle diffusion and fixed-bed column behaviors as emphasized in the succeeding chapters of this book.

Case I. Trace species

$$\frac{dq_A}{dC_A} = \text{constant and } \frac{d^2q_A}{d^2C_A} = 0$$

Case II. Rectangular isotherm/infinite selectivity

$$\frac{dq_A}{dC_A} = 0 \text{ and } \frac{d^2q_A}{d^2C_A} = 0$$

Case III. Non-trace species

For $K_{AB} > 1$ (i.e., favorable isotherm)

$$\frac{dq_A}{dC_A} > 0 \text{ and } \frac{d^2q_A}{d^2C_A} < 0$$

For $K_{AB} < 1$ (i.e., unfavorable isotherm)

$$\frac{dq_A}{dC_A} > 0 \text{ and } \frac{d^2q_A}{d^2C_A} > 0$$

3.3 Multi-Component Equilibrium

Let us consider ions $i,\ldots,n$ in a solution in equilibrium with an ion exchanger. The following assumptions are implied:

- Instantaneous mass transfer
- No shrinking or swelling of the resin
- Constant total exchange capacity
- Constant separation factors.

At equilibrium,

$$x_i = \frac{C_i}{C^0},\ x_j = \frac{C_j}{C^0},\ x_k = \frac{C_k}{C^0} \tag{3.19}$$

$$y_i = \frac{q_i}{Q},\quad y_j = \frac{q_j}{Q},\quad y_k = \frac{q_k}{Q} \tag{3.20}$$

Also,

$$C_i + C_j + C_k + \cdots = C^0 \tag{3.21}$$

$$x_i + x_j + x_k + \cdots = 1 \tag{3.22}$$

$$q_i + q_j + q_k + \cdots = Q \tag{3.23}$$

$$y_i + y_j + y_k + \cdots = 1 \tag{3.24}$$

Separation factor,

$$\alpha_{ij} = \frac{y_i x_j}{y_j x_i} = \frac{q_i C_j}{q_j C_i} = \frac{1}{\alpha_{ji}} \tag{3.25}$$

Now applying the conditions of equality in (3.25) into (3.24),

$$y_i + \frac{y_i x_j}{x_i}\alpha_{ji} + \frac{y_i x_k}{x_i}\alpha_{ki} + \cdots = 1 \tag{3.26}$$

$$y_i\left(1 + \frac{x_j}{x_i}\alpha_{ji} + \frac{x_k}{x_i}\alpha_{ki} + \cdots\right) = 1 \tag{3.27}$$

$$y_i = \frac{1}{1 + \sum_{j\neq i}\alpha_{ji}\frac{x_j}{x_i}} \tag{3.28}$$

$$y_i = \frac{x_i}{x_i + \sum_{j\neq i}\alpha_{ji}x_j} \tag{3.29}$$

In a similar vein,

$$y_j = \frac{x_j}{x_j + \sum_{n \neq i} \alpha_{nj} x_n} \tag{3.30}$$

Thus at equilibrium, exchanger phase composition for a given solution concentration can be determined for a multicomponent system based on constant separation factor values. Lastly, the assumption of "constant separation factor" is not valid when the total solution concentration (C^0 values) changes for heterovalent ion exchange. For such conditions, new separation factor values need to be computed prior to applying Eq. (3.29). Related solved problems have been included in Chapter 2.

Example 3.1 Fixed-Bed Process

A fixed-bed cation exchange column is considered to remove 5 mg/L of Cu^{2+} from an industrial wastewater also containing 300 mg/L of Na^+ and 100 mg/L of Ca^{2+} at pH 6.6. Consider the following two situations:

A. *A cation exchange resin with sulfonic acid functional groups has separation factor values of:*

$$\alpha_{Cu/Ca} = 1.1, \ \alpha_{Ca/Na} = 5.0$$

B. *A chelating cation exchanger with iminodiacetate functional group has separation factor values of:*

$$\alpha_{Cu/Ca} = 80.0, \ \alpha_{Ca/Na} = 10.0$$

Both resins have a capacity of 1.2 eq/L.

Questions:

1. Compute y_{Cu} values for both resins at equilibrium
2. How many liters, or bed volumes (BVs), of wastewater can be treated by one liter of each resin before copper breaks through from the column? Show the effluent histories.
3. Do the anions present in the wastewater have any impact on the results?

State assumptions if any.

Solutions:

1. The equilibrium conditions are

Name	Mass concentration (mg/L)	Equivalent concentration (meq/L)	Equivalent fraction (x_i)
Ca^{2+}	100	5	0.275
Cu^{2+}	5	0.16	0.009
Na^+	300	13	0.716

Using Eqs (3.29) and (3.30),

$$y_{Ca} = \frac{x_{Ca}}{x_{Ca} + \alpha_{Cu/Ca} \cdot x_{Cu} + \alpha_{Na/Ca} \cdot x_{Na}}$$

$$y_{Cu} = \frac{x_{Cu}}{x_{Cu} + \alpha_{Ca/Cu} \cdot x_{Ca} + \alpha_{Na/Ca} \cdot \alpha_{Ca/Cu} \cdot x_{Na}}$$

$$y_{Na} = \frac{x_{Na}}{x_{Na} + \alpha_{Ca/Na} \cdot x_{Ca} + \alpha_{Cu/Ca} \cdot \alpha_{Ca/Na} \cdot x_{Cu}}$$

Note:

$$\alpha_{Na/Ca} \cdot \alpha_{Ca/Cu} = \alpha_{Na/Cu}$$

$$\alpha_{Cu/Ca} \cdot \alpha_{Ca/Na} = \alpha_{Cu/Na}$$

For strong acid cation (SAC) exchange resin with the following properties,

$$\alpha_{Cu/Ca} = 1.1, \quad \alpha_{Ca/Na} = 5.0, \quad Q = 1.2\,\text{eq/L}$$

$$y_{Ca} = 0.643, \quad y_{Cu} = 0.022, \quad y_{Na} = 0.335$$

For iminodiacetate (IDA) chelating resin with the following properties:

$$\alpha_{Cu/Ca} = 80.0, \quad \alpha_{Ca/Na} = 10.0, \quad Q = 1.2\ \text{eq/L}$$

$$y_{Ca} = 0.265, \quad y_{Cu} = 0.666, \quad y_{Na} = 0.069$$

2. Sulfonic acid cation exchange resin has $y_{Cu} = 0.022$ at $Q = 1.2$ eq/L, or a capacity of 0.027 eq Cu^{2+}/L. Iminodiacetate cation exchange resin has $y_{Cu} = 0.666$ at $Q = 1.2$ eq/L, or 0.80 eq Cu^{2+}/L. With a feed of 0.16 meq/L, the resins can treat 170 BVs (sulfonic) and 5081 BVs (iminodiacetate). Considering instantaneous breakthrough with ideal plug-flow behavior, the breakthrough curves are

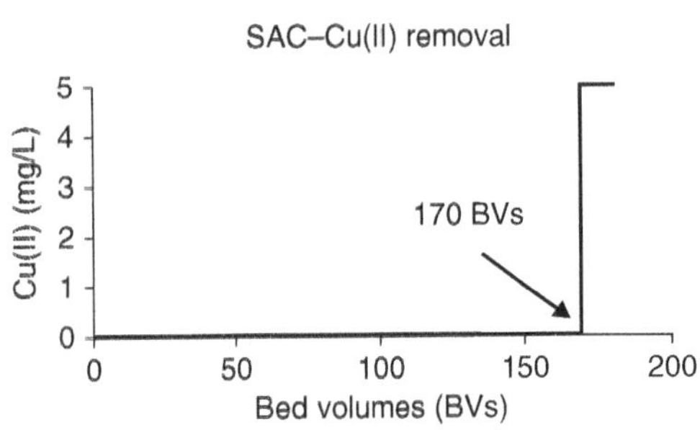

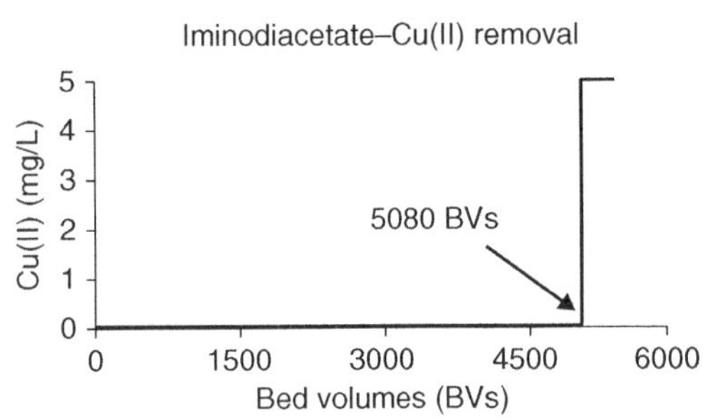

3. Anions present in the water will not influence the results. The separation factor and equivalent fraction calculations are only based on the cation concentrations in solution. Coion intrusion at lower TDS will not be significant and will not affect the intraparticle diffusion of counterions.

3.4 Agreement with Henry's Law

If "i" is a trace species in a multi-component system, the term "x_i" in the denominator of Eq. (3.29) can be ignored and we get,

$$y_i = \frac{x_i}{\sum_{j \neq i} \alpha_{ji} x_j} \tag{3.31}$$

During trace ion removal, since the composition of the solution phase remains virtually unchanged and the separation factor values are relatively constant, the denominator becomes a constant. Thus, if "c" is the inverse of the constant denominator,

$$y_i = c \cdot x_i, \text{ that is, } \frac{q_i}{Q} = c \cdot \frac{C_i}{C^0}, \text{ or,}$$

$$\frac{q_i}{C_i} = \text{constant} \tag{3.32}$$

This equation is identical to Eq. (3.17) and conforms to Henry's law, that is, linear dependence of solute distribution between two phases. The number of BVs treated during a fixed-bed column run prior to breakthrough of a specific trace contaminant "i" is proportional to q_i/C_i. Thus, the number of BVs treated does not change with a change in the concentration of the trace contaminant. Figure 3.2 shows effluent histories of As(V) anion (arsenate) at three different influent concentrations [5]. Note that while arsenic concentration in the influent changed 200× from 10 to 2000 μg/L, the number of BVs treated remained virtually constant, in accordance with the prediction of Eq. (3.32). Although the change in arsenic concentration in the influent was very large, arsenate was always a trace species in the presence of competing sulfate and chloride anions, thus satisfying the required condition presented in Eq. (3.32).

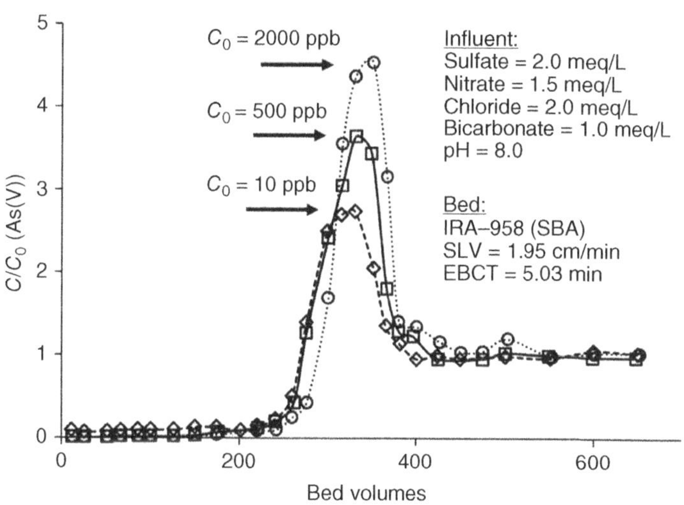

Figure 3.2 Effluent histories of As(V) for three different trace arsenic feed in fixed bed column runs using strong-base anion exchanger (IRA-958) under identical conditions. Reprinted with permission from SenGupta and Greenleaf 2001 [5].

It is worth noting that, for highly selective exchangers, the target solute isotherm, Figure 3.1, may deviate from linearity even at liquid-phase trace concentrations. For nonlinear isotherms, the number of BVs treated will change as the concentration in the influent changes. Example 3.2 attempts to demonstrate the relevant points.

Example 3.2 *Solve the problem from Example 3.1 when the copper concentration is 2 mg/L, all other conditions remaining the same. Appropriate comments welcome.*

Solutions:

1. The equilibrium conditions are

Name	Mass concentration (mg/L)	Equivalent concentration (meq/L)	Equivalent fraction (x_i)
Ca^{2+}	100	5	0.277
Cu^{2+}	2	0.063	0.003
Na^+	300	13	0.720

Applying the same process from Example 3.1, but with different equivalent fractions.

For the SAC resin,

$$y_{Ca} = 0.651, \quad y_{Cu} = 0.009, \quad y_{Na} = 0.340$$

For the IDA resin,

$$y_{Ca} = 0.441, \quad y_{Cu} = 0.444, \quad y_{Na} = 0.115$$

2. SAC resin has $y_{Cu} = 0.009$ at $Q = 1.2$ eq/L-resin, or a capacity of 0.011 eq Cu^{2+}/L. IDA resin has $y_{Cu} = 0.444$ at $Q = 1.2$ eq/L, or 0.533 eq Cu^{2+}/L. For resins with low selectivity for trace ions (e.g., SAC-Cu(II)), the ions remain trace species inside the resin. But, at high selectivity (e.g., IDA–Cu(II)), trace ions are major species inside. With a feed of 0.063 meq/L, the resins can treat 172 BVs (cationic) and 8466 BVs (iminodiacetate). The solution concentration was 40% of previous and the equivalent fraction of copper on SAC was 40% of Example 3.1: the number of BVs treated remained constant. The equivalent fraction on iminodiacetate resin was 67% of Example 3.1, that is, Y_{Cu}/X_{Cu} increased by 0.67/0.4 or 1.67. Thus, the number of bed volumes increased by 67% as shown below.

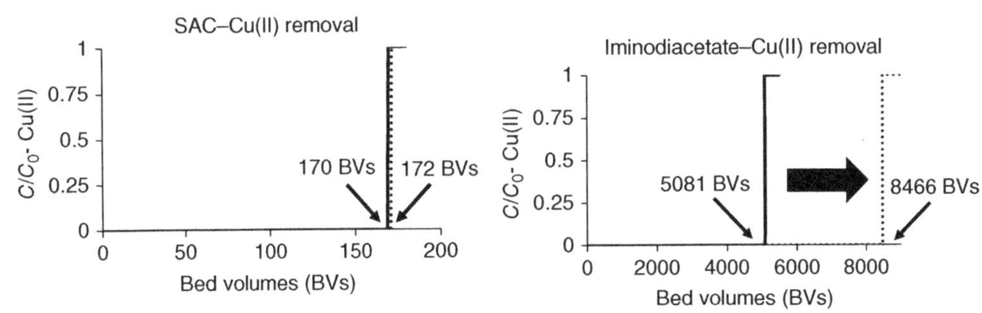

3. Anions present in the water will not influence the results. The separation factor and equivalent fraction calculations are based only on the cation concentrations in solution. Coion intrusion at lower TDS will not be significant and will not affect the intraparticle diffusion of counterions.

3.5 Multiple Trace Species: Genesis of Elution Chromatography

Let us consider the situation where i and j are trace species in the aqueous phase while k, l, m, and n are present at significantly greater concentrations. From Eq. (3.29),

$$y_i = \frac{x_i}{x_i + \sum_{j \neq i} \alpha_{ji} x_j} \tag{3.33}$$

Since j is also a trace species, we may rewrite the above equation as,

$$y_i = \frac{x_i}{x_i + \alpha_{ji} x_j + \sum_{k \neq i, j \neq i} \alpha_{ki} x_k} \tag{3.34}$$

Since both x_i and x_j are extremely small,

$$y_i = \frac{x_i}{\sum_{k \neq i, k \neq j} \alpha_{ki} x_k} \tag{3.35}$$

Thus, the exchanger phase concentration of i, y_i, is not influenced by the presence of trace species j. Similarly, y_j is not influenced by x_i.

$$y_j = \frac{x_j}{\sum_{k \neq i, k \neq j} \alpha_{kj} x_k} \tag{3.36}$$

In a multi-component system, individual trace species, therefore, behave as if they are oblivious to the existence of one another. The large excess of "k, l, m" thus "uncouple" the trace species i and j, that is, suppress their interactions with one another. This mathematical relationship, although not recognized intuitively, forms the heart of the analytical technique leading to the invention of elution or ion chromatography [20–22].

For prompt appreciation of the foregoing mathematical equalities, consider the ion chromatogram in Figure 3.3 for several anions where bicarbonate or bicarbonate–carbonate solution is used as the eluent or mobile phase. The order of elution is inversely related to the magnitude of the term, y_i/x_i, of the trace species, that is, the species with the lowest y_i/x_i value will appear first in the chromatogram due to its lowest affinity for the anion exchanger used as the stationary phase. However, the reproducibility or precision of ion chromatography rests on the properties related to trace ion exchange as presented in Eqs (3.35) and (3.36). Note that besides bicarbonate, all other ions are essentially trace species in the eluting stream, that is, y_i/x_i and y_j/x_j are essentially constant. Thus, the exact time of elution for any species is independent of the presence of others, that is, a nitrate peak will appear exactly at the same time in the chromatogram as if it were the only analyte, all other conditions remaining identical. In a way, elution times for different anions are independent of their concentrations in the sample and predestined if the sorbent (e.g., anion exchanger) and the eluent remain unchanged.

3.5.1 Determining Separation Factor from Elution Chromatogram

The elution times of different ions in the chromatogram can be used to determine their relative affinities (i.e., separation factor values). A typical elution chromatogram for a sample containing two analytes (ions A and B) is presented in Figure 3.4.

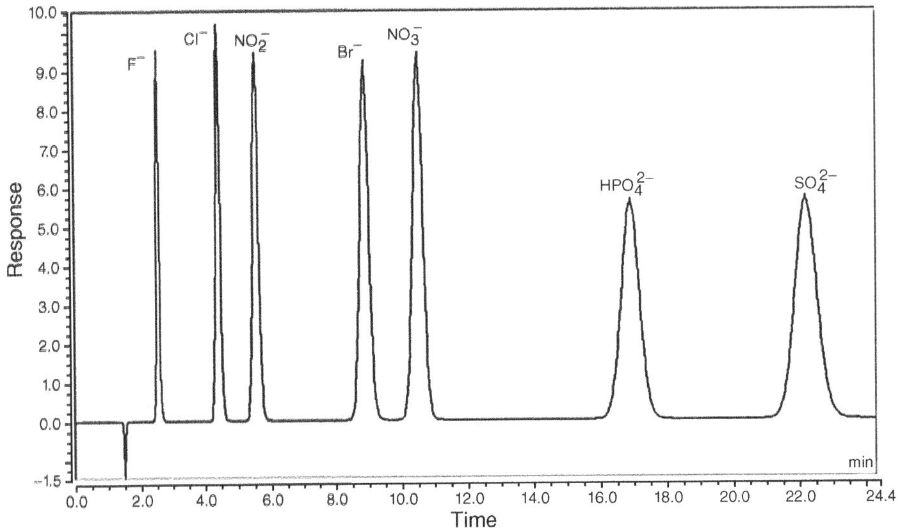

Figure 3.3 Typical ion chromatogram for some common anions.

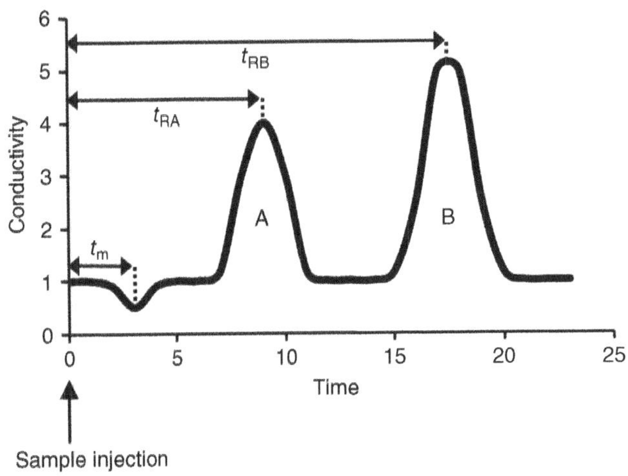

Figure 3.4 A schematic illustrating occurrence of chromatographic peaks of ions A and B. Note: The baseline conductivity is due to the eluent. The lower peak or "water dip" at t_m is caused by the lower conductivity solvent of the sample, for example, water, being pushed out of the column.

If Z is the length of the column pack, the average speed of carrier water solution,

$$u = \frac{Z}{t_m} \tag{3.37}$$

The average speed of ion A,

$$v_A = \frac{Z}{t_{RA}} \tag{3.38}$$

And that of B,

$$v_B = \frac{Z}{t_{RB}} \tag{3.39}$$

Understandably, ions A and B are retained in the stationary phase (ion exchanger) and that is why each proceeds with a velocity slower than that of carrier water. Of the two, B is retained longer than A. Let us consider ion A and its distribution or partitioning between water and the exchanger is

$$\lambda_A = \frac{\overline{C_A}}{C_A} \tag{3.40}$$

where $\overline{C_A}$ and C_A are equilibrium concentrations of A in the exchanger phase and the aqueous phase, respectively. Now we assume plug flow behavior of the ion exchange column and that the equilibrium of A between the two phases is instantly established. With these assumptions, the general mass balance or equation of continuity for A is given by the partial differential equation

$$\frac{\partial C_A}{\partial t} + \frac{(1-\varepsilon)}{\varepsilon}\frac{\partial \overline{C_A}}{\partial t} = -u\frac{\partial C}{\partial Z} \tag{3.41}$$

The first and second term on the left-hand side (LHS) of Eq. (3.41) imply accumulation of the solute A in the mobile and stationary phases, respectively. The term on the right-hand side (RHS) represents mass transport by convection. Now considering Eq. (3.40) and replacing $\frac{(1-\varepsilon)}{\varepsilon}$ with H

$$\frac{\partial C_A}{\partial t}[1 + H\lambda_A] = -u\frac{\partial C}{\partial Z} \tag{3.42}$$

The analytical solution for a pulse input of Eq. (3.42) is

$$\left(\frac{\partial Z}{\partial t}\right)_A = \frac{u}{1 + H\lambda_A} \tag{3.43}$$

Again,

$$\left(\frac{\partial Z}{\partial t}\right)_A = v_A = \frac{Z}{t_{RA}} \tag{3.44}$$

Thus,

$$t_{RA} = \frac{Z}{u}[1 + H\lambda_A] \tag{3.45}$$

From Eq. (3.37), $t_m = \frac{Z}{u}$, so

$$t_{RA} = t_m[1 + H\lambda_A] \tag{3.46}$$

Similarly, it may be shown that

$$t_{RB} = t_m[1 + H\lambda_B] \tag{3.47}$$

Combining Eqs (3.46) and (3.47)

$$\frac{t_{RA} - t_m}{t_{RB} - t_m} = \frac{\lambda_A}{\lambda_B} = \frac{\overline{C_A} C_B}{\overline{C_B} C_A} \tag{3.48}$$

The last term is essentially the separation factor between A and B, α_{AB}.

Thus,

$$\alpha_{AB} = \frac{t_{RA} - t_m}{t_{RB} - t_m} \tag{3.49}$$

Thus, α_{AB} can be computed by noting the retention times of A, B, and water from the chromatogram.

Example 3.3 From Elution Chromatogram to Separation Factor Values
Consider the ion chromatogram in Figure 3.3, that is, elution of the peaks of different anionic species at different times from an anion exchanger column.

Given: The dilute sodium carbonate and bicarbonate mixed solution eluent (carrier fluid) passes through the 15 cm long ion exchange column at a linear velocity of 4.8 m/h before it hits the ion detector at the end of the column. Calculate separation factor (α) for NO_3^-/Cl^-, HPO_4^{2-}/Cl^-, and SO_4^{2-}/Cl^- for the ion exchanger.

Solution:

Species	Peak time (min)
Cl^-	4.1
NO_3^-	8.2
HPO_4^{2-}	10.5
SO_4^{2-}	12.6

Note that the negative peak corresponding to the retention time of the mobile phase occurs at 1.8 min, that is, $t_m = 1.8$ min

$$\alpha_{AB} = \frac{t_{RA} - t_m}{t_{RB} - t_m}$$

$$\alpha_{N/C} = \frac{t_{RN} - t_m}{t_{RC} - t_m} = \frac{8.2 - 1.8}{4.1 - 1.8} = 2.8$$

Similarly,

$$\alpha_{P/C} = 3.8, \quad \alpha_{S/C} = 4.7$$

Supplementary Reading S3.1 Chromatography

Elution, displacement, and frontal chromatography are separation processes in which the solutes in the liquid (mobile phase) are separated during passage through a column containing a sorbent (stationary phase). In Greek, the word "chromatos" means "color." The term chromatography is derived from experiments carried out by a Russian Scientist, M.S. Tswett, over 100 years ago to separate pigments using a stationary column containing solid inorganic sorbents [23]. Here we will briefly discuss the underlying concept of three different types of chromatography using solutions containing three solutes (A, B, and C) with the following selectivity sequence A > B > C.

Elution chromatography is illustrated in Figure S3.1 where a solution containing A and B is introduced on top of the column containing the sorbent. The column is then washed with an eluent containing a high concentration of solute C with affinity lower than A and B. Since solute A is sorbed more strongly than B, A moves down the column more slowly than B and emerges at the bottom outlet after B.

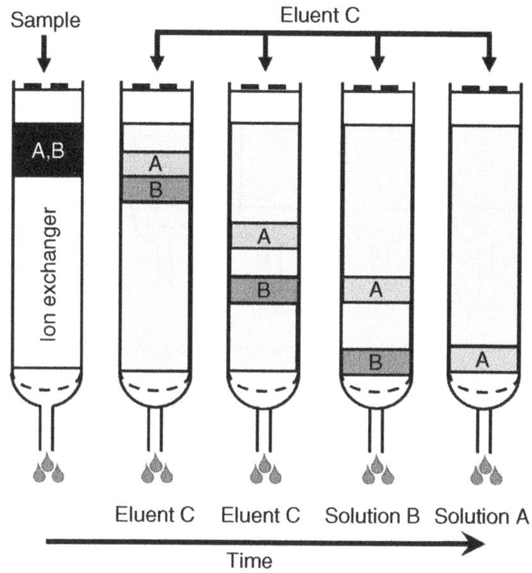

Figure S3.1. *Separation of A and B through elution chromatography using a stationary phase, for example, an ion exchange resin.*

Elution chromatography is commonly used for analytical purpose and the sorbent does not require a high capacity. Eluent C is normally used at significantly higher concentration than the concentrations of B and A in the analyte.

Displacement chromatography is a process where the mixture of solutes (B + C) is introduced into the top of the column and then pushed or displaced through the stationary phase with a solution of A, which has higher affinity than both B and C. With the passage of solution, B and C are separated and obtained at the outlet, Figure S3.2.

The primary application of displacement chromatography pertains to commercial or large-scale separation of B and C. High capacity of the stationary phase is desirable. At the end of the cycle, the mixture of B and C is introduced again at the top of the column followed by displacement with A. The displacing eluent, A, is often recovered with high purity and reused.

Frontal chromatography is the process in which the mixture of all solutes (A + B + C) to be separated is continuously passed through a column containing the sorbent (stationary phase). No eluent is needed. The solute, least preferred by the sorbent, appears first at the outlet followed by others in an increasing order of affinity, Figure S3.3.

Widely used fixed-bed or packed-bed treatment process is the most common form of frontal chromatography. Unlike elution and displacement chromatography, frontal

(Continued)

Supplementary Reading S3.1 (Continued)

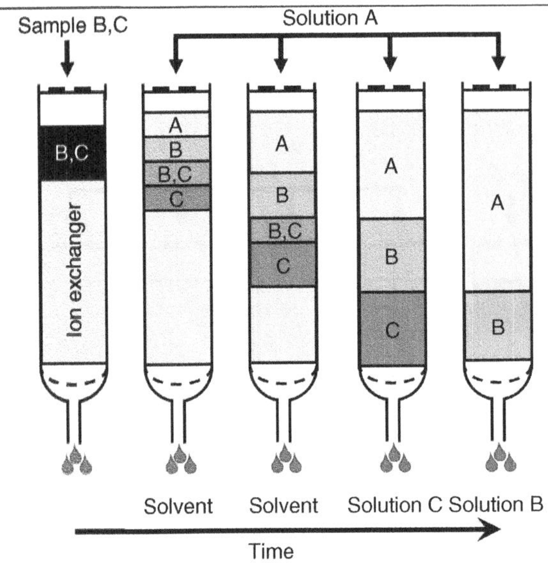

Figure S3.2. Separation of solutes B and C using displacement chromatography (affinity sequence A>B>C).

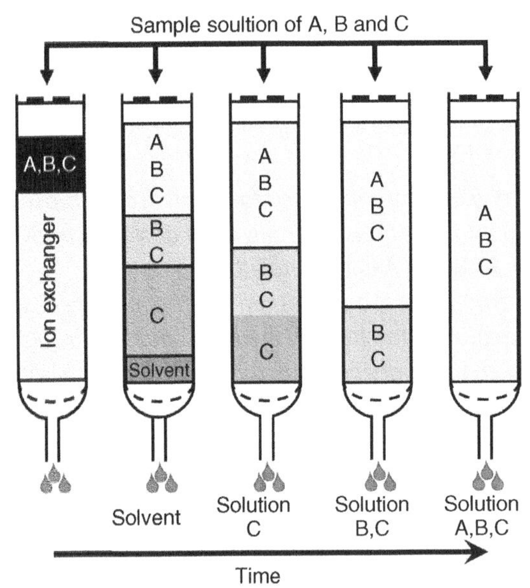

Figure S3.3. Separation with frontal chromatography with an affinity sequence A > B > C.

chromatography requires regeneration of the stationary phase after every cycle, mostly with a concentrated solution of C, that is, the solute with least affinity. For more in-depth and detailed discussion on the subject, the following two references are recommended:

Ion Exchange Chromatography of Proteins. S. Yamamato, K. Nakanishi, R. Matsuno, Marcel Decker. New York 1988.

Ion Chromatography (Modern Analytical Chemistry). H. Small. Springer. New York. 1989.

3.6 Uphill Transport of Trace Ions: Donnan Membrane Effect

In any solution, a solute spontaneously travels from a region of higher concentration to a region of lower concentration to level out any concentration gradient in accordance with the second law of thermodynamics. The process is referred to as *diffusion*. The presence of ion exchanger material in the solution as a barrier may drastically alter such transport behaviors in a way that may seem counterintuitive, that is, trace ions may spontaneously flow from a lower to a higher concentration. To make the concept readily comprehensible without any loss of generality, consider copper chloride ($CuCl_2$), a completely ionized 2:1 electrolyte and hydrochloric acid (HCl), a 1:1 electrolyte to be present in water. Consider Case I where a membrane (i.e., barrier) separates 0.001 M $CuCl_2$ in the solution from 0.1 M HCl as shown in Figure 3.5.

The membrane is permeable to both cations and anions and the solution volume in each side of the membrane is the same (say 1.0 L). At equilibrium, cations and anions will redistribute themselves across the permeable membrane and concentrations of various species at equilibrium will be the same in both sides of the membrane, that is,

$$\frac{[Cu^{2+}]_R}{[Cu^{2+}]_L} = \frac{[H^+]_R}{[H^+]_L} = \frac{[Cl^-]_R}{[Cu^-]_L} = 1 \tag{3.50}$$

The above equality is understandably trivial, as shown in Figure 3.5 (Case I). The distribution of Cu^{2+} in both sides of the membrane is, however, greatly altered if the barrier is replaced by cation exchanger membrane that is permeable only to cations and does not allow any passage of anion as shown in Figure 3.6 (Case II). At equilibrium, the electrochemical potential of copper ion, Cu^{2+}, (η_{Cu}) in the solution on the LHS of the membrane will be the same as that in the electrolyte solution on the RHS, that is,

$$\eta_{Cu}^{L} = \eta_{Cu}^{R} \tag{3.51}$$

$$\mu_{Cu}^{0} + RT\ln a_{Cu}^{L} + zF\varnothing^{L} = \mu_{Cu}^{0} + RT\ln a_{Cu}^{R} + zF\varnothing^{R} \tag{3.52}$$

where superscripts "0," "L," and "R" refer to standard state, LHS and RHS, and μ, a, F, and $\varnothing$ denote chemical potential, activity, Faraday's constant and electric potential,

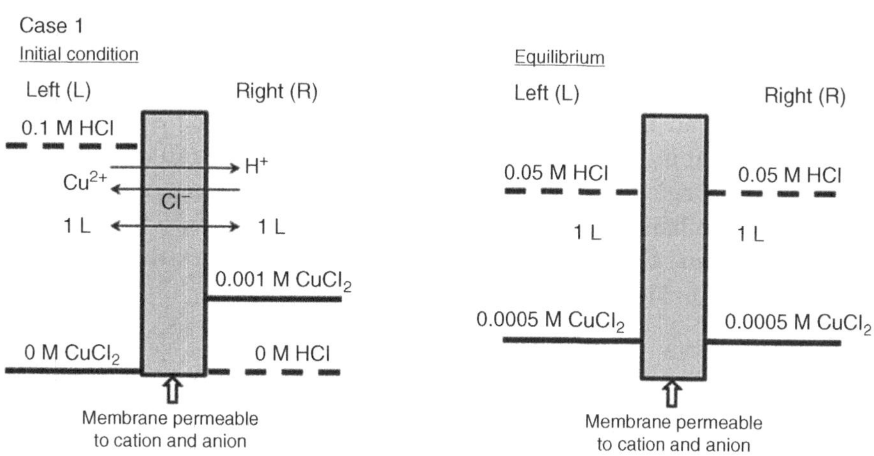

Figure 3.5 Illustration of separation of dilute $CuCl_2$ using a permeable membrane.

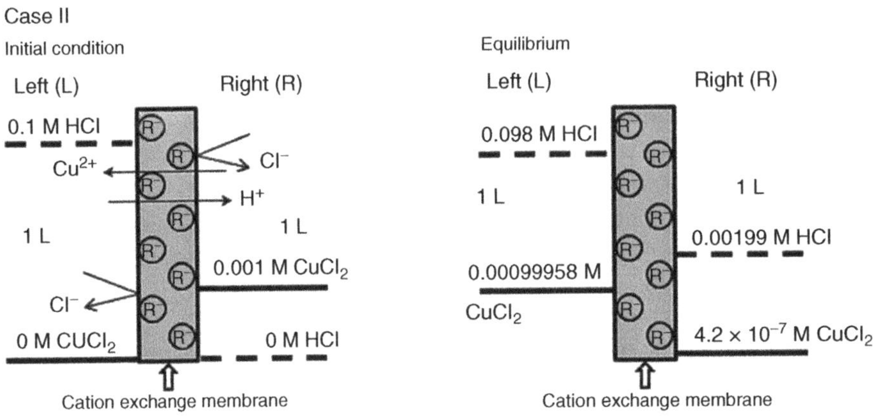

Figure 3.6 Illustration of separation of dilute $CuCl_2$ using a cation exchange membrane through Donnan dialysis (cation exchange membrane is impermeable to Cl^-).

respectively. *R* is the universal gas constant and "*z*" refers to the charge of the diffusing ion, which is +2 for Cu^{2+}. Equation (3.52) gives the following equality for the copper ions on two sides of the membrane:

$$\frac{F(\varnothing^R - \varnothing^L)}{RT} = \ln\left(\frac{a^L_{Cu}}{a^R_{Cu}}\right)^{1/2} \tag{3.53}$$

In a similar way, it can be shown for hydrogen ions that

$$\frac{F(\varnothing^R - \varnothing^L)}{RT} = \ln\left(\frac{a^L_{H}}{a^R_{H}}\right) \tag{3.54}$$

Considering ideality, Eqs (3.53) and (3.54) yield the following:

$$\left(\frac{C^L_{Cu}}{C^R_{Cu}}\right) = \left(\frac{C^L_{H}}{C^R_{H}}\right)^2 \tag{3.55}$$

If the ratio $\left(\frac{C^L_H}{C^R_H}\right) = 10$, it means C^L_{Cu} will be 100 times greater than C^R_{Cu}. Thus, by maintaining a relatively high hydrogen ion concentration on the LHS of the membrane, copper ions can be driven from the RHS to LHS, against a positive concentration gradient, that is, from a solution of lower concentration to a solution of higher concentration. Chloride ion is a coion in the process and its concentration in either side of the cation exchange membrane remains unchanged due to the Donnan coion exclusion effect. Thus, the electroneutrality condition provides the following equalities for the situation in Figure 3.6.

$$2[Cu^{2+}]_R + [H^+]_R = [Cl^-]_R = 0.002 \tag{3.56}$$

and,

$$2[Cu^{2+}]_L + [H^+]_L = [Cl^-]_L = 0.1 \tag{3.57}$$

Figure 3.6 (Case II) shows the equilibrium composition of copper and hydrogen ions for the example when the solutions are separated by a cation exchange membrane. It is the cation exchange barrier or more precisely, the Donnan potential gradient that allows concentration of ions of higher valence to move against its own concentration gradient. If Cu^{2+} in the example is replaced by Al^{3+}, greater concentrations can be achieved in accordance with the following:

$$\left(\frac{C_{Al}^{L}}{C_{Al}^{R}}\right) = \left(\frac{C_{H}^{L}}{C_{H}^{R}}\right)^{3} \tag{3.58}$$

The foregoing principle of concentrating ions of higher valence forms the foundation of the Donnan dialysis or Donnan membrane separation process [24–27]. Note that the valence of the permeating counterions, not their chemical make-up, is the sole driving force in the Donnan dialysis process. Replacing Cu^{2+} with any other divalent cation, for example, Ca^{2+}, Mg^{2+}, Zn^{2+}, Ni^{2+} would have yielded identical results.

Problem to be Solved

Carry out necessary calculations to confirm the equilibrium composition in the RHS of Figure 3.6.

Let us consider 100 L of dilute 0.01 M $AlCl_3$ solution from a wash stream in an industrial plant. The plan is to concentrate the aluminum with 10.0 L of 0.5 M H_2SO_4.

Draw a sketch of the Donnan dialysis process. Determine the aluminum concentration at equilibrium in the sulfuric acid compartment. What is the estimated pH in each chamber?

3.7 Trace Leakage

Contrary to the binary ion exchange reaction shown in Eq. (3.1), ion exchangers rarely exist in a single ionic form in any cyclic process due to the incompleteness of the regeneration process. Such an incomplete regeneration is often a deliberate practice for economic reasons. Also, the regenerant used is never pure; other counterions present in the regenerant along with B^- (or B^+ in case of cation exchange) may also be the source of *leakage* during the service cycle. The term "trace leakage" refers to the leaching of an ion of concern from the exchanger phase into a contacting solution without them. To illustrate the problem, let us consider a regenerated ion exchanger containing $i, j, k, \ldots$, that is,

$$y_i + y_j + y_k + \cdots = 1 \tag{3.59}$$

The ion exchanger is now brought in contact with a solution containing all other species except "k," that is

$$x_i + x_j + x_l + \cdots = 1 \tag{3.60}$$

Initial trace leakage of "k," that is, x_k value can be obtained from the following equality:

$$y_k = \frac{x_k}{x_k + \sum_{j \neq k} \alpha_{jk} x_j} \tag{3.61}$$

Considering "i" and "j" to be only two ions present in the contacting solution

$$y_k = \frac{x_k}{x_k + \alpha_{ik} x_i + \alpha_{jk} x_j} \tag{3.62}$$

For conditions related to trace leakage, $x_k \ll x_i$ and $x_k \ll x_j$.
Thus,

$$x_k = y_k(\alpha_{jk}x_j + \alpha_{ik}x_i) \tag{3.63}$$

Trace leakage depends greatly on separation factor or relative affinity, that is, lower the affinity of "k," greater will be its leakage. It is noteworthy that "trace leakage" is an equilibrium phenomenon and it should not be interchanged with "trace breakthrough" which occurs due to kinetic limitations. Secondly, trace leakage is an unsteady state phenomenon, that is, y_k value declines with trace leakage, thus causing continual reduction in the leakage of k. Lastly, the separation factor values in Eqs (3.61)–(3.63) may need to be recomputed with a change in total electrolyte concentrations for heterovalent ion exchange.

Example 3.4 *A strong-acid cation exchange softener on Na-cycle is used to remove* Ca^{2+} *from the water with the following composition:*

$Ca^{2+} = 1.5\,\text{meq/L}$	$SO_4^{2-} = 1.0\,\text{meq/L}$
$Na^+ = 9.5\,\text{meq/L}$	$Cl^- = 10.0\,\text{meq/L}$
$C_T = 11.0\,\text{meq/L}$	$C_T = 11.0\,\text{meq/L}$

For the cation exchanger,

Selectivity coefficient, $K_{Ca/Na} = 2.6$
Total exchange capacity, $Q = 2.0\,\text{eq/L}$.

In order to improve process economics, it was decided to regenerate the cation exchange column with excess seawater having the following composition:

$$Na^+ = 25{,}000\,\text{mg/L as } CaCO_3 \text{ or } 500\,\text{meq/L}$$

$$Ca^{2+} = 1000\,\text{mg/L as } CaCO_3 \text{ or } 20\,\text{meq/L}$$

Since calcium is present in the regenerant, there will always be calcium leakage in the treated water during the early stage of the service cycle.
Find the calcium leakage in milligram per liter as $CaCO_3$.

Solution
Regenerant composition:

$$[Na^+] = 500\,\text{meq/L}$$

$$[Ca^{2+}] = 20\,\text{meq/L}$$

Total, $C_T = 520\,\text{meq/L} = 0.52\,\text{eq/L}$
For the $Ca^{2+} - Na^+$ exchange,

$$K_{Ca/Na} = \frac{y_{Ca}}{(1-y_{Ca})^2}\frac{x_{Na}^2}{x_{Ca}}\frac{C_T}{Q}$$

Now $K_{Ca/Na}$, C_T, and Q are given

$$x_{Na} = \frac{500}{520} = 0.96$$

$$x_{Ca} = \frac{20}{520} = 0.04$$

Thus,

$$2.6 = \frac{y_{Ca}}{(1 - y_{Ca})^2} \frac{0.96^2}{0.04} \frac{0.52}{2.0}$$

After iterations of trial and error,

$$y_{Ca} = 0.242$$

$$y_{Na} = 1 - 0.242 = 0.758$$

Note that nearly 25% of the ion exchange sites are occupied by calcium. During the initial period of the service cycle, treated water will be in equilibrium with the regenerated resin, that is,

$$C_T = 11\,\text{meq/L} = 0.011\,\text{eq/L}$$

$$y_{Na} = 0.758$$

$$y_{Ca} = 0.242$$

Thus,

$$2.6 = \frac{0.242}{0.758^2} \frac{(1 - x_{Ca})^2}{x_{Ca}} \frac{0.011}{2.0}$$

$$x_{Ca} = 0.00089$$

$$[Ca^{2+}] = 0.00089 \times 11\,\text{meq/L} = 0.01\,\text{meq/L}$$

$$= 0.5\,\text{mg/L as}\,CaCO_3$$

Assuming attainment of equilibrium, 0.5 mg/L calcium as $CaCO_3$ will appear in the treated water after the column is restarted. For homovalent ion exchange, separation factor approach by Eq. (*3.63*) is more appropriate.

3.8 Trace Fouling by Natural Organic Matter

Fouling is a gradual and irreversible change in the generic properties of parent ion exchanger in the presence of a trace solute. Most importantly, fouling results in the steady loss of operating capacity. The fouling caused by natural organic matter, or NOM, onto anion exchange resins is of major concern during demineralization of water. Here we will discuss the underlying fundamentals leading to the genesis of NOM fouling. NOM is a large, high molecular weight organic compound, which is universally present in surface water due to gradual oxidation of leaves and decay of vegetation. NOM is classified into two groups, namely, humic and fulvic acids, but neither has a distinctive

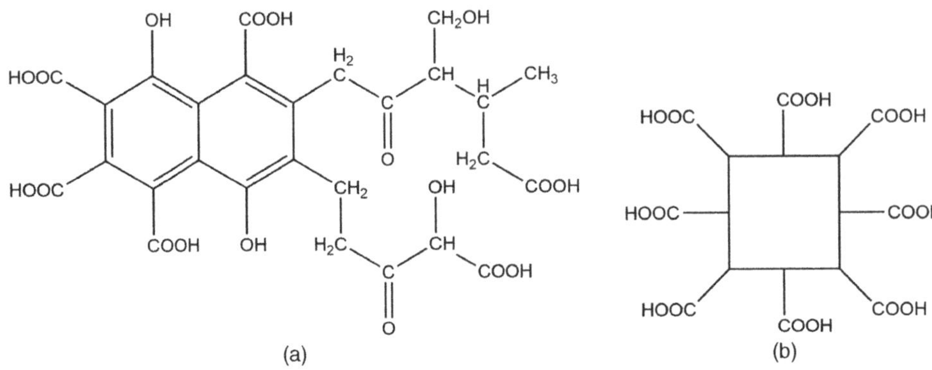

Figure 3.7 (a) An example structure of fulvic acid (L) and (b) a generalized structure of fulvic acid.

chemical formula. Between the two, while humic acids have lower solubility at acidic pH and high molecular weight, fulvic acids are soluble over a wide range of pH and have relatively low molecular weight. Due to their higher solubility, fulvic acids with a molecular mass of 500–10,000 Da and of size 5–200 nm are mostly responsible for organic fouling. NOM concentration in surface water typically ranges from 2 to 10 mg/L as DOC (dissolved organic carbon) and is seasonal, normally peaking during the Fall or Autumn period. In general, fouling-causing NOMs are polyanions with an aromatic core. In principle, it can interact with a sorbent phase through: hydrophobic, coulombic or electrostatic and Lewis acid–base interactions. Figure 3.7a illustrates the suggested structure of an NOM macromolecule; note that carboxylate group anions are prevalent along with aromatic hydrophobicity. Figure 3.7b provides a rendition of fulvic acid where the rectangle represents the hydrophobic core attached to pendant carboxylate groups on the surface. The ability to form micelles is extremely limited for fulvic acids due to the abundance of repulsive anionic charges.

Since NOMs exist as anions around neutral to slightly alkaline pH, they are rejected by the cation exchange resins via the Donnan exclusion effect. Cation exchange resins are thus immune to fouling by the NOM. In general, only the anion exchangers are susceptible to *organic fouling*. To elucidate the primary mechanisms of irreversible fouling and gradual deterioration in capacity, let us consider NOM with two major constituents: one hydrophobic core and negatively charged carboxylate groups. Due to both strong electrostatic and hydrophobic interactions, NOMs exhibit high affinity toward anion exchange resins, much higher than commonly present sulfate and chloride. Hence, during the conventional regeneration process with sodium hydroxide, NOMs are incompletely desorbed. Many pendant carboxylate groups of NOM attached to the anion exchanger act as cation exchange sites and pick up sodium ions during regeneration. During the rinsing process following regeneration, these carboxylate groups in sodium form undergo hydrolysis and slowly release sodium hydroxide as illustrated in Figure 3.8.

The conductivity in the rinse water remains unacceptably high for a long time due to presence of Na^+ and OH^-. During the same period, however, the capacity of the anion exchanger is significantly exhausted by other anions presented in rinse water. Thus,

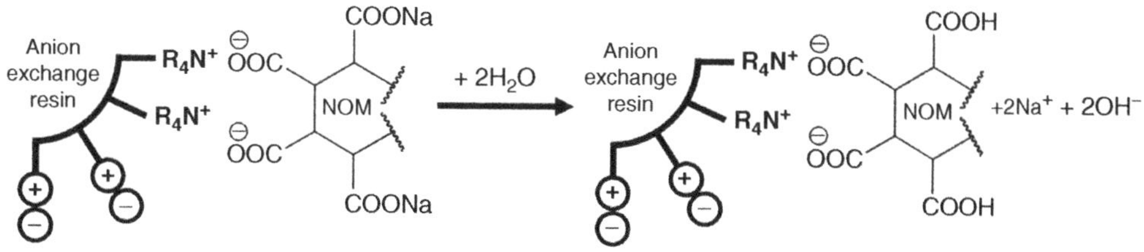

Figure 3.8 Gradual release of NaOH from anion exchange resin-bound NOM imparting high conductivity during the rinsing step.

the overall operating capacity of the anion exchanger is greatly decreased. Incomplete desorption of NOM during the regeneration step accompanied by slow hydrolysis, is responsible for the fouling, that is, loss in exchange capacity.

NOM has relatively large sizes and low diffusivity; they remain attached primarily to the outer periphery of an anion exchange resin and the carboxylate groups are only partly neutralized by the functional groups of the anion exchangers. Continued accumulation of NOM over multiple cycles due to incomplete regeneration thus creates an excess of negatively charged carboxylate groups that gradually transforms the outer periphery of the anion exchanger into a cation exchanger. So, anions tend to be rejected by the parent anion exchanger in accordance with the Donnan membrane principle. Figure 3.9 illustrates the phenomenon.

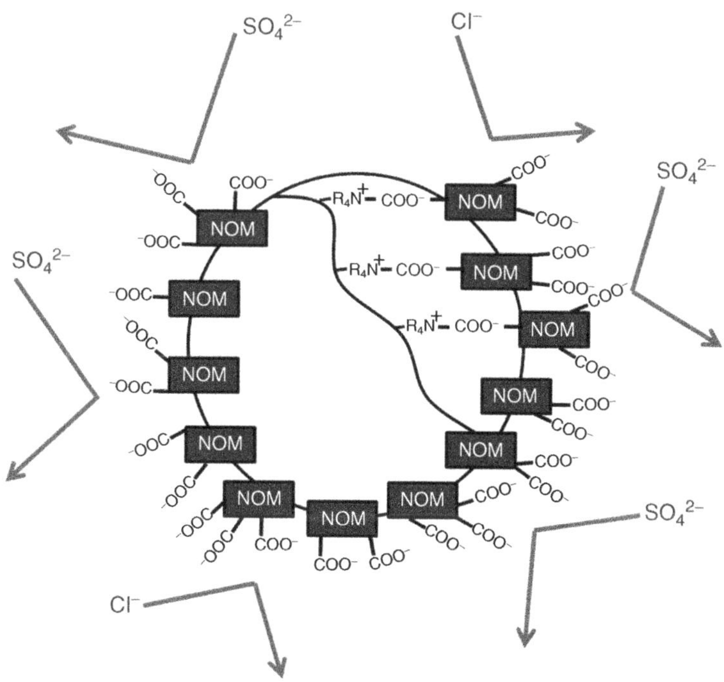

Figure 3.9 An illustration of the accumulation of the excess carboxylate groups in the outer periphery with concomitant cation exchange properties and anion rejection by the Donnan exclusion effect.

To rejuvenate an anion exchange resin fouled with NOM, a warm solution of NaCl and NaOH is used for a prolonged time. Elevated temperature (around 40 °C) increases NOM diffusivity facilitating desorption. The use of a non-polar cosolvent enhances the regeneration thermodynamics by reducing the dielectric constant of the solvent and hydrophilic repulsion as discussed earlier in Section 2.9. Macroporous resins have shorter diffusion path lengths and are, therefore, less susceptible to NOM fouling. For surface water with relatively high NOM content, macroporous anion exchange resins are routinely used to abate chronic fouling caused by NOM.

3.9 Ion Exchange Accompanied by Chemical Reaction

Equilibrium behaviors of ion exchange processes are altered by accompanying chemical reactions, such as precipitation, complexation, neutralization and redox reactions involving counterions. The overall process can be viewed as an extension of the Le Châtelier's principle where counterions are removed by other reactions in series, thus favorably bringing the ion exchange process to completion. Specific accompanying reactions of interest are as follows:

3.9.1 Precipitation

Let us consider the regeneration of a divalent metal cation (Me^{2+}) with sodium carbonate where the metal carbonate has very low solubility

$$CO_3^{2-} + 2Na^+ + \overline{(R^-)_2Me^{2+}} \leftrightarrow CO_3^{2-} + Me^{2+} + 2\overline{(R^-)Na^+}\,(K_{IX}) \tag{3.64}$$

$$Me^{2+} + CO_3^{2-} \leftrightarrow MeCO_3(s)\,(1/K_{sp}) \tag{3.65}$$

Overall,

$$CO_3^{2-} + 2Na^+ + \overline{(R^-)_2Me^{2+}} \leftrightarrow MeCO_3(s) + 2\overline{(R^-)Na^+}\,(K_{overall}) \tag{3.66}$$

Thus,

$$K_{overall} = \frac{K_{IX}}{K_{sp}} \tag{3.67}$$

For any sparingly soluble solid, K_{sp} is by several orders of magnitude lower than unity. Thus, the overall equilibrium constant, $K_{overall}$, is greatly enhanced due to the accompanying precipitation reaction. To get a quantitative feel for the effect of the precipitation reaction on ion exchange, consider the divalent metal cation to be Ca^{2+}. The value of K_{IX} from the general body of open literature is approximately 0.25 [6,28] while K_{sp} is $10^{-8.3}$. Thus, $K_{overall} = \frac{0.25}{10^{-8.3}} = 5 \times 10^7$. Similarly, for $CdCO_3(s)$, the K_{sp} value is 5.2×10^{-12}. Thus, the overall equilibrium constant will be even greater. From an equilibrium viewpoint, the forward reaction is thus highly favorable (irreversible) due to the accompanying precipitation reaction. In terms of kinetics, precipitation is significantly slower than ion exchange and requires nucleation to initiate the precipitation process. Even for fixed-bed ion exchange processes, the precipitation process can be carried out outside the column. The foregoing concept forms the basis of ion exchange isothermal supersaturation (IXISS) for environmentally benign application of ion exchange processes. In a similar vein, a trace target metal ion can first be removed by selective ion exchange, then regenerated and recovered outside the column as insoluble carbonate salt [6,28,29].

Also, note that following precipitation of $MeCO_3(s)$, the regenerant can be recycled and reused because it is mostly free of Me^{2+}. From a sustainability viewpoint, such hybrid ion exchange processes that allow reduction in the regenerant consumption and opportunities for simultaneous product recovery are receiving global acceptance.

3.9.2 Complexation

A complexing anionic ligand, L^{n-}, is excluded by a cation exchanger due to the Donnan effect. However, its ability to form stable complexes may shift the equilibrium favorably by replacing the precipitating agent (say $CO_3{}^{2-}$) (Eq. (3.64)) with an anionic ligand, L^{2-}.

$$L^{2-} + 2Na^{+} + \overline{(R^{-})_2Me^{2+}} \leftrightarrow L^{2-} + 2Me^{2+} + \overline{2(R^{-})Na^{+}} \quad (3.68)$$

$$L^{2-} + 2Me^{2+} \leftrightarrow (MeL^{0}) \quad (3.69)$$

Overall,

$$L^{2-} + 2Na^{+} + \overline{(R^{-})_2Me^{2+}} \leftrightarrow \overline{2(R^{-})Na^{+}} + (MeL^{0}) \quad (3.70)$$

Thus, a high value of metal–ligand stability constant (K_{st}) will greatly increase the favorable desorption of Me^{2+} from the cation exchanger by a sodium salt.

3.9.3 Redox Reaction

In previous two scenarios, ion exchange is followed by accompanying reactions, namely, precipitation and complexation. In engineered processes, such reactions can also be brought about prior to ion exchange to improve favorable sorption behaviors. Elemental mercury (Hg^0) and arsenite (H_3AsO_3) are electrically neutral and thus cannot participate in ion exchange processes. However, cation and anion exchangers appropriately dispersed with an oxidizing agent, namely, manganese dioxide nanoparticles, can oxidize the target species followed by selective ion exchange [30]. For elemental mercury,

$$Hg^{0} + MnO_2(s) + 4H^{+} \leftrightarrow Hg^{2+} + Mn^{2+} + 2H_2O \quad (3.71)$$

$$\overline{4R^{-}Na^{+}} + Hg^{2+} + Mn^{2+} \leftrightarrow \overline{(R^{-})_2Hg^{2+}} + \overline{(R^{-})_2Mn^{2+}} \quad (3.72)$$

Overall,

$$Hg^{0} + MnO_2(s) + \overline{4R^{-}Na^{+}} + 4H^{+} \leftrightarrow \overline{(R^{-})_2Hg^{2+}} + \overline{(R^{-})_2Mn^{2+}} + 4Na^{+} \quad (3.73)$$

In a similar way, for non-ionized arsenite,

$$HAsO_2 + MnO_2(s) + H^{+} \leftrightarrow H_2AsO_4^{-} + Mn^{2+} \quad (3.74)$$

$$\overline{(R^{+})Cl^{-}} + H_2AsO_4^{-} \leftrightarrow \overline{(R^{+})H_2AsO_4^{-}} + Cl^{-} \quad (3.75)$$

Overall,

$$HAsO_2 + MnO_2(s) + \overline{(R^+)Cl^-} + 4H^+ \leftrightarrow \overline{(R^+)H_2AsO_4^-} + Mn^{2+} + Cl^- \quad (3.76)$$

The foregoing examples demonstrate that by appropriately aligning other processes in tandem with ion exchange, the thermodynamic efficiency can be greatly enhanced. Ion exchange-assisted precipitation also yields new process routes for synthesis of pure chemicals [30].

3.10 Monovalent–Divalent Selectivity

From the earlier discussion on selectivity reversal during heterovalent ion exchange in Chapter 2, it was apparent that a divalent counterion becomes increasingly preferred to its monovalent competitor as the exchange capacity, Q, is increased. Equation (2.34) does not, however, reveal any scientific insight about the genesis of such a correlation. One easy-to-comprehend explanation is that as Q is increased, the number of charged functional groups per unit volume of the exchanger is increased, thus reducing the average distance between two neighboring ion exchange sites. So, the two charges of a divalent ion (cation or anion) can be satisfied with less amount of work and hence, higher selectivity. When coulombic or electrostatic interaction is the predominant ion exchange mechanism, previous research has demonstrated that the distance of fixed charge separation in the ion exchanger is the primary determinant of divalent/monovalent selectivity [31–33]. Cation exchangers with guaranteed close spacing of the negative functional groups are very divalent ion selective and fit nicely into the charge separation distance theory. The structures of these commercially available cation exchangers, namely, carboxylate, aminophosphonate and iminodiacetate are provided in Figure 3.10. They all are used to remove Ca^{2+} and Mg^{2+} from brines and take advantage of closely spaced negatively charged ion exchange sites. On the contrary, a commercial strong-acid cation exchanger with sulfonic-acid functional groups with greater charge separation distance offers significantly lower Ca^{2+}–Na^+ selectivity. It is worth noting that both Ca^{2+} and Na^+ are essentially hard cations with identical outer shell electronic configurations. Coulombic interaction is the primary ion exchange mechanism for both counterions.

3.10.1 Effect of Charge Separation: Mechanistic Explanation

The concept of fixed-charge separation offers a tool to improve monovalent ion selectivity. To develop an insight into the role of charge separation distance between two neighboring sites, let us first consider Case 1 in Figure 3.11 for a cation exchanger in sodium form with two exchange sites in close proximity to each other. The exchange of two monovalent Na^+ counterions with a divalent Ca^{2+} follows classical ion exchange where the coions (Cl^-) continue to exist only in the aqueous phase and electroneutrality is preserved in both phases.

For Case 2, we consider a cation exchanger with just one site, all other conditions remaining identical. For Ca^{2+} to displace Na^+ from the ion exchange site and maintain electroneutrality at the same time, one Cl^- must be brought from the aqueous to the

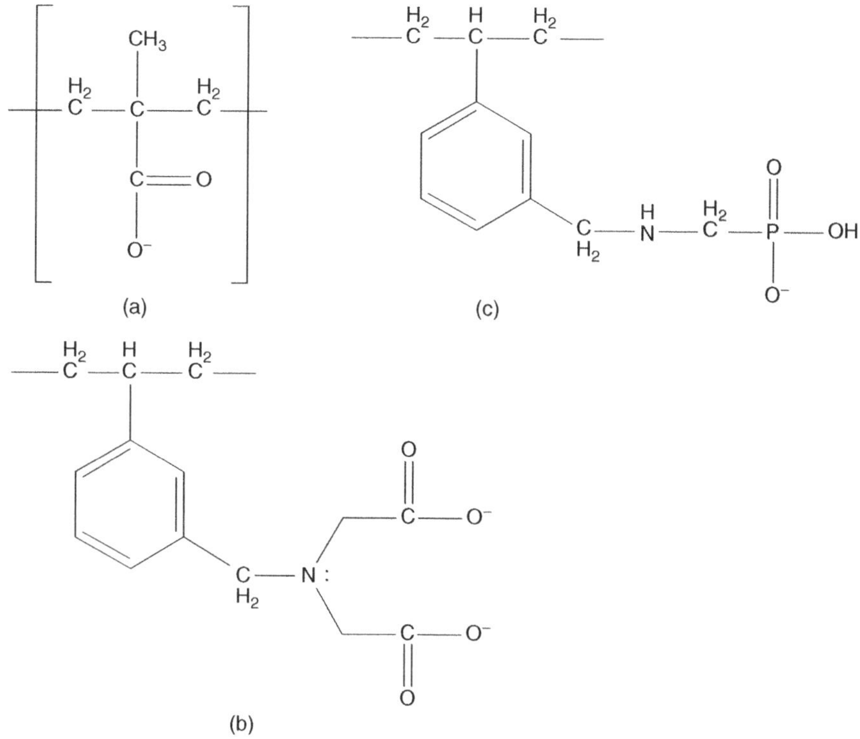

Figure 3.10 Functional groups of three commercially available cation exchangers, namely, (a) carboxylate, (b) aminophosphonate, and (c) iminodiacetate with high divalent cation selectivity in sodium form.

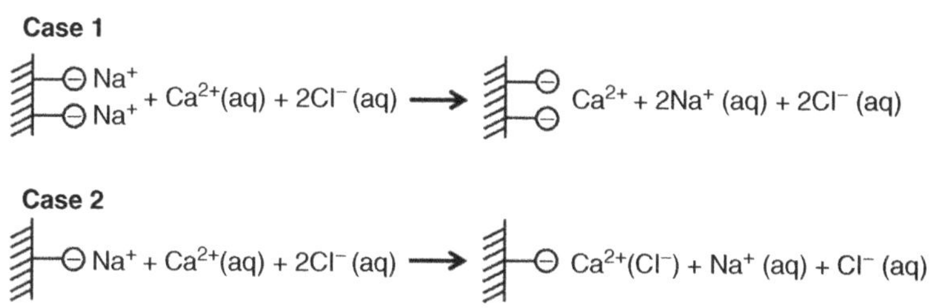

Figure 3.11 A cation exchanger in sodium form where in Case 1 there are two exchange sites in close proximity to each other; and in Case 2 there is only local exchange site.

exchanger phase. Binding of Ca^{2+} following exchange with Na^+ transforms the lone site into an anion exchanging one because of a residual positive charge. Following the same mathematical treatment using Coulomb's law, as illustrated in Eq. (2.40), the additional work or free energy change needed to bring Cl^- to the site at the standard state is

$$\Delta G^{o} = \frac{-e^2}{(r_{Ca} + r_{Cl})\varepsilon_D} \tag{3.77}$$

where r_{Ca} and r_{Cl} are the respective hydrated ionic radius of calcium and chloride, ε_D is the dielectric constant of the exchanger, and e is the charge of an electron.

Thermodynamically, the exchange reaction in Case 2 is less favorable compared to Case 1 due to the additional work or energy to be overcome as per Eq. (3.77). Thus, the calcium selectivity over sodium is significantly less for Case 2. From a broader perspective, as the distance between two neighboring sites is increased, the affinity for the divalent ion is reduced. Specifically, when that distance approaches or exceeds 1 nm or 10 Å, divalent ion selectivity is not observed and is essentially absent.

To validate the scientific premise of the hypothesis, consider typical strong-acid cation and strong-base anion exchange resins. In general, the cation exchange capacity per unit volume is about two times greater than its anion counterpart, implying shorter distance between two neighboring sites in a cation exchanger. Divalent/monovalent selectivity for a cation exchange (e.g., Ca^{2+}/Na^{+}) is significantly greater than that for an anion exchanger (e.g., SO_4^{2-}/Cl^{-}) under otherwise identical conditions.

3.10.2 Nitrate/Sulfate and Chloride/Sulfate Selectivity in Anion Exchange

Nitrate/sulfate selectivity is an important parameter in the design of an ion exchange process for nitrate removal from contaminated groundwater with low ionic strength, that is, the total dissolved solids significantly less than 500 mg/L. With electrostatic interaction as the primary binding mechanism for anion exchange resins, the dilute solution selectivity sequence for common anions stands as follows:

$$SO_4^{2-} > NO_3^{-} > Cl^{-}$$

Besides reduced capacity, the presence of sulfate in the feed solution during a fixed-bed column also leads to chromatographic elution of nitrate, that is, nitrate in the treated water can be significantly greater than its influent concentration for a prolonged period. Nitrate preference over sulfate for an anion exchange resin will obviously have specific advantages. Clifford [34–36] has investigated resin properties that influence nitrate/sulfate selectivity. Along the same vein, sulfate/chloride selectivity is of paramount significance in desulfation of seawater through anion exchange to avoid calcium and barium sulfate precipitation during desalination. The subject of sulfate/chloride selectivity at relatively high ionic strength has been studied by researchers [37–40].

Previous studies demonstrated that both the matrix and the functional group influence monovalent/divalent selectivity for anion exchange resins. These insightful findings were appropriately exploited to develop and synthesize new nitrate-, perchlorate-, and chromate-selective anion exchange resins. Figures 3.12–3.15 present salient experimental results from Clifford and Weber [31] pertaining to sulfate-nitrate isotherms at 5 meq/L or 0.005 N aqueous-phase electrolyte concentration. In particular, Figures 3.12 and 3.13 demonstrate the effects of resin matrix (styrene vs acrylic) on sulfate/nitrate isotherms and Figures 3.14 and 3.15 exhibit how different functional groups influence sulfate/nitrate selectivity. These experimental observations agree well with the concept of "charge separation distance" and other intrinsic properties of ion exchange resins discussed and emphasized earlier in the chapter pertaining to divalent/monovalent ion selectivity. Specifically, they demonstrate:

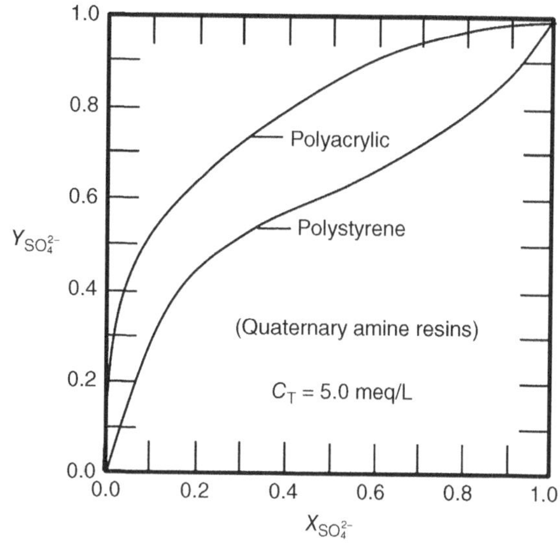

Figure 3.12 The effect of resin matrix on the divalent/monovalent (SO_4^{2-}/NO_3^-) selectivity of quaternary amine resins at 5.0 meq/L total aqueous phase concentration. *Source*: Clifford and Weber 1983 [31]. Reproduced with permission of Elsevier.

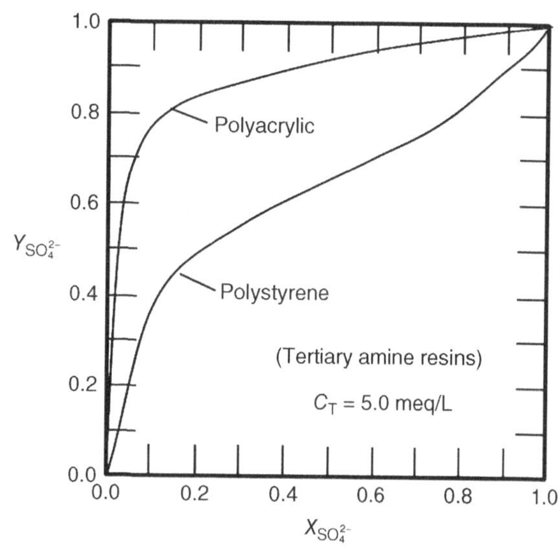

Figure 3.13 The effect of resin matrix on the divalent/monovalent (SO_4^{2-}/NO_3^-) selectivity of tertiary amine resins. *Source*: Clifford and Weber 1983 [31]. Reproduced with permission of Elsevier.

1. Between polystyrene and polyacrylic matrix, the latter is more hydrophilic and carries higher concentration of fixed functional groups per unit volume. So, the charge separation difference between the neighboring functional groups is lower for polyacrylic matrix, offering higher selectivity for divalent sulfate ions.
2. For identical matrix (polystyrene or polyacrylic), lower basicity (i.e., tertiary amine relative to quaternary ammonium) provides higher concentration of functional groups, that is, shorter charge separation distance between neighboring sites, thus offering greater sulfate selectivity.

The foregoing findings, although not amenable to quantitative treatment, led the foundation for nitrate-selective anion exchange resins.

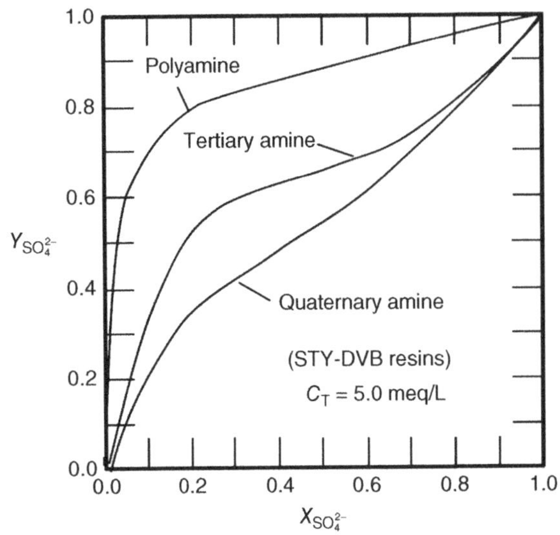

Figure 3.14 The effect of amine functionality on the divalent/monovalent (SO_4^{2-}/NO_3^-) selectivity of polystyrene resins. *Source*: Clifford and Weber 1983 [31]. Reproduced with permission of Elsevier.

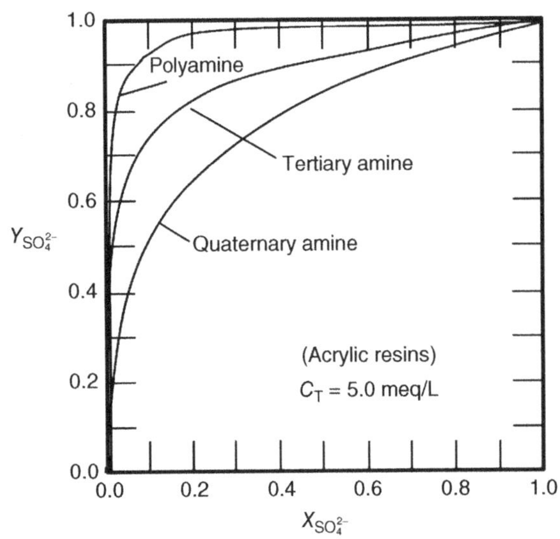

Figure 3.15 The effect of amine functionality on the divalent/monovalent (SO_4^{2-}/NO_3^-) selectivity of polyacrylic resins. *Source*: Clifford and Weber 1983 [31]. Reproduced with permission of Elsevier.

3.10.3 Genesis of Nitrate-Selective Resin

Nitrate-selective resins correspond to anion exchange resins that exhibit nitrate selectivity over sulfate (i.e., $\alpha_{N/S}$ greater than unity) in dilute solution at ionic strength less than 0.01 M. These anion exchange resins have been synthesized with polystyrene matrix by changing the alkyl groups of the quaternary ammonium functionality as presented in Figure 3.16. Table 3.3 includes capacity and selectivity coefficient data for nitrate–sulfate exchange as the alkyl group changes gradually from methyl to butyl [41] for the following exchange reaction:

$$\overline{(R^+)_2SO_4^{2-}} + NO_3^- \leftrightarrow \overline{2R^+NO_3^-} + SO_4^{2-} \tag{3.78}$$

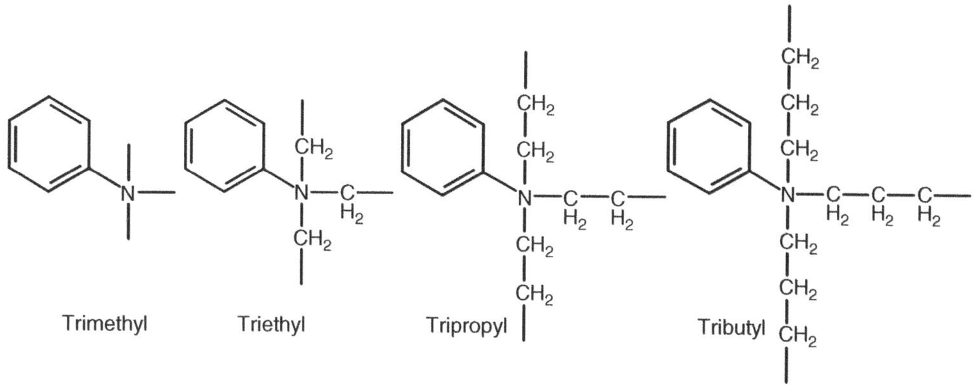

Figure 3.16 Strong-base anion exchange resins with trimethyl, triethyl, tripropyl, and tributyl functional groups.

Table 3.3 Selectivity coefficient values and capacity data for nitrate-selective anion exchangers [41].

Alkyl group in quaternary amine functional group	Moisture content (%)	Capacity (eq/L)	$K^{Se}_{N/S}$
Methyl	57.0	1.41	100
Ethyl	48.0	1.2	1,000
Propyl	30.4	0.84	1,100
Butyl	33.0	0.66	11,000

Note that nitrate selectivity increases steadily as the size of the alkyl group is increased from methyl to ethyl to propyl and then to butyl. Such an enhancement in nitrate selectivity is attributed to the following two reasons:

- Gradual but consistent decrease in exchange capacity enhances nitrate selectivity over sulfate in agreement with the concept of charge separation distance;
- More hydrophobic and bulky alkyl groups (e.g., propyl compared to methyl) exert greater steric hindrance to more hydrated divalent sulfate ion than monovalent nitrate.

Following a prolonged laboratory and field-scale investigation, Guter developed and subsequently commercialized a strong-base anion exchange resin with polystyrene matrix and tributyl quaternary ammonium functional groups [33]. Figure 3.17a and b shows comparison of breakthrough data for sulfate and nitrate during column runs of representative groundwater samples from McFarland, California. Sulfate and nitrate concentrations in the feed were about 375 and 110 mg/L, respectively. Note that while nitrate broke through at around 150 BVs for a commercial strong-base anion exchange resin with trimethyl alkyl groups, the tributyl functional groups enhanced the breakthrough BVs for nitrate to 250. For both column runs, resins used were in chloride form.

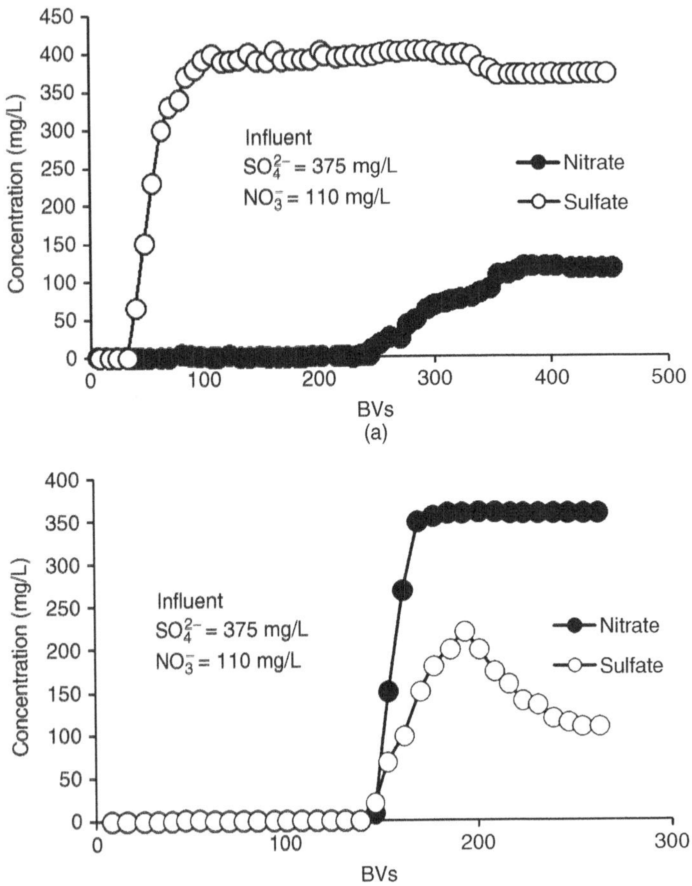

Figure 3.17 (a) Sulfate (50 BVs) and nitrate (>250 BVs) breakthrough for styrene–divinylbenzene resin with tributyl quaternary ammonium groups (nitrate-selective resin); (b) sulfate (150 BVs) and nitrate (150 BVs) breakthrough for styrene–divinylbenzene resin with trimethyl quaternary ammonium groups (Type I Anion exchange resin). *Source*: Guter 1995 [33]. Reproduced with permission of Taylor & Francis.

Also note that nitrate showed chromatographic elution for the commercial resin, that is, upon breakthrough, nitrate concentration became significantly greater than its influent concentration. For the nitrate-selective resin with tributyl functional group, no chromatographic elution was observed, further confirming higher nitrate selectivity over sulfate.

3.10.4 Chromate Ion Selectivity

Like nitrate, monovalent ion selectivity of chromate ($HCrO_4^-$) is very desirable over competing divalent sulfate ion for specific separations. Chromate chemistry is of particular interest because of its dissociation at near-neutral pH as follows:

$$HCrO_4^- \rightleftarrows H^+ + CrO_4^{2-}, \quad pK_a = 6.5 \tag{3.79}$$

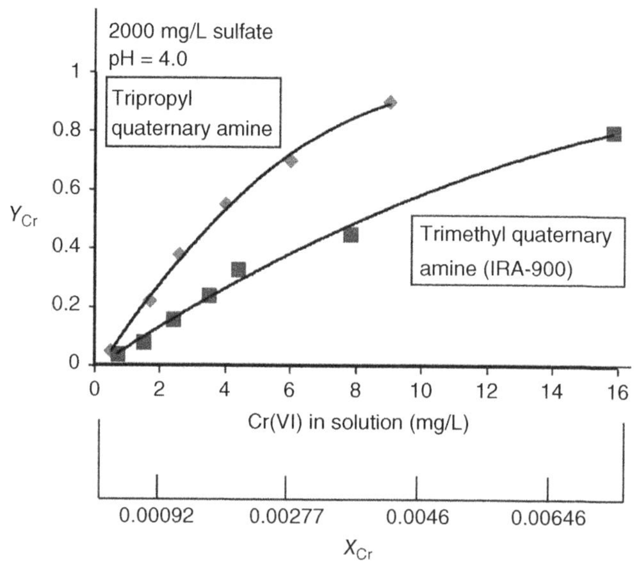

Figure 3.18 Comparison of chromate/sulfate isotherm (23 ± 2 °C) at pH 4.0 between the new resin (tripropyl quaternary ammonium functionality) and IRA-900 (trimethyl quaternary ammonium functionality) under identical conditions. *Source*: SenGupta *et al.* 1988 [42]. Reproduced with permission of Elsevier.

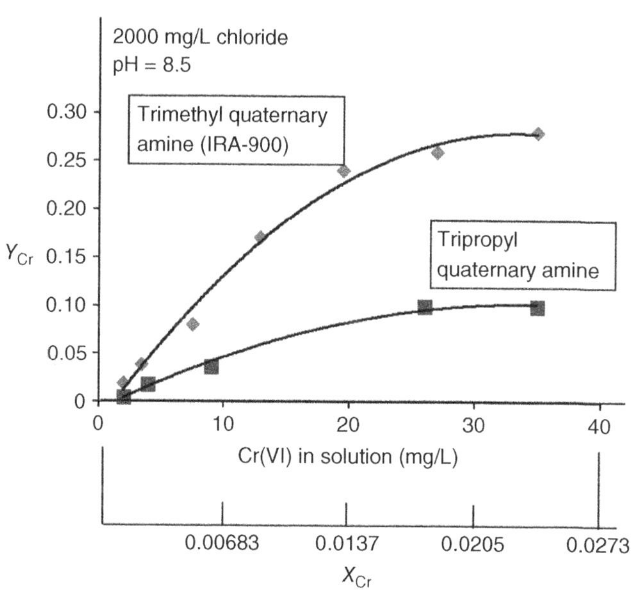

Figure 3.19 Comparison of chromate/chloride isotherms (23 ± 2 °C) at pH 8.5 between the new resin (tripropyl quaternary ammonium functionality) and IRA-900 (trimethyl quaternary ammonium functionality) under identical conditions. *Source*: SenGupta *et al.* 1988 [42]. Reproduced with permission of Elsevier.

While at above-neutral pH, divalent chromate (CrO_4^{2-}) predominates, monovalent chromate ($HCrO_4^-$) is the predominant species at acidic pH. Thus, chromate offers a unique opportunity to further confirm the scientific premise of the discussion in Section 3.10.2 regarding the effect of the alkyl group in quaternary ammonium functional group. Figure 3.18 shows the comparison of $HCrO_4^-/SO_4^{2-}$ isotherms at constant background sulfate concentrations for two strong-base anion exchange resins: one with tripropyl functional group and the other with commercially available trimethyl group (IRA-900 from Rohm and Haas Co., Philadelphia) [42]. In agreement with the findings of nitrate/sulfate selectivity, tripropyl functional groups offer greater monovalent $HCrO_4^-$ selectivity over sulfate.

To further confirm the role of alkyl group in determining the relative monovalent/divalent ion selectivity, CrO_4^{2-}/Cl^- isotherm tests were carried out with the same two anion exchange resins, but at alkaline pH where divalent CrO_4^{2-} predominates. Note that under the experimental conditions, chromate (CrO_4^{2-}) is a divalent anion while the competing chloride ion is monovalent. Figure 3.19 shows that trimethyl quaternary ammonium functional group exhibits greater selectivity for CrO_4^{2-} (divalent) relative to tripropyl groups in accordance with the general premise of the monovalent/divalent anion selectivity.

3.11 Entropy-Driven Selective Ion Exchange: The Case of Hydrophobic Ionizable Organic Compound (HIOC)

During a sorption process, solute molecules or ions are essentially transferred from the solvent phase to the sorbent phase. As the binding of a solute takes place at the sorption site, the rotational and translational freedom of the solute are reduced. Hence, the entropy change (ΔS) during sorption is negative. For the sorption to be favorable, Gibbs free energy change (ΔG) must be negative, which in turn requires the enthalpy change (ΔH) to be negative because $\Delta G = \Delta H - T\Delta S$. In general, all favorable sorption processes tend to conform to this stipulation, that is, they are exothermic and accompanied by an overall decrease in entropy. Figure 3.20 illustrates such enthalpy-driven sorption processes.

Ion exchange processes involve exchange of ions and for homovalent ion exchange based primarily on electrostatic interaction, this is followed nearly universally.

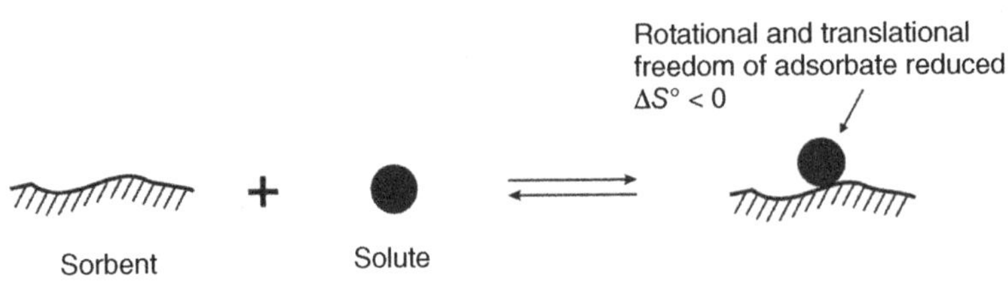

$$\Delta G^\circ = \Delta H^\circ - T\Delta S^\circ$$

Favorable sorption: $\Delta G^\circ < 0$

$\Delta H^\circ < 0$ (Exothermic and enthalpy-driven process)

Figure 3.20 A schematic illustrating an exothermic and enthalpy-driven sorption process.

For example, let us consider nitrate–chloride (NO_3^- : Cl^-) for a strong-base anion exchanger with polystyrene matrix as follows:

$$\overline{R^+NO_3^-} + Cl^- \leftrightarrow \overline{R^+Cl^-} + NO_3^- \tag{3.80}$$

At 25 °C, energetics of the NO_3^- : Cl^- exchange process at the standard state for 1 mol or equivalent of nitrate exchange are as follows, illustrating the exothermicity of the reaction, $\Delta G^0 = -3.4\,kJ$, $\Delta H^0 = -8.7\,kJ$, $T\Delta S^0 = -5.3\,kJ$.

By and large, all thermodynamically favorable homovalent ion exchange reactions pertaining to inorganic ions, both cations and anions, are exothermic.

Many synthetic aromatic compounds exhibit acidic characteristics due to the presence of carboxylic, phenolic and sulfonic acid moieties, and their acidities are often strengthened because of the electron-withdrawing effects of various substituent groups. For example, the pK_a value (i.e., negative logarithm of acid dissociation constants) for phenol is 9.3 while the same for pentachlorophenol or PCP is 4.75. As a result, PCP, which is extensively used in the wood preservation industry, exists as an anion in contaminated surface or groundwater at neutral pH. Contrary to other non-ionized hydrophobic aromatic compounds, pentachlorophenate or PCP^- is, therefore, more mobile in natural environment and not amenable to efficient removal by conventional hydrophobic sorbents like activated carbon. Like PCP^-, many other industrially significant aromatic compounds, namely, naphthalene sulfonates and quaternary ammonium compounds tend to exist as ions in the aqueous phase and are commonly referred to as HIOCs [43,44]. While the aromaticity imparts hydrophobic or non-polar characteristic, the ionic charge of these compounds enhances hydrophilicity through ion–dipole interaction with water molecules. The solubility of weak-acid type HIOC compounds, therefore, increases significantly at pH greater than pK_a values.

The aromatic anions have hydrophobic characteristics as well as ionic characteristics due to their non-polar moieties (NPM). Understandably, the sorption behaviors of such aromatic anions will be greatly influenced by both hydrophobic and ionic properties. Contrary to nonionized hydrophobic aromatic compounds, the sorption of these aromatic anions is not a physical sorption process. Such processes are characterized by equivalent exchange of ionic species between the liquid phase and ion-exchanger solid phase, but ion exchange selectivity is often determined by concurrent hydrophobic interactions other than electrostatic ones [10].

3.11.1 Focus of the Study and Related Implications

In this section, we will discuss favorable sorption or ion exchange behaviors of several environmentally significant HIOCs that are aromatic anions, such as pentachlorophenate, chlorophenate, benzene, and naphthalene sulfonates. Such favorable sorption equilibria are, however, distinctively unique because they are all endothermic processes and accompanied by highly positive entropy changes. Solvent dielectric constant, polarity or moisture content of the ion-exchanger matrix and the non-polar moiety (NPM) of the aromatic anion are the three fundamental process variables that govern the overall sorption equilibria. It is noteworthy that mechanistic understanding of the sorption behaviors of HIOCs, as discussed in the succeeding sections, can be extended to predict, model and quantify the sorption of NOM or many pharmaceutical and personal care products (PPCP) onto anion exchangers. In principle, NOM and many PPCP solutes are similar to HIOC compounds, that is, they all have hydrophobic aromatic core with negative charges.

Two types of aromatic anions were investigated: chlorophenols and sulfonated aromatic anions. Chlorophenols included PCP, 2,4,6-trichlorophenol, 2,6-dichlorophenol. Sulfonated aromatic anions included naphthalene-1-sulfonate, naphthalene-1,5-disulfonate, and benzene sulfonate. Tables 3.4 and 3.5 include salient information about chlorophenols and sulfonated aromatic acids. Because of the electron-withdrawing effect (or inductive effect) of Cl substituents, the pK_a values of the phenols decrease as more Cl substituents are introduced into benzene rings. In Table 3.4, the values of octanol/water partition coefficient ($K_{O/W}$) increase with increasing number of Cl substituents for undissociated acids. The $K_{O/W}$ is a measure of hydrophobicity. Note that the hydrophobicity of the phenols is enhanced with an increase in substituent Cl atoms. Naphthalenesulfonic and benzenesulfonic acids are strong acids and their pK_a values are very low. While naphthalene-1-sulfonate and benzene sulfonate are monovalent anions, naphthalene-1,5-disulfonate is a divalent anion.

Two types of ion exchange resins, namely, IRA-900 and IRA-958, were used. The salient properties of the ion exchangers are presented in Table 3.6. IRA-900 and IRA-958 are strong-base anion exchangers. Both anion exchangers have quaternary

Table 3.4 Properties of chlorophenols.

Name	Molecular formula	Molecular weight	pK_a	log $K_{O/W}$
Pentachlorophenol	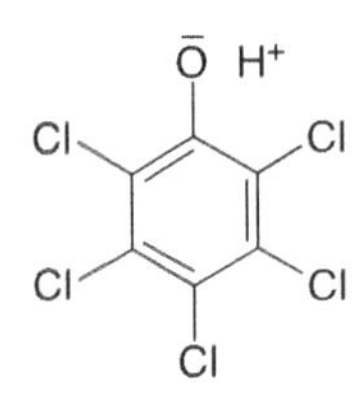	266.5	4.8	5.2
2,4,6-Trichlorophenol	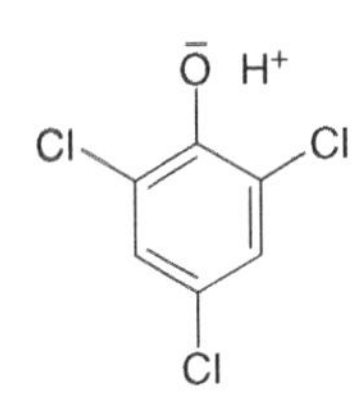	197.5	6.1	3.7
2,6-Dichlorophenol	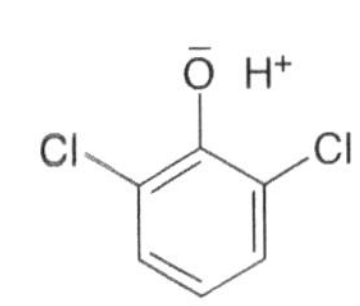	163	6.9	2.6

Source: Jafvert *et al.* 1990 [44]. Reproduced with permission of American Chemical Society.

Table 3.5 Properties of sulfonated aromatic acids.

Compounds	Molecular formula	Molecular weight	pK_a
Naphthalene-1-sulfonic acid	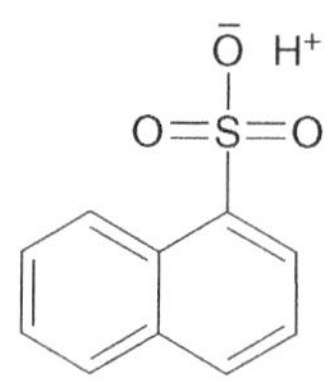	208	0.17
Naphthalene-1,5-disulfonic acid	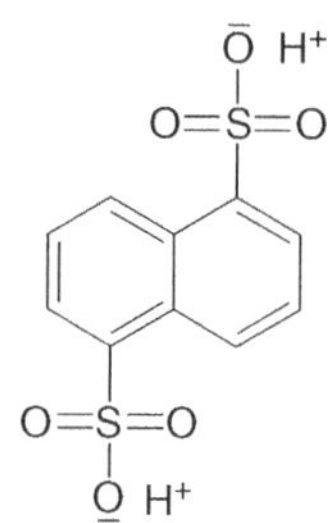	284	−3.37, −2.64
Benzenesulfonic acid	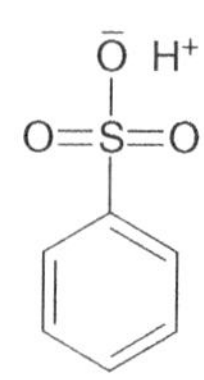	158	0.7

Source: Adapted from Stahl *et al*. 2008 [45].

ammonium functional groups. The matrix of IRA-900 is polystyrene, while the matrix of IRA-958 is polyacrylic. The polystyrene matrix is more non-polar and hydrophobic compared to the polyacrylic one.

3.11.2 Nature of Solute–Sorbent and Solute–Solvent Interactions

A polymeric anion exchanger with fixed positive charges will sorb aromatic anions like pentachlorophenate and naphthalene sulfonates. A typical anion exchange reaction between pentachlorophenate (PCP^-) and chloride (Cl^-) can be presented as follows:

$$\overline{R^+Cl^-} + PCP^- \leftrightarrow \overline{R^+PCP^-} + Cl^- \quad (3.81)$$

where overbar represents the exchanger phase and R^+ is an anion exchanger with fixed positive charges. Chloride (Cl^-) and pentachlorophenate (PCP^-) are identical electrostatically; they both have one negative charge. Strictly from an electrostatic or coulombic interaction viewpoint, the sorption of PCP^- onto a polymeric anion exchanger in the presence of competing chloride ion is unlikely to be a selective process. Previous studies have, however, shown very favorable sorption behaviors of chlorinated

Table 3.6 Salient properties of polymeric anion exchangers.

Resin	IRA-900	IRA-958
Structure (repeating unit)	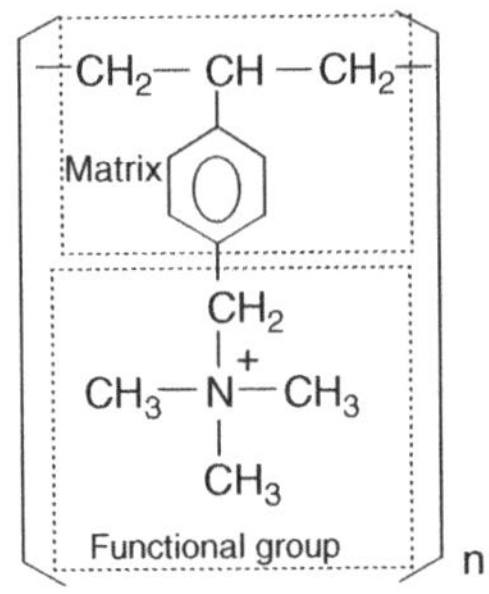	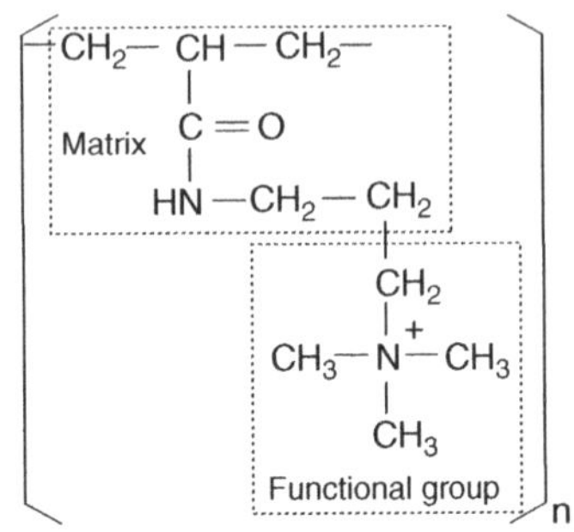
Functional group	Quaternary ammonium	Quaternary ammonium
Matrix	Polystyrene, macroporous	Polyacrylic, macroporous
Capacity (meq/g air-dried resin)	3.6	3.4
Manufacturer	Rohm and Haas Co., Philadelphia	Rohm and Haas Co., Philadelphia

phenols and aromatic anions onto polymeric exchangers in preference to chloride and other inorganic anions [46–48]. High ion exchange selectivity has also been reported for aliphatic anions with long alkyl chains [49,50]. Such high sorption affinities have, in general, been attributed to hydrophobic interactions resulting from the NPM of the aromatic anions. From a phenomenological viewpoint, the NPM–solvent and the NPM–matrix interactions are recognized as the two primary contributors toward high sorption affinity of aromatic ions in ion-exchange processes. The matrix represents the skeletal organic component in the polymeric ion exchanger besides the charged functional groups. Assuming insignificant change in the hydration of chloride ion between the aqueous and ion exchanger phases, the following ion exchange half reaction is the primary determinant of the overall equilibrium of the reaction in Eq. (3.81):

$$\overline{R^+} + PCP^-(aq) \leftrightarrow \overline{R^+PCP^-} + \text{water} \tag{3.82}$$

Since PCP^- sorption is favorable, the overall free energy change for Eq. (3.82) is negative. The free energy change at the standard state of choice (ΔG^0) is given by

$$\Delta G^0 = \Delta H^0 - T\Delta S^0 \tag{3.83}$$

Therefore, both enthalpic (ΔH^0) and entropic (ΔS^0) changes help decide the overall selectivity of the ion exchange process. Note that the definition of the standard state in the ion exchanger phase may alter the significance of exchanger-phase activity coefficients, but in no way alters the relative enthalpic and entropic contributions toward the overall equilibrium. To elucidate interactions associated with PCP^- sorption in Eq. (3.82), the sorption process can be broken down into two consecutive steps: first, desolvation of PCP^- and, second, PCP^- sorption onto the anion exchanger.

Interaction during Desolvation of PCP^-

A NPM is not capable of forming hydrogen bonds with polar water molecules. So, when an ion containing NPM is introduced into water (a polar solvent), the water molecules tend to turn away from NPM and reorganize themselves in clusters through hydrogen bonding. Hence, there is an overall entropy decrease in the system due to reduced degrees of freedom of these self-associated water molecules. The concept of cluster-like formation of structured water molecules around a hydrophobic solute was first discussed by Frank and Wen [51] and later elaborated by Némethy and Scheraga [52], and others [53–55]. As PCP^- leaves the aqueous phase during the ion exchange process, an overall increase in entropy will, therefore, result. Also, the solvent phase needs to absorb heat to break the highly associated cluster-like structure of water molecules, that is, the process is endothermic.

Interaction during PCP^- Sorption onto the Polymeric Exchanger

Once a PCP^- molecule enters the exchanger phase and binds to the fixed positive charge, its NPM tends to be in direct contact with the non-polar matrix of the ion exchanger. This results in expulsion of polar water molecules from the exchanger phase, which are present primarily due to the osmotic pressure difference between the exchanger phase and the solvent. Although thermal energy is required for such localized dehydration within the exchanger, the resulting increase in overall entropy due to the direct contact between these two non-polar substances (matrix and NPM of PCP^-) makes such a binding energetically advantageous [56].

Figure 3.21 illustrates a mechanistic interpretation of the foregoing two steps of the sorption process. Note that hydrophobic interactions energetically comprise both NPM–solvent and NPM–matrix interactions. Although not explicit, the effect of solvent–matrix interaction is also included in Figure 3.21. The weaker the solvent–matrix interaction, the smaller will be the energy required to expel the solvent molecules from the matrix and hence, more favorable will be the sorption process and vice versa.

For negligible swelling/shrinking of the polymeric exchanger, the overall free energy change for an ion exchange reaction involving a counterion with NPM is thus contributed by electrostatic (el), NPM–solvent and NPM–matrix interactions.

$$\Delta G^0_{\text{overall}} = \Delta G^0_{\text{el}} + \Delta G^0_{\text{NPM-solvent}} + \Delta G^0_{\text{NPM-matrix}} \tag{3.84}$$

When Eq. (3.84) is applied to homovalent PCP^-–Cl^- exchange in Eq. (3.81), the free energy change due to electrostatic interactions cancel out, and we get the following:

$$\begin{aligned}\Delta G^0_{\text{overall}} &= \Delta G^0_{\text{NPM-solvent}} + \Delta G^0_{\text{NPM-matrix}} \\ &= (\Delta H^0_{\text{NPM-solvent}} + \Delta H^0_{\text{NPM-matrix}}) - (T\Delta S^0_{\text{NPM-solvent}} + T\Delta S^0_{\text{NPM-matrix}})\end{aligned} \tag{3.85}$$

Note that only overall enthalpic and entropic changes during the sorption process can be determined experimentally. However, by changing the NPM of the solute, dielectric constant of the solvent and polarity of the matrix, one can assess the relative contributions of NPM–solvent and NPM–matrix interaction toward the overall free energy change. The overall free energy change is again related to the equilibrium constant, K, of the reaction in Eq. (3.81) as follows:

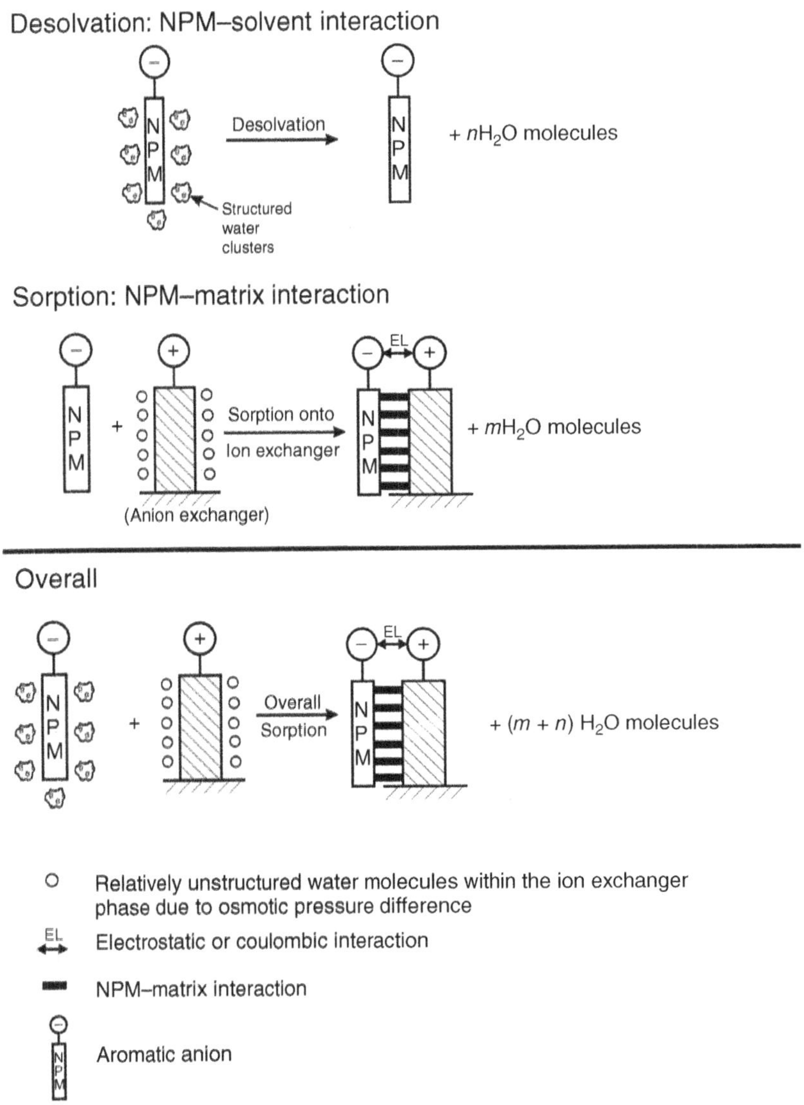

Figure 3.21 A schematic illustrating NPM–solvent, NPM–matrix, and electrostatic interactions during sorption of the aromatic anion from the aqueous phase. *Source*: Li and SenGupta 2001 [57]. Reproduced with permission of Elsevier.

$$\Delta G^0_{\text{overall}} = -RT \ln K \tag{3.86}$$

where R is the universal gas constant and T is the temperature in kelvin. For homovalent PCP^-–Cl^- exchange, the equilibrium constant, K, is given by

$$K_{\text{PCP/Cl}} = \frac{y_{\text{PCP}} \cdot f_{\text{PCP}}}{y_{\text{Cl}} \cdot f_{\text{Cl}}} \cdot \frac{x_{\text{Cl}} \cdot \gamma_{\text{Cl}}}{x_{\text{PCP}} \cdot \gamma_{\text{PCP}}} \tag{3.87}$$

where y_i and x_i represent equivalent fractions of counterion "i" in the exchanger phase and in the aqueous phase, respectively, while f_i and γ_i represent activity coefficients in the corresponding two phases. For ions with identical charges, the activity coefficients in dilute aqueous solutions tend to be equal, that is, γ_{PCP}/γ_{Cl} is unity [58]. The separation factor for PCP^-/Cl^- exchange can be determined experimentally at a particular resin loading and is given by

$$\alpha_{PCP/Cl} = \frac{y_{PCP} \cdot x_{Cl}}{y_{Cl} \cdot x_{PCP}} \tag{3.88}$$

The variation in exchanger-phase loading for PCP^-–Cl^- exchange is, however, contained between $y_{PCP} = 0$ and $y_{PCP} = 1.0$. For homovalent ion exchange, the equilibrium constant can then be approximated as the average separation factor value, integrated over the entire exchanger-phase composition, that is,

$$\ln K_{PCP/Cl} = \frac{\int_0^1 \ln \alpha_{PCP/Cl} \mathrm{d}y_{PCP}}{\int_0^1 \mathrm{d}y_{PCP}} = \int_0^1 \ln \alpha_{PCP/Cl} \mathrm{d}y_{PCP} \tag{3.89}$$

The overall free energy change for PCP^-:Cl^- exchange is now

$$\Delta G^0_{\text{overall}} = -RT \ln K = -RT \int_0^1 \ln \alpha_{PCP/Cl} \mathrm{d}y_{PCP}$$

$$= -RT \int_0^1 \ln \frac{y_{PCP} \cdot (1 - x_{PCP})}{(1 - y_{PCP}) \cdot x_{PCP}} \mathrm{d}y_{PCP} \tag{3.90}$$

The above integral can now be computed from the binary sorption isotherm data.

If the equilibrium constant values are determined at different temperatures around 298 K where standard enthalpy change ($\Delta H^0_{\text{overall}}$) may be assumed to be considered constant, the van't Hoff equation gives

$$\frac{\mathrm{d}(\log K)}{\mathrm{d}(1/T)} = -\frac{\Delta H^0_{\text{overall}}}{2.3R} \tag{3.91}$$

where T is the absolute temperature in K. The standard enthalpy change can be computed from the slope of $\log K$ versus $1/T$ plot. Similar approaches have been successfully used earlier to determine ΔH^0 values during sorption processes at ambient temperature [49,59]. Enthalpic changes thus determined agreed well with the values obtained independently using microcalorimetric technique [59]. The standard entropic contribution at 298 K [$T\Delta S^0_{\text{overall}}$] can subsequently be determined from the following relationship:

$$T\Delta S^0_{\text{overall}} = \Delta H^0_{\text{overall}} - \Delta G^0_{\text{overall}} \tag{3.92}$$

High selectivity of counterions with NPM results from hydrophobic interactions, which are again manifested in enthalpic and entropic changes [60,61]. Altogether, there are three independent process variables: hydrophobicity of the solute, polarity of the ion exchanger matrix and the dielectric constant of the solvent influencing the selectivity of a specific aromatic anion.

3.11.3 Experimental Observations: Stoichiometry, Affinity Sequence, and Cosolvent Effect

Figure 3.22 shows the plot of a stepwise PCP^- uptake onto IRA-900 versus the corresponding stepwise release of chloride ions into the aqueous phase in milliequivalent

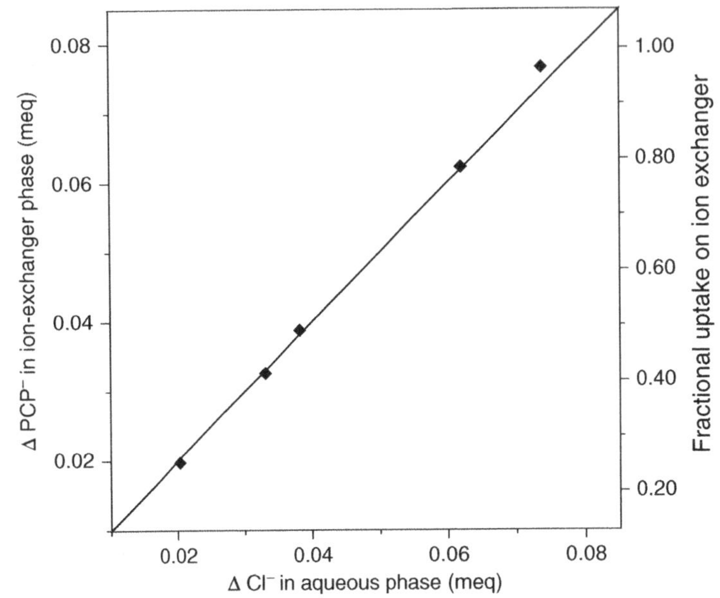

Figure 3.22 Milliequivalents of PCP^- uptake onto the anion exchanger versus the corresponding release of Cl^- into the aqueous phase. *Source*: Li and SenGupta 2001 [57]. Reproduced with permission of Elsevier.

or meq units. Note that for a wide range of ion exchange site coverage, the plot is essentially a perfect straight line passing through the origin with a slope equal to unity. Thus, an uptake of PCP^- by the exchanger is always accompanied by desorption of an equivalent amount of chloride ions. Similar stoichiometry of equivalent ion exchange between naphthalenesulfonate (NS) and Cl^- was also observed [10,57].

To understand pH effects on the sorption of aromatic anions, batch equilibrium test results were obtained at different pH values for sorption of PCP onto an ion exchanger (IRA-900) and a synthetic adsorbent (XAD-2). Both IRA-900 and XAD-2 have an identical macroporous polystyrene matrix and divinylbenzene cross-linking, but XAD-2 does not have anion exchange functional groups. Figure 3.23a presents the results of the batch equilibrium tests for the sorption onto IRA-900 and XAD-2, while Figure 3.23b presents the theoretical speciation of the neutral species (PCP^0) and the anionic species (PCP^-) as a function of pH. PCP is a weak acid with a pK_a value (negative logarithm of acid dissociation constants) of 4.75. As shown in Figure 3.23b, PCP^- is the predominant species when pH in aqueous solutions is higher than the pK_a value, while PCP^0 is the predominant species when pH in aqueous solutions is lower than the pK_a value. It is noted that the ion exchanger attains high sorption capacity when the anionic species PCP^- is predominant in the aqueous phase, but it significantly loses sorption capacity when the neutral species PCP^0 is predominant. Conversely, the synthetic adsorbent attains high sorption capacity when the neutral species PCP^0 is predominant in the liquid phase. The experimental results imply distinctly different mechanisms of the PCP sorption onto these two types of sorbents, namely, ion exchanger IRA-900 and synthetic adsorbent XAD-2. The mechanism of PCP sorption onto IRA-900 is ion exchange while the mechanism of PCP sorption onto XAD-2 is

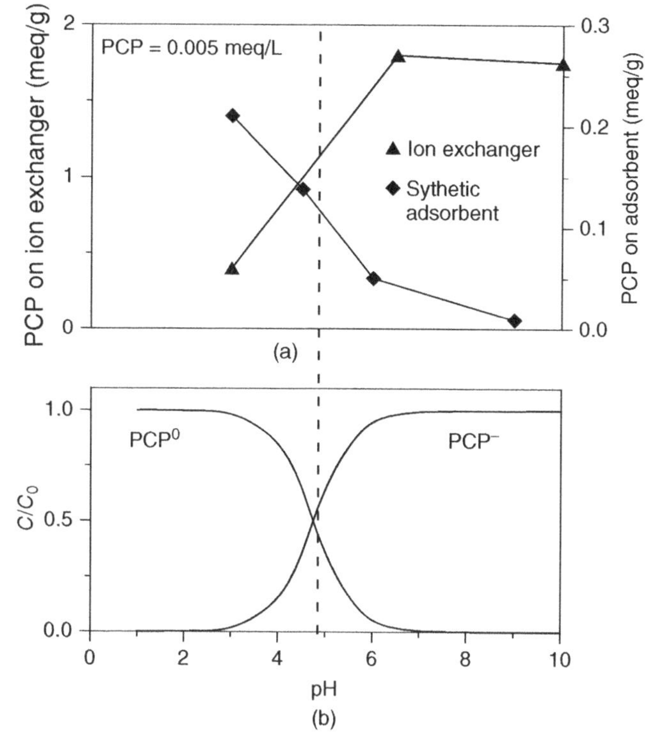

Figure 3.23 (a) Sorption of pentachlorophenol onto ion exchanger and synthetic adsorbent under different pH. (b) Theoretical speciation of the neutral species (PCP^0) and the anionic species (PCP^-) as a function of pH. *Source*: Li and SenGupta 2001 [57]. Reproduced with permission of Elsevier.

physical adsorption. Also, ion exchange mechanism predominates when pH values in liquid phases are higher than pK_a values of parent aromatic acids.

Figure 3.24 shows the complete effluent history of a fixed-bed column run using IRA-900 (polystyrene matrix, quaternary ammonium functional group) for an influent containing trace concentration of dissolved PCP^- (2.7 mg/L or 0.01 mmol/L) along with much higher concentrations of competing bicarbonate, chloride, and sulfate ions. Note that while the inorganic anions including divalent sulfate broke through early, monovalent PCP^- was completely removed well over 10,000 BVs and the column run lasted for several months. The higher preference of monovalent PCP^- over divalent sulfate demonstrates that the electrostatic or coulombic interaction is not the primary determinant of relative selectivity in such an ion exchange process.

Figure 3.25 presents the separation factor ($\alpha_{PCP/Cl}$) in a logarithm scale versus the volume fraction (f_c) of various organic solvents with low dielectric constants. The organic solvents include methanol, acetone and dioxane. For each of the organic solvents, log($\alpha_{PCP/Cl}$) decreases with increasing f_c, and the linear relationship between log($\alpha_{PCP/Cl}$) and f_c can be noted. In Chapter 2, it was demonstrated how the use of a cosolvent can enhance the efficiency of regeneration of PCP^- for anion exchange processes.

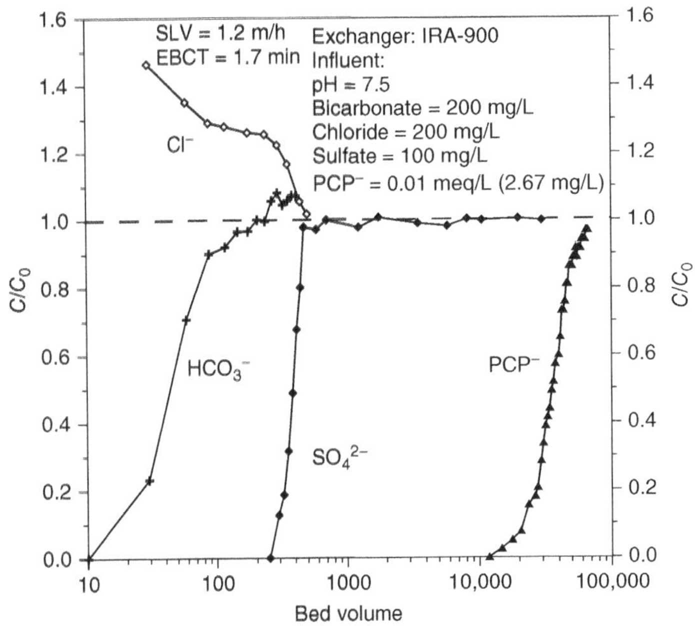

Figure 3.24 A complete effluent history of PCP^- and other competing inorganic anions during a fixed-bed column run with IRA-900 in chloride form. *Source*: Li and SenGupta 1998 [10]. Reproduced with permission of American Chemical Society.

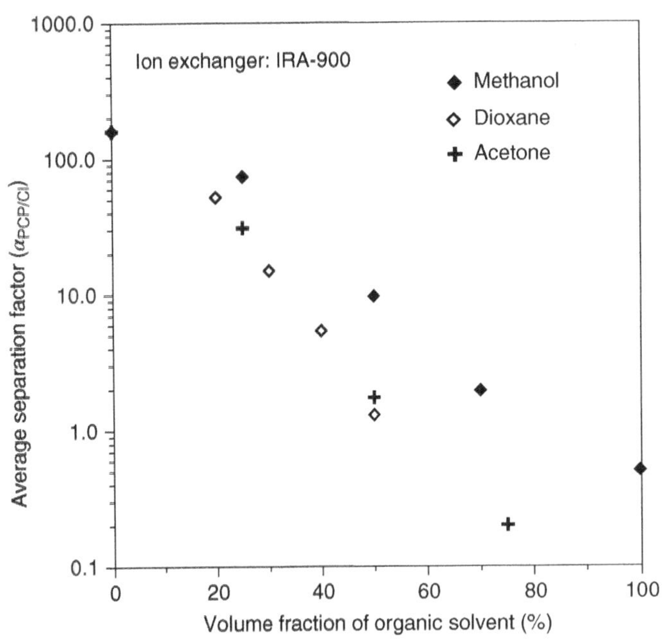

Figure 3.25 The average separation factor ($\alpha_{PCP/Cl}$) versus the volume fraction of organic solvents in water. *Source*: Li 1999 [6]. Reproduced with permission of American Chemical Society.

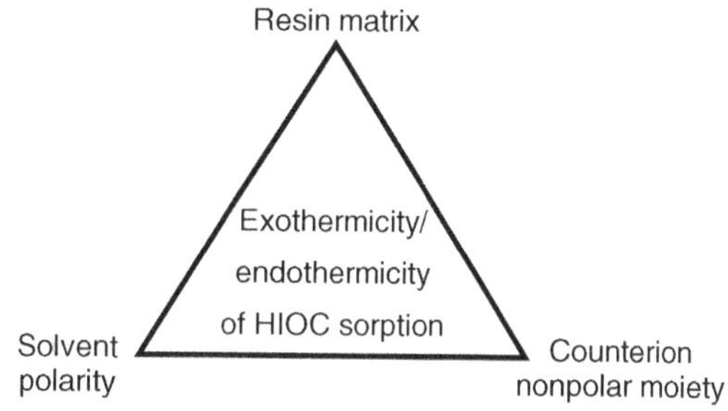

Figure 3.26 Three major factors that govern the thermodynamics of HIOC sorption onto an anion exchanger.

3.11.4 Energetics of the Sorption Process

Exothermicity or endothermicity of HIOC sorption onto an anion exchanger is governed by three variables, namely, resin matrix, solvent polarity, or dielectric constant and the NPM of the solute or HIOC as illustrated in Figure 3.26.

Figure 3.27a–c provides binary PCP^-/Cl^- isotherms (the equivalent fraction of PCP^- in the ion-exchanger phase, y_{PCP}, versus the equivalent fraction of PCP^- in the liquid phase, x_{PCP}) at three different temperatures for (i) IRA-900 and water, (ii) IRA-958 and water, and (iii) IRA-900 and methanol–water (50%/50% by volume) solvent. Note that when pure water is the solvent, PCP^- uptake increases with an increase in temperature strongly for IRA-900 and moderately for IRA-958. However, as water is replaced by methanol–water solvent for IRA-900, the effect of temperature is reversed.

Equilibrium constant (K) and free energy change (ΔG^0) values were subsequently determined for each individual isotherm using Eqs (3.89) and (3.90) for the three systems in Figure 3.27a–c. With the van't Hoff equation (Eq. (3.91)), values of enthalpy change (ΔH^0) were determined from a plot of $\ln K$ versus $1/T$. From the slopes of the straight lines ($-\Delta H^0/R$), values of ΔH^0 were computed for abovementioned binary ion exchanges. Figure 3.28 shows van't Hoff plots ($\ln K$ vs $1/T$) for the three abovementioned isotherms. Values of entropy change (ΔS^0) were then estimated by using Eq. (3.92). Estimated ΔG^0, ΔH^0, and $T\Delta S^0$ values are given in Figure 3.28.

The following observations are noteworthy:

1. For all the three systems, PCP^- sorption onto the exchanger is preferred to Cl^-, that is, $\ln K$ is greater than zero, and ΔG^0 values are negative. In pure water system, IRA-900 gives rise to high positive ΔH^0 value while the same for IRA-958 is marginally greater than zero. Favorable ion exchange type sorption behaviors with positive enthalpy changes (endothermic) are very unusual but have previously been reported for long-chain alkanesulfonates and quaternary ammonium compounds [49,62–64]. Note that the entropy contribution ($T\Delta S^0$) for IRA-900 with non-polar polystyrene matrix is significantly larger than that for IRA-958 with a relatively polar matrix, all other conditions remaining identical. High endothermicity of the exchange reaction with IRA-900 makes PCP^- sorption gradually less favorable as the temperature decreases. The van't Hoff plots of IRA-900 and IRA-958 intersect at 11 °C, as may be seen in Figure 3.28, that is, PCP^- sorption onto IRA-958 is thermodynamically more favorable than with IRA-900 at temperature lower than 11 °C.

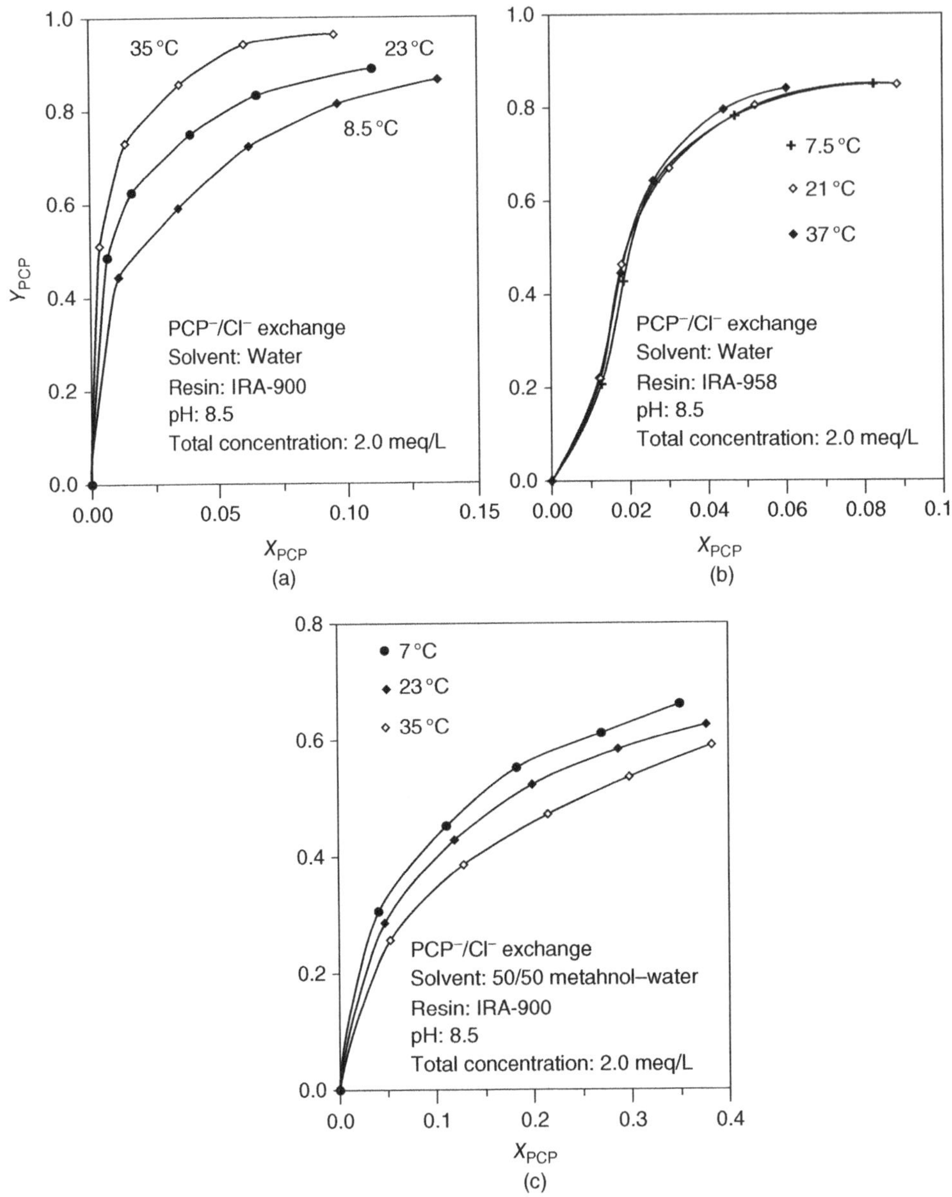

Figure 3.27 PCP^-/Cl^- isotherms at three different temperatures for (a) IRA-900 and water; (b) IRA-958 and water; and (c) IRA-900 and methanol/water systems. *Source*: Li and SenGupta 1998 [10]. Reproduced with permission of American Chemical Society.

2. Contrary to the pure water system, the van't Hoff plot for IRA-900 in the presence of a cosolvent (50% methanol + 50% water) has a positive slope, that is, PCP^-–Cl^- exchange is exothermic and accompanied by a negative enthalpy change. In pure water systems, negative ΔG^0 values result from positive entropic contributions, that is, favorable PCP^- sorption is an entropy-driven process. In contrast, the negative free energy change for methanol–water solvent is an enthalpy-driven process.

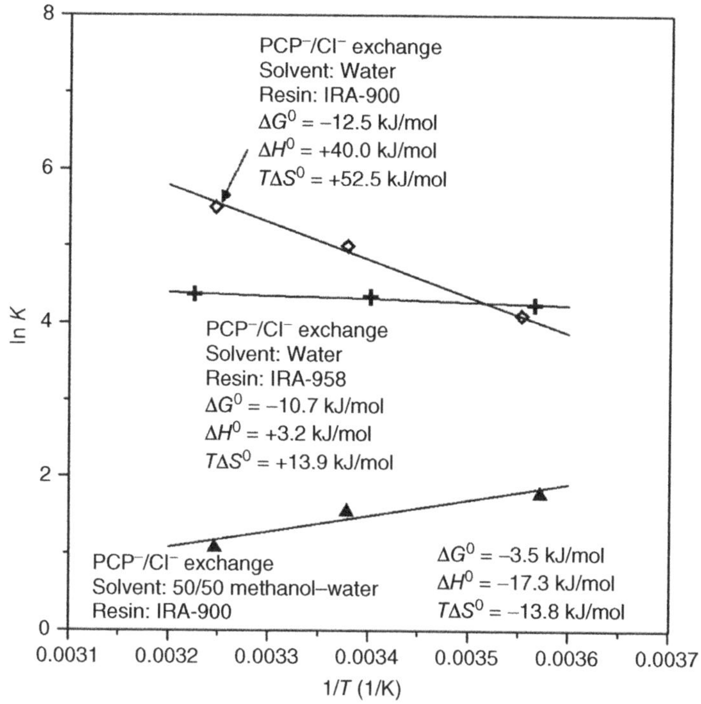

Figure 3.28 van't Hoff plots (ln *K* vs 1/*T*) for three different types of isotherms reported in Figure 3.27. *Source*: Li and SenGupta 1998 [10]. Reproduced with permission of American Chemical Society.

In addition, van't Hoff plots ($\ln K$ vs $1/T$) were also plotted in Figure 3.29 for three other binary systems using HIOC solutes, namely, benzenesulfonate/chloride (BS^-/Cl^-), naphthalene-1-sulfonate/chloride (NS^-/Cl^-), and naphthalene-1,5-disulfonate/chloride (NDS^{2-}/Cl^-) at three temperatures. While IRA-900 with polystyrene matrix was the anion exchange resin used, water was the only solvent. Estimated ΔG^0, ΔH^0, and $T\Delta S^0$ values are given in Figure 3.29. Note that all three sulfonates are selectively sorbed by anion exchange resins over chloride and the favorable equilibrium is driven solely by positive entropy changes.

3.11.5 Unifying Hydrophobic Interaction: From Gas–Liquid to Liquid–Solid System

To interpret the significance of various experimentally determined ΔH^0 and $T\Delta S^0$ values during the sorption of aromatic anions, we will consider the classical work pertaining to the dissolution of gaseous non-polar methane between cyclohexane (a non-polar, non-associated solvent) and water (a polar, self-associated solvent) [65]. ΔH^0 and $T\Delta S^0$ values are provided for the transfer of methane from water to cyclohexane in Figure 3.30a where subscripts "S" and "W" represent solvent cyclohexane and water, respectively. Note that the enthalpic and entropic changes for transfer of methane from polar water to non-polar cyclohexane are positive. Since the overall free energy change ($\Delta G^0_{W\to S}$) is negative, the methane transfer is a favorable, endothermic,

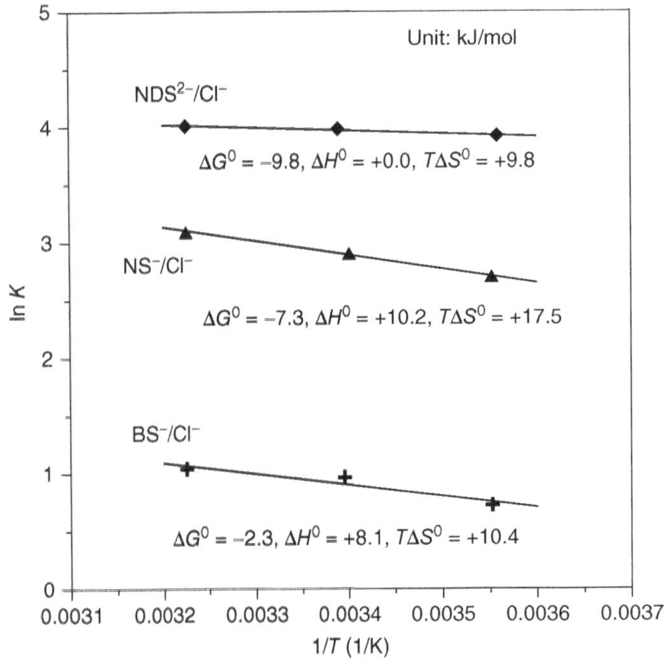

Figure 3.29 van't Hoff plot (ln K vs $1/T$) for sulfonated aromatic anions. *Source*: Li 1999 [6]. Reproduced with permission of American Chemical Society.

and an entropy-driven process. In order to draw an analogy and assess the relative magnitude of different interactions, ΔH^0 and $T\Delta S^0$ values of ion exchange processes involving aromatic anions and chloride under different experimental conditions are included in the same figure. The following provides a generic analysis attempting to unify experimentally determined enthalpic and entropic changes under varying conditions:

- In Figure 3.30b, IRA-900 has a highly hydrophobic polystyrene matrix and is analogous to non-polar cyclohexane in Figure 3.30a. Similar to methane transfer in Figure 3.30a, PCP^- sorption onto IRA-900 is thermodynamically favorable (i.e., negative ΔG^0) and endothermic, and involves positive entropy changes.
- In Figure 3.30c, IRA-958 has a more polar matrix, that is, it is equivalent to replacing cyclohexane in Figure 3.30a with a more polar solvent. As a result, although favorable, PCP^- sorption is much less endothermic (ΔH^0 is nearly zero) and positive entropic contribution is relatively low.
- In Figure 3.30d, water is replaced by 50/50 methanol–water solvent with a significantly lower dielectric constant ($\varepsilon = 55$). This is analogous to using a high non-polar solvent in the place of water in the methane transfer process. Understandably, PCP^- desolvation in such a solvent, as illustrated in Figure 3.28, no longer involves a significant structure breaking of solvent molecules. Positive entropy change associated with the desolvation step, therefore, diminishes sharply. Also, less heat needs to be absorbed because the structure breaking of solvent molecules is unwarranted. All in all, the overall equilibrium becomes much less favorable for PCP^- sorption (i.e., lower negative ΔG^0 value) and the process is exothermic (i.e., negative ΔH^0).

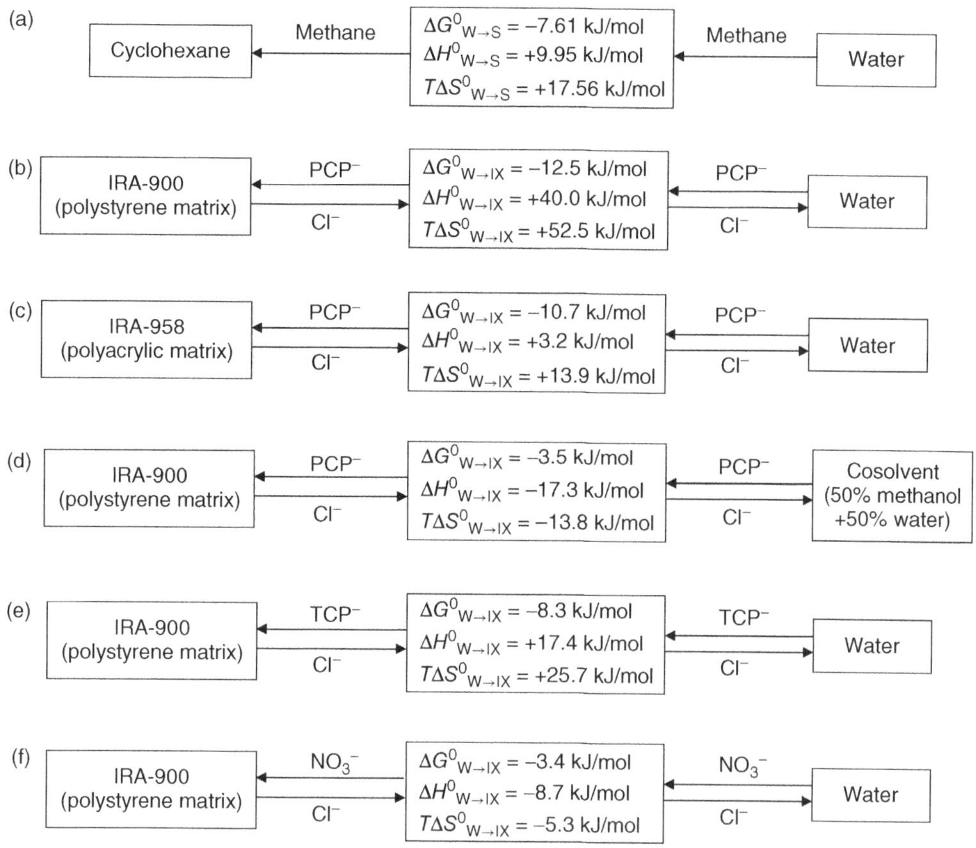

Figure 3.30 (a) Energetics of methane transfer between water and cyclohexane as a control in comparison to transfer of HIOCs between water–ion exchange resin. Enthalpic and entropic changes during PCP^- or TCP^- sorption under varying conditions (b–e) and their relationships to methane transfer between cyclohexane and water; (f) represents nitrate–chloride exchange in water with an anion exchanger. *Source*: Li and SenGupta 1998 [10]. Reproduced with permission of American Chemical Society.

- Figure 3.30e shows the results of a sorption process very similar to Figure 3.30b excepting that the solute PCP^- has been replaced with trichlorophenol or TCP^-. The NPM of TCP^- is less hydrophobic than that of PCP^-, as reflected in their K_{OW} values. As a result, TCP^- sorption is favorable and the sign of ΔH^0 and $T\Delta S^0$ remain unchanged, that is, both are positive, but their absolute values are lower compared to those obtained with PCP^-.
- To distinguish the difference between strictly inorganic ion exchange and the exchange involving aromatic anions, results of nitrate–chloride (NO_3^-/Cl^-) exchange are included in Figure 3.30f. Other conditions, namely, the ion exchanger and solvent, are essentially the same as those shown in Figure 3.30b. Nitrate sorption is favorable, that is, free energy change is negative. But contrary to sorption of PCP^- or TCP^-, the favorable equilibrium in this case is driven by negative enthalpy changes, that is, nitrate–chloride exchange is essentially an exothermic process accompanied by an overall decrease in entropy. The foregoing observation is true

for typical inorganic ion exchange processes, both cationic and anionic, where the energy of solvation is the primary determinant of the relative selectivity of exchanging counterions [7,66]. In contrast, cluster-like formation of structured water molecules around HIOC in the aqueous phase and their breakdown following sorption are absent [67–70].

3.11.6 Effect of Polymer Matrix and Solute Hydrophobicity

Figure 3.31 shows the average PCP^-/Cl^- separation factor values ($\alpha_{PCP/Cl}$) for the two polymeric anion exchangers used in the study, namely, IRA-900 and IRA-958. The high $\alpha_{PCP/Cl}$ values (well over unity) for both anion exchangers clearly demonstrate their high preference toward PCP^- to Cl^-. However, for IRA-958 with a more polar polyacrylic matrix, $\alpha_{PCP/Cl}$ is significantly lower compared to IRA-900 with polystyrene matrix. This observation corroborates that the solute–matrix, that is, NPM–matrix interaction, as illustrated in Figure 3.21, contributes toward the relative selectivity of PCP^-. For both the exchangers, the fixed positive functional groups reside in the gel phase and that is where PCP^- sorption is predominant. Between polyacrylic and polystyrene matrices, the former is more polar (i.e., less hydrophobic) due to its open-chain aliphatic structure containing carbonyl groups. Polyacrylic resins, therefore, tend to imbibe more water molecules within the exchanger phase; expelling water molecules imbibed into the more polar matrix of IRA-958 is thus energetically more difficult. So, the affinity of PCP^- toward IRA-958 is significantly lower compared to IRA-900, which has a relatively non-polar polystyrene matrix.

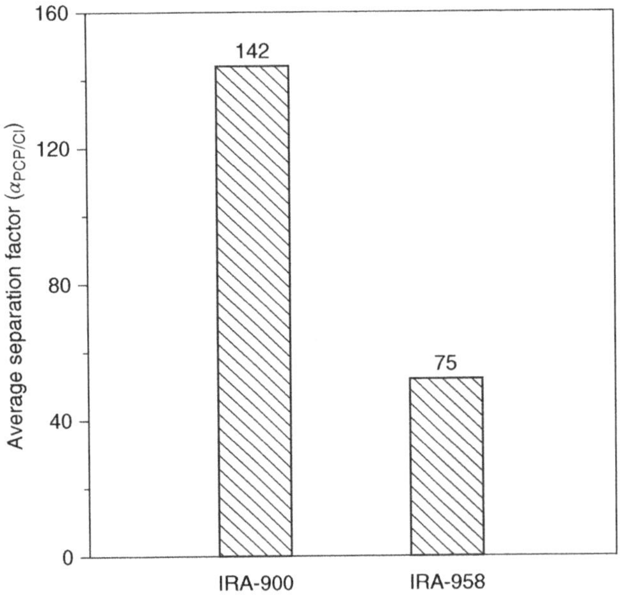

Figure 3.31 Average PCP^-/Cl^- separation factor values for two anion exchangers, IRA-900 and IRA-958. *Source*: Li 1999 [6]. Reproduced with permission of American Chemical Society.

For a given solvent (i.e., water) and a given polymeric anion exchanger (i.e., IRA-900), the solute affinity should be strongly correlated to the hydrophobicity of its NPM. The octanol–water partition coefficient (K_{OW}) of an undissociated chlorophenol may be a representative measure of the NPM's hydrophobicity. The derivative of the free energy change with respect to $\ln K_{OW}$ for a monovalent hydrophobic anion during exchange with chloride should then be a constant, that is,

$$\frac{d\Delta G^0}{d \ln K_{OW}} = \text{constant} \tag{3.93}$$

$$-RT\frac{d \ln K}{d \ln K_{OW}} = \text{constant} \tag{3.94}$$

or,

$$\frac{d \ln K}{d \ln K_{OW}} = \text{constant} \tag{3.95}$$

Therefore, theoretically, a linear relationship exists between $\ln K$ and $\ln K_{OW}$. Figure 3.32 shows a plot of experimentally determined K values of three different chlorophenols for IRA-900 and their corresponding K_{OW} values. The plot, in general, recognizes the strong agreement between Eq. (3.95) and the experimental data. It is noteworthy that as the $\ln K_{OW}$ value drops to near 2.0, the hydrophobic interaction is no longer predominant over electrostatic interaction and $\ln K$ tends to be zero or K is close to unity. Under such conditions, the basic premises of NPM–solvent and NPM–matrix interactions as presented earlier are no longer valid. When $\log K_{OW}$ values are, say, below 2.0 (or $\ln K_{OW}$ below 4.6), ΔH^0 for favorable sorption will change from positive to negative and the exchange will be exothermic.

3.12 Linear Free Energy Relationship and Relative Selectivity

According to linear free energy relationships (LFERs), standard state free energy changes (ΔG^0) for identical reactions in two different phases (e.g., water and ion exchange resin) are linearly dependent on each other [71]. Thus, the existing knowledge of the equilibrium constant data available for the aqueous phase for different reactions can readily be extrapolated to predict or validate the equilibrium relationship in the ion exchanger phase. Consider the following metal–ligand reaction in the aqueous phase:

$$M^{2+}(aq) + L^{2-}(aq) \leftrightarrow ML^0(aq) \tag{3.96}$$

A similar reaction in the ion exchanger phase where the ligand has been covalently attached, is:

$$M^{2+}(aq) + \overline{L^{2-}}(IX) \leftrightarrow ML^0(IX) \tag{3.97}$$

From LFER,

$$\frac{\Delta G^0_{IX}}{\Delta G^0_{aq}} = a(\text{constant}) \tag{3.98}$$

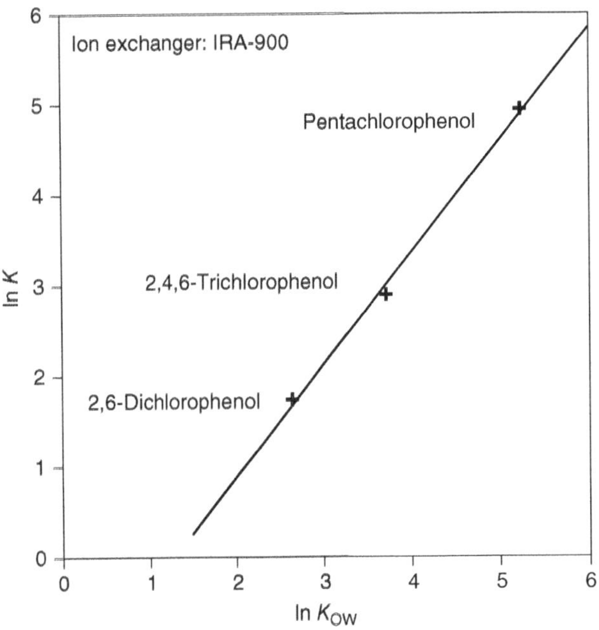

Figure 3.32 A plot of experimentally determined ln *K* values versus log K_{OW} for three chlorophenols. *Source*: Li and SenGupta 1998 [10]. Reproduced with permission of American Chemical Society.

Thus, $-RT\ln K_{IX} = -aRT\ln K_{aq}$ and

$$\log K_{IX} = a\log K_{aq} \tag{3.99}$$

The following experimental data underscore how the aqueous phase equilibrium relationship can be extended to predict exchanger phase selectivity.

Figure 3.33 shows the relationship between copper/calcium separation factor values for three commercial chelating exchangers and the corresponding aqueous phase stability constant values for representative ligands [4]. Noteworthy is the fact that as the composition of the functional groups in Figure 3.20 changes from hard oxygen donor atoms (e.g., carboxylate) to relatively soft nitrogen donor atoms (bispicolylamine), the affinity of Cu(II), a borderline Lewis acid is greatly enhanced over the hard cation, Ca^{2+}. Understandably, the composition of the functional groups in chelating exchangers can be judiciously tailored to improve specific affinities toward target metal ions.

Along the same vein, Figure 3.34 shows the separation factor values of five different heavy metal cations for a weak-acid cation exchange resin with carboxylate functional groups (IRC DP-1, Rohm and Haas Co., Philadelphia). Note that the sequence and relative affinity of dissolved heavy metals are strongly correlated to their aqueous phase metal-acetate stability constant values [72].

In the case of polymeric ligand exchange (PLE), the ligands or the Lewis bases are the exchangeable anions while metals (e.g., Cu^{2+}) are the Lewis acids anchored onto the polymer phase. Figure 3.35 [72] plots the experimentally determined binary separation factor values for divalent anionic ligands, namely, succinate, maleate, oxalic,

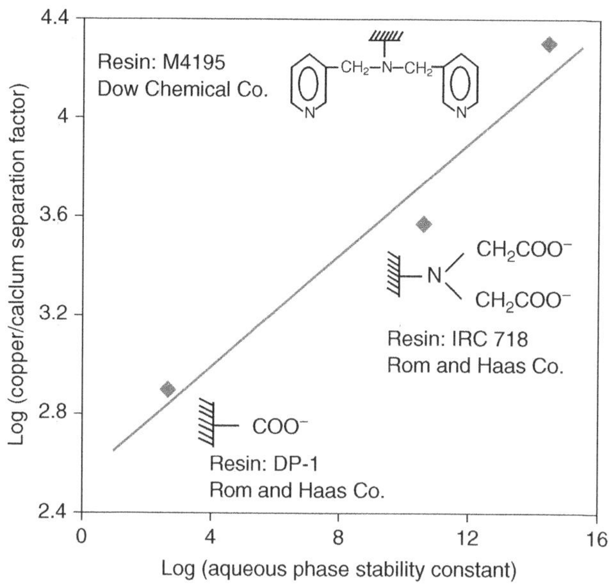

Figure 3.33 Plot of copper/calcium separation factors for three commercial chelating exchangers versus aqueous phase stability constant of representative ligands of chelating species. *Source*: SenGupta 2001 [4]. Reproduced with permission of Taylor & Francis.

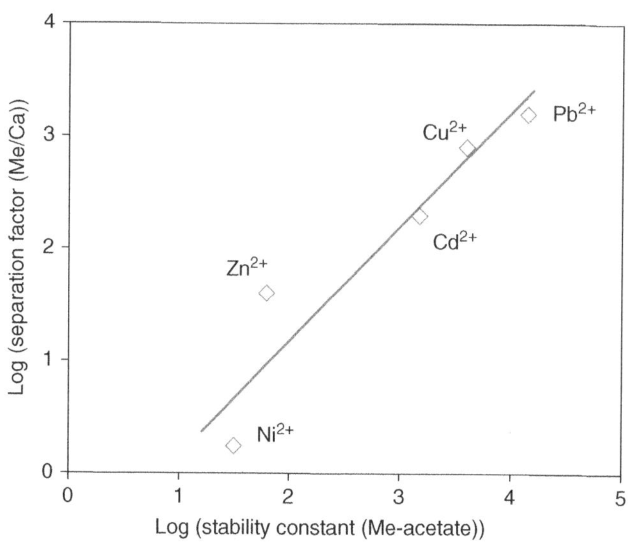

Figure 3.34 Plot of metal/calcium separation factors for a weak-acid cation exchanger (carboxylate functionality) versus aqueous phase metal acetate stability constants. *Source*: SenGupta 2001 [4]. Reproduced with permission of Taylor & Francis.

and phosphate with respect to sulfate versus the first copper–ligand stability constant values. Again, a linear relationship is observed.

As stated in the previous sections, aromatic anions often exhibit high selectivity towards polymeric anion exchangers due to concurrent hydrophobic interaction between the aromatic portion of the ion and the nonpolar matrix. Thus, greater the

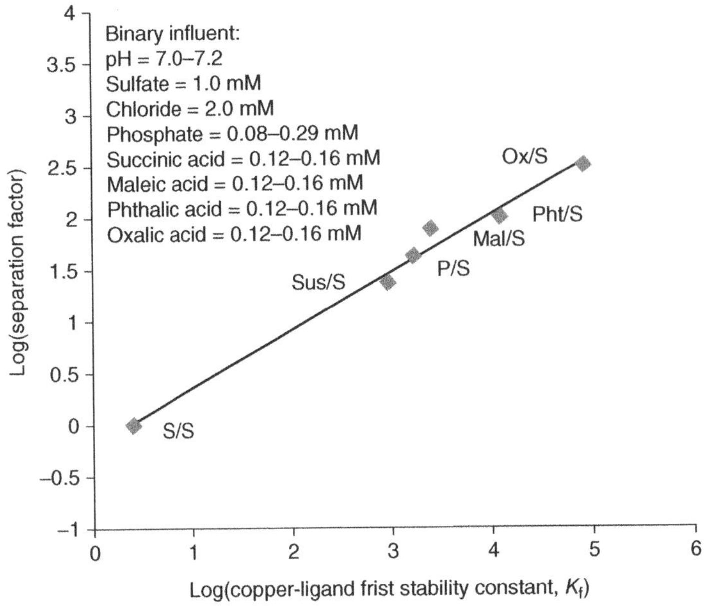

Figure 3.35 Plot showing relationship of binary separation factors of divalent anionic ligand versus corresponding copper ligand stability constant values. *Source*: Zhao and SenGupta 2000 [72]. Reproduced with permission of American Chemical Society.

hydrophobicity of the counterion, greater will be its selectivity. As the hydrogen atom from phenol is gradually replaced by more electron-withdrawing chlorine atoms, the hydrophobicity of the resulting chlorophenols is enhanced. Figure 3.31 in the previous section demonstrates the linearity between the separation factor values of different chlorophenols and their K_{OW} values.

3.13 Simultaneous Removal of Target Metal Cations and Anions

In principle, an ion exchanger cannot simultaneously remove both cations and anions due to Donnan coion exclusion effect and nearly all ion exchangers conform to that. Some specially functionalized polymers may carry electrically neutral chelating groups with nitrogen, oxygen or sulfur donor atoms which can form coordinate bonds with transition metal cations, namely, Cu^{2+}. Once anchored into the polymer, the transition metal cations can serve as sites for anion exchange. Obviously, the anions with high ligand strength such as arsenate, phosphate and chromate will be preferred over others. Figure 3.36 shows the various constituents of the conceptualized polymeric sorbent for sequential removals of transition metal cations and anionic ligands.

Commercially available chelating polymers with bispicolylamine functional groups containing nitrogen donor atoms (M-4195, Dow Chemical Co., Michigan) satisfy the requisite criteria. Figure 3.37 shows the effluent history of a column run for an influent containing both Cu(II) and Cr(VI) along with other competing cations and anions [72]. Note that over 95% of Cu(II) and Cr(VI) are simultaneously removed for well

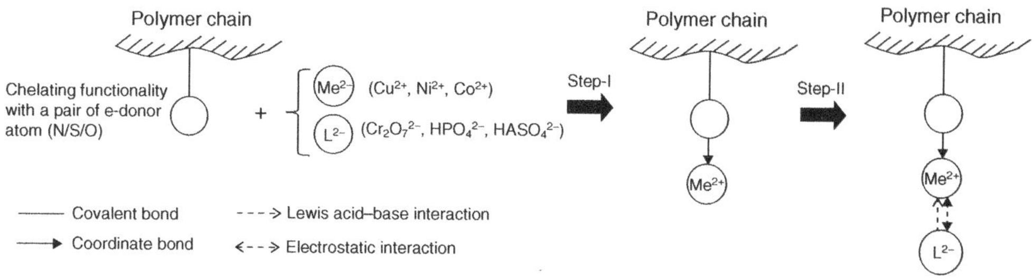

Figure 3.36 Various constituents of the conceptualized polymeric sorbents for sequential removal of transitional metal cations and anionic ligands. *Source*: Reprinted with permission from Zhao *et al.* 1998 [71].

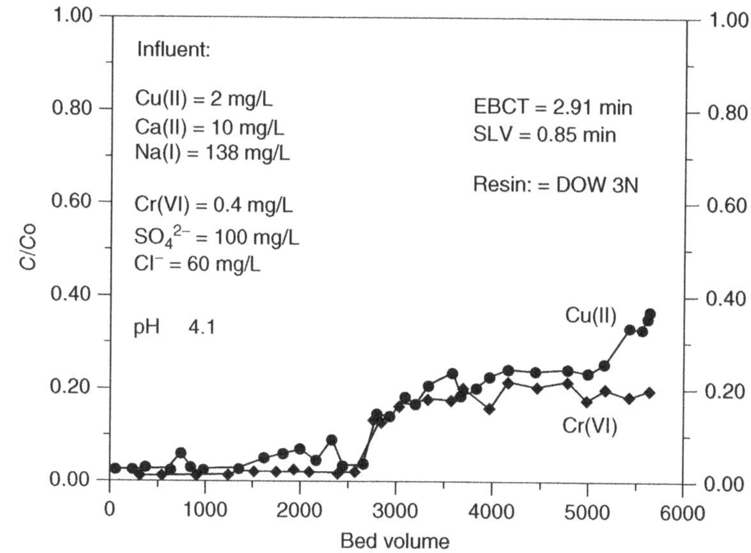

Figure 3.37 Effluent history of Cu(II) and Cr(VI) with a background of other competing cations and anions from a column run using conceptualized polymeric sorbent. *Source*: Zhao *et al.* 1998 [71]. Reproduced with permission of American Chemical Society.

over 2500 BVs. Other cations and anions, namely, calcium, sodium, chloride, and sulfate break through quite early. Mechanistically, such a phenomenon can be explained in accordance with Figure 3.36 as follows: first, Cu(II) is selectively sorbed onto the chelating polymer in preference to other competing cations; and second, once sorbed onto chelating polymers, Cu(II), acts as selective anion exchange sites for chromate in preference to chloride and sulfate. Note that other anionic ligands, namely, arsenate, phosphate, oxalate may also be removed selectively in a manner similar to chromate [72].

Other commercially available chelating exchangers with iminodiacetate, thiol, aminophosphonate and carboxylate functional groups are unable to provide dual removal of transition metal cations and chromate. The functional groups of these metal-selective chelating exchangers are negatively charged, and hence, they reject chromate anions through Donnan coion exclusion effect.

3.14 Deviation from Henry's Law

Trace ion sorption, as validated earlier conforms to Henry's law, that is, aqueous-phase and exchanger-phase concentrations are linearly dependent on each other.

$$q_A = \lambda C_A \tag{3.100}$$

Considering A to be a preferred species over other competing ions, with increased loading the sorption isotherm gradually changes from linear to convex upward (favorable) in accordance with Langmuir behavior. At low q_A values, species A in the exchanger are distant and independent from each other and, hence, there is no lateral interaction. Figure 3.38 illustrates gradual progress of the sorption process for a favorably sorbed species. The foregoing phenomenon is widely observed in selective ion exchange for nearly every trace ion. This section presents examples that deviate conspicuously from the Henry's law behavior.

3.14.1 Ions Forming Polynuclear Species

At high concentrations and in a certain pH range, chromates, molybdates, tungstates, bismuth and a few other metal ions can form polynuclear species, that is, multiple metal atoms can be present in a single ionic species, as shown below [73]:

$$2HCrO_4^- \leftrightarrow Cr_2O_7^{2-} + H_2O \tag{3.101}$$

$$7MoO_4^{2-} + 8H^+ \leftrightarrow Mo_7O_{24}^{6-} + 4H_2O \tag{3.102}$$

$$6WO_4^{2-} + 7H^+ \leftrightarrow HW_6O_{21}^{5-} + 3H_2O \tag{3.103}$$

$$6Bi^{3+} + 12H_2O \leftrightarrow Bi_6(OH)_{12}^{6+} + 12H^+ \tag{3.104}$$

The foregoing reactions have two things in common: (i) at high concentration of metal ions, the formations of polynuclear species are preferred and (ii) polynuclear species have higher charges than their mononuclear counterparts. An ion exchanger can be viewed as a polyelectrolyte gel of ionized cations and anions having concentrations as high as 1.2 eq/L or 1.2 N. Thus, during ion exchange, polynuclear species of these metals are preferred by the exchanger phase, while mononuclear species are their predominant representative in the aqueous phase. We will consider chromate ion exchange in this regard and the following are the important equilibrium reactions [71,74]:

Reaction	log *k* (25 °C)	
$H_2CrO_4^- \leftrightarrow H^+ + HCrO_4^-$	−0.8	(3.105)
$HCrO_4^- \leftrightarrow H^+ + CrO_4^{2-}$	−6.5	(3.106)
$2HCrO_4^- \leftrightarrow Cr_2O_7^{2-} + H_2O$	1.52	(3.107)

The distribution of chromate species is dependent on both pH and total chromate or Cr(VI) concentration and Figure 3.39 is a predominance diagram for various Cr(VI)

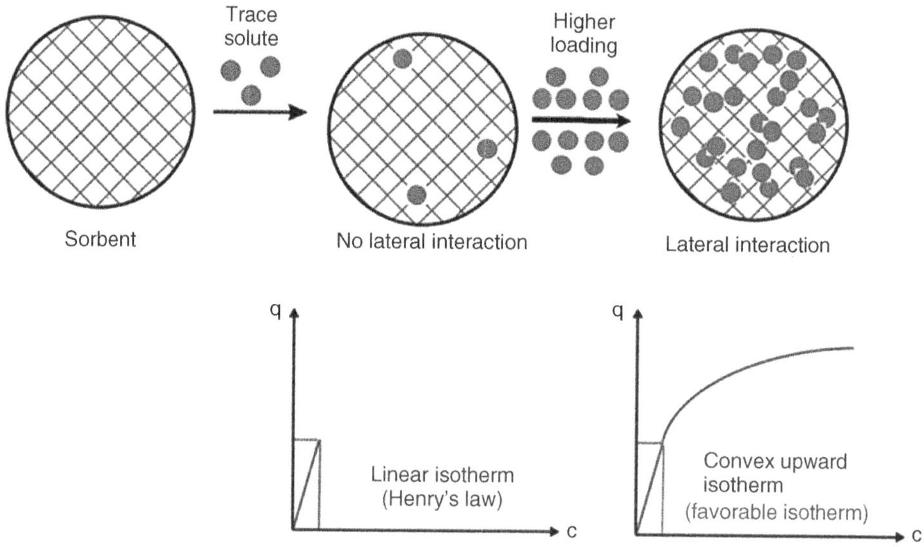

Figure 3.38 Illustration of gradual progress of sorption process for a favorably sorbed species.

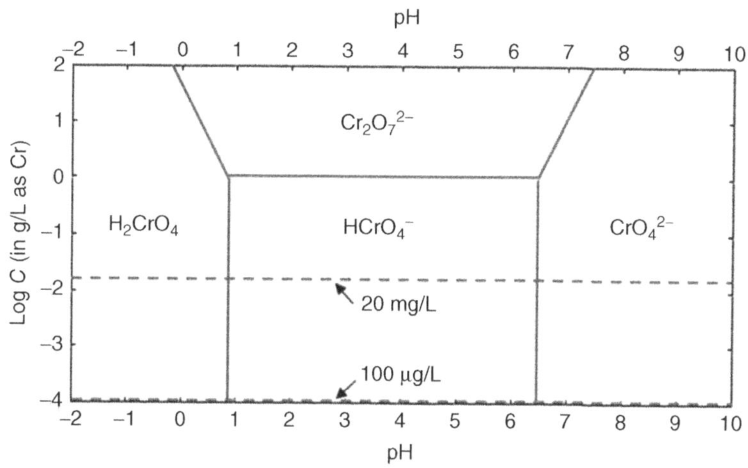

Figure 3.39 Predominance diagram for various Cr(VI) species. *Source*: SenGupta and Clifford 1986 [67]. Reproduced with permission of Elsevier.

species [15,71]. The area between the horizontal dashed line and the abscissa indicates the range of chromate concentration between 100 μg/L and 20 mg/L as Cr. Note that $HCrO_4^-$ and CrO_4^{2-} are, by far, the most predominant species at this total Cr(VI) concentration. Also, dimerization of $HCrO_4^-$ into $Cr_2O_7^{2-}$ is possible only at acidic pH and anion exchanger prefers divalent $Cr_2O_7^{2-}$ to monovalent $HCrO_4^-$.

Figure 3.40 illustrates the qualitative partitioning of different species between the two phases and highlights the fact that, in spite of being almost absent in the aqueous phase, $Cr_2O_7^{2-}$ is significantly present inside the anion exchanger.

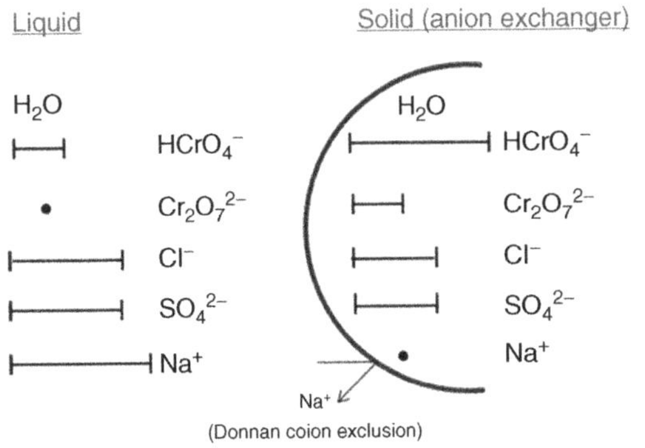

Figure 3.40 Qualitative partitioning of different Cr(VI) species along with other background cations and anions in anion exchanger (fixed positive charge) and in liquid phase, horizontal bars representing relative concentrations.

When chromate or Cr(VI) is a trace solute in the presence of high concentration of chloride or sulfate in the aqueous phase, the exchanger and the aqueous phase chromate concentrations show the following relationship:

$$q_{\mathrm{cr}} = A_1 C_{\mathrm{cr}} + A_2 C_{\mathrm{cr}}^2 \tag{3.108}$$

or,

$$y_{\mathrm{cr}} = A_1 C_{\mathrm{cr}} + A_2 C_{\mathrm{cr}}^2 \tag{3.109}$$

where q_{cr} and y_{cr} represent exchanger phase concentrations in milliequivalent per gram and equivalent fractions, respectively. A_1 and A_2 are constants under trace conditions for a given anion exchanger and solution composition. Equations (3.108) and (3.109) are conspicuous because they are parabolic and not linear, as predicted by Henry's law under trace conditions. The derivation of Eq. (3.108) is available in open literature [15], but being avoided here for its peripheral role in substantiating the primary theme of this section, that is, departure from Henry's law under trace conditions. Figures 3.41 and 3.42 show experimental results of chromate isotherms at trace conditions for several anion exchange resins in the presence of high concentrations of competing sulfate or chloride anions [15,67]. It is quite apparent that the chromate isotherm is always parabolic, that is, concave upward even at the trace Cr(VI) concentration in accordance with the predictions of Eqs (3.108) and (3.109). For isotherms in Figures 3.41 and 3.42 chromate is well preferred over competing sulfate and chloride, that is, chromate/chloride and chromate/sulfate separation factor values are much greater than unity. Thus, these isotherms are unfavorable and, hence, they would exhibit gradual chromate breakthrough during fixed-bed column runs [15,67,75]. Other studies have shown that, like chromate, tungstate also exhibits an unfavorable isotherm at acidic pH values.

At alkaline pH, however, chromate exists solely as CrO_4^{2-}, which cannot dimerize within an anion exchanger, but is still preferred over competing sulfate and chloride. Chromate isotherms are, therefore, favorable at alkaline pH and they conform to Henry's law at trace concentrations. Figure 3.43 [67] shows chromate/chloride isotherm, both at acidic and alkaline pH; the contrasts in the curvatures of the two isotherms can be readily noted [75,76]. Since the isotherms are favorable at alkaline pH,

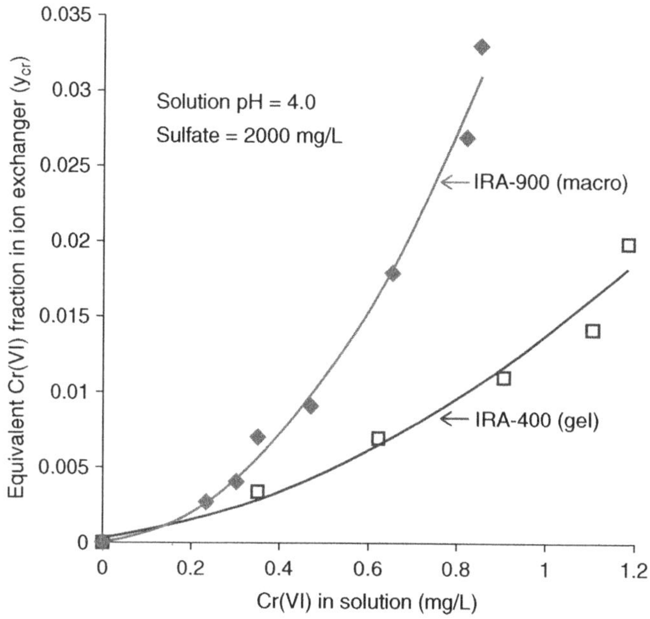

Figure 3.41 Isotherms for trace chromate in the background of high sulfate concentration for macroporous and gel-type anion exchangers. *Source*: SenGupta *et al.* 1988 [75]. Reproduced with permission of American Society of Civil Engineers.

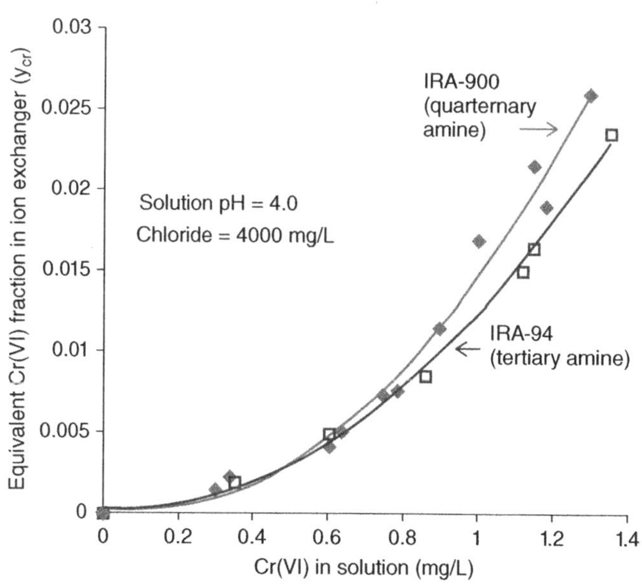

Figure 3.42 Isotherms for trace chromate in the background of high chloride concentration for macroporous and gel-type anion exchangers. *Source*: SenGupta *et al.* 1998 [75]. Reproduced with permission of American Society of Civil Engineers.

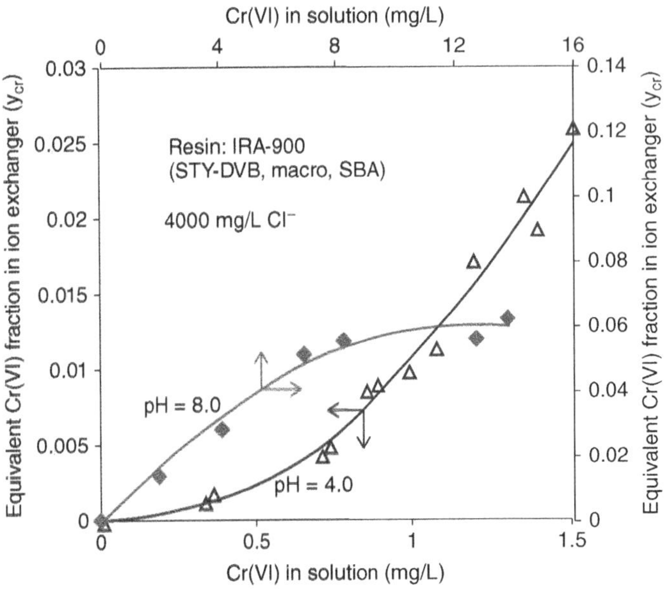

Figure 3.43 Isotherms for trace chromate in the presence of high chloride concentration for a macroporous strong-base anion exchanger at acidic and alkaline conditions. *Source*: SenGupta *et al.* 1988 [75]. Reproduced with permission of American Society of Civil Engineers.

chromate breakthrough during fixed bed column runs is of a self-sharpening type. Predicting effluent histories under such conditions is relatively straightforward [75].

3.15 Tunable Sorption Behaviors of Amphoteric Metal Oxides

Several polyvalent metal oxides, namely, oxides of Fe(III), Zr(IV), Ti(IV), and Al(III) are environmentally benign, inexpensive and readily available with surface sorption properties. These metal oxide particles exhibit amphoteric sorption behaviors, that is, they can selectively bind Lewis acids or transition metal cations (e.g., Cu^{2+}) as well as Lewis bases or anionic ligands (e.g., arsenate or $HAsO_4^{2-}$) through formation of inner sphere complexes [76–78]. Thus, Cu^{2+}, a Lewis acid and an environmentally regulated heavy metal, is sorbed in preference to other competing but innocuous alkaline and alkaline-earth metal cations, namely, Na^+, Ca^{2+}, Mg^{2+}. Similarly, sorption of arsenate, an anionic ligand with oxygen donor atoms, is preferred to commonly encountered anion namely, sulfate, chloride, and bicarbonate. The point of zero charge of crystalline or amorphous iron oxide nanoparticles in an inert electrolyte (e.g., sodium nitrate, sodium perchlorate or equivalent) resides within a pH range of 7.0–8.5 [9,77]. At circumneutral pH, iron oxide nanoparticles sorb both Cu^{2+} and $HAsO_4^{2-}$ simultaneously and selectively in the presence of commonly occurring competing ions that can form only outer-sphere complexes through coulombic interaction. Oxides of zirconium(IV), titanium(IV), and aluminum(III) also exhibit similar favorable sorption behaviors toward Lewis acids and Lewis bases [8,79–82]. Since sorption or binding sites reside only on the surface, nanoscale metal oxide particles with a very high surface area to volume ratio offer significantly enhanced sorption capacity. However, such metal oxide nanoparticles are unable to separate transition metal cations from anionic ligands. Since nanoparticles cause unusually high pressure drops in fixed-bed columns or any flow-through system, attempts were made to dope activated carbon, alginate, chitosan, cellulose, and polymeric sorbents with metal oxide nanoparticles [16–19,80]. These host materials improved permeability in flow-through systems but were unable to alter or influence the sorption behaviors of metal oxides including separation of Lewis acids from Lewis bases.

It was conceptualized that if HFO or other metal oxide nanoparticles were dispersed within a cation or anion-exchanger, the anions or cations will be rejected by the respective ion exchanger in accordance with the Donnan coion exclusion effect. Thus, in principle, an amphoteric metal oxide nanoparticle can be tailored to behave either as a strictly metal-selective sorbent or as a ligand-selective exchanger, Figure 3.44a–c illustrates the same [83].

For experimental validation, three separate fixed-bed column runs were carried out using: (i) Commercially available granulated ferric hydroxide (GFH) from US Filter Co. without any ion exchanger support material; (ii) hydrated Fe(III) oxide or HFO nanoparticles dispersed in a cation exchanger, referred to as hybrid cation exchanger (HCIX-Fe); and (iii) HFO dispersed in an anion exchanger, referred to as hybrid anion exchanger (HAIX-Fe). The feed composition was identical in all three cases and trace concentrations of both anionic As(V) and cationic Cu^{2+} were present as target solutes along with other electrolytes. Figure 3.45a–c shows the column-run results and note that (i) GFH removed both As(V) anions and Cu^{2+} quite significantly; (ii) HCIX removed only Cu^{2+} very selectively for well over 2000 BVs but rejected As(V) anions completely; and (iii) HAIX showed extraordinary As(V) sorption with no significant breakthrough for nearly 5000 BVs while Cu^{2+} broke through immediately [83]. It is worth mentioning that the parent cation exchanger with sulfonic acid functional groups can sorb cations only through electrostatic interaction and thus offers no specific selectivity toward Cu^{2+} in the presence of higher concentration of other competing cations, namely, calcium and sodium. Likewise, anion exchangers with quaternary ammonium functional groups do not exhibit specific selectivity toward anionic arsenate in the presence of competing sulfate anions.

Similar to HFO, the amphoteric sorption behavior of zirconium oxide (ZrO_2) nanoparticles may also be tuned through ion exchanger support. One gel-type cation exchanger was dispersed with zirconium oxide nanoparticles. The resulting hybrid cation exchanger (HCIX-Zr) was used in a batch sorption study where both copper and arsenate were present in trace concentrations. Figure 3.46 shows that while copper concentration dropped to nearly zero within an hour, As(V) concentration remained essentially unchanged [84].

The tunability of amphoteric HFO or zirconium oxide nanoparticles in the forgoing examples resulted from the Donnan membrane effect exerted by the ion exchanger support. The gel phase of an ion exchanger can be viewed as a polyelectrolyte where the functional groups (quaternary ammonium groups for anion exchanger and sulfonic acid groups for cation exchanger) are covalently attached and, hence,

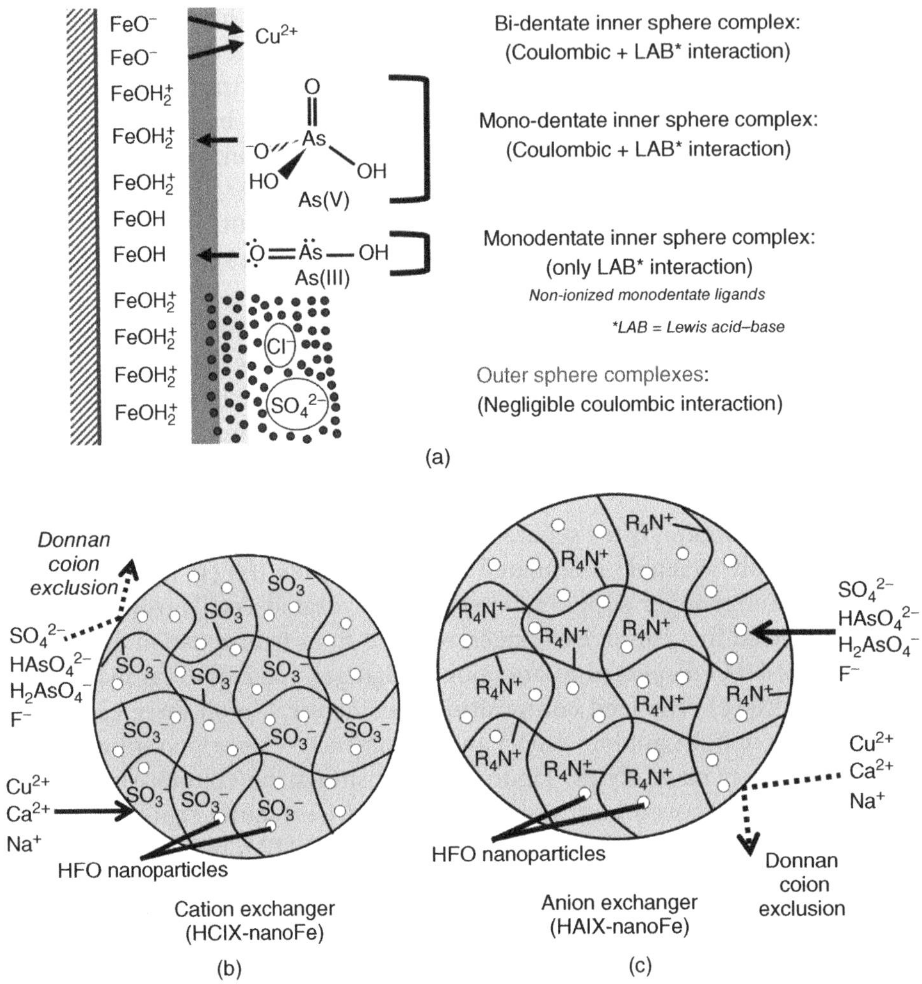

Figure 3.44 (a) Illustration of binding of heavy metal cations and ligands by HFO functionalities at different pH condition; (b) selective binding of metal cations (e.g., Cu^{2+}) by HFO doped cation exchanger; (c) selective binding of ligands (e.g., $HAsO_4^{2-}$) by HFO doped anion exchanger. *Source*: Puttamraju and SenGupta 2006 [84]. Reproduced with permission of American Chemical Society.

non-diffusible. The manifestation of the Donnan effect in two types of ion exchangers can be explained as follows: (i) a high concentration of fixed, non-diffusible negatively charged sulfonic acid functional groups in a cation exchanger disallows permeation of anions including arsenate into the gel phase and hence arsenate sorption by HFO nanoparticles is negligible; (ii) conversely, a high concentration of positively charged non-diffusible quaternary ammonium groups in an anion exchanger imbibes arsenate into the exchanger phase but rejects Cu^{2+}. It is worth mentioning that the conditions leading to the Donnan membrane equilibrium do not arise from the physical existence of a semi-permeable membrane or externally charged surfaces.

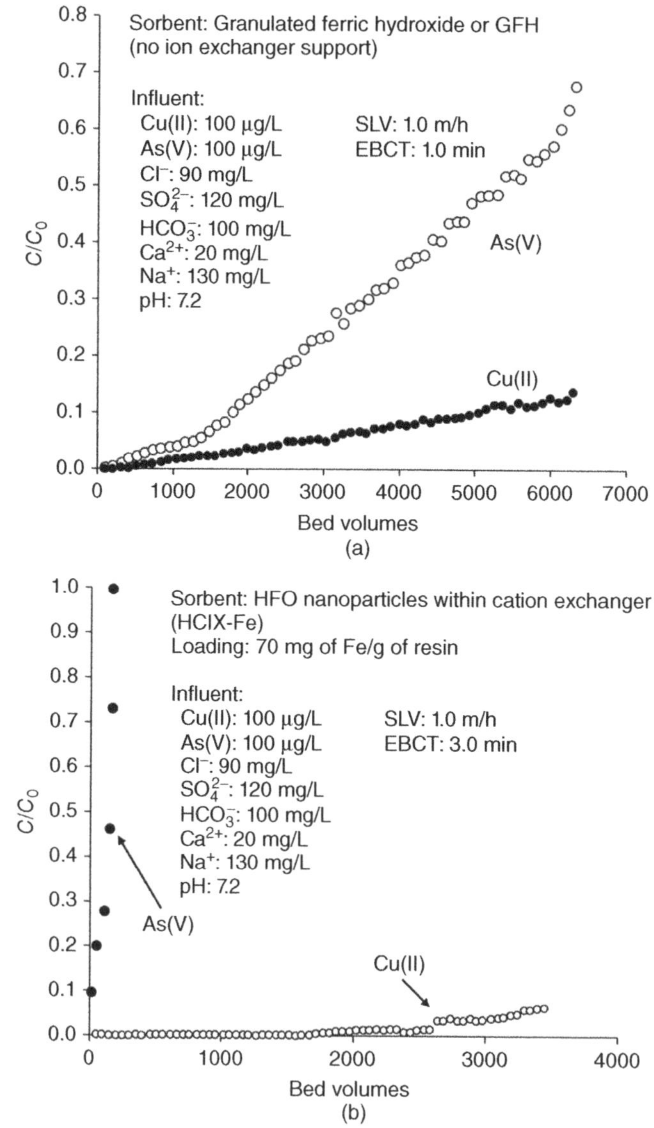

Figure 3.45 Results of fixed bed column runs under identical conditions with (a) GFH, (b) HCIX-NanoFe, and (c) HAIX-NanoFe. *Source*: Puttamraju and SenGupta 2006 [84]. Reproduced with permission of American Chemical Society.

3.16 Ion Sieving

Sieving or separations based upon ionic size differences are quite striking when one of the ions is of such size that it cannot diffuse into the interior of the exchanger. The choice of an exchanger, whose structure is such that only the exchange of the smaller counterions is feasible, will enable separation by rejecting macro-counterions from smaller-sized ones. The natural zeolite, chabazite, will sorb ammonium ions (NH_4^+), but, as demonstrated in Table 3.7, the percentage sorption capacity diminishes as hydrogen in NH_4^+ is substituted with methyl ($—CH_3$) groups [89].

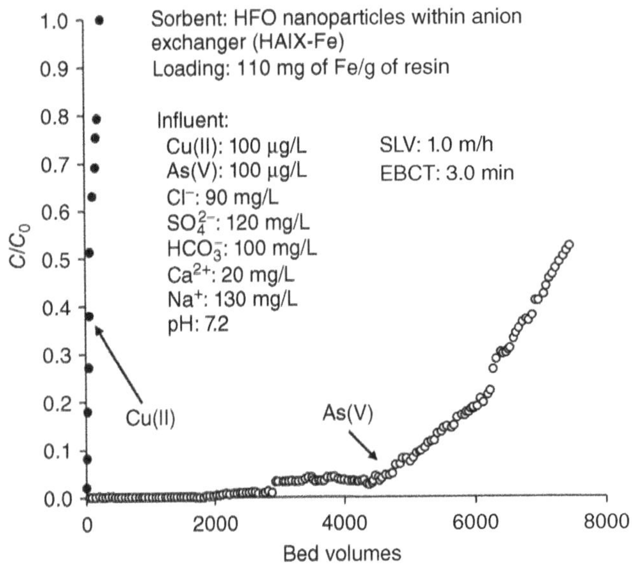

Figure 3.45 *(Continued)*

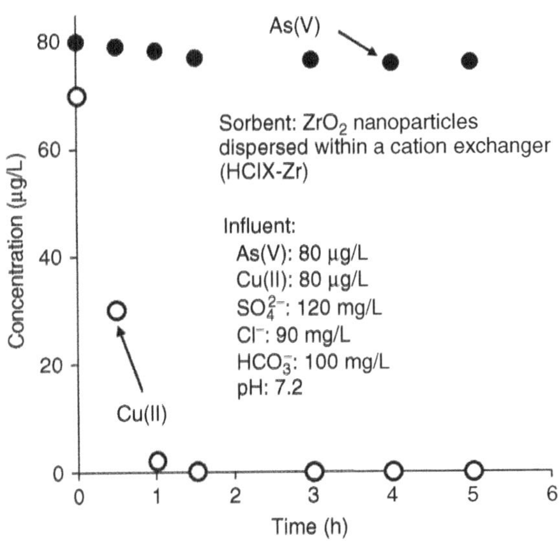

Figure 3.46 Result of batch sorption study of As(V) and Cu(II) onto ZrO_2-doped cation exchanger (HCIX-NanoZr). *Source:* Puttamraju and SenGupta 2006 [84]. Reproduced with permission of American Chemical Society.

The decrease in the cation exchange capacity of the sulfonic acid exchanger, Amberlite IRC 120, for large cations may be readily noted in Figure 3.47.

These data indicate that ion exchangers may be used for the separation of relatively small inorganic ions from complex organic ions. Since the large ions maybe sorbed and exchanged with ions at the exchanger surface, it is obvious that the particle size of the exchanger that is to be used, must be large enough so that the surface contribution is only negligible; ion exchange particle sizes of 500 μm or greater are normally satisfactory. Earlier, in Section 3.8, trace fouling of anion exchangers caused by NOM was discussed. Due to the large ionic radius, NOM sorption is restricted only onto the surface of the anion exchange resin.

Table 3.7 Relationship between percentage capacity of chabazite utilized and varying ionic diameter.

Ion	Ionic diameter (nm)	Total capacity realized (%)
NH_4^+	0.290	100
$CH_3NH_3^+$	0.318	21
$(CH_3)_2NH_2^+$	0.594	9
$(CH_3)_3NH^+$	0.654	9
$(CH_3)_4N^+$	0.698	4

Source: Adapted from Kunin and Myers 1950 [89].

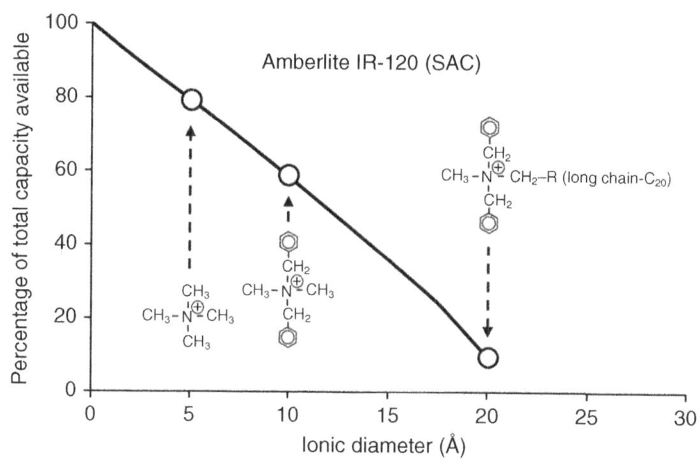

Figure 3.47 Effect of ionic radius of large cations on the exchange capacity. Note: For Na^+ and Ca^{2+} the cation exchange capacity is 100%. *Source*: Adapted from Kunin and Myers 1950 [89].

Supplementary Reading S3.2 Reactor Configuration in Ion Exchange: CSTR versus PFR

The focus of this book has been deliberately limited to the science of ion exchange and its relevance to broad areas of engineered environmental processes. Equipment design and process configuration of ion exchange processes often has close resemblance to those of other sorption/adsorption processes. Detailed discussions in such common overlapping areas already exist in the literature [84,85]. That is the key reason why we have avoided any major discussion in these areas. Nevertheless, it is imperative that we distinguish between two chemical reactor configurations – namely, continuous stirred tank reactors (CSTRs) and plug flow reactor (PFR) – for application of ion exchange processes. The utilization of ion exchange capacity is strongly dependent on the reactor configuration for continuous flow systems, all other conditions remaining identical.

Supplementary Reading S3.3 Capacity Utilization

Figures S.3.4a and b illustrate an ideal CSTR (mixer-settler) and an ideal PFR (fixed bed), respectively, while Figure S3.4c provides the isotherm for the target contaminant "i." Note that CSTR (i.e., the mixer) operates at the exit concentration and, thus, ion exchange resin beads attain the equilibrium capacity, q_e, corresponding to C_e. On the contrary, ion exchange resin beads in a PFR always attain equilibrium at the influent concentration, C_I. Hence, ion exchange resin beads attain the equilibrium capacity of q_I corresponding to C_I. Since C_I is always greater than C_e, $q_I > q_e$. Thus, in principle, PFR configuration always requires lesser amount of ion exchange resins for identical removal of the contaminant "i." The following example illustrates the point.

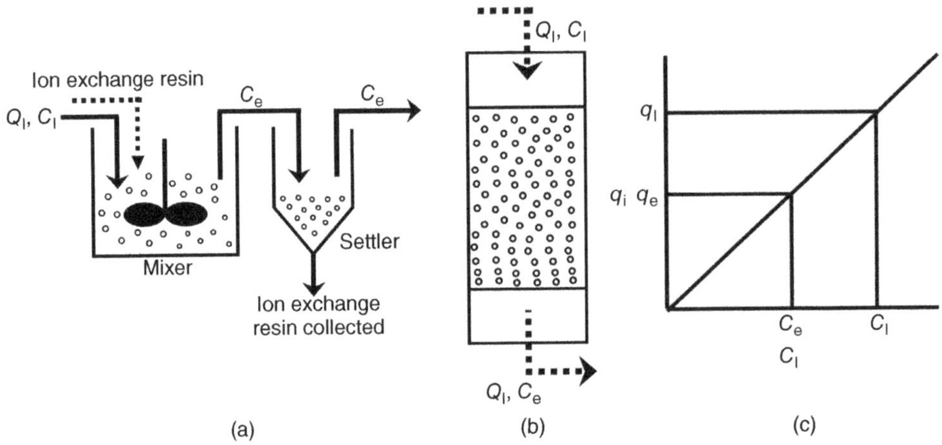

Figure S3.4. *Illustration of (a) CSTR, (b) PFR, and (c) prevailing isotherm for the contaminant of interest "i."*

Example S3.1

Trichlorophenol (TCP) is present as an anion in a wastewater stream at 5 mg/L. Find out the volume of anion exchange resins needed per hour to treat 5 m^3/h of wastewater for 80% TCP removal by the following two treatment processes: (i) mixer-settler and (ii) fixed bed.

The linear ion exchange isotherm for TCP in the wastewater is given as

$$q_{TCP}\left(\frac{\text{mg TCP}}{\text{g resin}}\right) = 0.1 \cdot C_{TCP}\left(\frac{\text{mg TCP}}{\text{L}}\right)$$

State assumptions, if any.

Solution:

Let us consider mixer-settler to be an ideal CSTR and fixed-bed to be an ideal PFR. After 80% removal, the desired TCP concentration in the treated water is 1 mg/L.

Mixer-Settler (CSTR)
TCP removal capacity corresponding to 1 mg/L concentration is

$$q_{TCP}\left(\frac{\text{mg TCP}}{\text{g resin}}\right) = 0.1 \cdot 1\,\text{mg/L} = 0.1\,\text{mg/g}$$

From the mass balance across the mixer,

$$Q(C_I - C_e) = P \cdot q_e$$

where Q is the hourly flow rate of wastewater and P is the hourly dosage of ion exchanger. Thus,

$$P = \frac{Q(C_I - C_e)}{q_e}$$

$$= \frac{5\,\text{m}^3/\text{h} \cdot 1000\,\text{L/m}^3\,(5\,\text{mg/L} - 1\,\text{mg/L})}{0.1\,\text{L/g} \cdot 1\,\text{mg/L}}$$

$$= 200{,}000\,\text{g/h} = 200\,\text{kg/h}$$

Assuming the specific gravity of the ion exchanger is 1.0, the volume of ion exchanger needed is 200 L/h. Note that the settling tank only physically separates ion exchange resin from the solution without any additional TCP uptake.

IX resin dosage $= P/Q = 200\,\text{kg}/5\,\text{m}^3 = 40$ *g of IX/L of wastewater.*

Fixed-Bed (PFR)
From the mass balance

$$QC_I = P \cdot q_I$$

$$P = \frac{QC_I}{0.1 \cdot C_I}$$

$$= \frac{5\,\text{m}^3/\text{h} \cdot 1000\,L/\text{m}^3\,(5\,\text{mg/L})}{0.1\,\text{L/g} \cdot 5\,\text{mg/L}}$$

$$= 50{,}000\,\text{g/h} = 50\,\text{kg/h}$$

The volume needed is 50 kg/h.

IX resin dosage $= P/Q = 50/5\,\text{kg/m}^3 = 10$ *g of IX/L of wastewater.*

Comment: Note that the volume needed for the fixed bed is much smaller than that for the mixer-settler. Both reactors are assumed to have attained equilibrium, that is, there is no kinetic limitation.

Example S3.2

Compute the amount of ion exchanger needed for a two-stage mixer-settler with recycle of ion exchange resin from the second settler, as illustrated. Note $q_2 = q_e$.

From the mass balance across both mixer 1 + 2,

$$Q(C_I - C_e) = P \cdot q_1$$

$$Q(C_I - C_e) = P \cdot 0.1 \cdot C_1 \quad (1)$$

(Continued)

Supplementary Reading S3.3 (Continued)

Q, C_I, and C_2 are known, but P and C_1 are both unknown and a second equation is required. From the mass balance across mixer 2, a second equation can be derived,

$$Q(C_1 - C_e) = P \cdot q_e$$

$$Q(C_1 - C_e) = P \cdot 0.1 \cdot C_e$$

$$C_1 = \frac{C_e(P \cdot 0.1 + Q)}{Q}$$

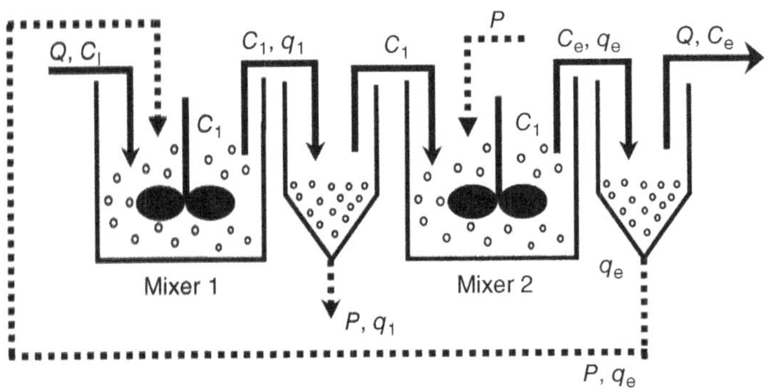

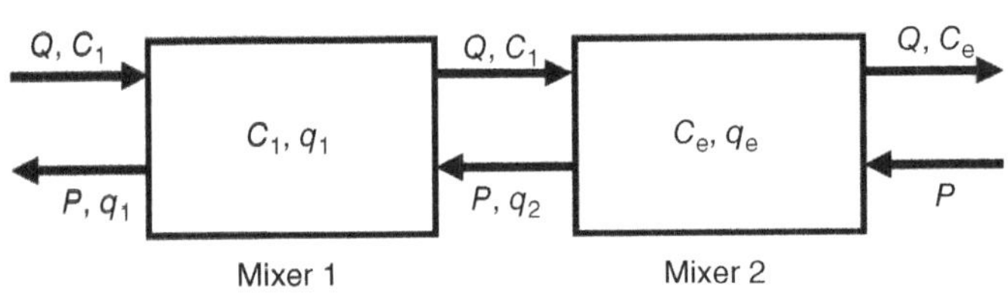

Inserting C_1 into the total mass balance in Eq. (1),

$$Q(C_I - C_e) = P \cdot 0.1 \cdot \frac{C_e(P \cdot 0.1 + Q)}{Q}$$

$$\frac{Q^2(C_I - C_e)}{0.1 \cdot C_e} = P \cdot (P \cdot 0.1 + Q)$$

$$0 = 0.1 \cdot P^2 + QP - \frac{Q^2(C_I - C_e)}{0.1 \cdot C_e}$$

$$P = \frac{-5000\,\text{L/h} + \sqrt{(5000\,\text{L/h})^2 + 4 \cdot 0.1 \cdot \frac{(5000\,\text{L/h})^2(5-1\ \text{mg/L})}{0.1 \cdot 1\ \text{mg/L}}}}{0.2\,\text{L/g}}$$

$$P = 78{,}078\,\text{g/h} = 78\,\text{kg/h}$$

$$\text{IX resin dosage} = \frac{P}{Q} = \frac{78\,\text{kg/h}}{5\,\text{m}^3/\text{h}} = 15.6\,\text{g/L}$$

$$C_1 = \frac{C_e(P \cdot 0.1 + Q)}{Q} = \frac{1\,\text{mg/L}(78{,}078\,\text{g/h} \cdot 0.1\,\text{L/g} + 5000\,\text{L/h})}{5000\,\text{L/h}}$$

$$C_1 = 2.56\,\text{mg/L}$$

Note that the requirement of IX resin has dropped significantly (from 200 to 78 kg/h) by increasing the number of mixer-settler stages from one to two. Ideally, an infinite number of stages will make the dosage equal to that of a PFR.

3.17 Trace Ion Removal

It is only appropriate that we briefly discuss in this chapter the challenges and underlying fundamentals for removal of trace ions of major environmental significance.

3.17.1 Uranium(VI)

Uranium may exist in different oxidation states in water, but U(VI), is the most stable under normally encountered pH and redox environment. It is the heaviest naturally occurring radioactive element and has high affinity for oxygen. Under acidic conditions, uranyl ion (UO_2^{2+}) is the most stable state of uranium in water with low total dissolved solids. However, at neutral to slightly alkaline pH, UO_2^{2+} complexes with hard anions, namely CO_3^{2-}, to form strong labile uranyl–carbonate complexes.

$$UO_2^{2+} + 2CO_3^{2-} \leftrightarrow [UO_2(CO_3)_2]^{2-},\ K_{a1} = 1.67 \times 10^{16} \tag{3.110}$$

$$[UO_2(CO_3)_2]^{2-} + CO_3^{2-} \leftrightarrow [UO_2(CO_3)_3]^{4-},\ K_{a2} = 3.0 \times 10^{5} \tag{3.111}$$

Many groundwater sources in the Western United States are contaminated with naturally occurring radioactive uranium. The US EPA introduced a maximum contaminant level (MCL) of 20 μg/L and a maximum contaminant level goal (MCLG) of zero for uranium due to health concerns associated with carcinogenicity and kidney chemotoxicity [89–94]. Nearly 2000 communities in the USA are confronted with the risk of drinking uranium-contaminated water from underground.

Of all the technologies currently available, the fixed-bed anion exchange process is particularly effective because it can selectively remove trace uranium in the presence of other electrolytes and the process is operationally simple. The high affinity of anionic uranyl carbonate counterions to strong-base anion exchange resins through coulombic interactions is attributed to its preferred uptake in accordance with the following:

$$2\overline{R^+Cl^-} + UO_2(CO_3)_2^{2-} \leftrightarrow \overline{R_2UO_2(CO_3)_2} + 2Cl^- \tag{3.112}$$

$$4\overline{R^+Cl^-} + UO_2(CO_3)_3^{4-} \leftrightarrow \overline{R_4UO_2(CO_3)_3} + 4Cl^- \tag{3.113}$$

Clifford and co-workers carried out extensive laboratory and field-scale studies on uranium removal [95,96]. Figure 3.48 represents uranium removal under field conditions at two different pH values, namely, pH = 8.0 and pH = 4.3.

Note that while at pH = 8.0, uranium breakthrough was not observed until well after 25,000 BVs, at pH = 4.3 uranium breakthrough occurred almost in the beginning and in less than 5000 BVs, uranium breakthrough was complete. Besides, the difference in pH values, the two influents were identical in every respect. At pH = 4.3, nearly all

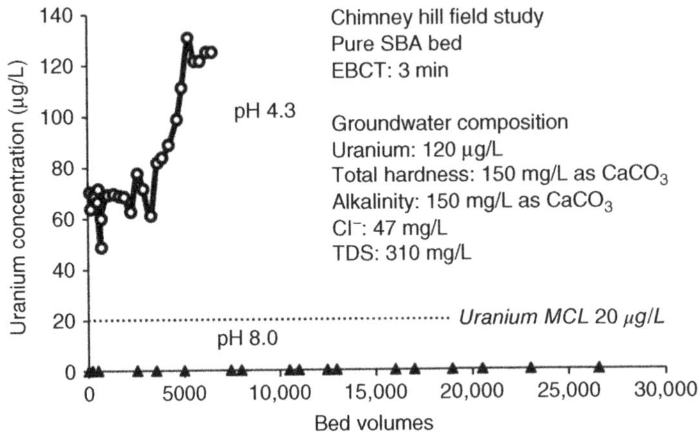

Figure 3.48 Effluent uranium levels during the virgin exhaustions of a bed of pure SBA resin at pH 4.3. *Source*: Adapted from Clifford and Zhang 1995 [96].

the bicarbonate (HCO_3^-) is converted to H_2CO_3 and stripped out of water. So, anionic uranyl carbonate complexes are essentially absent in the feed, thus greatly reducing removal of uranium by the anion exchanger.

SBA resins, loaded with U(VI), may be easily regenerated with brine or NaCl solutions. Figure 3.49 shows efficient uranium elution of the exhausted bed with 2.0 N and 3.0 N NaCl solution. Absent carbonate in the regenerant solution, uranyl carbonate, $[UO_2(CO_3)_2{}^-]^{2-}$ undergoes hydrolysis, thus favoring the desorption process as follows:

$$\overline{(R^+)_2[UO_2(CO_3)_2]^{2-}} + 2Cl^- \leftrightarrow \overline{2R^+Cl^-} + [UO_2(CO_3)_2]^{2-} \tag{3.114}$$

$$[UO_2(CO_3)_2]^{2-} + 2H_2O \leftrightarrow UO_2^{2+} + 2HCO_3^- + 2OH^- \tag{3.115}$$

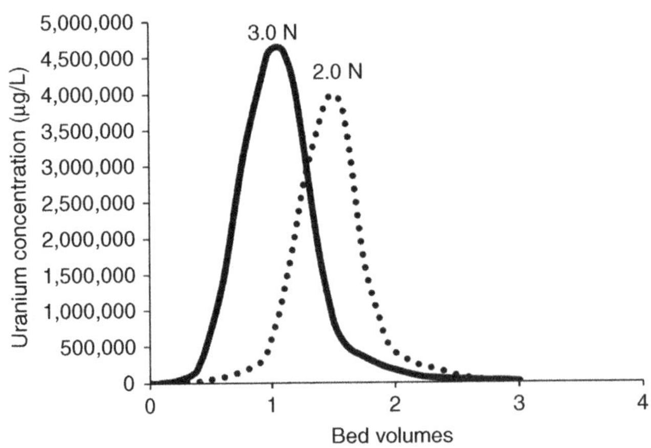

Figure 3.49 Uranium elution by 2.0 N and 3.0 N NaCl during the first concurrent regeneration of a pure SBA resin bed exhausted to 30,000 BV; regeneration level 10 eq Cl−/eq resin (36 lb NaCl/ft^3 resin). *Source*: Adapted from Clifford and Zhang 1995 [96].

Note that one major uranium species that exists at acidic conditions, UO_2^{2+}, is cationic and, hence, completely rejected by the anion exchanger due to the Donnan coion exclusion effect. That is why regeneration is very efficient and is completed in less than five BVs.

3.17.2 Radium

Radium is a radioactive alkaline earth metal and present in Group IIA of the periodic table, along with calcium, magnesium, and barium. Of all the radioactive isotopes, Ra-226 has the longest half-life ($t_{1/2} = 1600$ years) and is the most significant in the context of contaminating the potable water supply. Like other alkaline earth metal ions, radium exists in the aqueous phase as a divalent cation, Ra^{2+}, and is seemingly amenable to removal by cation exchange processes. In fact, Ra^{2+} is the most preferred alkaline earth metal cation by the synthetic cation exchange resins containing sulfonic acid functional groups in the following order of preference: $Ra^{2+} > Ba^{2+} > Sr^{2+} > Ca^{2+} > Mg^{2+}$. However, with very high concentrations of competing Ca^{2+} and/or Mg^{2+} ions in the groundwater, the run length of the fixed-bed column operation in Na-cycle is short and NaCl regeneration is inefficient. Also, disposal of spent regenerant laden with radium always poses environmental challenges.

To enhance radium removal capacity and eliminate spent regenerant disposal, Dow Chemical Co., developed a cation exchanger doped with barium sulfate or $BaSO_4$ [97,98]. The material is referred to as a radium-selective complexer or RSC, which takes advantage of the lower solubility product value (K_{sp}) of $RaSO_4$ ($K_{sp} = 4.2 \times 10^{-11}$) compared to $BaSO_4$ ($K_{sp} = 1.08 \times 10^{-10}$).

Radium or Ra^{2+} removal takes place in two consecutive steps inside the cation exchange resins in a fixed-bed column: (i) ion-exchange; (ii) precipitation/dissolution. First, Ra^{2+} is sorbed onto the cation exchange sites through ion exchange with very fast kinetics. Subsequently, as Ra^{2+} counterions are eluted from the ion exchange sites, they precipitate and replace Ba^{2+} in the solid phase because the formation of $RaSO_4$(s) is thermodynamically favorable. The stepwise radium removal can be presented as follows:

Step 1. Ion exchange

$$2\overline{(RSO_3^-)Na^+} + Ra^{2+} \leftrightarrow \overline{(RSO_3^-)_2Ra^{2+}} + 2Na^+ \quad (3.116)$$

Step 2. Desorption followed by precipitation

$$\overline{(RSO_3^-)_2Ra^{2+}} + Ca^{2+} \leftrightarrow \overline{(RSO_3^-)_2Ca^{2+}} + Ra^{2+} \quad (3.117)$$

$$Ra^{2+} + BaSO_4(s) \leftrightarrow RaSO_4(s) + Ba^{2+} \quad (3.118)$$

$$2\overline{(RSO_3^-)Na^+} + Ba^{2+} \leftrightarrow \overline{(RSO_3^-)_2Ba^{2+}} + 2Na^+ \quad (3.119)$$

Overall:

$$4\overline{(RSO_3^-)Na^+} + BaSO_4(s) + Ra^{2+} + Ca^{2+} \leftrightarrow \overline{(RSO_3^-)_2Ba^{2+}} + \overline{(RSO_3^-)_2Ca^{2+}} + RaSO_4(s) + 4Na^+ \quad (3.120)$$

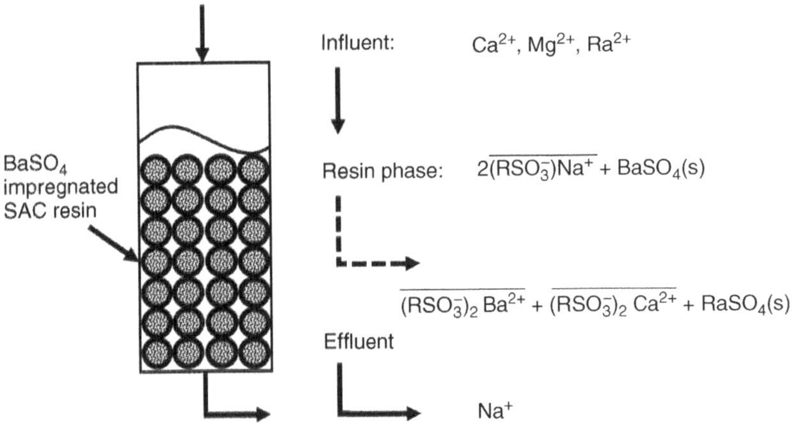

Figure 3.50 Schematic of radium removal by a strong-acid cation exchanger loaded in sodium form with barium sulfate precipitates.

Figure 3.50 provides a schematic illustrating Ra^{2+} removal.

Since the run length is greatly enhanced, RSC is discarded without being regenerated. Primarily due to the limited market warranting treatment of radium-contaminated groundwater, use of RSC has been rather insignificant in the USA. However, scientifically, RSC is considered to be the first sorbent that can accomplish both ion exchange and precipitation in the same bed, but the process becomes increasingly inefficient at high total dissolved solids or TDS.

3.17.3 Boron

Boron is widely distributed in the environment, mainly in the form of boric acid or borate salts. The recommended limit for the boron concentration from the World Health Organization (WHO) is 0.5 mg/L for drinking water [99]. Over intake of boron may cause acute boron toxicity with nausea, headache, diarrhea and kidney damage. Also, most crops are sensitive to high boron levels in irrigation water. So, it is generally recommended that boron levels remain less than 1.5 mg/L for irrigated crops, especially for citrus plants.

Boron is widely distributed throughout the lithosphere and due to its high affinity toward oxygen, boron exists mostly as boric acid or $B(OH)_3$ or H_3BO_3. Boric acid is a weak acid that dissociates into borate anion ($pK_a = 9.1$).

$$B(OH)_3 + H_2O \rightarrow B(OH)_4^- + H^+, \; pK_a = 9.1 \tag{3.121}$$

At neutral pH, boric acid or $B(OH)_3$ is the predominant borate species. It has long been known that boric acid or borate forms very stable complexes rapidly with polyalcohols to form an acid that is considerably stronger than boric acid [99]. Glycerol is commonly used analytically as the polyalcohol and the resulting strong acid can then be titrated by aqueous NaOH. Kunin and Preuss at Rohm and Haas Company used this principle from analytical chemistry in synthesizing a boron-selective resin (BSR) by aminating a chloromethylated styrene–divinylbenzene copolymer with

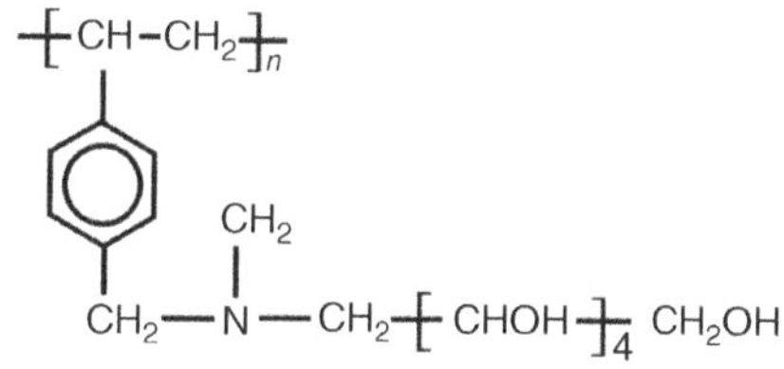

Figure 3.51 The structure of boron selective resins. *Note*: The repeating polyol structure is used for boron chelation.

N-methyl-D-glucamine (NMDG) functional groups [100–102]. This BSR was subsequently produced commercially as Amberlite-743 from Rohm and Haas Co.; many similar products are now available from other manufacturers. Figure 3.51 provides the typical composition of the boron selective Amberlite-743 resin with polystyrene matrix and divinylbenzene cross-linking:

In contrast to standard ion exchange processes, the NMDG moieties of BSR capture boron via coordination complexation and not by coulombic interaction. The maximum boron removal capacity occurs at pH ~8.0, but boron can be removed over a wide range of pH values. Mechanistically, boric acid of $B(OH)_3$ is complexed with two sorbitol groups of the NMDG functional group, and the proton released is retained by the tertiary amine site of the weakly basic anion exchanger. The exhausted BSR is regenerated first by using acid to desorb boric acid through hydrolysis. Subsequently, the weak-base resins are deprotonated with NaOH. Figure 3.52 illustrates the schematic of the cyclic process [100,101].

Simonnot *et al.* demonstrated that the boron removal capacity of BSR remains virtually the same in the pH range of 5.5–8.0 and is not influenced by NaCl concentration in the aqueous phase [103]. Since the boron uptake mechanism is not based on exchange of ions, the process is kinetically slow. The practical ion exchange capacity depends on the flow rate or contact time: the operating boron removal capacity drops rapidly as the flow rate is increased or contact time is decreased. Information in the open literature provides various modes of operation and recent progresses using hybrid process for boron removal [103,104].

In seawater, boron is present at concentrations around 5 mg/L. Since boron exists predominantly as non-ionized boric acid, boron rejection is significantly lower than chloride for reverse osmosis (RO) desalination processes. Thus, boron is present in the treated water exceeding the drinking water limit. Boron rejection by RO membrane also decreases with an increase in temperature of the feed seawater in the range of 10–45 °C. Use of BSR is the most effective treatment for treating boron-containing RO permeate.

3.17.4 Perchlorate (ClO_4^-)

Perchlorate is a monovalent anion. In dilute aqueous solutions, perchlorate is stable, extremely nonreactive and cannot be removed by coprecipitation. In the United States, more than 30 states have detected perchlorate-contaminated water due to past mishandling of ammonium perchlorate fuel. The USEPA added perchlorate to the contaminant candidate list for drinking water due to its adverse health impact, but has not set a MCL or MCLG as of 2016 [105]. Many wells in California have been idled due to the presence of perchlorate and the Cal/EPA's Office of Environmental Health Hazard

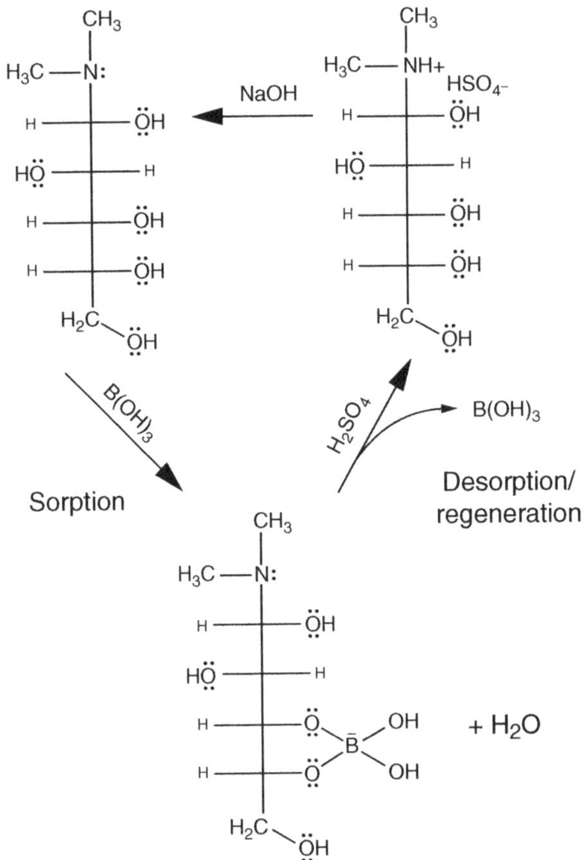

Figure 3.52 A schematic illustrating the uptake into and desorption from a boron-selective resin with polyol functional groups.

Assessment (OEHHA) established a public health goal (PHG) of 1 μg/L in 2015 [106]. Perchlorate is very mobile with negligible sorption onto soil or NOM. The oxidation number of chlorine in perchlorate is +VII. However, reducing perchlorate abiotically is a kinetically slow process. Selective sorption onto anion exchange resins is a viable treatment option for contaminated groundwater and other wastewaters for perchlorate removal.

From an anion exchange perspective, nitrate and perchlorate are quite similar. Both are monovalent anions, hydrophobic and do not form complexes with cations. Thus, the rules or criteria governing ion exchange selectivity, as described for nitrate in Section 3.10, are also applicable for perchlorate. Table 3.8 summarizes how different composition variables of anion exchange resins influence perchlorate/chloride separation factor or selectivity.

In general, all other conditions remaining identical, perchlorate is preferred to nitrate. Figure 3.53 demonstrates how increasing the size of the alkyl group in a strong-base anion exchanger enhances perchlorate selectivity [87].

Anion exchange resin with triethyl- (Amberlite IRA-996) and tripropyl-amine (Ionac SR-7) are now commercially available, and removing trace perchlorate selectively from

Table 3.8 Effect of resin composition on ClO_4^-/Cl^- selectivity.

Composition variable	$\alpha_{ClO_4^-/Cl^-}$
Polyacrylic (more hydrophilic) to polystyrene (more hydrophobic) matrix	Increases
Increased degree of cross-linking	Increases
Trimethyl quaternary amine to triethyl quaternary amine functionality	Increases

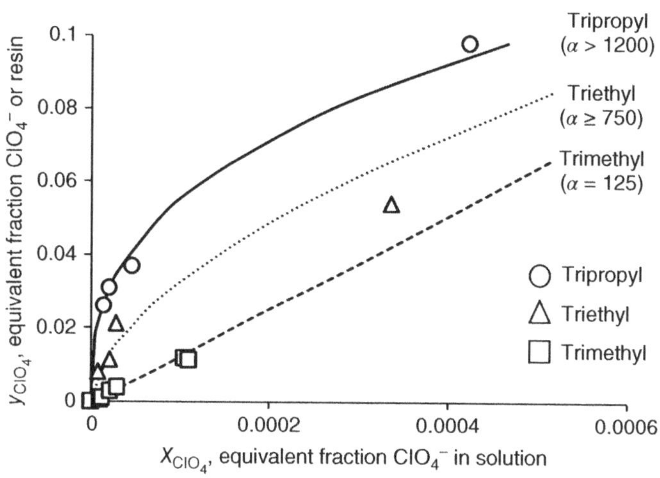

Figure 3.53 Twenty-four hours perchlorate–chloride binary isotherms at 20 °C comparing three polystyrene resins with varying alkyl chain lengths of the quaternary amine functional group. The perchlorate separation factors are in parentheses. *Source*: Adapted from Tripp and Clifford 2004 [87].

contaminated groundwater is not a challenging separation problem. Traditional anion exchange resins with polystyrene matrix and trimethyl quaternary amine functional groups are, in general, quite effective. However, the challenge lies in efficient regeneration and reuse of the anion exchanger. Figure 3.54 shows the elution/desorption of perchlorate from a strong-base anion exchanger at three different temperatures using 1 N NaCl as the regenerant.

Like other favorable monovalent inorganic ion exchange processes, perchlorate–chloride exchange is an exothermic process. Thus, perchlorate selectivity diminishes with an increase in temperature and the efficiency of regeneration is enhanced as the temperature is increased from 23 to 60 °C.

A multifunctional strong-base anion exchange resin was developed using alkyl groups of different lengths to improve perchlorate selectivity without compromising sorption kinetics [107,108]. The multifunctional perchlorate-selective resin is now commercially available, for example, Purolite A-530E, from Purolite Co. in the USA. A new technique has been developed to efficiently regenerate multifunctional and other Type I strong-base anion exchange resins loaded with perchlorate. The new method for multifunctional resin involves a mixed solution of $FeCl_3$ and HCl, in which

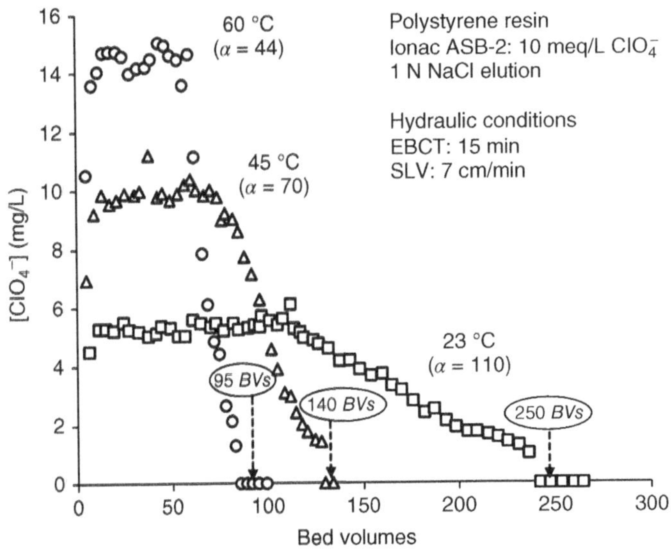

Figure 3.54 Effect of temperature on elution of perchlorate from a polystyrene resin using 1 N chloride solutions. Perchlorate–chloride separation factors, α values, are given for each temperature. *Source*: Adapted from Tripp and Clifford 2004 [87].

tetrachloroferrate ($FeCl_4^-$) ion forms in the presence of excess Cl^-:

$$FeCl_3(aq) + Cl^- \rightarrow FeCl_4^- \tag{3.122}$$

Like ClO_4^-, $FeCl_4^-$ is a large, poorly hydrated anion that can more effectively displace ClO_4^- from the anion exchange resin than can Cl^- or most other counterions. In practice, a mixed solution of 1 M $FeCl_3$ and 4 M HCl was effective in eluting ClO_4^- from an exhausted resin bed. The sorbed $FeCl_4^-$ can subsequently be removed just by passing water accompanied by hydrolysis:

$$FeCl_4^- \rightarrow Fe^{3+} + 4Cl^- \tag{3.123}$$

Resulting Fe^{3+} is readily desorbed due to the Donnan exclusion effect and the anion exchange resin returns to chloride form. Further details of the regeneration process are available in the open literature [109,110].

3.17.5 Emerging Contaminants of Concern and Multi-Contaminant Systems

Traces of active pharmaceuticals, personal care products and their metabolites from agricultural and human applications and wastes, have been found in all aquatic environments, including drinking water [111–117]. Understandably, measures are underway to intercept them before they find access to natural waterways. One of the treatment approaches is to capture and concentrate them first, followed by biological or chemical oxidation using a combination of highly reactive oxygen species, for example, ozone, hydrogen peroxide, catalytic metal oxides, and high energy source, for example, UV. Figure 3.55 includes a few of the leading pharmaceutical compounds present in waterways globally because of their widespread use across all sections of the human population.

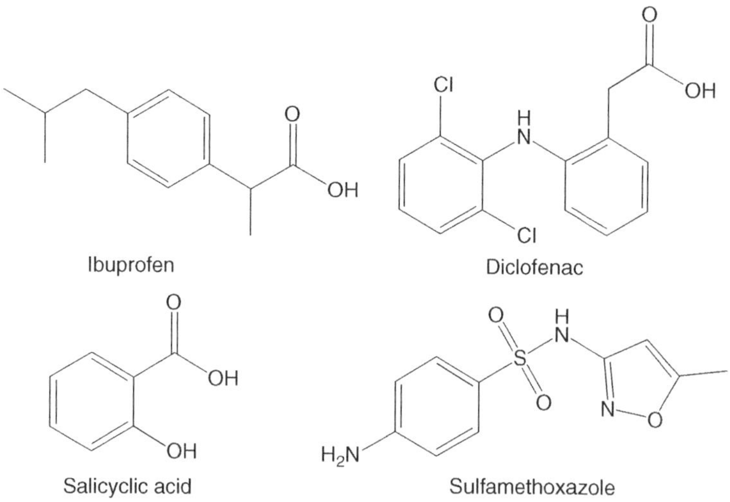

Figure 3.55 Structures of ibuprofen (e.g., tylenol), diclofenac (e.g., NSAID), salicylic acid (e.g., topical skin products), and sulfamethoxazole (e.g., antibiotic, Bactrim).

Since these are either weak-acid or weak-base compounds, they also exist as cations or anions depending on pH. Thus, they are essentially HIOCs with NPMs. Their sorption affinity and desorption regenerability are characteristically similar to the discussion in Section 3.11 pertaining to chlorophenates, benzoates and naphthalene sulfonates. A generic review for a broad range of such ionic compounds and their physical properties is available in the open literature [111,112,116,117].

Urine source separation has been proposed as a more sustainable approach to wastewater management than treating combined wastewater streams [113,118,119]. This methodology can remove and recover phosphate while simultaneously removing pharmaceutical residuals, for example, diclofenac (DCF), using commercially available HAIX-NanoFe that is essentially a strong-base anion exchanger within which HFO nanoparticles have been irreversibly dispersed [18]. Note that HAIX-NanoFe has dual functional groups: quaternary ammonium groups with high affinity toward hydrophobic anions and HFO sorption sites with high affinity toward ligands. Figure 3.56 demonstrates removal of phosphate and diclofenac by HAIX-NanoFe.

Note that both phosphate and diclofenac are removed simultaneously and effectively. There is no significant competition between diclofenac removal and phosphate concentration. Other negatively-charged pharmaceuticals, for example, ibuprofen, naproxen, and ketoprofen, are also expected to be removed effectively by HAIX-NanoFe.

In Lake Isabella, California, groundwater is naturally contaminated with both uranium and arsenic. Due to severe water paucity caused by drought, attempts are underway in California to appropriately use impaired sources of water. Under the auspices of the US Environmental Protection Agency (EPA), field scale trials were carried out to evaluate the performance of HAIX-NanoFe for simultaneous removal of uranium and arsenic from the contaminated water. Figure 3.57a and b shows the demonstration plant on-site and the results of a prolonged column run on-site.

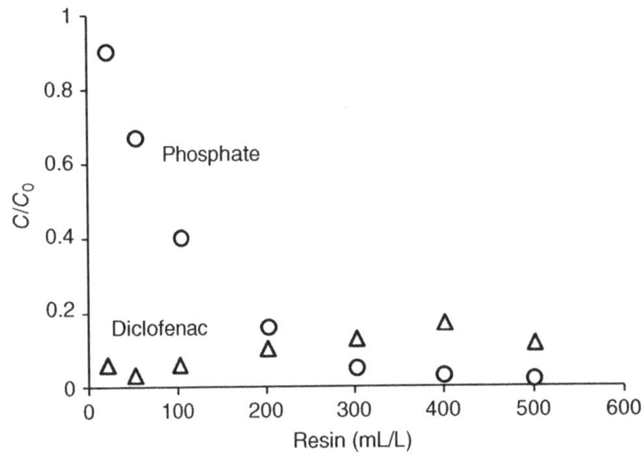

Figure 3.56 Coremoval of diclofenac and phosphate from fresh urine using hybrid anion exchange resin. Mixing conditions: 2 h at 200 rpm. Initial concentrations: 0.204 mmol/L diclofenac, 704 mg P/L phosphate. *Source*: Sendrowski and Boyer 2013 [120]. Reproduced with permission of Elsevier.

At groundwater conditions on-site, uranium was present as uranyl tricarbonate ($UO_2(CO_3)_3^{4-}$), a tetravalent anion with very high affinity for strong-base anion exchange resins. Nearly all arsenic on site was as As(V), which as a Lewis base ligand, has high sorption capacity on the iron oxide nanoparticles. Uranium and arsenic had separate mechanisms of removal by HAIX-NanoFe and were concurrently removed below their MCLs for greater than 30,000 BVs. Chapter 6 is devoted to the concept, underlying science and performance of HIX-nanotechnology.

3.17.6 Arsenic and Phosphorus: As(V), P(V), and As(III)

Of all the naturally occurring contaminants present in groundwater, arsenic is by far the most toxic, and quite prevalent. Although not known 30 years ago, nearly 100 million people in more than 50 countries are routinely exposed to arsenic poisoning by drinking contaminated groundwater. According to both the WHO and the USEPA, the MCL of arsenic in drinking water is 10 μg/L. Both arsenic and phosphorus are in Group V of the periodic table and their chemistries are quite similar. However, phosphorus is not toxic and, from a broader environmental perspective, it is a limiting nutrient responsible for algal blooms and eutrophication in ponds, lakes and waterways. Thus, phosphorus, often present as phosphate, needs to be avoided even at trace concentrations. While arsenic exists primarily in +V and +III oxidation states, phosphorus is prevalent in the +V oxidation state. Both arsenic and phosphorus exist as oxyacids or oxyanions that are ligands or Lewis bases with an ability to donate a lone pair of electrons. Table 3.9 summarizes the salient properties of commonly occurring As(V), P(V), and As(III) compounds at below-neutral and above-neutral pH values and highlights the key differences.

In Section 3.15, sorption of various anionic ligands onto polyvalent metal oxides has been discussed. Iron oxide has been widely used for removing arsenate, arsenite and phosphate at trace concentrations. In Chapter 6, a new class of hybrid anion exchangers (HAIX-NanoFe) have been discussed where iron oxide nanoparticles are dispersed

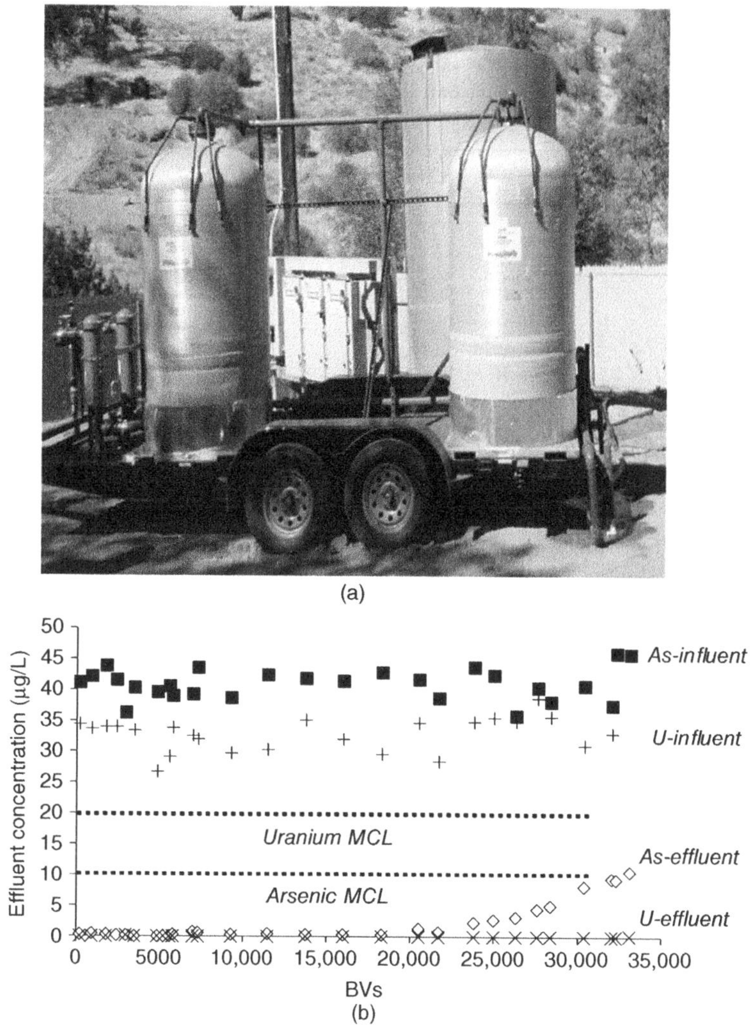

Figure 3.57 (a) Skid-mounted HAIX-NanoFe treatment columns at Lake Isabella, CA for EPA evaluation; (b) influent and effluent data for concurrent uranium and arsenic removal by HAIX-NanoFe from the skid-mounted HAIX-NanoFe columns. *Source*: After: Wang *et al.* 2010 [121].

within polymeric anion exchangers. Sorption properties of phosphate and arsenate are nearly identical: HAIX sorbents have been used for removal of both arsenate and phosphate [82,121–124].

Lewis acid–base interactions are the primary sorption mechanism for such selective separations, while ion exchange or coulombic interactions have only a very minor effect. Of all the commonly present competing anions, sulfate is divalent and expected to offer significant competition through enhanced coulombic or electrostatic interactions. Figure 3.58 shows that doubling the sulfate concentration from 120 to 240 mg/L has practically no effect on phosphate sorption capacity of HAIX.

In fixed-bed column runs also, HAIX demonstrated its ability to remove phosphate from diluted secondary wastewater from the Bethlehem wastewater treatment plant

Table 3.9 As(V) and As(III) oxyacids and conjugate anions.

Oxyacid	pK_a values	Predominant dissolved species at pH 5.5	Predominant dissolved species at pH 8.5
As(V): H_3AsO_4	$pK_{a1} = 2.2$ $pK_{a2} = 6.98$ $pK_{a3} = 11.6$	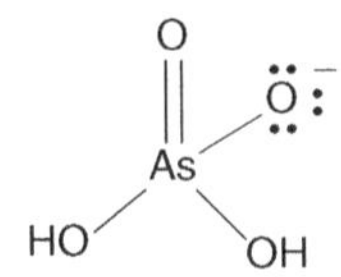 Monovalent monodentate ligand	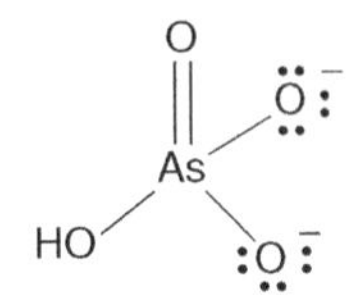 Divalent bidentate ligand
P(V): H_3PO_4	$pK_{a1} = 2.12$ $pK_{a2} = 7.21$ $pK_{a3} = 12.67$	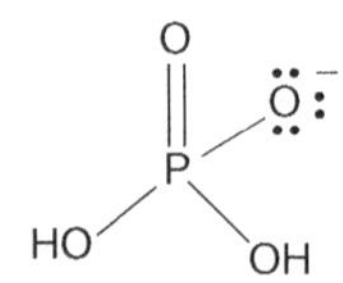 Monovalent monodentate ligand	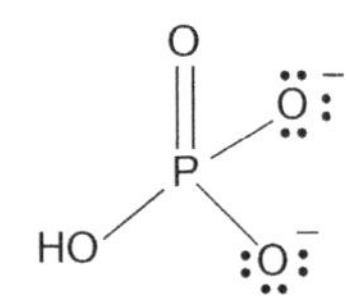 Divalent bidentate ligand
As(III): $HAsO_2$	$pK_{a1} = 9.2$	$O{=}\ddot{As}{-}OH$ Non-ionized monodentate ligand	$O{=}\ddot{As}{-}OH$ Non-ionized monodentate ligand

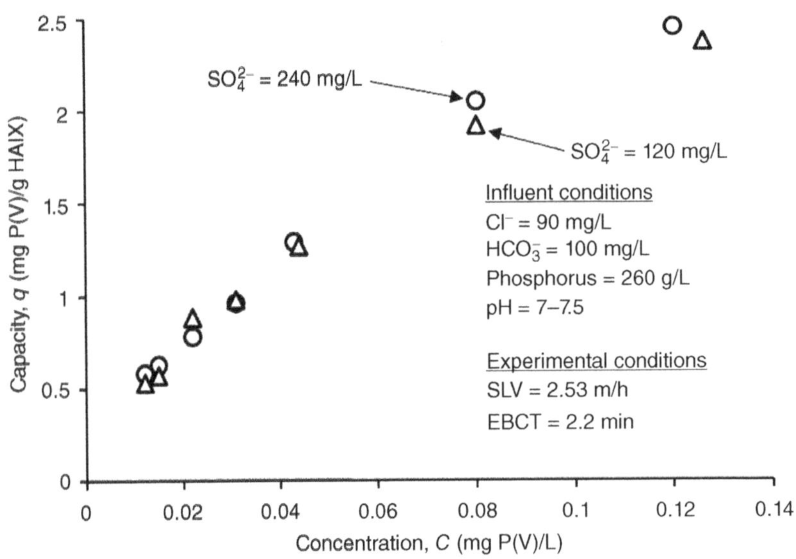

Figure 3.58 Comparison of phosphate isotherms for HAIX at two different background sulfate concentrations, all other conditions remaining identical. *Source*: Blaney *et al*. 2007 [82]. Reproduced with permission of Elsevier.

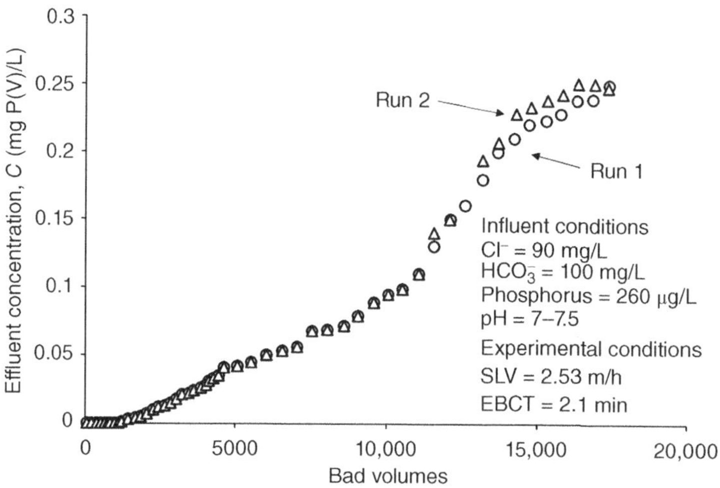

Figure 3.59 Phosphate effluent histories during two consecutive runs with secondary wastewater from the Bethlehem WWTP (Bethlehem, PA, USA) using "virgin" HAIX (Run 1) and "regenerated" HAIX (Run 2). *Source*: Blaney *et al.* 2007 [82]. Reproduced with permission of Elsevier.

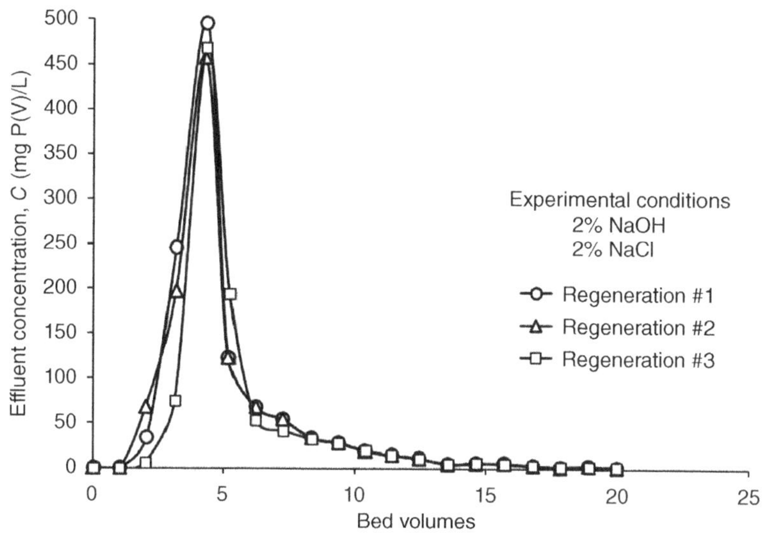

Figure 3.60 Phosphate elution profiles during regeneration of HAIX resin with high phosphate recovery (>95%) in 12 bed volumes. *Source*: Blaney *et al.* 2007 [82]. Reproduced with permission of Elsevier.

(WWTP) for consecutive cycles. In between the repeat treatment cycles, the bed was efficiently regenerated with 2% NaCl/2% NaOH achieving over 95% phosphate recovery from the bed. Figures 3.59 and 3.60 include the experimental results for column runs and intermittent regeneration [122].

The chemistry of P(V) and As(V) is nearly the same and so is their removal by sorption, with As(V) having slightly greater affinity than P(V), in accordance with its greater molecular weight and smaller hydrated ionic radius. Selective removal of

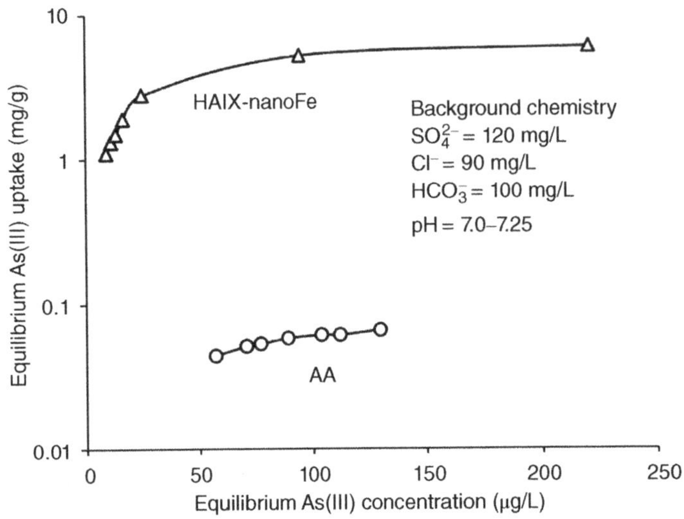

Figure 3.61 Isotherm of HAIX and activated alumina (AA) for sorption of As(III) species. *Source*: Sarkar *et al*. 2007 [124]. Reproduced with permission of Elsevier.

As(III), although through Lewis acid–base interactions, is significantly different from P(V) and As(V) sorption due to two reasons: (i) As(III) or arsenite is non-ionized and (ii) As(III) is a relatively soft Lewis base compared to arsenate or phosphate. Between iron oxide and aluminum oxide, the former is a relatively soft Lewis acid. Consequently, arsenite exhibits much higher sorption affinity onto Fe(III) oxide than Al(III) oxide. Figure 3.61 demonstrates that, all other conditions remaining identical, iron oxide-based HAIX-NanoFe offers nearly two orders of magnitude greater As(III) sorption capacity than activated alumina (AA).

It is noteworthy that the redox environment for groundwater often favors the formation of arsenite or As(III) over arsenate or As(V). Thus, arsenic remediation must include pre-oxidation of As(III) to As(V) or a sorbent with high As(III) removal capacity [125–127].

3.17.7 Fluoride (F^-)

Due to natural geochemical soil leaching, fluoride is present in many underground aquifers around the world, particularly in the continents of Asia and Africa. Although not as toxic as arsenic from a life-threatening viewpoint, consumption of fluoride-contaminated groundwater causes mottled teeth and bone deformations, commonly known as dental and skeletal fluorosis, respectively. In many remote villages in Africa and Asia, fluoride-contaminated groundwater is the only viable potable water source and the majority of the people continue to suffer from health impairments caused by contaminated groundwater. A fluoride concentration of 1.5 mg/L is the drinking water limit recommended by the WHO, but thousands of groundwater wells currently in use exceed the limit. Mitigation of the global crisis resulting from fluoride-contaminated groundwater remains mostly unsolved. A new class of hybrid ion exchangers that has been conceived, field-tested and commercialized for fluoride removal is discussed in Chapter 6 [128]. Key attributes of this new sorbent are: (i) high fluoride selectivity; (ii) regenerability for multiple cycles without loss in sorption capacity, and (iii) partial desalination with over 95% treated water recovery. The interested reader is encouraged to read further in Chapter 6.

Summary

- Selective ion exchange is more than mere exchange of ions with coulombic or electrostatic interactions. Lewis acid–base and/or hydrophobic interactions are often concurrently present and enhance selectivity.
- Ions are considered trace when their concentrations are relatively insignificant compared to competing ions in both the aqueous and the exchanger phase. Trace ions yield linear isotherms and do not influence the removal capacities for other trace ions.
- Elution ion chromatography of trace species is based on the principle of trace ion exchange. Having a linear isotherm, the chromatographic peak of 2.0 mg/L NO_3^- elutes at the same time and exactly double the height of 1.0 mg/L NO_3^-.
- Trace ions can be transported through ion exchange membranes against a negative concentration gradient without an applied electric filed. This is a thermodynamically favorable process known as Donnan dialysis.
- In dilute solutions, divalent anions exhibit higher selectivity than monovalent anions for anion exchange resins due to the electroselectivity effect, for example, for SO_4^{2-} vs NO_3^-, $\alpha_{S/N} > 1$. By increasing the size of the alkyl group in the quaternary ammonium functional group of a strong-base anion exchanger, sulfate-nitrate selectivity can be reversed, even in dilute solutions, that is, by replacing methyl groups with butyl groups, $\alpha_{S/N} < 1$.
- HIOCs, for example, pentachlorophenate, have NPM due to the presence of aromatic groups. HIOCs exhibit very high sorption affinity for polymeric anion exchange resins. Such favorable sorption processes are endothermic and associated with positive entropy changes.
- Linear force energy relationships (LFERs) can be used to predict the ion exchange affinity of target solutes from the knowledge of aqueous-phase stability constants or association constants.
- Oxides of Fe(III), Zr(IV), and Ti(IV) have amphoteric surfaces that can remove both transition metal cations, for example, Cu(II) and Pb(II), and anionic ligands, for example, arsenate, oxalate and phosphate. By appropriately selecting either cation exchangers or anion exchangers as the host material, the ion selectivity of amphoteric metal oxides can be tuned to either of the specific groups of solutes.
- All other conditions remaining identical, the plug-flow-reactor (PFR) configuration offers higher ion exchange capacity than a continuous stirred-tank reactor (CSTR) system.
- Uranium in groundwater exists as the uranyl carbonate tetravalent anion ($UO_2(CO_3)_3^{4-}$) and is very selectively removed by most anion exchangers. Removal of boron as borate onto a BSR takes place through formation of stable

complexes with polyalcohol functional groups. However, boron sorption kinetics are much slower compared to other ion exchange processes.

References

1 Pearson, R.G. (1968) Hard and soft acids and bases, HSAB, Part I: fundamental principles. *Journal of Chemical Education*, **45** (9), 581.

2 Pearson, R.G. (1968) Hard and soft acids and bases, HSAB, Part II: underlying theories. *Journal of Chemical Education*, **45** (10), 643.

3 Nieboer, E. and Richardson, D.H. (1980) The replacement of the nondescript term 'heavy metals' by a biologically and chemically significant classification of metal ions. *Environmental Pollution. Series B: Chemical and Physical*, **1** (1), 3–26.

4 SenGupta, A.K. (2001) Chapter 1: Principles of heavy metal separation: an introduction, in *Environmental Separation of Heavy Metals: Engineering Processes* (ed. A.K. SenGupta), CRC Press, Boca Raton, FL, pp. 1–14.

5 SenGupta, A.K. and Greenleaf, J.E. (2001) Chapter 8: arsenic in subsurface water: its chemistry and removal by engineered processes, in *Environmental Separation of Heavy Metals: Engineering Processes* (ed. A.K. SenGupta), CRC Press, Boca Raton, FL, pp. 265–306.

6 Li, P. (1999) *Sorption of Synthetic Aromatic Anions onto Polymeric Ion Exchangers: Genesis of Selectivity and Effects of Equilibrium Process Variables on Sorption Kinetics*, PhD dissertation, Lehigh University, Bethlehem, PA.

7 Helfferich, F.G. (1962) *Ion Exchange*, Dover Publications, New York.

8 Cotton, F.A., Wilkinson, G., Murillo, C.A. *et al.* (1999) *Advanced Inorganic Chemistry*, Wiley, New York.

9 Stumm, W. and Morgan, J.J. (1996) *Aquatic Chemistry: Chemical Equilibria and Rates in Natural Waters*, 4th edn, John Wiley & Sons, New York.

10 Li, P. and SenGupta, A.K. (1998) Genesis of selectivity and reversibility for sorption of synthetic aromatic anions onto polymeric sorbents. *Environmental Science & Technology*, **32** (23), 3756–3766.

11 Bolto, B., Dixon, D., Eldridge, R. *et al.* (2002) Removal of natural organic matter by ion exchange. *Water Research*, **36** (20), 5057–5065.

12 Zhao, D. and SenGupta, A.K. (1998) Ultimate removal of phosphate from wastewater using a new class of polymeric ion exchangers. *Water Research*, **32** (5), 1613–1625.

13 Wang, L., Chen, A.S., Lewis, G.M., Sorg, T.J., and Supply W. (2008) Arsenic and Uranium Removal from Drinking Water by Adsorptive Media: US EPA Demonstration Project at Upper Bodfish in Lake Isabella, CA: Final Performance Evaluation Report. National Risk Management Research Laboratory, Office of Research and Development, US Environmental Protection Agency.

14 Drever, J.I. (1988) *The Geochemistry of Natural Waters*, Prentice Hall, New Jersey.

15 SenGupta, A.K. (1995) Chapter 3: Chromate ion exchange, in *Ion Exchange Technology: Advances in Pollution Control* (ed. A.K. SenGupta), Technomic Publishing Co., Lancaster, PA, pp. 115–148.

16 Jang, M., Chen, W., and Cannon, F.S. (2008) Preloading hydrous ferric oxide into granular activated carbon for arsenic removal. *Environmental Science & Technology*, **42** (9), 3369–3374.

17 DeMarco, M.J., SenGupta, A.K., and Greenleaf, J.E. (2003) Arsenic removal using a polymeric/inorganic hybrid sorbent. *Water Research*, **37** (1), 164–176.

18 Cumbal, L. and SenGupta, A.K. (2005) Arsenic removal using polymer-supported hydrated iron(III) oxide nanoparticles: role of Donnan membrane effect. *Environmental Science & Technology*, **39** (17), 6508–6515.

19 SenGupta, A.K., Cumbal, L.H., inventors. SenGupta A.K., assignee (2007) Hybrid anion exchanger for selective removal of contaminating ligands from fluids and method of manufacture thereof. US Patent 7,291,578 B2. Nov. 6, 2007.

20 Fritz, J.S. and Gjerde, D.T. (2009) *Ion Chromatography*, 4th edn, John Wiley & Sons.

21 Skoog, D.A., Holler, F.J., and Crouch, S.R. (2007) *Principles of Instrumental Analysis*, 6th edn, Brooks Cole, Canada.

22 Small, H., Stevens, T.S., and Bauman, W.C. (1975) Novel ion exchange chromatographic method using conductimetric detection. *Analytical Chemistry*, **47** (11), 1801–1809.

23 Tswett, M.S. (1905) "О новой категории адсорбционных явлений и о применении их к биохимическому анализу" (O novoy kategorii adsorbtsionnykh yavleny i o primenenii ikh k biokkhimicheskomu analizu (On a new category of adsorption phenomena and on its application to biochemical analysis)), Труды Варшавского общества естествоиспытателей, отделении биологии. *Trudy Varshavskago Obshchestva Estestvoispytatelei, Otdelenie Biologii (Proceedings of the Warsaw Society of Naturalists [i.e., Natural Scientists], Biology Section)*, **14** (6), 20–39.

24 Donnan, F.G. (1911) Theorie der Membrangleichgewichte und Membranpotentiale bei Vorhandensein von nicht dialysierenden Elektrolyten. Ein Beitrag zur physikalisch-chemischen Physiologie. *Zeitschrift für Elektrochemie und Angewandte Physikalische Chemie*, **17** (14), 572–581.

25 Donnan, F. and Guggenheim, E. (1932) Exact thermodynamics of membrane equilibrium. *Zeitschrift fur Physikalische Chemie A*, **162**, 346–360.

26 Donnan, F.G. (1934) The thermodynamics of membrane equilibria. *Zeitschrift fur Physikalische Chemie A*, **A168**, 369–380.

27 Donnan, F.G. (1995) Theory of membrane equilibria and membrane potentials in the presence of non-dialysing electrolytes. A contribution to physical–chemical physiology. *Journal of Membrane Science*, **100** (1), 45–55.

28 C.E. Harland. *Ion Exchange: Theory and Practice.* : Royal Society of Chemistry; 1994.

29 Muraviev, D.N. and Khamizov, R. (2004) Ion-exchange isothermal supersaturation: concept, problems, and applications, in *Ion Exchange and Solvent Extraction. A Series of Advances*, vol. **16** (eds A.K. SenGupta and Y. Marcus), Marcel Dekker, New York, pp. 119–210.

30 Greenleaf, J.E., Cumbal, L., Staina, I., and SenGupta, A.K. (2003) Abiotic As(III) oxidation by hydrated Fe(III) oxide(HFO) microparticles in a plug flow columnar configuration. *Process Safety and Environmental Protection*, **81** (2), 87–98.

31 Clifford, D. and Weber, W.J. (1983) The determinants of divalent/monovalent selectivity in anion exchangers. *Reactive Polymers, Ion Exchangers, Sorbents*, **1** (2), 77–89.

32 Li, P. and SenGupta, A.K. (2004) Sorption of hydrophobic ionizable organic compounds (HIOCs) onto polymeric ion exchangers. *Reactive and Functional Polymers*, **60**, 27–39.

33 Guter, G.A. (1995) Nitrate removal from contaminated groundwater by anion exchange, in *Ion Exchange Technology – Advances in Pollution Control* (ed. A.K. SenGupta), Technomic Publishing Co., Inc., Lancaster, PA, pp. 61–113.
34 Clifford, D. and Liu, X. (1993) Ion exchange for nitrate removal. *Journal American Water Works Association*, **85** (4), 135–143.
35 Subramonian, S. and Clifford, D. (1988) Monovalent/divalent selectivity and the charge separation concept. *Reactive Polymers, Ion Exchangers, Sorbents*, **9** (2), 195–209.
36 Horng, L. and Clifford, D. (1997) The behavior of polyprotic anions in ion-exchange resins. *Reactive and Functional Polymers*, **35** (1), 41–54.
37 Liberti, L. and Passino, R. (1974) Chloride-sulphate exchange on anion-exchange resins: kinetic investigations. I. *Journal of Chromatography A*, **102**, 155–164.
38 Sarkar, S. and SenGupta, A.K. (2009) A hybrid ion exchange-nanofiltration (HIX-NF) process for energy efficient desalination of brackish/seawater. *Water Science and Technology: Water Supply*, **9** (4), 369–377.
39 Aveni, A., Boari, G., Liberti, L. *et al.* (1975) Sulphate removal and dealkalization on weak resins of the feed water for evaporation desalting plants. *Desalination*, **16** (2), 135–149.
40 Boari, G., Liberti, L., Merli, C., and Passino, R. (1974) Exchange equilibria on anion resins. *Desalination*, **15** (2), 145–166.
41 Guter GA, inventor (1984) The United States of America as Represented by the Administrator Environmental Protection Agency, assignee. Removal of nitrate from water supplies using a tributyl amine strong base anion exchange resin. US Patent 4,479,877 A. Oct. 30, 1984.
42 SenGupta, A.K., Roy, T., and Jessen, D. (1988) Modified anion-exchange resins for improved chromate selectivity and increased efficiency of regeneration. *Reactive Polymers, Ion Exchangers, Sorbents*, **9** (3), 293–299.
43 Matzner, R.A., Hunter, D.R., and Bales, R.C. (1991) The effect of pH and anions on the solubility and sorption behavior of acridine, in *Organic Substances and Sediments in Water, Volume 2: Processes and Analytical* (ed. R.A. Baker), Lewis Publishers, Chelsea, MI, p. 365.
44 Jafvert, C.T., Westall, J.C., Grieder, E., and Schwarzenbach, R.P. (1990) Distribution of hydrophobic ionogenic organic compounds between octanol and water: organic acids. *Environmental Science & Technology*, **24** (12), 1795–1803.
45 Stahl, P.H. and Wermuth, C.G. (2008) Monographs on acids and bases, in *Handbook of Pharmaceutical Salts Properties, Selection, and Use*, 2nd revised edn (eds P.H. Stahl and C.G. Wermuth), John Wiley & Sons, New York, pp. 327–422.
46 Gustafson, R. and Lirio, J. (1968) Adsorption of organic ions by anion exchange resins. *Industrial & Engineering Chemistry Product Research and Development*, **7** (2), 116–120.
47 Hinrichs, R. and Snoeyink, V. (1976) Sorption of benzenesulfonates by weak base anion exchange resins. *Water Research*, **10** (1), 79–87.
48 Lee, K. and Ku, Y. (1996) Removal of chlorophenols from aqueous solution by anion-exchange resins. *Separation Science and Technology*, **31** (18), 2557–2577.
49 Janauer, G.E. and Turner, I.M. (1969) Selectivity of a polystyrene-benzyltrimethyl-ammonium-type anion-exchange resin for alkanesulfonates. *The Journal of Physical Chemistry*, **73** (7), 2194–2203.

50 Gregory, J. and Semmens, M. (1972) Sorption of carboxylate ions by strongly basic anion exchangers. *Journal of the Chemical Society, Faraday Transactions 1: Physical Chemistry in Condensed Phases*, **68**, 1045–1052.
51 Frank, H.S. and Wen, W. (1957) Ion–solvent interaction. Structural aspects of ion–solvent interaction in aqueous solutions: a suggested picture of water structure. *Discussions of the Faraday Society*, **24**, 133–140.
52 Némethy, G. and Scheraga, H.A. (1962) Structure of water and hydrophobic bonding in proteins. I. A model for the thermodynamic properties of liquid water. *The Journal of Chemical Physics*, **36** (12), 3382–3400.
53 Huque, E.M. (1989) The hydrophobic effect. *Journal of Chemical Education*, **66** (7), 581.
54 Israelachvili, J.N. (1985) *Intermolecular and Surface Forces with Applications to Colloidal and Biological Systems*, Academic Press, New York.
55 Valsaraj, K.T. and Melvin, E.M. (1995) *Elements of Environmental Engineering: Thermodynamics and Kinetics*, Lewis Publishers, Boca Raton.
56 Lüning, U. (1989) Chr. Reichardt: solvents and solvent effects in organic chemistry, second, completely revised and enlarged edition, VCH Verlagsgesellschaft, Weinheim, Basel, Cambridge, New York 1988. 534 Seiten, Preis: DM 148. *Berichte der Bunsengesellschaft für Physikalische Chemie*, **93** (3), 416.
57 Li, P. and SenGupta, A.K. (2001) Entropy-driven selective ion exchange for aromatic ions and the role of cosolvents. *Colloids and Surfaces A: Physicochemical and Engineering Aspects*, **191** (1), 123–132.
58 Stumm, W. and Morgan, J.J. (1996) *Aquatic Chemistry: Chemical Equilibria and Rates in Natural Water*, 4th edn, Wiley-Interscience Publication, New York.
59 Maity, N., Payne, G.F., Ernest, M.V., and Albright, R.L. (1992) Caffeine adsorption from aqueous solutions onto polymeric sorbents: the effect of surface chemistry on the adsorptive affinity and adsorption enthalpy. *Reactive Polymers*, **17** (3), 273–287.
60 Mackay, D., Shiu, W.Y., and Ma, K. (1997) *Illustrated Handbook of Physical–Chemical Properties of Environmental Fate for Organic Chemicals*, CRC Press, Boca Raton.
61 Dean, J.A. and Lange, N.A. (1992) *Lange's Handbook of Chemistry*, 14th edn, McGraw Hill, New York.
62 Boyd, G.E. and Larson, Q.V. (1967) Binding of quaternary ammonium ions by polystyrenesulfonic acid type cation exchangers. *Journal of the American Chemical Society*, **89** (24), 6038–6042.
63 Ide, M., Maeda, Y., and Kitano, H. (1997) Effect of hydrophobicity of amino acids on the structure of water. *The Journal of Physical Chemistry B*, **101** (35), 7022–7026.
64 Feitelson, J. (1969) in *Ion Exchange, Version 2* (ed. J.A. Marinsky), Marcel Dekker, New York, pp. 135–166.
65 Franks, F., Mathias, S.F., Galfre, P. *et al.* (1983) Ice nucleation and freezing in undercooled cells. *Cryobiology*, **20** (3), 298–309.
66 Eisenman, G. (1962) Cation selective glass electrodes and their mode of operation. *Biophysical Journal*, **2** (2 Pt 2), 259–323.
67 SenGupta, A.K. and Clifford, D. (1986) Some unique characteristics of chromate ion exchange. *Reactive Polymers, Ion Exchangers, Sorbents*, **4** (2), 113–130.
68 Akhadov, Y.Y. (1981) *Dielectric Properties of Binary Solutions: A Data Handbook*, Pergamon Press, Oxford.

69 Burger, K. (1983) *Solvation, Ionic and Complex Formation Reactions in Non-aqueous Solvents – Experimental Methods for Their Investigation*, Elsevier, Amsterdam.
70 Krygowski, T.M., Wrona, P.K., Zielkowska, U., and Reichardt, C. (1985) Empirical parameters of Lewis acidity and basicity for aqueous binary solvent mixtures. *Tetrahedron*, **41** (20), 4519–4527.
71 Zhao, D., SenGupta, A.K., and Stewart, L. (1998) Selective removal of Cr(VI) oxyanions with a new anion exchanger. *Industrial and Engineering Chemistry Research*, **37** (11), 4383–4387.
72 Zhao, D. and SenGupta, A.K. (2000) Ligand separation with a copper(II)-loaded polymeric ligand exchanger. *Industrial and Engineering Chemistry Research*, **39** (2), 455–462.
73 Roy, T.K. (1989) *Chelating Polymers: Their Properties and Applications in Relation to Removal, Recovery and Separation of Toxic Metals*, MS thesis, Lehigh University.
74 Butler, J.N. (1964) *Ionic Equilibrium: A Mathematical Approach*, Addison-Wesley, Reading, MA.
75 SenGupta, A.K., Subramonian, S., and Clifford, D. (1988) More on mechanism and some important properties of chromate ion exchange. *Journal of Environment Engineering*, **114** (1), 137–153.
76 SenGupta, A.K., Clifford, D., and Subramonian, S. (1986) Chromate ion-exchange process at alkaline pH. *Water Research*, **20** (9), 1177–1184.
77 Clifford, D.A. (1999) Chapter 9: Ion exchange and adsorption of inorganic contaminants, in *Water Quality & Treatment Handbook*, 5th edn (ed. R.D. Letterman), McGraw-Hill Professional, pp. 9.1–9.91.
78 Dzombak, D.A. and Morel, F.M. (1990) *Surface Complexation Modeling: Hydrous Ferric Oxide*, John Wiley & Sons.
79 Dutta, P.K., Ray, A.K., Sharma, V.K., and Millero, F.J. (2004) Adsorption of arsenate and arsenite on titanium dioxide suspensions. *Journal of Colloid and Interface Science*, **278** (2), 270–275.
80 Lieser, K.H. (1991) Non-siliceous inorganic ion exchangers, in *Ion Exchangers* (ed. K. Dorfner), Walter de Gruyter, pp. 519–546.
81 Yuchi, A., Ogiso, A., Muranaka, S., and Niwa, T. (2003) Preconcentration of phosphate and arsenate at sub-ng ml^{-1} level with a chelating polymer-gel loaded with zirconium(IV). *Analytica Chimica Acta*, **494** (1), 81–86.
82 Blaney, L.M., Cinar, S., and SenGupta, A.K. (2007) Hybrid anion exchanger for trace phosphate removal from water and wastewater. *Water Research*, **41** (7), 1603–1613.
83 Sarkar, S., Blaney, L.M., Gupta, A. *et al.* (2008) Arsenic removal from groundwater and its safe containment in a rural environment: validation of a sustainable approach. *Environmental Science & Technology*, **42** (12), 4268–4273.
84 Puttamraju, P. and SenGupta, A.K. (2006) Evidence of tunable on–off sorption behaviors of metal oxide nanoparticles: role of ion exchanger support. *Industrial and Engineering Chemistry Research*, **45** (22), 7737–7742.
85 Yang, R.T. (2013) *Gas Separation by Adsorption Processes*, Butterworth-Heinemann, Oxford.
86 Suzuki, M. (1990) *Adsorption Engineering*, vol. **551**, Kodansha, Tokyo, pp. 128–132.

87 Tripp, A.R. and Clifford, D.A. (2004) Selectivity considerations in modeling the treatment of perchlorate using ion-exchange processes, in *Ion Exchange and Solvent Extraction: A Series of Advances* (eds A.K. SenGupta and Y. Marcus), CRC Press, New York, pp. 267–338.
88 Bonnesen, P.V., Brown, G.M., Alexandratos, S.D. *et al.* (2000) Development of bifunctional anion-exchange resins with improved selectivity and sorptive kinetics for pertechnetate: batch-equilibrium experiments. *Environmental Science & Technology*, **34** (17), 3761–3766.
89 Kunin, R. and Myers, R.J. (1950) *Ion Exchange Resins*, John Wiley & Sons, Inc., New York.
90 Cothern, R.C. and Lappenbusch, W.L. (1983) Occurrence of uranium in drinking water in the US. *Health Physics*, **45** (1), 89–99.
91 Kurttio, P., Auvinen, A., Salonen, L. *et al.* (2002) Renal effects of uranium in drinking water. *Environmental Health Perspectives*, **110** (4), 337.
92 Zamora, M.L., Tracy, B.L., Zielinski, J.M. *et al.* (1998) Chronic ingestion of uranium in drinking water: a study of kidney bioeffects in humans. *Toxicological Sciences*, **43** (1), 68–77.
93 Kurttio, P., Komulainen, H., Leino, A. *et al.* (2005) Bone as a possible target of chemical toxicity of natural uranium in drinking water. *Environmental Health Perspectives*, **1**, 68–72.
94 Sorg, T.J. (1988) Methods for removing uranium from drinking water. *Journal American Water Works Association*, **80**, 105–111.
95 Zhang, Z. and Clifford, D.A. (1994) Exhausting and regenerating resin for uranium removal. *Journal of the American Water Works Association*, **86** (4), 228–241.
96 Clifford, D.A. and Zhang, Z. (1995) Chapter 1: Removing uranium and radium from groundwater by ion exchange resins, in *Ion Exchange Technology: Advances in Pollution Control* (ed. A.K. SenGupta), Technomic Publishing Co., Lancaster, PA, pp. 1–60.
97 Rozelle, R.E. and Ma, K.W. (1983) *A New Potable Water Radium/Radon Removal System*, Dow Chemical Co., Midland, MI.
98 Hatch, M, inventor (1984) The Dow Chemical Company, assignee. Removal of metal ions from aqueous medium using a cation-exchange resin having water-insoluble compound dispersed therein. EP Patent 0,071,810 A1. Aug. 3, 1981.
99 World Health Organization (2004) *Guidelines for Drinking-Water Quality: Recommendations*, World Health Organization. REF Cotton, Wilkinson. Inorganic Chemistry Handbook.
100 Cotton, A.F., Wilkinson, G., and Gaus, P.L. (1995) *Basic Inorganic Chemistry*, 3rd edn, Wiley, New York, NY.
101 Xu, Y. and Jiang, J.Q. (2008) Technologies for boron removal. *Industrial & Engineering Chemistry Research*, **47** (1), 16–24.
102 Vanhoorne, P., Schelhaas, M., inventors. Lanxess Deutschland Gmbh, assignee (2010) Boron-selective resins. US Patent 20,110,108,488 A1. May 12, 2011.
103 Simonnot, M., Castel, C., Nicolai, M. *et al.* (2000) Boron removal from drinking water with a boron selective resin: is the treatment really selective? *Water Research*, **34** (1), 109–116.

104 Ipek, I., Guler, E., Kabay, N., and Yuksel, M. (2016) Removal of boron from water by ion exchange and hybrid processes, in *Ion Exchange and Solvent Extraction: A Series of Advances*, vol. **22** (ed. A.K. SenGupta), CRC Press, Boca Raton, pp. 33–64.
105 Beauvais, J. (2016) Request for nominations for peer reviewers for EPA's draft biologically based dose–response (BBDR) model for perchlorate, draft model support document and draft approach for deriving a maximum contaminant level goal (MCLG) for perchlorate in drinking water. FR Doc. 2016-12724.
106 State of California Environmental Protection Agency. (2015) *Public Health Goal for Perchlorate in Drinking Water*. OEHHA, Pesticide and Environmental Toxicology Branch.
107 Gu, B., Brown, G.M., Bonnesen, P.V. *et al.* (2000) Development of novel bifunctional anion-exchange resins with improved selectivity for pertechnetate sorption from contaminated groundwater. *Environmental Science & Technology*, **34** (6), 1075–1080.
108 Barrett, J., Lundquist, E., Miers, J., Pafford, M., Carlin, W., inventors. Henry, B.J., Gustave, L.E., Alfred, M.J., Mary, P.M., and Harris, C.W., assignees (2004) High selectivity perchlorate removal resins and methods and systems using same. US Patent 20,040,256,597 A1. Dec. 23, 2004.
109 Gu, B., Bohlke, J.K., Sturchio, N.C. *et al.* (2011) Applications of selective ion exchange for perchlorate removal, recovery, and environmental forensics, in *Ion Exchange and Solvent Extraction: A Series of Advances*, vol. **20** (ed. A.K. SenGupta), CRC Press, Boca Raton, pp. 117–144.
110 Gu, B., Brown, G.M., and Chiang, C. (2007) Treatment of perchlorate-contaminated groundwater using highly selective, regenerable ion-exchange technologies. *Environmental Science & Technology*, **41** (17), 6277–6282.
111 Bhattarai, B., Muruganandham, M., and Suri, R.P. (2014) Development of high efficiency silica coated β-cyclodextrin polymeric adsorbent for the removal of emerging contaminants of concern from water. *Journal of Hazardous Materials*, **273**, 146–154.
112 Chimchirian, R.F., Suri, R.P., and Fu, H. (2007) Free synthetic and natural estrogen hormones in influent and effluent of three municipal wastewater treatment plants. *Water Environment Research*, **79** (9), 969–974.
113 Qu, W., Suri, R.P., Bi, X. *et al.* (2010) Exposure of young mothers and newborns to organochlorine pesticides (OCPs) in Guangzhou, China. *The Science of the Total Environment*, **408** (16), 3133–3138.
114 He, K., Soares, A.D., Adejumo, H. *et al.* (2015) Detection of a wide variety of human and veterinary fluoroquinolone antibiotics in municipal wastewater and wastewater-impacted surface water. *Journal of Pharmaceutical and Biomedical Analysis*, **106**, 136–143.
115 Van Epps, A. and Blaney, L. (2016) Antibiotic residues in animal waste: occurrence and degradation in conventional agricultural waste management practices. *Current Pollution Reports*, **2**, 135–155.
116 Hopkins, Z.R. and Blaney, L. (2016) An aggregate analysis of personal care products in the environment: identifying the distribution of environmentally-relevant concentrations. *Environment International*, **92**, 301–316.

117 Jadbabaei, N. and Zhang, H. (2014) Sorption mechanism and predictive models for removal of cationic organic contaminants by cation exchange resins. *Environmental Science & Technology*, **48** (24), 14572–14581.
118 O'Neal, J.A. and Boyer, T.H. (2013) Phosphate recovery using hybrid anion exchange: applications to source-separated urine and combined wastewater streams. *Water Research*, **47** (14), 5003–5017.
119 Landry, K.A. and Boyer, T.H. (2013) Diclofenac removal in urine using strong-base anion exchange polymer resins. *Water Research*, **47** (17), 6432–6444.
120 Sendrowski, A. and Boyer, T.H. (2013) Phosphate removal from urine using hybrid anion exchange resin. *Desalination*, **322**, 104–112.
121 Wang, L., Chen, A.S.C., and Lewis, G.M. (2010) Arsenic and Uranium Removal from Drinking Water by Adsorptive Media. Final performance evaluation report for contract no. 68-C-00-185. *Battelle*; EPA/600/R-10/165.
122 Sengupta, S. and Pandit, A. (2011) Selective removal of phosphorus from wastewater combined with its recovery as a solid-phase fertilizer. *Water Research*, **45** (11), 3318–3330.
123 Cumbal, L., Greenleaf, J., Leun, D., and SenGupta, A.K. (2003) Polymer supported inorganic nanoparticles: characterization and environmental applications. *Reactive and Functional Polymers*, **54** (1), 167–180.
124 Sarkar, S., Blaney, L.M., Gupta, A. *et al.* (2007) Use of ArsenXnp, a hybrid anion exchanger, for arsenic removal in remote villages in the Indian subcontinent. *Reactive and Functional Polymers*, **67** (12), 1599–1611.
125 Sankar, M.U., Aigal, S., Maliyekkal, S.M., Chaudhary, A., Kumar, A.A., Chaudhari, K., and Pradeep, T. (2013) Biopolymer-reinforced synthetic granular nanocomposites for affordable point-of-use water purification. *Proceedings of the National Academy of Sciences*. **110** (21), 8459–8464.
126 Ravenscroft, P., Brammer, H., and Richards, K. (2009) *Arsenic Pollution: A Global Synthesis*, John Wiley & Sons.
127 De, S. and Maiti, A. (2012) *Arsenic Removal from Contaminated Groundwater*, The Energy Resources Institute (TERI), New Delhi.
128 German, M. (2017) *Hybrid anion exchange nanotechnology (HAIX-Nano) for concurrent trace contaminant removal with partial desalination*, PhD dissertation, Lehigh University, Bethlehem, PA.

4

Ion Exchange Kinetics: Intraparticle Diffusion

Similar to other sorption processes, equilibrium or near-equilibrium capacity is unattainable in real-life ion exchange processes, regardless of the equipment configuration, due to kinetic limitations. The rate at which ion exchange reactions proceed is a complex function of several inter-related processes that may be influenced by the individual or combined effects of:

(1) Ion exchanger properties (e.g., capacity, functional groups, porosity)
(2) External fluid dynamics (e.g., Reynolds number)
(3) Fluid properties (e.g., concentration, pH, temperature)
(4) Counterion properties (e.g., diffusivity, sorption affinity, hydrophobicity)
(5) Concentration gradients in both phases
(6) Electric charge gradient in both phases.

Ion exchange is a coupled process: sorption of one counterion is always accompanied by desorption of equivalent amounts of other counterions. The time-dependent coupling of mass and charge transfer for real-life ion exchange processes is complex and mathematically cumbersome, and more so for heterovalent ion exchange. Their application to real-life systems is often viewed as unduly time-consuming and not worth the effort. Yet, an insightful understanding of the kinetics of ion exchange may be meaningful in two major ways: (i) to elucidate the mechanisms that control or contribute to the overall reaction rate and (ii) to identify opportunities to improve ion exchange kinetics.

4.1 Role of Selectivity

To illustrate the relationship between selectivity and kinetics in ion exchange processes, let us consider our familiar exchange of counterions A^+ and B^+ as follows:

$$\overline{R^-B^+} + A^+(aq) \leftrightarrow \overline{R^-A^+} + B^+(aq) \tag{4.1}$$

The entire ion exchange process, uptake of A^+ by the ion exchanger and the release of B^+ into the aqueous solution, can be broadly divided into six consecutive steps:

1. Transport of the counter ion A^+ from the bulk phase into the ion exchange liquid film layer.
2. Diffusion of A^+ across the film layer to the surface of the ion exchanger.

Ion Exchange in Environmental Processes: Fundamentals, Applications and Sustainable Technology,
First Edition. Arup K. SenGupta.

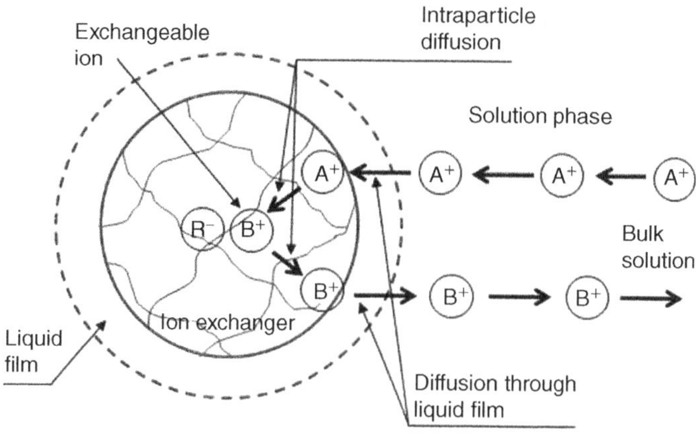

Figure 4.1 Illustration of ion exchange reaction and transport of counterions between the solution (liquid phase) and the ion exchanger (solid phase) where counterion B^+ is being gradually replaced by A^+.

3. Intraparticle diffusion of A^+ within the ion exchanger to the functional group $\overline{R^-}$.
4. Ion exchange at the fixed sites of the ion exchanger.
5. Intraparticle diffusion of B^+ to the film layer of the ion exchanger.
6. Transport of B^+ from the film layer to the bulk solution phase.

Figure 4.1 illustrates the process where the flux of A^+ is equal to, but directionally opposite to that of B^+, and electroneutrality is conserved. The rate-limiting step for selective ion exchange, almost without any exception, is the transport of counterions within the exchanger, that is, *intraparticle diffusion*. Thus, any improvement or tailoring of the ion exchange process, from a kinetic perspective, must address intraparticle diffusion.

In agreement with the previous chapters, let us emphasize the physical reality of intraparticle diffusion in selective ion exchange through a set of easy-to-comprehend experimental data, as presented in Figure 4.2a and b. Both figures present the results of anion exchange in the same format: normalized fractional uptake versus time. Fractional uptake varies from zero to unity, where unity represents the equilibrium capacity or capacity after infinite time under the experimental conditions. Figure 4.2a represents the isotopic exchange of sulfate ions with different sulfur isotopes, S^{32} and S^{34}, for two different anion exchange resins. Isotopic exchange allows the study of ion exchange between two otherwise identical counterions with separation factor equal to unity, that is, an ideal case of non-selective ion exchange [1]. On the contrary, Figure 4.2b presents the results of a kinetic study for the selective ion exchange between pentachlorophenol (PCP^-) and chloride (Cl^-). PCP^-, an aromatic anion, has much higher affinity than Cl^-; the PCP^-/Cl^- separation factor ($\alpha_{PCP/Cl}$), is well over 100 [2,3]. The two anion exchange reactions are shown below:

$$\overline{(R^+)_2SO_4^{2-}} + S^*O_4^{2-}(aq) \leftrightarrow \overline{(R^+)_2S^*O_4^{2-}} + SO_4^{2-}(aq) \tag{4.2}$$

$$\overline{(R^+)Cl^-} + PCP^-(aq) \leftrightarrow \overline{(R^+)PCP^-} + Cl^-(aq) \tag{4.3}$$

asterisk (*) represents sulfur isotopes of different atomic mass.

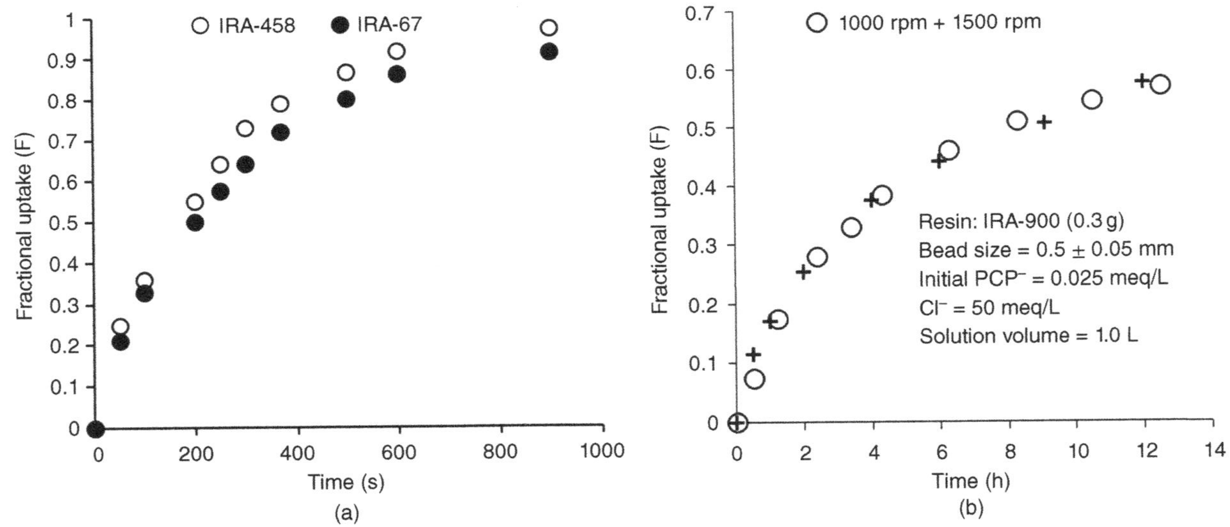

Figure 4.2 (a) Fractional uptake versus time plot for isotopic exchange of $S^*O_4^{2-}/SO_4^{2-}$ on anion exchanger IRA-458 and IRA-67 under identical experimental and hydrodynamic conditions. *Source:* Liberti 1983 [1]. Reproduced with permission of Springer. (b) Fractional uptake versus time plot for PCP^-/Cl^- exchange on anion exchanger IRA-900 for two different stirring speeds. *Source:* Li and SenGupta 2000 [2]. Reproduced with permission of Elsevier.

Both of these reactions are cases of homovalent anion exchange, but with a great difference in counterion selectivity. Note that in 15 min, more than 80% equilibrium capacity (fractional uptake, $F > 0.8$) is attained for the isotopic $S^*O_4^{2-}/SO_4^{2-}$ exchange (Figure 4.2a). In contrast, only 55% of equilibrium capacity is gained for PCP^-/Cl^- exchange even after 12 h, all other experimental conditions nearly being the same (Figure 4.2b). Increasing the stirrer speed in the batch kinetic study from 1000 to 1500 revolutions per minute (rpm) showed no noticeable increase in the uptake rate in either case.

Intraparticle diffusion, that is, the transport process within the anion exchanger, is the rate-limiting step in both cases. However, PCP^-/Cl^- exchange is markedly slower than the isotopic exchange because of the difference in sorption affinity between PCP^- and Cl^- toward the exchange sites of the anion exchanger. The scientific question is: *how does higher selectivity of an incoming counterion impede the rate of progress, or kinetics, of ion exchange*? Thus, intraparticle diffusion bears a special significance in selective ion exchange. Besides selectivity, water content, porosity and cross-linking also influence intraparticle diffusivity.

During the course of this chapter, we will first present experimental observations (i.e., physical realities) in relation to individual process variables. Only then will we attempt to provide mathematical models with reference to selective ion exchange. For now, it is advisable that the reader keeps in mind the most distinctive elements of Figure 4.2: the difference in sorption affinity of exchanging counterions strongly influences the kinetics of ion exchange and electroneutrality is to be conserved.

The supplementary materials, S4.1, presented below provide relevant background materials and examples. However, a reader with sufficient exposure to ion exchange or sorption processes may skip it without loss of continuity.

Supplementary Reading S4.1 Batch Kinetic Test and Construction of Fractional Uptake Curves

Figure S4.1 illustrates the batch kinetic test apparatus that was originally developed by Kressman and Kitchener [4,5]. A solution of an electrolyte of known concentration and volume is placed in the batch reactor. An ion exchange material of known mass is placed inside a polypropylene-wired (fine mesh) cage, which constitutes the center part of a centrifugal stirrer. The stirrer is immersed in the electrolyte solution and then stirred in the batch reactor. The ion exchange materials placed inside the cage are subjected to a rapid circulating flow of solution because fresh solution is sucked in through the bottom of the cage, forced out radially through the sorbent and the openings provided in the casing. Aliquots from the solution are taken at different time intervals and analyzed for the concentration of solute to record the ion exchange rate.

Figure S4.2 describes the results of a kinetics batch test for selective Zn(II) uptake by a specially prepared hybrid inorganic material (HIM) in the presence of competing calcium and sodium ions; experimental conditions are in figure [6]. The solution was continuously agitated using a stirrer at a speed of 1600 rpm. Note that the aqueous zinc concentration rapidly decreases from an initial concentration of 0.25 mg/L (250 μg/L) to 0.10 mg/L, and gradually decreases over time until it reaches an equilibrium

(Continued)

Supplementary Reading S4.1 (Continued)

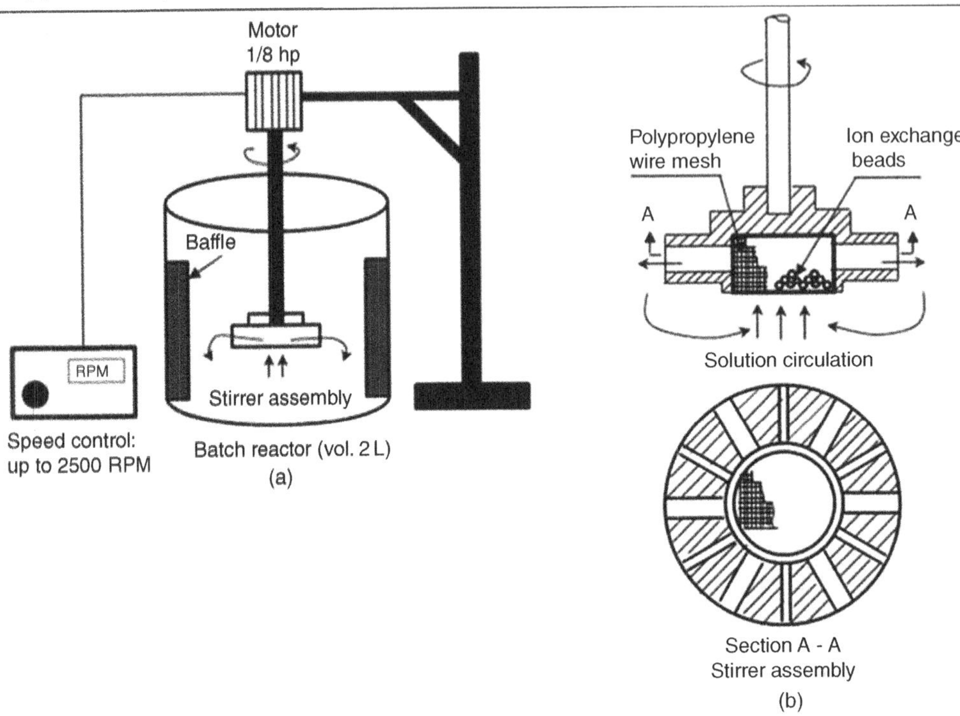

Figure S4.1. *(a) Kinetics batch test set-up; (b) illustration of stirrer assembly showing cross-sectional view. Source: Kressman and Kitchener 1949 [4]. Reproduced with permission of Royal Society of Chemistry.*

concentration of 0.081 mg/L. Fractional uptake by an ion exchanger is related, but expressed as the ratio of mass of solute taken up by the ion exchange material at any time "t" since the start of experiment and the equilibrium uptake after infinite time.

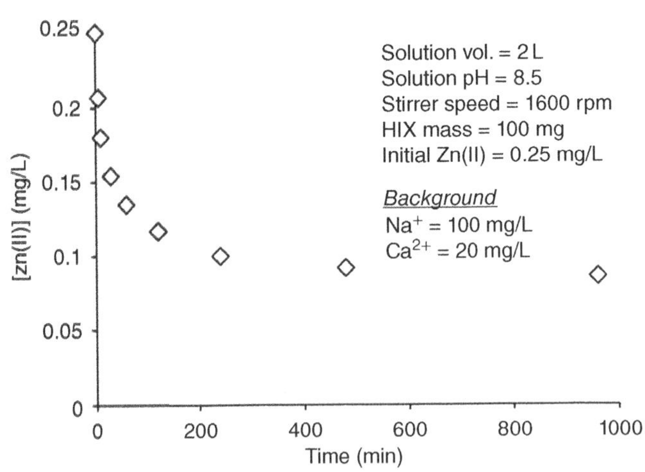

Figure S4.2. *Concentration versus time plot of batch kinetic test for Zn(II) sorption by HIM at stirrer speed 1600 rpm. Source: Reprinted with permission from Chatterjee 2011 [6].*

Example S4.1 Construction of a Fractional Uptake Curve

The batch kinetic test is performed for selective zinc uptake by HIM with 2 L solution having an initial Zn(II) concentration 0.25 mg/L along with 100 mg/L of Na^+ and 20 mg/L of Ca^{2+}. About 100 mg of HIM granules were placed inside the stirrer cage rotating at 1600 rpm. The solution Zn concentration versus time is provided in Table S4.1. The equilibrium concentration (i.e., concentration of Zn in solution taken after 72 h) is 0.081 mg/L [6]. Plot the Zn fractional uptake curve and find out the half time ($t_{1/2}$). Half time or $t_{1/2}$ is defined as the time at which the ion exchanger has attained 50% of its equilibrium capacity.

Table S4.1. *Zn(II) concentration over time during a kinetic batch test with HIM.*

Time (min)	Zn(II) concentration, mg/L (C_t)
0	0.250
5	0.207
10	0.180
30	0.158
60	0.135
120	0.117
240	0.100
480	0.092

Solution: *For a batch kinetic test with mass of ion exchanger "m," solution volume "V," initial solute concentration "C_o," and solute concentration "C_t" at time "t," the mass concentration of solute uptake "q_t" (mass of solute/mass of ion exchange, e.g., mg solute/g ion exchanger) by an ion exchanger is given by the following mass balance.*

$$q_t = \frac{V}{m}(C_o - C_t) \qquad (S4.1)$$

At equilibrium, solute concentration in solution is C_e (i.e., concentration after infinite time) and the corresponding mass of solute uptake by the ion exchanger after infinite time or equilibrium uptake q_∞ is

$$q_\infty = \frac{V}{m}(C_o - C_e) \qquad (S4.2)$$

Fractional uptake, F_t, by the ion exchanger is defined as the ratio of the mass uptake of the target solute (e.g., zinc) at time "t" since the start of the experiment versus equilibrium solute uptake capacity after infinite time.

Therefore, fractional uptake at time t

$$F_t = \frac{q_t}{q_\infty} = \frac{C_o - C_t}{C_o - C_e} \qquad (S4.3)$$

In accordance with Eq. (S4.3), the fractional uptake at different times can be computed for the present problem with $C_0 = 0.25$ mg/L, $C_e = 0.081$ mg/L, and C_t from Table S4.1. Computed "F_t" at time "t" is given in the following table. The plot of F_t versus t is shown

(Continued)

Supplementary Reading S4.1 (Continued)

in Figure S4.3 from where the half time ($t_{1/2}$) is readily obtained for $F_t = 0.5$ and equal to about 20 min.

Table S4.2. *Fractional uptake of Zn(II) at time t by HIM.*

Time (min)	$F_t = (C_o - C_t)/(C_o - C_e)$
0	0
5	0.254
10	0.414
30	0.568
60	0.680
120	0.787
240	0.887

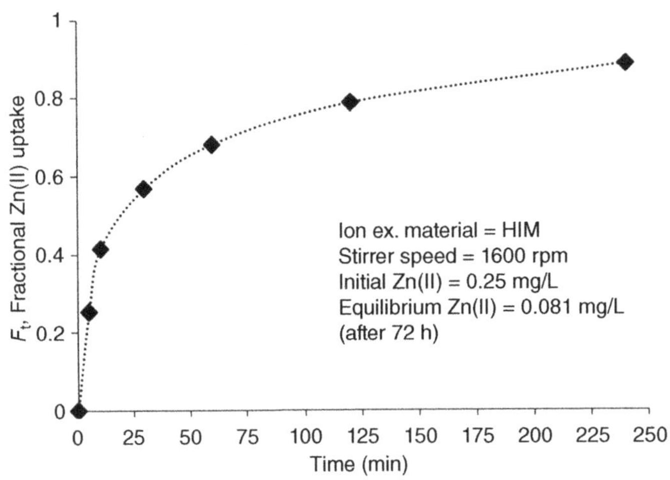

Figure S4.3. *Fractional uptake (F_t) versus time plot for Zn(II) sorption by HIM over time. Source: Reprinted with permission from Chatterjee 2011 [6].*

Example S4.2 Stirrer Speed and Rate Limiting Phenomena

Figure S4.4 shows concentration of zinc versus time plots for batch kinetic tests of zinc sorption by HIM at different stirrer speeds using the previously described experimental set-up (Figure S4.1). Besides stirrer speeds, all other conditions remained unchanged. Note that the rate of uptake is faster with higher stirrer speeds. The analysis of the data in Figure S4.4 raises the two following questions:

- *Why did the concentration plot (i.e., zinc uptake by ion exchanger) change with the change in stirrer speed from 500 to 800 to 1200 rpm?*
- *Why did the increase in stirrer speed from 1200 to 1600 rpm have no effect on the Zn concentration plot, that is, no effect on zinc uptake rate?*

Explain conceptually using words and figures.

Solution: *The governing or the rate-limiting step of ion exchange is determined by the slower of the two transport processes that occur in series, that is, external liquid-phase*

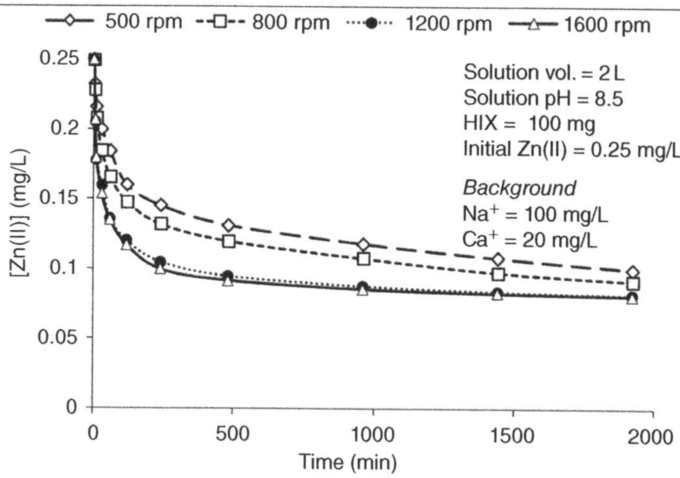

Figure S4.4. *Concentration versus time plots at different stirrer speeds for kinetic test of Zn(II) sorption by HIM. Source: Reprinted with permission from Chatterjee 2011 [6].*

film diffusion or intraparticle diffusion. Later, we will discuss and confirm that the "chemical reaction" of ion exchange is extremely fast and never the rate-limiting step. For the current batch reactor setup, lower stirrer speeds (e.g., 500 and 800 rpm) offer lower agitation rendering external "film diffusion" as the rate-limiting step. The term film diffusion is generally used to describe the condition when the resistance to mass transfer (or transport) lies across the liquid film at the solid–liquid interface. Therefore, the concentration gradient exists only across the liquid film surrounding every particle. In comparison, diffusion within the particle (i.e., intraparticle diffusion) is significantly faster than film diffusion, so that the concentration differences within the beads are instantly leveled out along the radius as illustrated in Figure S4.5.

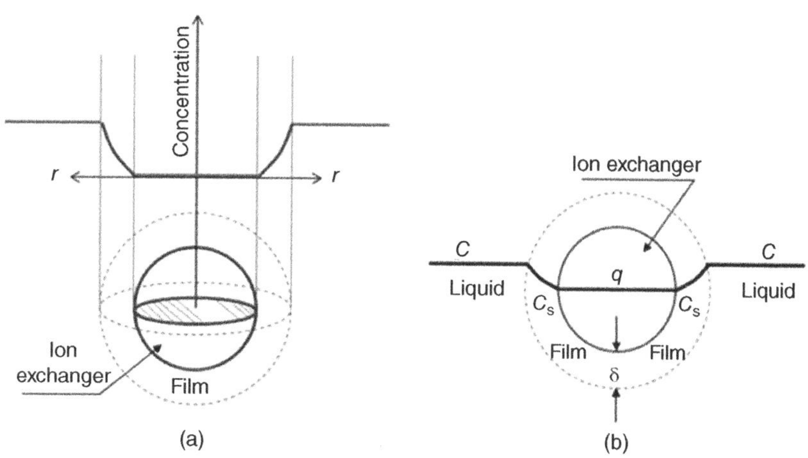

Figure S4.5. *(a) Radial concentration profile of an ion exchanger under film diffusion control kinetics; (b) illustration of a concentration gradient across the film surrounding the ion exchanger particle. C is the solute (i.e., dissolved zinc) concentration in the bulk; C_s is the concentration at the solid–liquid interface; and q is the solid phase concentration of zinc in equilibrium with C_s.*

(Continued)

Supplementary Reading S4.1 (Continued)

An increase in stirrer speed creates more agitation or turbulence in the water of the stirrer chamber (i.e., increase in Reynolds number), thus reducing film thickness (δ), which in turn increases the rate of transport. This phenomenon explains that external film diffusion is the rate-limiting step as the stirrer speed is gradually increased from 500 to 800 and then to 1200 rpm with concomitant increase in the uptake rate with stirrer speed.

However, as the stirrer speed was increased from 1200 to 1600 rpm, no change in concentration profiles (i.e., uptake rate) was observed. Beyond 1200 rpm stirrer speed, the resistance to the transport process was larger within the HIM particle than in the external liquid film. So, the concentration gradient existed only within the particle and intraparticle diffusion became the rate-limiting step. The prevailing concentration gradient under this condition is presented in Figure S4.6. Intraparticle diffusion is not influenced by the stirrer speed or agitation in the solution phase. Thus, the concentration versus time plots remains almost identical at stirrer speeds of 1200 and 1600 rpm.

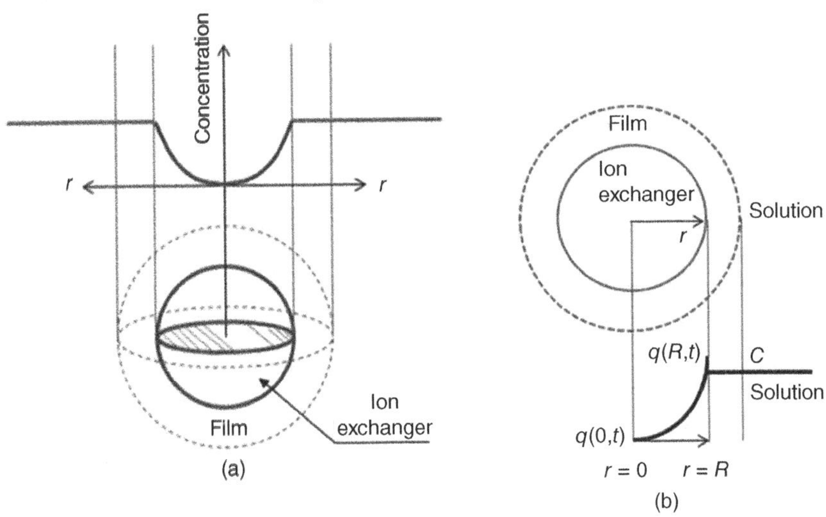

Figure S4.6. *(a) Radial concentration profile for intraparticle diffusion control; (b) illustration of concentration gradient within an ion exchanger for intraparticle diffusion-controlled process.*

4.2 State of Water Molecules inside Ion Exchange Materials

Even for isotopic exchange, that is, exchange with no relative selectivity between counterions, the intraparticle diffusivities within an ion exchanger are consistently lower than corresponding aqueous phase values and are greatly influenced by the exchanger's water content. The intraparticle diffusion coefficient ($\overline{D}_i$) of a species "i" is related to its solvent or aqueous phase diffusion coefficient (D_i). The two most common models describing such a relationship are as follows [7,8]:

$$\overline{D}_i = \frac{D_i \cdot \epsilon}{2} \tag{4.4}$$

$$\overline{D}_i = D_i\left(\frac{\epsilon}{2-\epsilon}\right)^2 \tag{4.5}$$

where "ϵ" is the dimensionless porosity or fractional pore volume within the exchanger and ϵ increases with an increase in free water molecules within the hydrated ion exchanger. The underlying commonality among the above two models is that with an increase in ϵ, $\overline{D}_i$ always increases. Thus, kinetics of selective ion exchange is greatly determined by the water content and swelling of the exchanger. Understandably, the properties of the ion exchanger that control swelling/shrinking directly influence the rate of intraparticle diffusion, for example, capacity and cross-linking.

Water molecules inside a typical ion exchanger that are in contact with an aqueous solution are not all identical, they exist in four different states:

(1) Water molecules are in spheres of hydration around fixed coions and diffusing counterions, primarily through ion–dipole interactions.
(2) Water is sorbed to the polymer matrix through hydrogen bonding or dipole–dipole interactions (significant for a polyacrylic matrix, but negligible for a polystyrene matrix).
(3) Water molecules form the hydration spheres of electrolytes that entered into the gel phase of the ion exchanger in violation of the Donnan coion exclusion.
(4) The bulk water molecules present inside the ion exchanger due to osmosis, that is, the difference in osmotic pressure between the ion exchanger and the external solution.

Let us make a note that inside the macropores of macroporous ion exchangers, water has the same activity as that of water in the external solution. Of all types of water inside an ion exchanger, the fourth type, caused by osmotic pressure difference, is the least structured (i.e., most mobile) and by far the most predominant within the exchanger. They fill up the conduits and channels within the ion exchanger through which exchanging counterions move or swim from site to site. For both selective and non-selective ion exchange, intraparticle diffusivity is significantly influenced by these free water molecules. The illustration in Figure 4.3 attempts to depict different types of water molecules within an ion exchanger.

At this point, the reader may appropriately refer to Example 2.1 and note that in a typical cation exchange resin with sulfonic acid functional groups, free water constitutes over 90% of the water present within an ion exchanger. Figure 4.4 shows how the $\overline{D}_i/D_i$ ratio, rapidly decreases with a decrease in ϵ value. Since the difference in osmotic pressure governs the amount of free water molecules in the resin phase, intraparticle diffusion can be enhanced by intelligently exploiting this phenomenon during selective ion exchange. Such an example is provided at the end of the chapter. By introducing non-selective strong-acid sulfonic acid functional groups along with weak-acid phosphonic acid chelating groups, the effective intraparticle diffusion rate is greatly improved [9,10].

Besides capacity and cross-linking, the swelling of a chelating ion exchanger is influenced by the relative affinity of the metal counterion, that is, lower affinity leads to greater swelling. Representative experiments were carried out using a chelating ion exchanger with iminodiacetate functional groups (IRC-718, Rohm and Haas Co.). Two spherical beads (H-form) with identical sizes were immersed in two separate solutions: (i) 200 mg/L $CaCl_2$ and (ii) 200 mg/L $CuCl_2$ [6,11–13]. The swelling rate was

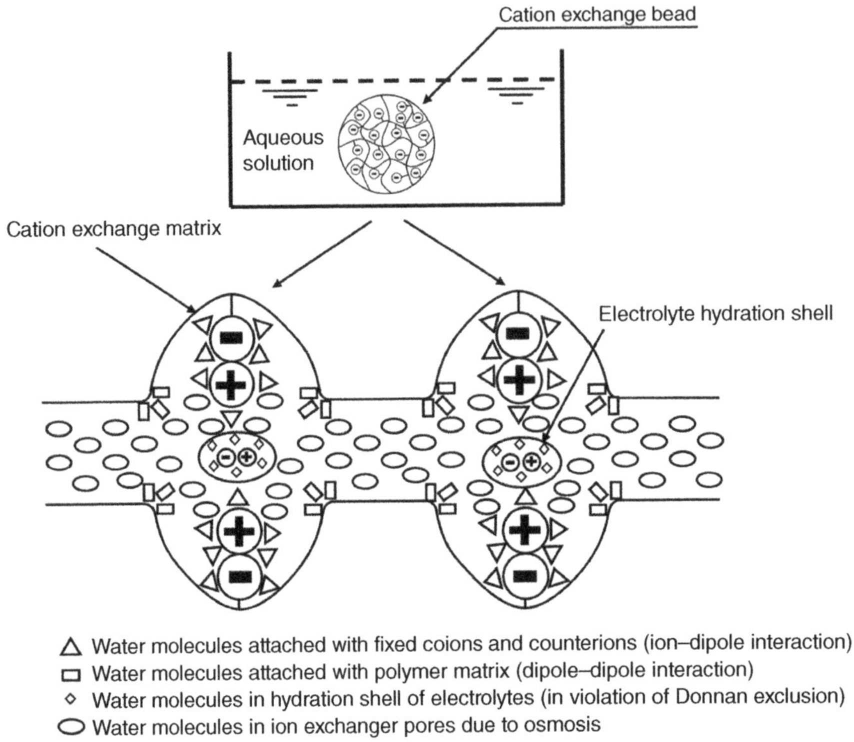

Figure 4.3 Schematic representation of different types of water molecules inside a cation exchanger.

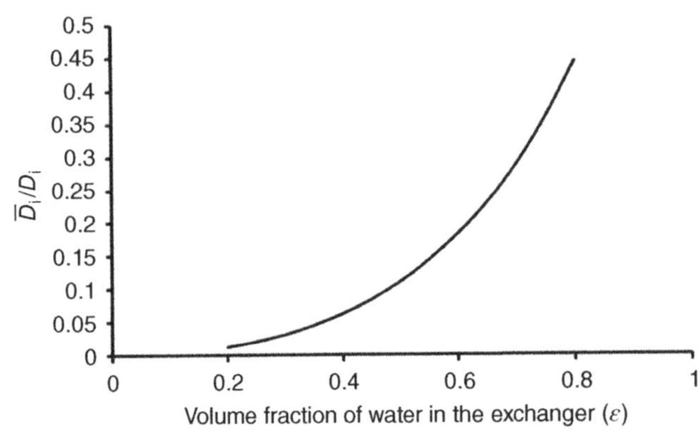

Figure 4.4 Plot showing the typical relation between intraparticle diffusivity and internal porosity of an ion exchanger bead.

monitored under a high-resolution microscope. Although both copper and calcium have the same valence, IRC-718 has a much higher affinity toward copper through Lewis acid-base interactions. In Figure 4.5, the bead swelled noticeably less in copper solution than in calcium solution under otherwise identical conditions. The formation of inner sphere complexes between the resin and copper releases waters of hydration from inside the chelating exchanger back into water. On the contrary, calcium ions only

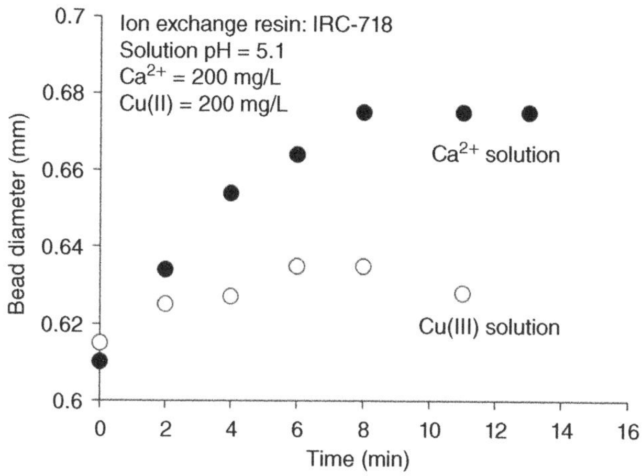

Figure 4.5 Experimental results of resin swelling of a chelating ion exchanger with iminodiacetate functional groups in the presence of Ca(II) or Cu(II) solutions. *Source:* Reprinted with permission from Chatterjee 2011 [6].

form outer-sphere complexes and retain a high degree of hydration. So, the osmotic pressure of a chelating exchanger loaded with copper is significantly less than that with calcium and, hence, there is less swelling. Figure 4.6 provides a schematic in this regard showing relative changes in osmotic pressure from H-form to Ca-form. In principle, any interaction or phenomenon that leads to swelling, and consequent increase in free water molecules in the exchanger phase, enhances intraparticle diffusion rates.

4.3 Activation Energy Level in Ion Exchangers: Chemical Kinetics

Although the process is reversible and no electron transfer takes place between participating ions, ion exchange has the appearance of a chemical process. During the early evolution of ion exchange technology around the time of the Second World War, chemical reactions were accepted, understandably, as the primary mechanism of ion exchange kinetics [14]. Subsequently, the role of diffusion-controlled transport processes, namely, external film diffusion and intraparticle diffusion were duly recognized [15]. The approach to determine the rate limiting step for a specific ion exchange reaction, however, remained empirical and the stepwise procedure entailed the following:

(1) Assume one of the three limiting mechanisms of ion exchange (e.g., external film diffusion, intraparticle diffusion, and chemical reaction).
(2) Deduce approximate mathematical solutions and linearize them in terms of concentration versus time plots.
(3) Carry out ion exchange testing.
(4) Identify the prevailing rate mechanism based on the best fit between the experimental data and the model prediction [16,17].

In accordance with this methodology, experiments are always carried out at a single temperature and minor imprecisions in data collection and subsequent trend fitting

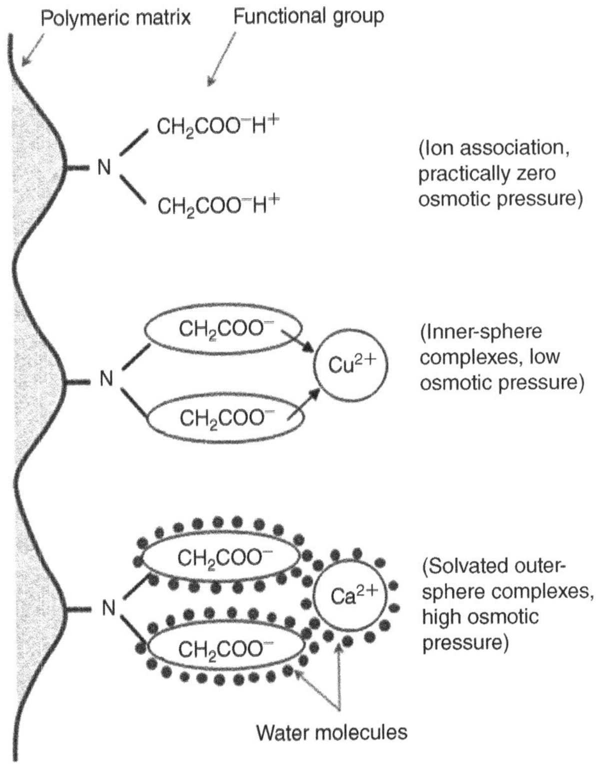

Figure 4.6 Schematic illustrating the relative increase in osmotic pressure of the chelating exchanger as it transitions from H-form to Cu-form and to Ca-form.

may lead to incorrect conclusions – such examples abound in the open literature. The knowledge of activation energy value – kilojoule per mole of the reactant – in a chemical process is the single most important determinant in identifying the rate-limiting step [18].

4.3.1 Activation Energy Determination from Experimental Results

Price *et al.* [19,20] studied nickel–sodium ion exchange using a chelating ion exchanger of macroporous structure containing aminophosphonic functional groups.

$$2\overline{RNa} + Ni^{2+} \leftrightarrow \overline{R_2Ni} + 2Na^{+} \tag{4.6}$$

The second order bimolecular reaction rate model for Eq. (4.6) is

$$r = -\frac{d[Ni^{2+}]}{dt} = k_2[\overline{RNa}][Ni^{2+}] \tag{4.7}$$

Experimental kinetic data were determined over a temperature range from 5 to 55 °C, as shown in Figure 4.7. At each temperature, the results were analyzed in accordance with the bi-molecular rate model of Eq. (4.7) to obtain the best-fitted specific rate constant k_2.

From the Arrhenius rate law,

$$k_2 = A\, e^{-E_a/RT} \tag{4.8}$$

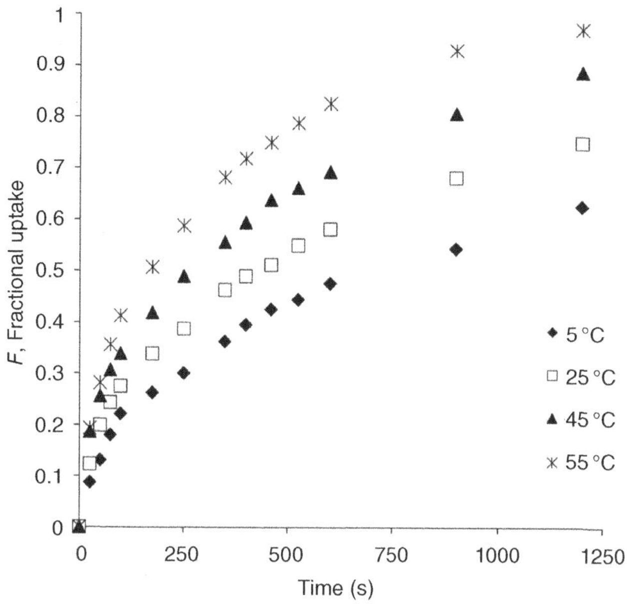

Figure 4.7 Effects of temperature on the fractional uptake profile of Ni(II) by a chelating ion exchanger in Na-form containing aminophosphonic functional groups. *Source:* Adapted from Price *et al.* 1988 [19].

Equation (4.8) can be linearized as follows:

$$\ln k_2 = \ln A - \frac{E_a}{RT} \tag{4.9}$$

where E_a is the activation energy, R the ideal gas constant, T the absolute temperature, and A the frequency factor constant characteristic to the chemical reaction. Specific rate constants were subsequently used to construct an Arrhenius plot, that is, $-\ln k_2$ versus $1/T$ as shown in Figure 4.8.

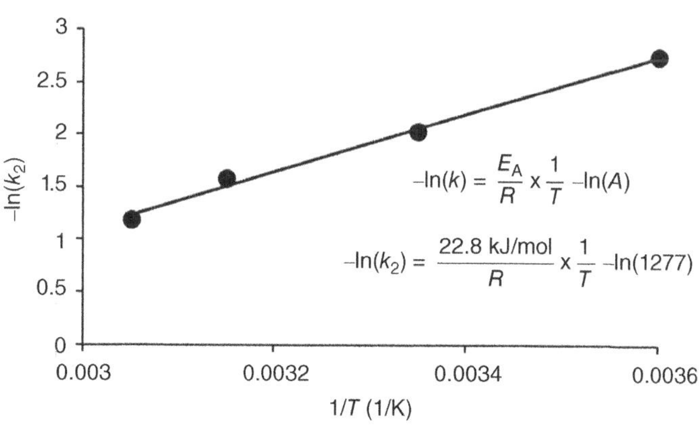

Figure 4.8 Arrhenius plot for Na/Ni exchange showing influence of temperature on ion exchange rate constant. *Source:* Adapted from Price *et al.* 1988 [19].

The activation energy for Ni^{2+}–Na^{+} exchange was found to be 22.8 kJ/mol. This relatively low value (≤100 kJ/mol) is characteristic of diffusion-controlled processes and is, therefore, compatible with a diffusion rate-limited mechanism. It is worth noting that highly selective ion exchange processes are not reaction kinetics limited unless they are accompanied by redox reactions. In the early history of ion exchange, it took a rather long time to dispel the notion that a "chemical reaction" is *not* the rate-limiting step in ion exchange processes. The following example deals with selective sorption of an anionic ligand, arsenate, onto a hybrid ion exchanger and provides a stepwise procedure to compute the activation energy from kinetic data.

Example 4.1 *The kinetic test of arsenate, As(V), sorption on to a ferric oxide based sorbent (hybrid anion exchanger (HAIX)) was conducted at three different temperatures: 5, 22, and 35 °C under controlled laboratory conditions. The mass of sorbent used was 0.04 g with diameter of sorbent beads 0.5 ± 0.05 mm. Initial arsenate concentration was around 100 μg/L with sulfate at 200 mg/L, a competing background ion. The volume of solution was 1.0 L and pH of the solution during the test was maintained at 7.0 ± 0.5. At different time intervals, 1 mL aliquots were withdrawn for analysis. Arsenic concentration versus time for tests at different temperatures is shown in Table 1. The equilibrium concentration was obtained after five days (120 h) (Experimental data are from unpublished work at Lehigh University).*

Determine if the Sorption of Arsenic is Controlled by Diffusion or Chemical Reaction. Provide Necessary Steps

Solution

Step 1. Calculate fractional uptake at different times for experiments conducted at different temperatures.

From Eq. (S4.3), fractional uptake (F_t) at any time t is given by:

$$F_t = \frac{q_t}{q_e} = \frac{V_t(C_o - C_t)}{V_e(C_o - C_e)} \qquad (1)$$

Ignoring the aliquots removed for sample analysis, V_t and V_e are identical. Where, C_o is initial solute concentration (at t = 0), C_t is concentration at time t, and C_e is concentration at equilibrium, that is, after five days.

From Table 1 for sorption at 5 °C and t = 2 h, $C_o = 111.0$ μg/L, $C_t = 92.6$ μg/L, and $C_e = 30.9$ μg/L

$$F_t = \frac{111.0 - 92.6}{111.0 - 30.9} = 0.230 \qquad (2)$$

Similarly, for 35 °C at t = 6 h, $C_o = 103.0$ μg/L, $C_t = 51.9$ μg/L, and $C_e = 10.2$ μg/L

$$F_t = \frac{103.0 - 51.9}{103.0 - 10.2} = 0.551 \qquad (3)$$

Table 1. Arsenic concentration versus time at three different temperatures.

Temperature (°C)	Time (h)	As(V) concentration in liquid phase, μg/L (ppb)
5	0	111.0
	0.5	105.0
	1.0	100.0
	1.5	95.2
	2.0	92.6
	3.0	87.1
	4.0	82.7
	5.0	80.3
	6.5	75.1
	120 (final)	30.9
22	0	105.0
	0.5	93.0
	1.0	87.1
	1.5	82.3
	2.0	76.9
	3.0	71.0
	4.0	65.7
	5.0	58.3
	6.0	55.8
	120 (final)	7.6
35	0	103.0
	0.25	97.6
	0.5	91.9
	1.0	86.1
	1.5	77.2
	2.0	75.0
	3.0	65.1
	4.0	60.1
	6.0	51.9
	120 (final)	10.2

The calculated fractional uptake is indicated in Table 2.

Data from Table 2 is plotted in Figure 1.

Step 2. Find out values of specific rate constants for bi-molecular reaction of As sorption onto the ferric oxide based surface.

The sorption of As(V) onto a ferric oxide based sorbent is represented as:

$$\mathrm{As(V)} + \overline{\mathrm{Fe}} \leftrightarrow \overline{\mathrm{Fe} : \mathrm{As(V)}} \tag{4}$$

The symbol overbar represents solid phase. The bi-molecular rate of this sorption reaction can be represented as:

$$r = -\frac{\mathrm{d[As(V)]}}{\mathrm{d}t} = k[\mathrm{As(V)}][\overline{\mathrm{Fe}}] \tag{5}$$

Table 2. Calculated fractional uptakes at three different temperatures.

Time (h)	F_t (5 °C)	F_t (22 °C)	F_t (35 °C)
0.25	–	–	0.058
0.5	0.075	0.123	0.120
1.0	0.137	0.184	0.182
1.5	0.197	0.233	0.278
2.0	0.230	0.289	0.302
3.0	0.298	0.349	0.408
4.0	0.353	0.403	0.462
5.0	0.383	0.479	–
6.0	–	0.505	0.551
6.5	0.448	–	–

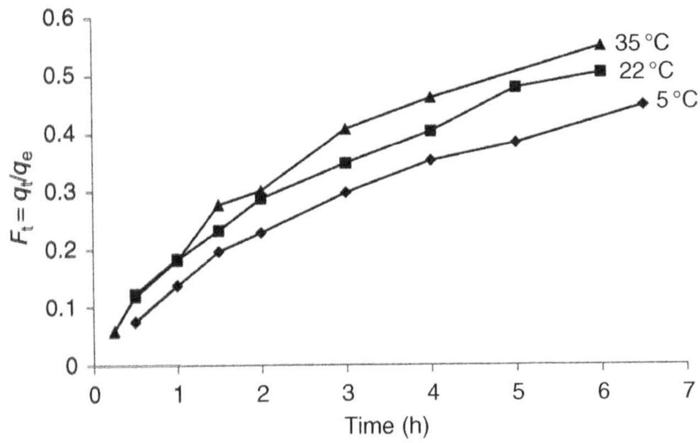

Figure 1. Fractional uptake of arsenic versus time at three different temperatures.

In Eq. (5), k is the specific reaction rate constant, [As(V)] is the concentration of arsenate in aqueous solution at time t and $[\overline{\text{Fe}}]$ represents the available sorption capacity of the sorbent for arsenate. It is assumed that the available sorption capacity, $[\overline{\text{Fe}}]$, is very high relative to the arsenic loading considered in this problem. Thus, $[\overline{\text{Fe}}]$ is constant and can be lumped together with the reaction constant to form the pseudo-first order reaction:

$$\frac{d[\text{As(V)}]}{dt} = -k_{\text{obs}}[\text{As(V)}] \tag{6}$$

where k_{obs} is the observed reaction rate coefficient. Solving this expression gives the solution

$$\ln[\text{As(V)}] = -k_{\text{obs}}t + \ln\,[\text{As(V)}]_0 \tag{7}$$

$[As(V)]_0$ is the arsenic concentration at time zero or the time the ferric oxide based sorbent is added to the arsenate containing solution. Plotting ln [As(V)] versus time should result in a straight line with negative slope k_{obs}.

For the experimental data at 5 °C, the plot is shown in Figure 2.

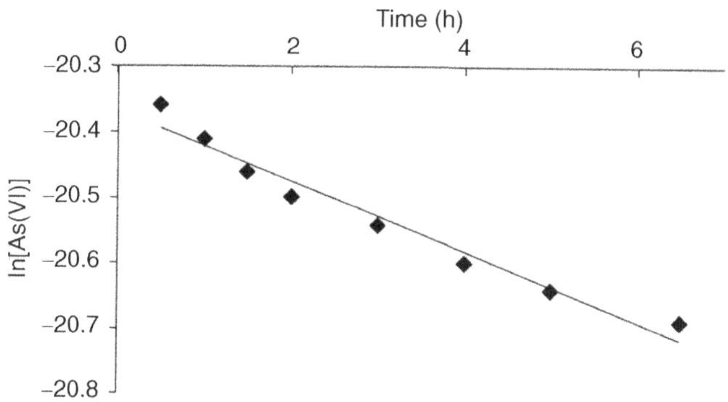

Figure 2. Plot of ln[As(V)] versus time for 5 °C test.

The slope of the line is equal to $-k_{obs}$, which was computed to be 0.056 h^{-1} at 5 °C. Subsequently, the value of k_{obs} was computed for 22 and 35 °C and shown in Table 3.

Table 3. Calculated values of k_{obs} for three different temperatures.

T (°C)	k_{obs} (h^{-1})
5	0.056
22	0.099
35	0.105

Step 3. Computing specific rate constant, k.

The specific rate constant, k, is computed by dividing k_{obs} by $[\overline{Fe}]$ which is constant for all batch studies and is equal to 0.04 g/L. The k at three different temperatures is listed in Table 4.

Table 4. Calculated k values for all temperatures.

T (°C)	k_{obs} (h^{-1})	k (L/g h)
5	0.056	1.40
22	0.099	2.47
35	0.105	2.63

Step 4. Calculate the value of activation energy for the sorption reaction.

From the Arrhenius rate law,

$$k = A\ e^{-E_a/RT} \tag{8}$$

where E_a is the activation energy for the reaction, A is the frequency factor, R the universal gas constant, and T is the absolute temperature. Eq. (8) can be further expressed in linearized form,

$$\ln k = \ln A - \frac{E_a}{RT} \tag{9}$$

A plot of ln k versus 1/T will produce a straight line with a slope of (E_a/R) and intercept of ln A, as shown in Figure 3.

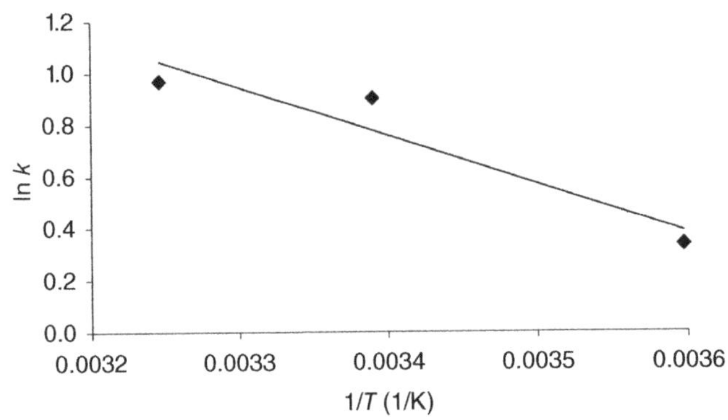

Figure 3. Determination of activation energy.

From the linear correlation, the activation energy, E_a, of As(V) sorption onto the ferric oxide based sorbent or HAIX is determined to be 15.5 kJ/mol (3.72 kcal/mol).

Step 5. Comments on the nature of the reaction.

Such a low activation energy, that is, well below 50 kJ/mol, signifies that the kinetics of sorption, or the rate-limiting process, is diffusion controlled. Such a sorption process does not involve any permanent chemical change and the sorption is reversible in nature. In practice, the ferric oxide based sorbent or HAIX can be regenerated and reused for multiple cycles by swinging pH from near neutral to alkaline pH to desorb and release As(V).

4.4 Physical Anatomy of an Ion Exchanger: Gel, Macroporous and Fibrous Morphology

Since intraparticle diffusion or transport within the exchanger is often the rate-limiting step, the morphology or the physical configuration of the ion exchanger plays a significant role in influencing the rate of ion exchange. Of them, gel resin beads, macroporous resin beads and fibers are the most common morphologies. We will discuss them individually with an intent to highlight their individual distinctiveness.

4.4.1 Gel-Type Ion Exchanger Beads

The gel structure or gel-type ion exchanger beads was the first one to be synthesized and is still the most widely used type due to its high capacity and low cost of production. A gel-type ion exchanger, also referred to as microporous or isoporous exchanger, may be viewed analogous to a condensed cross-linked polyelectrolyte. Structurally, an extremely high number of charged functional groups have been covalently attached to a three-dimensional crosslinked matrix. Gel structure is also a significant constituent for other morphologies, including macroporous beads and fibrous resins. Different from most adsorbents, gel-type ion exchangers or gel resins do not possess any internal surface area; a stark contrast to activated carbon, the most widely used adsorbent for non-polar organic solutes. The Brunauer–Emmet–Teller (BET) surface area according to the nitrogen adsorption protocol is essentially absent inside a gel-type ion exchanger, while it is a defining property of different activated carbons. It is the free water molecules within the ion exchanger, due to osmosis, that create the pore volume within a gel-type ion exchanger. These water molecules form a continuum through which counterions transport from one fixed site to the next during the sorption–desorption processes.

It is worth noting that for both ion exchangers and activated carbon, kinetics or the rate of reaction is often intraparticle diffusion-controlled. But activated carbon is a hydrophobic sorbent that barely swells or shrinks. The huge interconnected network of macropores, mesopores, and micropores of activated carbon provides the transport pathways for solutes: the water content rarely influences the prevailing surface diffusion rate of activated carbon. The intraparticle diffusion for gel-type ion exchangers, on the contrary, is affected by the water content. In fact, the parameter "surface area" does not bear any significance for gel-type ion exchangers and is nearly impossible to determine experimentally by conventional techniques because the beads, or granules, will collapse in the absence of water. Compared to macroporous resins and fibers, the intraparticle diffusion path length, as we will see later, is longer in gel-type resins and, hence, the sorption–desorption rate is slower.

4.4.2 Macroporous Ion Exchanger Beads

A macroporous ion exchanger particle is an ensemble of millions of tiny microgels with interconnected networks of pores. These macroporous ion exchangers are manufactured using the process of suspension polymerization; the resulting microgels become microspheres. While the sizes of macroporous beads vary from 0.2 to 1.0 mm, the sizes of microgels (<100 nm) are always much smaller [21–29]. The entire ion exchange capacity of the particle essentially resides within the microgels. Thus, gel phases are present inside the macroporous exchanger beads, but they are fragmented with continuous pores (10–100 nm) in between. Transmission electron microphotographs (TEMs) of a gel and macroporous anion exchanger are provided in Figure 4.9 along with representative illustrations. Note that the macroporous anion exchanger (e.g., IRA-900 from Rohm and Hass Co., Philadelphia) is truly a biphasic agglomerate of microgels with a continuous network of pores.

In principle, the solute transport inside a macroporous (biphasic) exchanger can proceed in parallel through both the pore and gel phases. For selective ion exchange, let us consider an anion exchange reaction where B^- in the exchanger is being replaced

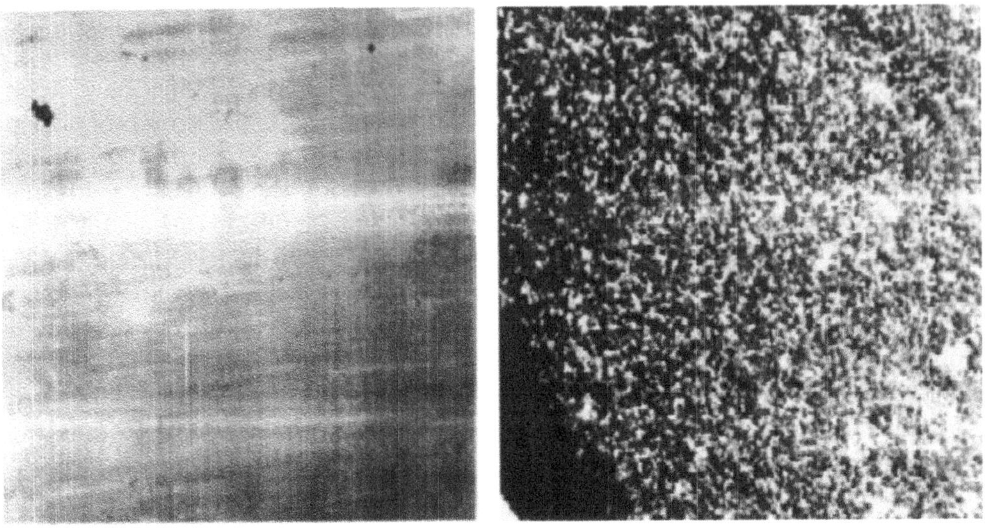

Figure 4.9 Tunneling electron microphotographs (TEM) of a gel (L) and macroporous (R) anion exchanger. *Source:* Li and SenGupta 2000 [2]. Reproduced with permission of Elsevier.

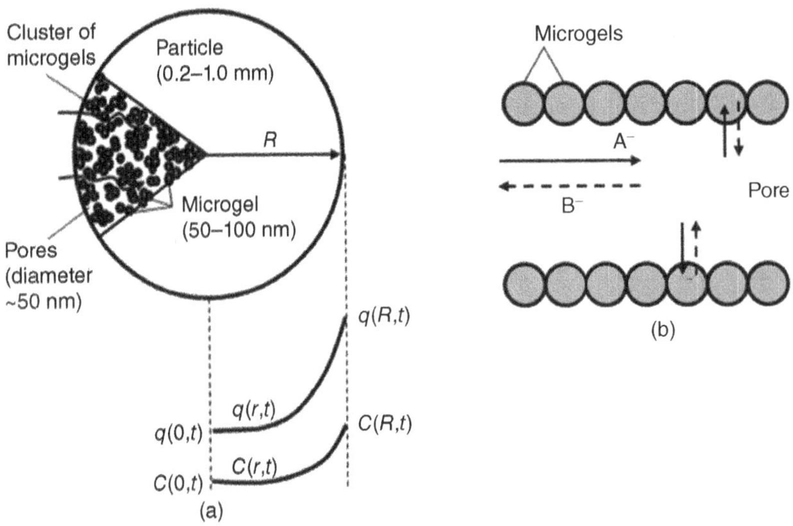

Figure 4.10 (a) Schematic representation of a macroporous particle containing microgels, where intraparticle diffusion is the rate limiting step and (b) explanation of diffusion of counterions A^+ and B^+ in parallel through microgels and macropores. *Source:* Li and SenGupta 2000 [2]. Reproduced with permission of Elsevier.

by a preferred A^- in accordance with reaction (4.10)

$$\overline{R^+B^-} + A^-(aq) \leftrightarrow \overline{R^+A^-} + B^-(aq) \qquad (4.10)$$

Figure 4.10a depicts a macroporous particle containing microgels and the concentration gradient of A^- at any time inside the exchanger under conditions when intraparticle diffusion is the rate-limiting step. Figure 4.10b shows how counterions A^- and B^- diffuse in parallel through microgels and macropores.

Example 4.2 Number of Microgels and Average Distance between Neighboring Sites

Let us consider an air-dried macroporous anion exchanger with polystyrene-DVB matrix and quaternary ammonium functional groups (IRA-900 from Rohm and Haas Co.). The average capacity of the exchanger is 3.6 meq/g and the dry bead density is 1100 kg/m^3. For a spherical bead of 0.5 mm (500 µm) diameter:

Estimate the number of microgels inside the resin bead and the average distance between neighboring ion exchange sites.

Solution:

1. *Number of microgels in an anion exchanger bead.*
   The volume of a single ion exchanger bead with a diameter of 0.5 mm is:

$$\frac{4}{3}\cdot\pi\cdot\left(\frac{0.5\times10^{-3}\,\text{m}}{2}\right)^3 = 6.5\times10^{-11}\,\text{m}^3 \tag{1}$$

While the volume of a microgel with an approximate diameter of 0.07 µm is:

$$\frac{4}{3}\cdot\pi\cdot\left(\frac{0.07\times10^{-6}\,\text{m}}{2}\right)^3 = 1.8\times10^{-22}\,\text{m}^3 \tag{2}$$

Thus, the number of closely packed microgels in a single bead (assuming a porosity of 0.3) can be calculated as:

$$\frac{6.5\times10^{-11}\,\text{m}^3\cdot(1-0.3)}{1.8\times10^{-22}\,\text{m}^3} = 2.5\times10^{11} = 250\,\text{billion} \tag{3}$$

2. *Number of charges or functional groups in a microgel.*
   For a bead density of 1100 kg/m^3, the weight of a single bead is:

$$1100\times10^3\,\frac{\text{g}}{\text{m}^3}\times6.5\times10^{-11}\,\text{m}^3 = 7.2\times10^{-5}\,\text{g} \tag{4}$$

Then the number of charges in a single bead can be calculated as:

$$3.6\times10^{-3}\,\frac{\text{eq}}{\text{g}}\times7.2\times10^{-5}\,\text{g}\times6.02\times10^{23}\,\text{charge/eq} = 1.6\times10^{17}\,\text{charges} \tag{5}$$

Therefore, number of charges in a single microgel is obtained:

$$\frac{1.6\times10^{17}\,\text{charges}}{2.5\times10^{11}\,\text{microgels}} = 6.4\times10^5 = 640{,}000\,\text{charges per microgel} \tag{6}$$

3. *Average distance between functional groups in a single microgel.*
   Let us assume that 640,000 charged spheres are packed in a microgel with a void space of about 0.3. The average radius, r, of this sphere is:

$$r = \sqrt[3]{\frac{3}{4\pi}\cdot\left[\frac{1.8\times10^{-22}\,\text{m}^3\cdot(1-0.3)}{640{,}000}\right]} = 3.6\times10^{-10}\,\text{m} = 0.36\,\text{nm} \tag{7}$$

Therefore, the average distance between two neighboring charges is: $2\times3.6\times10^{-10}$ m or 7.2 Å units or 0.72 nm. Note that the average distance between two neighboring ion exchange sites is on the same order of magnitude as the hydrated ionic radius of typical inorganic ions. But, even a single ion exchange bead is not homogeneous, that is, the distance between neighboring sites varies widely within the gel phase, and this can impart non-ideal behaviors.

4.4.3 Ion Exchange Fibers

Reducing diffusion path length enhances kinetics for intraparticle diffusion-controlled processes. Such a reduction can be achieved by reducing the particle diameter of a spherical resin bead. But the pressure drop or head loss in a typical fixed-bed column increases inversely with the square of the particle diameter. Thus, it is impractical and highly energy-intensive to reduce resin particle sizes. That is why most commercial ion exchange resins don't have particle sizes smaller than 500 μm. A cylindrical configuration of ion exchange materials with relatively small diameters (10–50 μm), more commonly referred to as ion exchange fibers or IX fibers, tends to offer some unique advantages in this regard. In a typical fixed-bed, the resultant void fraction when using IX fibers is relatively high and, hence, the pressure drop does not increase significantly. Figure 4.11 shows visual comparison between spherical resin beads and ion exchange fibers. Both polypropylene and glass have been used as parent substrates on which functional groups are covalently attached [30–32]. Table 4.1 provides salient properties of weak-acid IX fibers versus their resin bead counterparts; note that the chemical makeup and the exchange capacity are nearly the same.

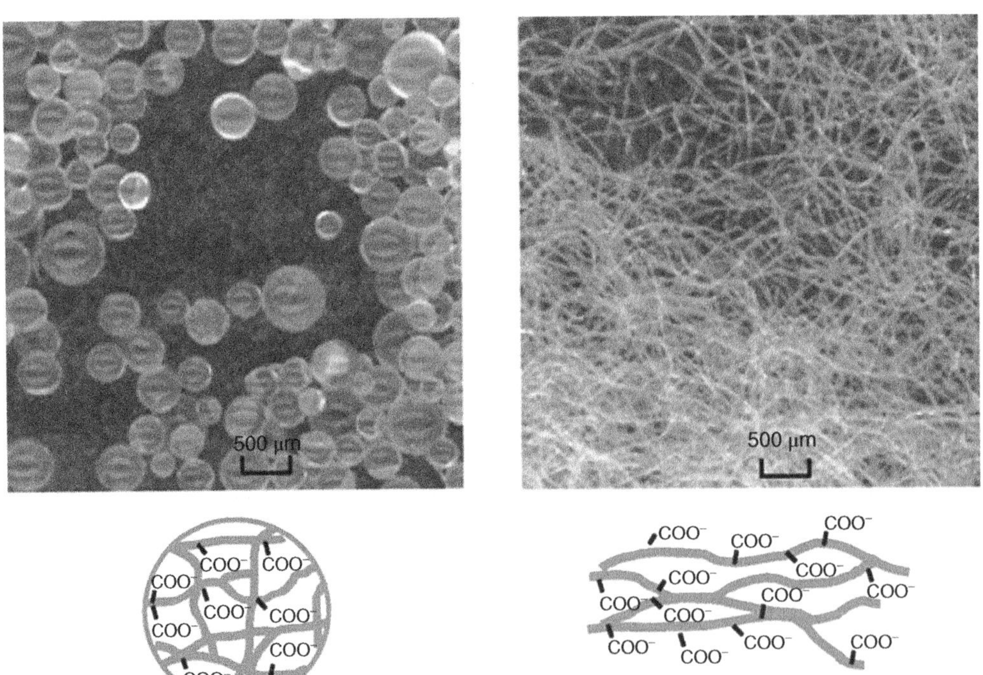

Figure 4.11 Visual comparison between spherical ion exchange resin beads and ion exchange fibers. *Source:* Greenleaf *et al.* 2006 [30]. Reproduced with permission of John Wiley & Sons.

Table 4.1 Salient properties of weak-acid ion exchange fiber and weak-acid ion exchange resin beads.

Description	Weak-acid ion exchange fibers (Fiban K-4)	Weak-acid ion exchange resin beads
Diameter	10–50 μm	500–1200 μm
Physical shape	Cylindrical	Spherical
Functionality	Carboxylate (COO^-)	Carboxylate (COO^-)
Capacity (air-dried)	4–5 meq/g	5–8 meq/g
Equipment configuration	Fixed-bed	Fixed-bed

To compare intraparticle diffusion phenomena between fiber and bead configurations, let us recognize the term, half-diffusion path length or $d_{1/2}$, that corresponds to the distance from the periphery toward the center (for both spherical and fibrous configurations) at which the exchange capacity encompassed equals half of the total exchange capacity. Figure 4.12 shows both spherical and cylindrical configurations of resins and fibers, illustrating the radial conversion during counterion uptake.

Assuming a uniform charge distribution, half-capacity of a cylindrical fiber material will be reached when the outer shell (converted periphery) volume, V_s, equals half the total particle volume, when r_0 is the fiber radius and r is the radius of unconverted resin fiber of length h

$$V_s = \frac{1}{2}\pi r_0^2 h \tag{4.11}$$

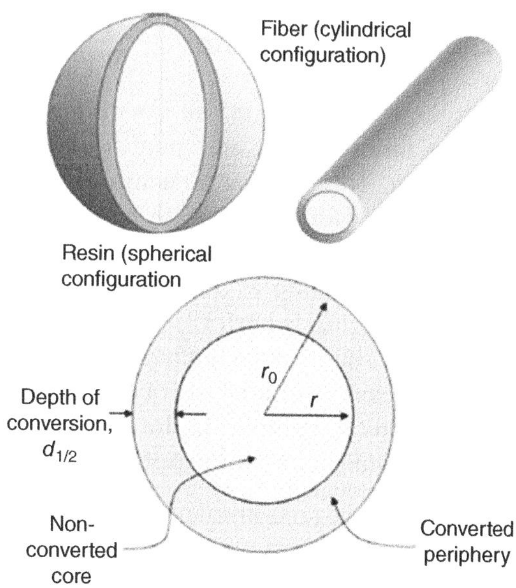

Figure 4.12 Schematic illustration of radial conversion during uptake for a spherical resin bead and a cylindrical ion exchange fiber.

V_s is also the difference between the total fiber volume and the unconverted core of the fiber,

$$V_s = \pi r_0^2 h - \pi r^2 h \tag{4.12}$$

Combining Eqs (4.11) and (4.12)

$$r = r_0 \sqrt{\frac{1}{2}} \tag{4.13}$$

A similar analysis for a spherical resin particle geometry yields

$$r = r_0 \sqrt[3]{\frac{1}{2}} \tag{4.14}$$

When the half-capacity has been reached for spherical or cylindrical geometries, the shell depth at the point of half conversion or half-diffusion path length, $d_{1/2}$, may be expressed as follows:

$$d_{1/2} = r_0 - r \tag{4.15}$$

Given the typical radii of fibers and resin particles as 25 and 500 μm, respectively, and applying the condition of half capacity, the computed half-diffusion path lengths are as follows:

$d_{1/2}$ (cylindrical fiber) = 7.3 μm
$d_{1/2}$ (spherical bead) = 103 μm

Since $d_{1/2}$ is more than an order of magnitude lower, the sorption/desorption kinetics are significantly faster with IX fibers compared to spherical resin beads, as reported in various independent studies [33–35].

4.5 Column Interruption Test: Determinant of Diffusion Mechanism

The column interruption test is a more definitive stand-alone technique to determine the rate limiting step and may be used, if necessary, in conjunction with other approaches. The test was first introduced by Kressman and Kitchener [4] and it is convenient because no separate experimental endeavor is necessary. All that is needed is that during an ongoing column run with ion exchange resins as the stationary phase and the feed solution as the mobile phase under conditions representative of normal use, the following is performed: (i) an interruption of the liquid flow (e.g., 12–24 h), where the column is virtually idle; (ii) subsequent restart of the flow of the liquid phase as before; (iii) analysis of liquid-phase effluent samples for the counterions of interest; and (iv) comparing the effluent concentration profile after flow interruption with the profile before interruption. The two following scenarios may emerge:

1. If the rate is predominantly controlled by external film-diffusion, no concentration gradient exists within the ion exchanger; the resistance to the transport of ions resides in the liquid film. Thus, upon restart of the column following interruption,

the concentration gradient is instantly reestablished. So, the rate of uptake remains unchanged, all other flow conditions remaining identical. The concentrations of samples, before and after interruption, therefore, remain essentially the same, that is, interruption in the flow of the mobile phase shows no effect.

2. If the rate is controlled by intraparticle diffusion, a concentration gradient exists within the exchanger, that is, the exchanger-phase concentration at the surface (q_S) in contact with the liquid phase, is greater than at the core. Thus, during the time of interruption, although there is no flow of solution, the concentration gradient within the exchanger tends to level out. Hence the exchanger-phase concentration of the counterion being loaded drops near the periphery. Immediately after the restart of the solution flow following interruption, the concentration gradient at the exchanger-water interface is at its highest. Hence, the solute uptake rate significantly increases resulting in a sharp decrease in the counterion concentration in the solution phase.

Thus, a sharp drop in liquid-phase counterion concentration as a result of the interruption test is a distinctive indicator that intraparticle diffusion is the rate-limiting step, while no change in liquid phase concentration confirms that external film diffusion is the predominant rate-limiting step [36]. Most importantly, the test can be carried out during fixed-bed columnar operations with no need to withdraw ion exchange particles from the column. Figure 4.13a–d illustrates the changes in the concentration gradient in the ion exchanger during various stages of the interruption step when intraparticle diffusion is the rate-limiting step. Note that while the gradient inside the exchanger gradually levels out during the interruption, it is the highest right after the restart of the column run, that is, dq/dr is a maximum in Figure 4.13c. The anticipated change in dq/dr at various stages of the protocol is depicted in Figure 4.13e. When external film diffusion is the rate-limiting step, the concentration gradient remains essentially the same before and after an interruption within the exchanger particle, as shown in Figure 4.13f. No significant drop in concentration is observed following restart of the column with flow of solution [37].

Figure 4.14 shows experimental results of an interruption test during a column run for removal of pentachlorophenate (PCP^-), an anion, in the presence of much higher concentrations of chloride and sulfate. The influent composition and the hydrodynamic conditions of the column run (e.g., EBCT or empty bed contact time; SLV or superficial liquid velocity) are included in the figure. A significant drop in PCP^- concentration is observed immediately after the restart of the column following interruption for 24 h, which demonstrates that intraparticle diffusion is the rate-limiting step. The following points are worth noting:

- Although there is a sharp drop in concentration following an interruption test when kinetics are intraparticle diffusion-controlled, the overall capacity of the exchanger, which is strictly an equilibrium phenomenon, does not increase.
- The operating capacity up to a specific breakthrough concentration can, however, be enhanced by resorting to intermittent interruptions for intraparticle diffusion-controlled processes.

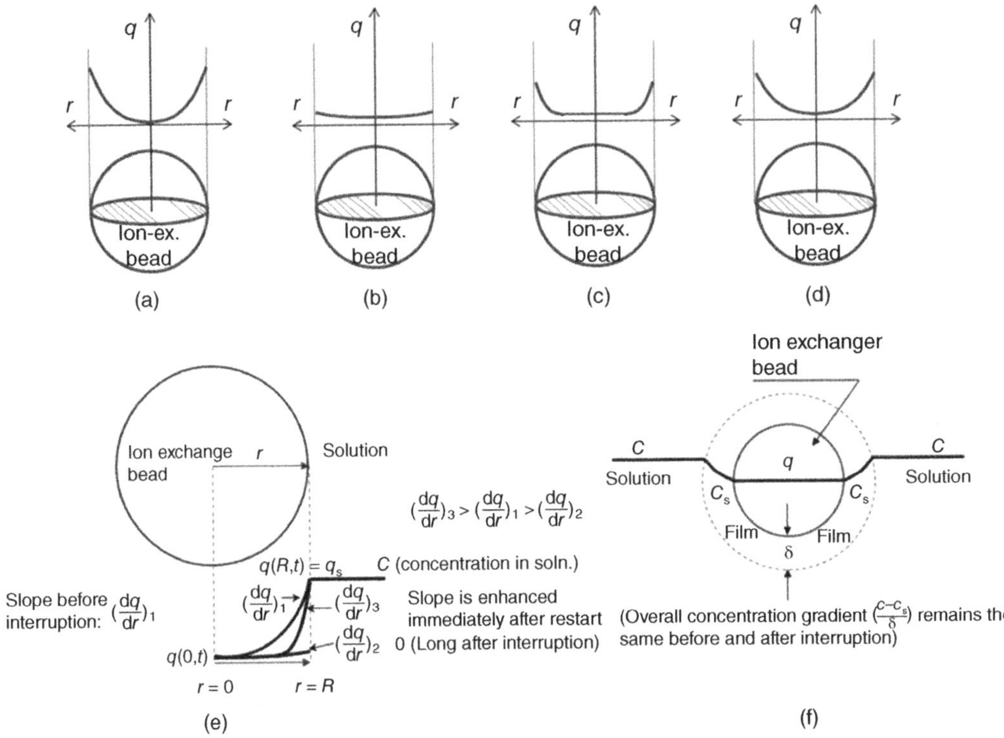

Figure 4.13 A schematic illustration of intraparticle diffusion control through an interruption test presenting concentration profile changes within an ion exchanger bead (a) before interruption; (b) after interruption; (c) immediately after restart; (d) long after restart; (e) their slopes at different stages of interruption test; (f) concentration profiles for liquid-film diffusion control step (i.e., no change in concentration gradient within the exchanger bead before and after interruption). *Source:* Li and SenGupta 2000 [37]. Reproduced with permission of American Chemical Society.

- The resistance to intraparticle diffusion decreases very rapidly with a decrease in diffusion path length. Thus, for pellicular resins (functional groups located primarily in the periphery), ion exchange nanofibers and nanoparticles, liquid film diffusion tends to be the rate-limiting step even for selective ion exchange.

4.6 Observations Related to Ion Exchange Kinetics

In general, the field of ion exchange kinetics still remains quite empirical and is often not amenable to appropriate quantitative interpretation. This shortcoming is primarily due to the inherent complexity of ion exchange processes caused by simultaneously occurring phenomena: coupled transport, Donnan exclusion effect, law of electroneutrality, multiple counterions and their difference in relative selectivity. However, even absent rigorous mathematical models, logical and scientific analyses are often appropriate. In this section, we will include half a dozen examples from different studies concerning ion exchange kinetics that are seemingly counterintuitive and demand scientific validation.

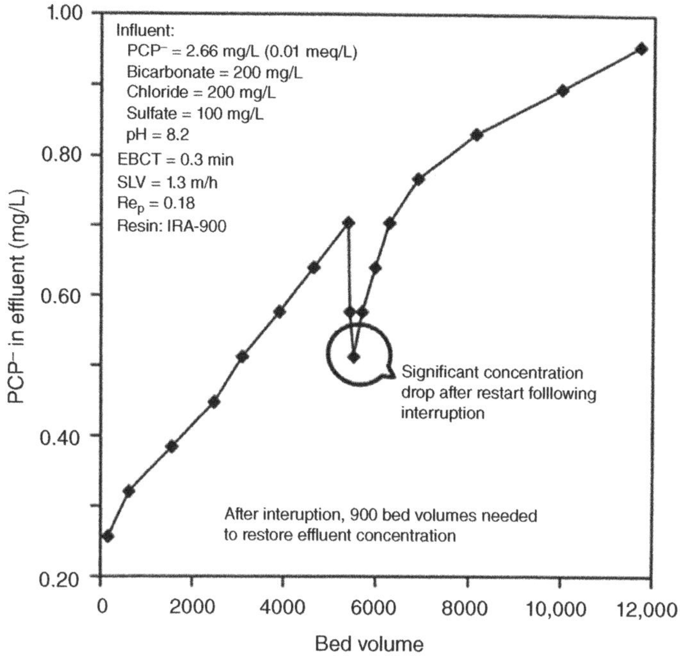

Figure 4.14 Plot of PCP^- concentration versus bed volumes. There is a significant drop in PCP^- concentration immediately after restart following a 24-h column interruption. This concentration profile is demonstrative of intraparticle diffusion being the rate-limiting step. *Source:* Li and SenGupta 2000 [37]. Reproduced with permission of American Chemical Society.

We will subsequently develop quantitative models in succeeding sections with particular emphasis on intraparticle diffusion and trace ion exchange. Finally, later in the chapter, we will return to validate each experimental observation using simple-to-follow mathematical models enabling the reader to recognize the usefulness of the quantitative approach.

4.6.1 Effect of Concentration on Half-time ($t_{1/2}$)

Half-time or $t_{1/2}$, as explained before, is the time required to attain half of the equilibrium capacity; $t_{1/2}$ and the kinetics of exchange are inversely related. Under a given set of experimental conditions with intraparticle diffusion as the sole rate-limiting step, $t_{1/2}$ should be independent of the solute concentration in the aqueous phase, that is, intraparticle diffusivity should remain constant. Figure 4.15 presents $t_{1/2}$ values for nickel (Ni^{2+}) uptake by a chelating ion exchanger from solutions for different equilibrium nickel concentrations in the presence of much higher competing Na^+ concentration. Experimental conditions with high stirrer speeds confirmed that intraparticle diffusion remains the rate-limiting step. Note that the $t_{1/2}$ values dropped significantly from 600 to 80 s as the equilibrium Ni concentration (C_{Ni}) increased from 0.006 to 0.1 M. This observation implies that the intraparticle diffusivity of Ni^{2+} increased with an increase in aqueous-phase nickel concentration. How is such an observation amenable to scientific explanation?

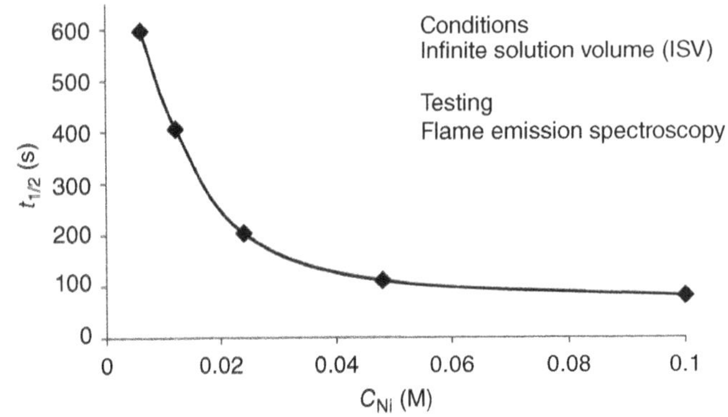

Figure 4.15 Plot of $t_{1/2}$ versus equilibrium Ni concentration for Ni uptake by a chelating ion exchanger from solution containing Ni in presence of much higher Na^+ concentration. *Source:* Adapted from Price *et al.* 1988 [19].

4.6.2 Major Differences in Ion Exchange Rate

Experimental observations made by Kunin and Barry [38] with weak- and strong-acid cation (SAC) exchange resins in H-form and Na-form are included in Table 4.2. Note that the neutralization of weak-acid cation (WAC) exchange resin (R-COOH) by KOH takes the longest time (i.e., 7 days) while other cation exchange reactions reach equilibrium in 2 min. It is well known that the neutralization reaction or a routine titration between a weak acid (say CH_3COOH) and a strong base (say KOH) occurs almost instanty in the aqueous phase. The obvious question is: Why does acid-base neutralization mediated by a WAC exchange resin take such an unusually long time?

4.6.3 Chemically Similar Counterions with Significant Differences in Intraparticle Diffusivity

Chlorophenols are synthetic aromatic compounds that exist primarily as monovalent anions in the aqueous phase at neutral to alkaline pH conditions (pH > 7). They have been widely used as pesticides in agricultural and industrial applications, for example, lumber decay by insects. As the number of substituted chlorine atoms is increased,

Table 4.2 Approximate rate data for weak-acid and strong-acid cation exchange.

Equilibrium reaction	Equilibrium attainment time	Apparent density (g/mL)
RCOOH + KOH	7 days	0.4
RSO_3H + KOH	2 min	0.435
RCOONa + $CaCl_2$	2 min	0.3
RSO_3Na + $CaCl_2$	2 min	0.5

Source: Kunin and Barry 1949 [38]. Reproduced with permission of American Chemical Society.

its potency as a pesticide becomes stronger, but the likelihood of the chlorophenol to entering the food chain through biomagnification is also enhanced due to higher octanol/water partition coefficient values (K_{OW}). However, being monovalent anions, their aqueous-phase self-diffusion coefficient values are nearly the same and close to chloride's (Cl^-). Table 4.3 includes the formula, acid dissociation constant (K_a), and octanol–water partition coefficients ($\log K_{OW}$) of three chlorophenol anions, for example, di-chloro, tri-chloro, and penta-chloro phenols or DCP^-, TCP^-, and PCP^- [3,39]. Their estimated self-diffusion coefficient values are both on the order of 10^{-6} cm^2/s.

Intraparticle diffusivities of DCP^-, TCP^-, and PCP^- were independently determined using chloride as a bulk competing ion for an anion exchange resin with a polystyrene matrix and quaternary ammonium functional groups. Figure 4.16 shows a plot of effective intraparticle diffusivity ($\overline{D}_{eff}$) versus octanol–water partition coefficients (K_{ow}) of three chlorophenols. The $\overline{D}_{eff}$ value of PCP^- is well over an order of magnitude lower than DCP^- and the intraparticle diffusivities of the three chlorophenols (e.g., DCP^-, TCP^-, and PCP^-) are inversely correlated to K_{ow}. What is the genesis of such a correlation for chemically similar compounds with nearly equal self-diffusion coefficient values in water?

Table 4.3 Pertinent properties of hydrophobic aromatic anionic chlorophenol.

Chlorophenols	Molecular formula	Molecular weight	pK_a (dissociation constant)	$\log K_{O/W}$
Pentachlorophenol (PCP)	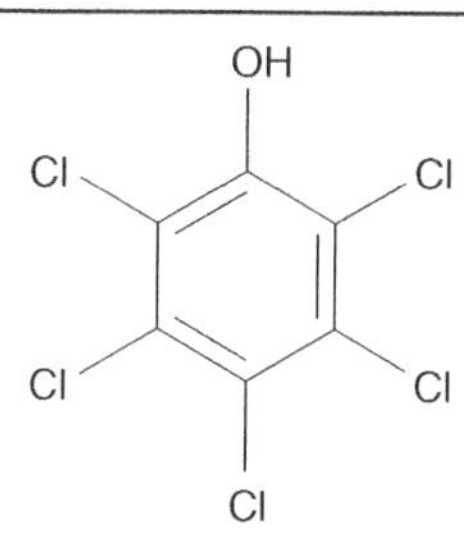	266.5	4.8	5.2
2,4,6-Trichlorophenol (TCP)	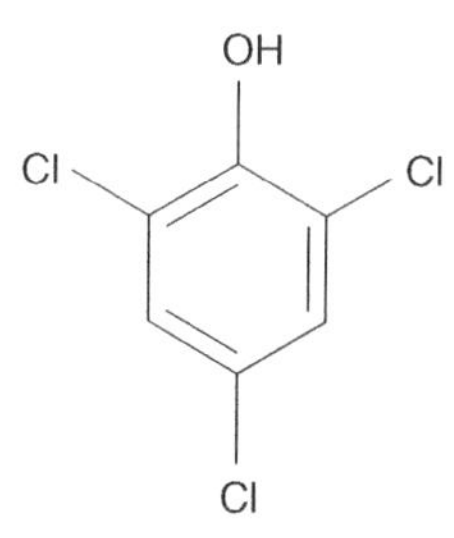	197.5	6.1	3.7
2,6-Dichlorophenol (DCP)	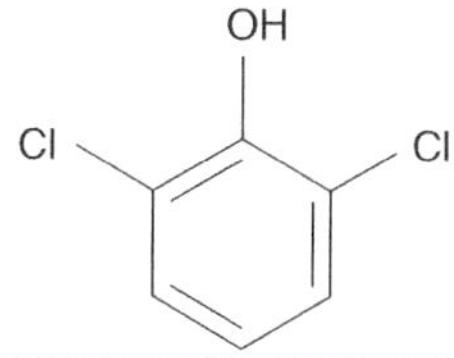	163	6.9	2.6

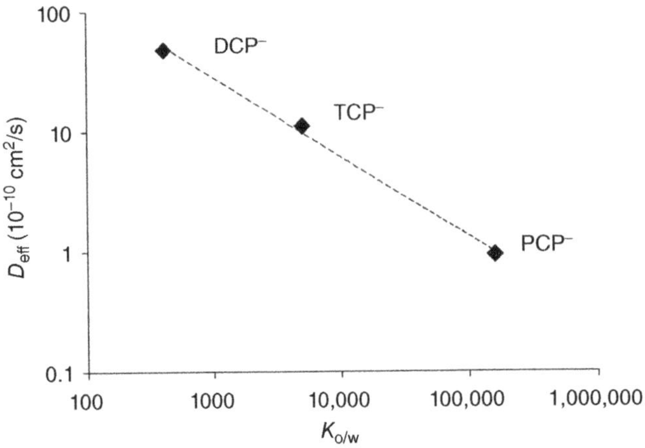

Figure 4.16 Plot of effective intraparticle diffusivity versus octanol–water partition coefficient for three chlorophenols. *Source:* Li and SenGupta 2000 [2]. Reproduced with permission of Elsevier.

4.6.4 Effect of Competing Ion Concentrations: Gel versus Macroporous

Batch kinetic tests were carried out under conditions where PCP^- was always a trace anion in the presence of varying competing chloride concentrations where the anion exchangers used were (i) gel (Biorad AG1-X8) and (ii) macroporous (Amberlite IRA-900). Figures 4.17 and 4.18 show the results of the kinetic tests (fractional uptake of PCP^- vs time) for these two anion exchangers at two background chloride concentrations (50 and 100 meq/L), with all other conditions remaining identical [37]. Note that the rate curves for IRA-900 and AG 1-X8 are characteristically different. An increase in competing chloride concentration from 50 to 100 meq/L had a negligible effect on the PCP^- uptake rate for the gel anion exchanger (AG 1-X8), but the PCP^-

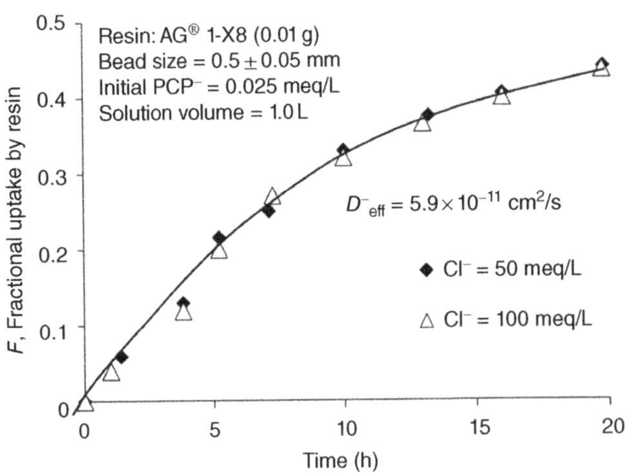

Figure 4.17 Fractional uptake profiles of PCP^- versus time for gel-type anion exchanger at two background chloride concentrations. *Source:* Li and SenGupta 2000 [37]. Reproduced with permission of American Chemical Society.

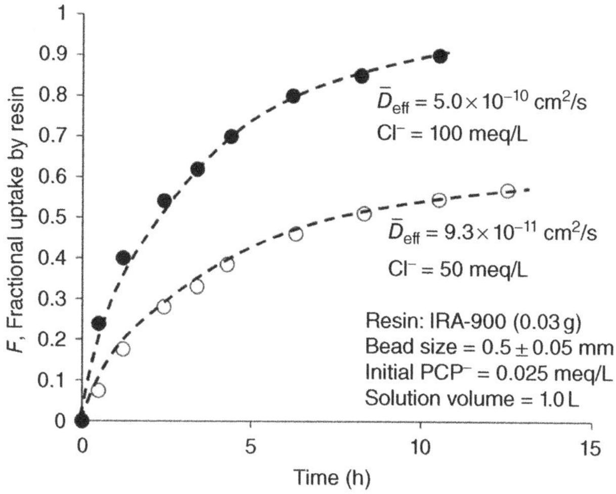

Figure 4.18 Fractional uptake profiles of PCP^- versus time for the macroporous anion exchanger at two background chloride concentrations. *Source:* Li and SenGupta 2000 [37]. Reproduced with permission of American Chemical Society.

uptake rate increased significantly for macroporous IRA-900 from 50 to 100 meq/L. Effective intraparticle diffusivity ($\overline{D}_{eff}$) values, as computed from the experimental data, showed similar behaviors, that is, no change was observed for the gel resin while $\overline{D}_{eff}$ increased for the macroporous resin with an increase in competing chloride ion concentration. The question is: What is the interplay between the ion exchanger morphology (gel vs macroporous) and competing ion concentration influencing intraparticle diffusivity?

4.6.5 Intraparticle Diffusion during Regeneration

Ion exchange is predominantly employed as a cyclic process, that is, every service cycle is followed by a relatively short regeneration process. The electrolyte concentration in the aqueous phase during the regeneration step is often very high to hasten the process. For a strong-acid gel-type cation exchanger in Na^+-form, Figure 4.19 shows how experimentally determined intraparticle self-diffusion coefficient values of sodium ($\overline{D}_{Na}$) decreased nearly two times as the concentration of sodium chloride in the solution phase increased from 0.5 M to nearly 5 M [40]. So, the goal of using high regenerant concentration is partly impaired due to the decrease in $\overline{D}_{Na}$ and a scientific explanation is warranted.

4.6.6 Shell Progressive Kinetics versus Slow Diffusing Species

The phenomenon of shrinking core or shell progressive kinetics is a special case of intraparticle diffusion with sharp moving boundaries for equilibrium conditions represented by rectangular isotherms. Hydrogen ions have a very high affinity for WAC exchange resins compared to other divalent alkaline-earth metal cations and approximate a scenario similar to rectangular isotherm as discussed in Chapter 3:

$$\overline{(R-COO^-)_2Ca^{2+}} + 2H^+ \rightarrow 2\overline{(R-COOH)} + Ca^{2+} \tag{4.16}$$

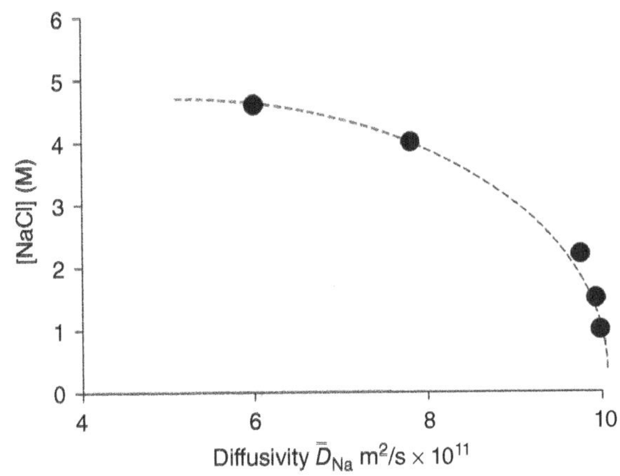

Figure 4.19 Plot of intraparticle diffusion coefficient of sodium ($\bar{D}_{Na}$) versus concentration of sodium chloride (NaCl) solution for a strong-acid gel type cation exchanger in Na^+-form. *Source:* Slater 1991 [40]. Reproduced with permission of Elsevier.

One gel-type WAC exchanger was originally loaded with calcium ions and then placed in an acid solution; Ca^{2+} ions from the gel phase were gradually eluted by hydrogen ions, progressing from the circumference toward the center. With time, the calcium-loaded core shrank and eventually disappeared. Figure 4.20 provides a photographic testimony, as developed by Höll, for shrinking core ion exchange kinetics [41]. For copper uptake onto chelating ion exchangers, similar shrinking core phenomenon was observed by Phelps and Ruthven [42].

Shell-progressive phenomenon may also be observed during gradual uptake of a slow diffusing species due to the low overall diffusion coefficient. Streat [43] investigated alpha particle emission rates on an autoradiographic film during exchange of plutonium(IV) nitrate on a weak-base anion exchange resin presaturated with nitrate at an acidic pH. Figure 4.21 shows the autoradiographs of the weak-base anion exchange resin during sorption of plutonium from 7.5 M nitric acid solution. Note that until 48 h, plutonium uptake followed shrinking-core like phenomenon. However, the sharp boundary between the shell and the unreacted core gradually blurred between 48 and

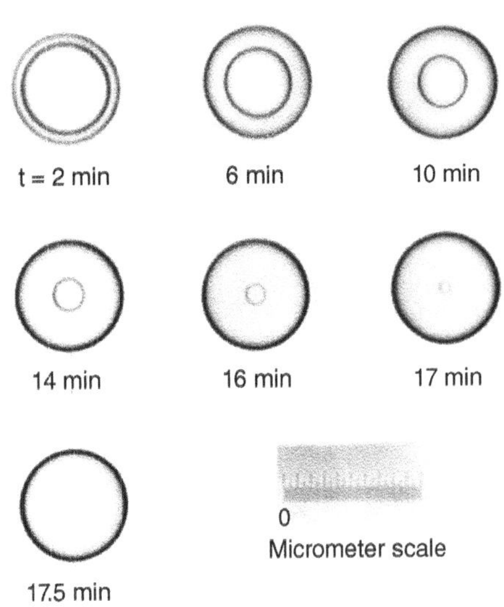

Figure 4.20 Photographic testimony of shrinking core ion exchange kinetics for Ca^{2+}–H^+ exchange for Amberlite IRC-84 (weak-acid cation exchanger) in 1 M HNO_3 solution. *Source:* Höll 1984 [41]. Reproduced with permission of Elsevier.

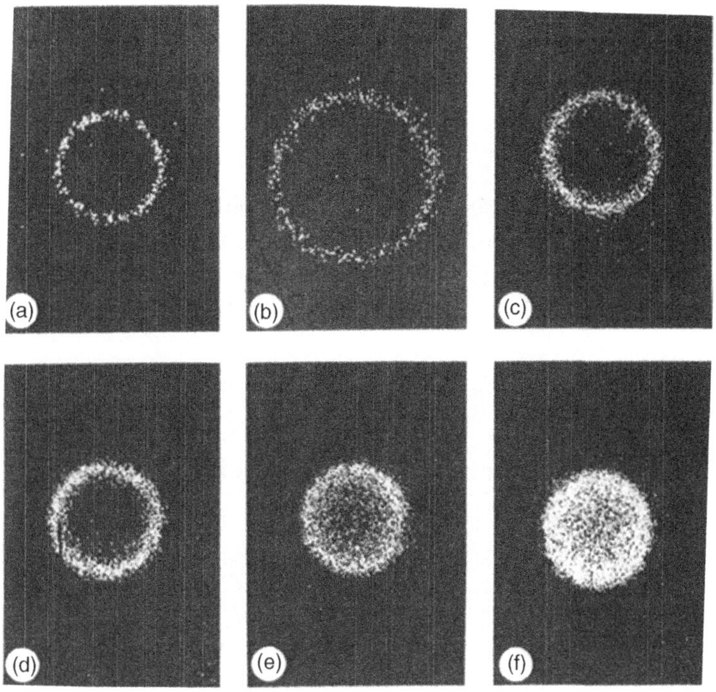

Figure 4.21 Autoradiograph of Ionac XAX 1284 as a function of time during sorption of plutonium(IV) from 7.5 M nitric acid: (a) 1 h, (b) 7 h, (c) 24 h, (d) 48 h, (e) 336 h, (f) 547 h. *Source:* Streat 1984 [43]. Reproduced with permission of Elsevier.

336 h, representing switchover from shell progressive kinetics to regular intraparticle diffusion. The questions were: Is the shrinking core kinetics of protonation of WAC exchange resin characteristically similar to plutonium uptake in Figure 4.21? Did the mechanism of plutonium uptake kinetics change with its progression from periphery to the center?

So far in the chapter, we have deliberately avoided mathematical interpretation of ion exchange kinetics controlled by intraparticle diffusion. Instead, we highlighted observations that raise meaningful inquiries and scientific questions about the ion exchange rate processes. We will now gradually delve into quantitative models and finally in Section 4.10, we will attempt to use the model predictions to explain the genesis of the foregoing observations made in this section.

4.7 Interdiffusion Coefficients for Intraparticle Diffusion

The physical mechanism of coupled transport of counterions forms the basis of intraparticle diffusion in ion exchange kinetics. Any difference in the diffusive flux of two counterions produces an electric potential gradient in a direction to eliminate the difference and maintain electroneutrality. The most important effect from the generation of an electric potential gradient is that the exchanger phase diffusivity of individual exchanging counterions is influenced by the transport of other counterions.

The flux of a counterion, i, within an ion exchanger needs to account for its charge and possible electric potential gradient and it is aptly described by the Nernst–Planck equation:

$$\bar{J}_i = -\bar{D}_i \ \mathrm{grad}\, q_i - \bar{D}_i Z_i q_i \left(\frac{F}{RT}\right) \mathrm{grad}\, \varnothing \tag{4.17}$$

where F is Faraday's constant, R is universal gas constant, T is absolute temperature, $\varnothing$ is electric potential. For intraparticle diffusion, we may ignore the presence of coions inside the exchanger and a constant exchange capacity is assumed. For exchange of counterions A and B, these constraints lead to the following equalities:

Maintenance of electroneutrality:

$$Z_A q_A + Z_B q_B = nQ \tag{4.18}$$

where, n = sign of fixed charge (−1 for cation exchangers and +1 for anion exchangers); Q = capacity of the ion exchanger.

With no electric current:

$$Z_A J_A + Z_B J_B = 0 \tag{4.19}$$

The two equations for counterions A and B in Eq. (4.17) can be combined after eliminating the electric potential by use of Eqs (4.18) and (4.19). The resulting solution for J_A is

$$J_A = -\frac{\bar{D}_A \bar{D}_B (Z_A^2 q_A + Z_B^2 q_B)}{Z_A^2 q_A \bar{D}_A + Z_B^2 q_B \bar{D}_B} \mathrm{grad}\, q_A \tag{4.20}$$

Considering this equation as a special form of traditional Fick's first law, we may express the coupled interdiffusion coefficient in terms of one combined diffusion coefficient as follows:

$$\bar{D}_{AB} = \frac{\bar{D}_A \bar{D}_B (Z_A^2 q_A + Z_B^2 q_B)}{Z_A^2 q_A \bar{D}_A + Z_B^2 q_B \bar{D}_B} \tag{4.21}$$

Interestingly, however, the interdiffusion coefficient $\bar{D}_{AB}$, is not constant, even for the counterions A and B for the same exchanger. Instead, it depends on the composition, or relative distribution of A and B in the exchanger, that is, q_A and q_B. The extreme scenarios can be readily noted:

For $q_A \gg q_B$ (i.e., when B is a trace species)

$$\bar{D}_{AB} = \bar{D}_B \tag{4.22}$$

And for $q_B \gg q_A$

$$\bar{D}_{AB} = \bar{D}_A \tag{4.23}$$

The general premise of intraparticle diffusion is that the ion present at lower concentrations tends to control the interdiffusion coefficient. Figure 4.22 shows how interdiffusion coefficient varies with the composition of the exchanger and as already stated for

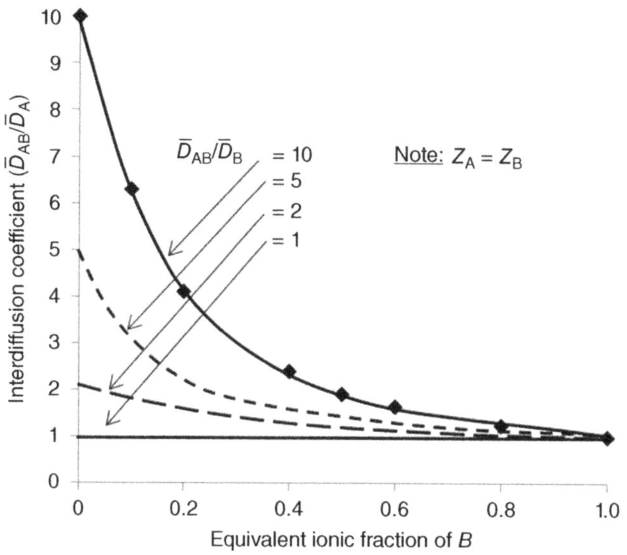

Figure 4.22 Graphical presentation for variations in interdiffusion coefficients with respect to the composition of the ion exchanger; when $q_B \gg q_A$, $\bar{D}_{AB}$ tends to be equal to $\bar{D}_A$.

$q_B \gg q_A$, $\bar{D}_{AB}$ approaches toward $\bar{D}_A$, that is, exchanger phase self-diffusion coefficient of A dictates the interdiffusion coefficient. In a similar vein, $\bar{D}_{AB} = \bar{D}_B$ when $q_A \gg q_B$. Another noteworthy point: for $q_A = q_B$, $\bar{D}_{AB}$ is closer to the diffusivity of the slower counterion, that is, the ion with lower diffusivity value. For selective ion exchange, and especially for trace solutes, intraparticle diffusion is often, if not always, the predominant rate-limiting step. As we proceed with this chapter, the foregoing observations will provide the basic foundation for the quantitative treatment of ion exchange kinetics governed by intraparticle diffusion.

Supplementary Reading S4.2 Determining the Effective Intraparticle Diffusivity from Batch Kinetic Experiments

Example S4.3 *Figure S4.7 presents results of Zn(II) sorption by a hybrid inorganic material (HIM) of average particle diameter of 250 μm for a batch kinetic study at 1600 and 2000 rpm, keeping all other conditions identical. Other pertinent experimental details are provided in the figure. Determine the effective intraparticle diffusivity coefficient value and show the individual steps. State all assumptions.*

The concentration versus time plots remain identical at both stirrer speeds of 1600 and 2000 rpm during the batch kinetic study, confirming intraparticle diffusion is the rate-limiting step. With the assumption of spherical geometry of ion exchange particles, where diffusion is occurring radially, as shown in Figure S4.6, the model diffusion equation with constant effective intraparticle diffusivity takes the following form:

$$\frac{\partial q}{\partial t} = \bar{D}_{eff}\left(\frac{\partial^2 q}{\partial r^2} + \frac{2}{r}\frac{\partial q}{\partial r}\right) \tag{S4.4}$$

(Continued)

Supplementary Reading S4.2 (Continued)

where "r" is the radial space coordinate (related to the radius of the particle) and q is the concentration of the solute within the ion exchanger phase at any time "t."

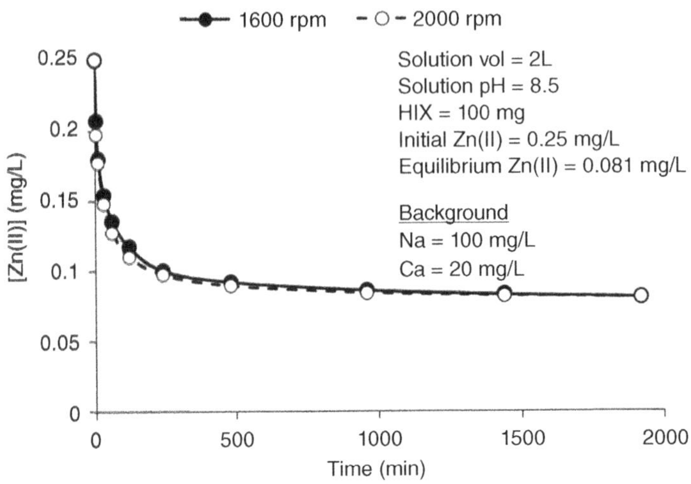

Figure S4.7. *Concentration versus time plots for Zn(II) sorption by HIM under intraparticle diffusion controlled conditions. Source: Reprinted with permission from Chatterjee 2011 [6].*

Step 1: Experimental data in Figure S4.7 was converted into fractional uptake, F_t, versus time plot using Eqs (S4.1)–(S4.3) and plotted in Figure S4.8.

Step 2: Considering a spherical sorbent particle of radius "r" which is initially free from solute, the total amount of solute mass absorbed by the sorbent after time t (q_t) is expressed as a fraction of solute uptake (F_t) of the corresponding quantity after infinite time (q_∞) by the following relationship [5].

$$q_t = \frac{V}{m}(C_o - C_t) \tag{S4.5}$$

$$q_\infty = \frac{V}{m}(C_o - C_\infty) \tag{S4.6}$$

$$F_t = \frac{q_t}{q_\infty} = 1 - \sum_{n=1}^{\infty} \frac{6\omega(\omega+1)\exp\left(-\frac{\overline{D_{eff}}\beta_n{}^2 t}{r^2}\right)}{9 + 9\omega + \omega^2\beta_n^2} \tag{S4.7}$$

C_t is the concentration of solute in the aqueous phase at time t and can be obtained from Figure S4.7; "m" is the mass of sorbent used for the test, m = 0.1 g; β_n is the non-zero root expressed in radians from the following equality:

$$\tan \beta_n = \frac{3\beta_n}{3 + \omega\beta_n^2} \tag{S4.8}$$

The parameter ω is expressed in terms of the final fractional uptake of solute by the ion exchange particles according to the following relation:

$$\frac{q_\infty}{VC_o} = \frac{C_o - C_\infty}{mC_o} = \frac{1}{1+\omega} \tag{S4.9}$$

q_∞ is determined from the aqueous phase concentration after equilibration for 72 h (C_e = 0.081 mg/L) which is essentially the amount of zinc (target ion) in the ion exchanger when equilibrium is attained.

$$q_\infty = \frac{V}{m}(C_o - C_e) = \frac{2\,\text{L}}{0.1\,\text{g}}(0.25\,\text{mg/L} - 0.081\,\text{mg/L}) = 3.38\ \text{mg/g} \qquad (S4.10)$$

From Eq. (S4.9), $\omega = 0.478$.

β_n values are further solved from Eq. (S4.8) using $\omega = 0.478$.

Table S4.3. Computed β values.

n	β
1	3.987
2	6.956
3	9.959
4	13.001
5	16.07
6	19.16
7	22.26
8	25.37
9	28.48
10	31.61

Note: β may be calculated for higher n and used in Eq. (S4.7) unless its contribution toward summation in the equation becomes insignificant.

Step 3: Fractional uptakes (F_t) at different times can be calculated using Eq. (S4.7) for a specific assumed value of diffusivity ($\overline{D}_{eff}$) with a particle radius r = 125 μm, $\omega = 0.478$ and computed roots of β_n as tabulated during Step 1. With different assumed effective diffusivity values, fractional uptake versus time (F_t vs t) plots can now be constructed and superimposed on the experimentally determined F_t versus t plot of Figure S4.8.

Step 4: Figure S4.8 illustrates the fractional zinc uptake versus time plots; the dotted plots represent the model predictions with different effective intraparticle diffusivity values. From the best-fit curve, the effective intraparticle diffusivity is 7.5×10^{-10} cm²/s.

Point to note: For the situation of infinite solution volume, the concentrations at the exchanger surface are the same as in the bulk solution and they don't change with time. In this case, Eq. (S4.4) has a simpler analytical solution for the fractional uptake as follows [5]:

$$F_t = 1 - \frac{6}{\pi^2}\sum_{n=1}^{\infty}\frac{1}{n^2}\exp\left(-\frac{\overline{D}_{eff}tn^2}{r^2}\right) \qquad (S4.11)$$

where F_t = fractional uptake at any time t, $\overline{D}_{eff}$ = intraparticle diffusivity, and r = radius of particle.

(Continued)

Supplementary Reading S4.2 (Continued)

Table S4.4. *Model predicted and experimentally determined F_t values.*

Time (min)	*F_t from model equation with $\bar{D}_{eff} = 7.5 \times 10^{-10}$ cm²/s* $F_t = 1 - \sum_{n=1}^{\infty} \frac{6\omega(\omega+1)\exp\left(-\frac{\bar{D}_{eff}\beta_n^2 t}{R^2}\right)}{9 + 9\omega + \omega^2\beta_n^2}$	*F_t from experimental data* $F_t = \frac{C_o - C_t}{C_o - C_e}$
2.5	0.24	–
5	0.32	0.313
10	0.41	0.430
20	0.515	–
30	0.59	0.600
40	0.64	–
60	0.71	0.727
90	0.77	–
120	0.82	0.828
180	0.875	–
240	0.91	0.905

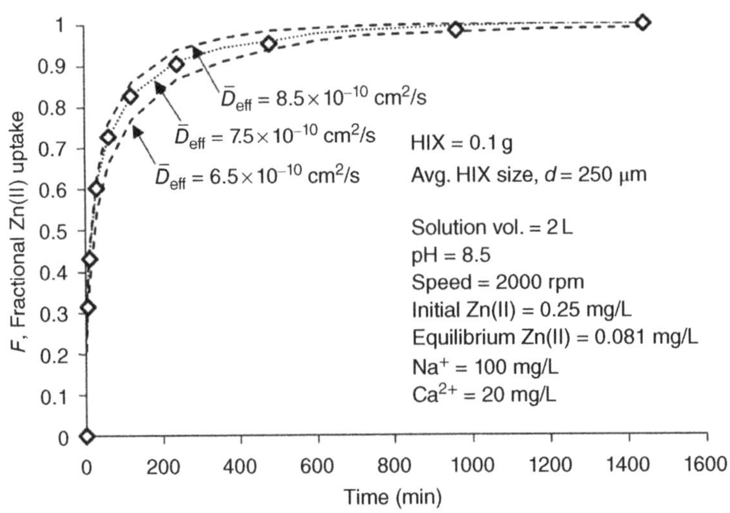

Figure S4.8. *Fractional uptake (F_t) versus time plots from experimental data and predicted results from model equation. Source: Reprinted with permission from Chatterjee 2011 [6].*

Example S4.4 Effect of Exchanger Particle Size

Consider the effective intraparticle diffusivity value (7.5×10^{-10} cm²/s) computed for selective zinc exchange in Example S4.3. Now consider different particle sizes (diameter) 100, 200, 300, and 500 μm. Develop and plot the F_t versus t curve for each particle size for the situation of infinite solution volume.

Question 1: Compute $t_{1/2}$ value for each particle size.

Question 2: Plot $t_{1/2}$ versus r (radius of the ion exchanger particle) and comment on how particle size influences the rate of intraparticle diffusion.

Solution:

1. *In case of infinite solution volume F_t is obtained from Eq. (S4.11). Based on effective diffusivity $\overline{D}_{eff} = 7.5 \times 10^{-10}$ cm²/s, F_t is calculated for different particle sizes at different times as explained in the following table:*

Table S4.5. *Computed fractional uptake values (F_t) for different particle sizes.*

	Fractional uptake (F_t)			
Time (s)	r = 50 μm = 0.005 cm (diameter = 100 μm)	r = 100 μm = 0.01 cm (diameter = 200 μm)	r = 150 μm = 0.015 cm (diameter = 300 μm)	r = 250 μm = 0.025 cm (diameter = 500 μm)
0	0	0	0	0
5	0.099	0.051	0.037	0.029
20	0.193	0.099	0.067	0.042
60	0.321	0.168	0.114	0.069
100	0.402	0.214	0.146	0.089
140	0.464	0.251	0.171	0.105
168	0.500	0.273	0.187	0.114
300	0.627	0.354	0.245	0.151
500	0.748	0.442	0.310	0.193
670	0.816	0.500	0.353	0.221
900	0.879	0.562	0.402	0.254
1200	0.929	0.627	0.454	0.290
1510	0.959	0.681	0.500	0.322
2000	0.983	0.748	0.559	0.365
4190	0.999	0.908	0.732	0.500

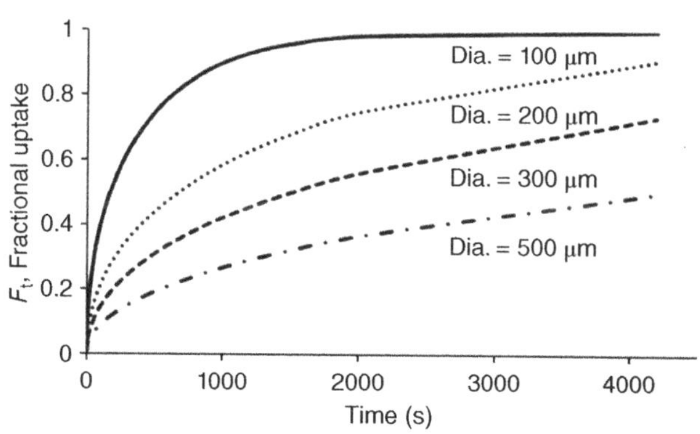

Figure S4.9. *Plot of fractional uptake versus time for different particle sizes in an infinite volume solution.*

(Continued)

Supplementary Reading S4.2 (Continued)

Figure S4.9 shows fractional uptake (F_t) versus time plot for different particle sizes. The plot clearly explains that uptake rate increases with a decrease in particle size.

2. *For different particle sizes, computed $t_{1/2}$ values, that is, time at which 50% uptake is complete are as follows:*

Table S4.6. *Half time values for different particle sizes.*

Particle radius (μm)	$t_{1/2}$ *(s)*
50	168
100	670
150	1510
250	4190

Figure S4.10 shows plot of $t_{1/2}$ versus particle radius on a log–log plot. Slope close to 2.0 suggests that half-time uptake values increase with the square of the particle radius.

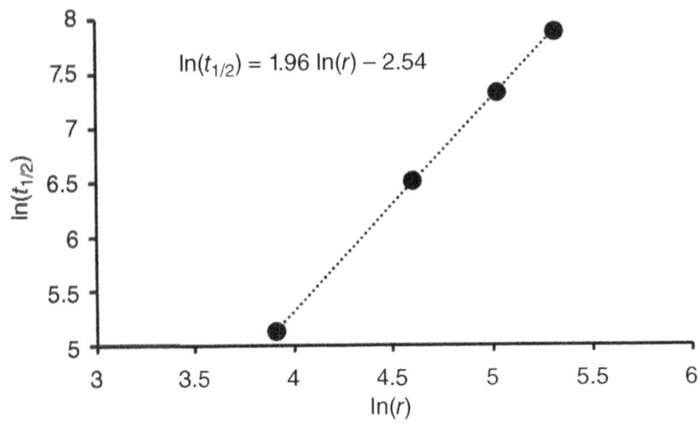

Figure S4.10. *Plot showing influence of particle size (radius of ion exchange particles) on $t_{1/2}$, that is, time required to attain 50% equilibrium capacity.*

4.8 Trace Ion Exchange Kinetics

Earlier, we devoted Chapter 3 to trace ion exchange equilibria. In this section, we will develop mathematical models for trace ion exchange kinetics controlled by intraparticle diffusion. Concurrently, we will present experimental results for validation of the models and for insightful understanding of the mathematical predictions.

4.8.1 Chlorophenols as the Target Trace Ions

One of the key experimental parameters associated with trace ion exchange studies stems from the fact that the affinity of trace ions of interest, namely, transition metals (cations) and ligands (anions), is strongly dependent on pH. This phenomenon is caused by the formidable competing effect of H^+ or OH^- on metal or ligand sorption during selective ion exchange. Thus, the experimental data collection needs to be carried out at specific pH values under carefully controlled experimental conditions. In order to circumvent this difficulty, we identified a group of environmentally significant synthetic organic compounds, chlorophenols that exist as monovalent anions over a wide range of pH. Chlorophenols are often referred to as hydrophobic ionizable organic compounds (HIOCs). The chlorophenate anions exhibit high sorption affinity onto polymeric anion exchangers over competing chloride or sulfate anions and pH has no noticeable effect on the equilibrium. Table 4.3, presented earlier in this chapter, included pertinent properties of three hydrophobic aromatic anions: pentachlorophenate (PCP^-), trichlorophenate (TCP^-), and dichlorophenate (DCP^-). Related studies as described in Chapter 3, have shown that the uptake of chlorophenol anions (e.g., PCP^-) by an anion exchanger is truly an anion exchange reaction, that is, it is accompanied by the desorption of an equivalent amount of competing ions (say chloride) in accordance with the following stoichiometry:

$$\overline{(R^+)Cl^-} + PCP^-(aq) \leftrightarrow \overline{(R^+)PCP^-} + Cl^-(aq) \tag{4.24}$$

Assuming ideality in both the aqueous phase and the ion exchanger phase, the pseudo-equilibrium constant ($\alpha_{PCP/Cl}$) or separation factor for the exchange reaction in (4.24) is given by:

$$\alpha = \frac{q_{PCP}C_{Cl}}{C_{PCP}q_{Cl}} \tag{4.25}$$

where q_i and C_i are the concentrations of solute i in the exchanger phase (meq/g) and aqueous phase (meq/L), respectively. The total exchange capacity of the ion exchanger, Q, and the total aqueous phase concentration, C_T, however remain unchanged during the binary sorption process, that is, $Q = q_{PCP} + q_{Cl}$ and $C_T = C_{PCP} + C_{Cl}$. After applying these equalities, Eq. (4.25) can be rearranged as follows:

$$q_{PCP} = \frac{\alpha C_{PCP}}{C_T + (\alpha - 1)C_{PCP}} Q \tag{4.26}$$

The interaction between the aromatic moiety of PCP^- and the anion exchanger's matrix contributes much greater selectivity of PCP^- over Cl^-, that is, $\alpha \gg 1$. Thus,

$$q_{PCP} = \frac{\alpha C_{PCP}}{C_T + \alpha C_{PCP}} Q \tag{4.27}$$

Or,

$$\frac{q_{PCP}}{Q} = \frac{\alpha x_{PCP}}{1 + \alpha x_{PCP}} \tag{4.28}$$

Or,

$$y_{PCP} = \frac{\alpha x_{PCP}}{1 + \alpha x_{PCP}} \tag{4.29}$$

where x_{PCP} = fractional PCP^- concentration in the aqueous phase or C_{PCP}/C_T and y_{PCP} = fractional PCP^- concentration in the exchanger phase or q_{PCP}/Q.

Note that the sorption behavior of PCP^- according to Eqs (4.27)–(4.29) is non-linear and conforms to the Langmuir isotherm. In environmental separation processes, the concentration of the target contaminant is very low and under trace contaminant conditions, $\alpha x_{PCP} \ll 1$. Thus,

$$y_{PCP} = \alpha x_{PCP} \tag{4.30}$$

Or,

$$q_{PCP} = \frac{\alpha Q C_{PCP}}{C_T} \tag{4.31}$$

Or,

$$q_{PCP} = K C_{PCP} \tag{4.32}$$

that is, the isotherm is linear and the partition coefficient, K, is equal to:

$$K = \frac{\alpha Q}{C_T} \tag{4.33}$$

As this section unfolds, we will gradually present how the partition coefficient (K) and competing ion concentration ($C_T \sim C_{Cl}$) influences intraparticle diffusion rates during trace ion exchange. It is worth noting that the mathematical treatment and resulting conclusions are not limited to $PCP^- \leftrightarrow Cl^-$ exchange and will remain identical for other trace systems pertaining to selective ion exchange.

4.8.2 Intraparticle Diffusion inside a Macroporous Ion Exchanger

Based on the physical morphology of the macroporous ion exchanger as described in Section 4.4.2, the solute transport inside a macroporous (biphasic) sorbent particle can proceed in parallel through both interconnected pores and adjacent gel phases. Figure 4.23a depicts a macroporous particle containing microgels under conditions where intraparticle diffusion is the rate-limiting step, and Figure 4.23a shows how coupled counterions PCP^- and Cl^- diffuse through microgels and macropores. Figure 4.23 is similar to Figure 4.10 where PCP^- and Cl^- are exchanging counterions.

With the assumptions that (i) local equilibrium exists between the microgels and adjoining pore liquid, (ii) counterions diffuse in parallel through microgels and pores, as shown in Figure 4.23b, and (iii) PCP^- is a trace solute, and its diffusivity is equal to the interdiffusion coefficient, the transport of PCP^- in a single macroporous particle may be written as:

$$\epsilon_P \frac{\partial C_{PCP}}{\partial t} + \rho_P \frac{\partial q_{PCP}}{\partial t} = \epsilon_P D_P \left(\frac{\partial^2 C_{PCP}}{\partial r^2} + \frac{2}{r} \frac{\partial C_{PCP}}{\partial r} \right) + D_g \rho_P \left(\frac{\partial^2 q_{PCP}}{\partial r^2} + \frac{2}{r} \frac{\partial q_{PCP}}{\partial r} \right) \tag{4.34}$$

where the first terms on both sides represent the pore diffusion and the second term corresponds to solid phase diffusion of PCP^-; ϵ_P and ρ_P denote the void fraction and the density of the wet macroporous particle while D_P and D_g are the diffusivities in the pores and in the gel phase, respectively. C_{PCP} represents the liquid phase concentration (mass per unit volume of liquid) in equilibrium with the gel phase concentration, q_{PCP} (mass per unit mass of ion exchanger).

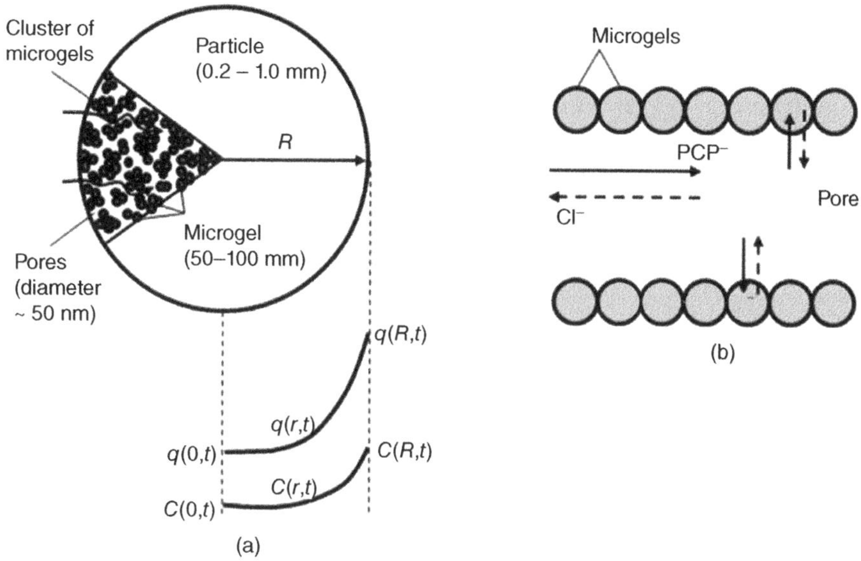

Figure 4.23 (a) Schematic representation of a macroporous particle containing microgels, where intraparticle diffusion is the rate limiting step, (b) explanation of coupled diffusion of counterions PCP^- and Cl^- in parallel through microgels and macropores. *Source:* Li and SenGupta 2000 [37]. Reproduced with permission of American Chemical Society.

Note: For an air-dried macroporous particle, an additional $(1 - \epsilon_P)$ will appear for the second term of both the LHS and the RHS.

When PCP^- is a trace species, from Eq. (4.32)

$$q_{PCP} = KC_{PCP} \tag{4.35}$$

Hence,

$$\frac{\partial q_{PCP}}{\partial t} = K\frac{\partial C_{PCP}}{\partial t} \tag{4.36}$$

and

$$\frac{\partial^2 q_{PCP}}{\partial r^2} = K\frac{\partial^2 C_{PCP}}{\partial r^2} \tag{4.37}$$

Applying these equations to Eq. (4.34) and rearranging,

$$\frac{\partial q_{PCP}}{\partial t} = \frac{D_P\epsilon_P + KD_g\rho_P}{\epsilon_P + K\rho_P}\left(\frac{\partial^2 q_{PCP}}{\partial r^2} + \frac{2}{r}\frac{\partial q_{PCP}}{\partial r}\right) = \overline{D}_{\text{eff}}\left(\frac{\partial^2 q_{PCP}}{\partial r^2} + \frac{2}{r}\frac{\partial q_{PCP}}{\partial r}\right) \tag{4.38}$$

Where the effective intraparticle diffusivity is

$$\overline{D}_{\text{eff}} = \frac{D_P\epsilon_P + KD_g\rho_P}{\epsilon_P + K\rho_P} \tag{4.39}$$

Note that Eq. (4.38) is characteristically similar to the solid phase diffusion with spherical geometry as presented in Example S4.3 earlier.

For selective ion exchange, $K\rho_P \gg \epsilon_P$.

Thus,

$$\overline{D}_{\text{eff}} = D_g + \frac{D_P \epsilon_P}{K \rho_P} \tag{4.40}$$

From Eq. (4.33), $K = \frac{\alpha Q}{C_T}$.

And the pore phase diffusivity of PCP^- is equal to

$$D_P = \frac{D_P^0}{\tau} \tag{4.41}$$

where D_P^0 and τ represent the solute's liquid phase diffusivity and the tortuosity factor of the macroporous sorbent particle, respectively. Thus,

$$\overline{D}_{\text{eff}} = D_g + \frac{D_P^0 \epsilon_P C_T}{\tau \rho_P \alpha Q} \tag{4.42}$$

For parallel transport, if gel-phase diffusion is significantly faster than the pore-phase diffusion, that is, for a strictly gel-type ion exchanger,

$$\overline{D}_{\text{eff}} = D_g \tag{4.43}$$

Similarly, when pore diffusion is the rate-limiting step,

$$\overline{D}_{\text{eff}} = \frac{D_P^0 \epsilon_P C_T}{\tau \rho_P \alpha Q} \tag{4.44}$$

4.8.3 Effect of Sorption Affinity on Intraparticle Diffusion

We will use anionic chlorophenols as trace counterions to validate the relationship between sorption affinity and intraparticle diffusion. Figure 4.24 shows binary chlorophenol–chloride isotherms for three different anionic chlorophenols, namely, pentachlorophenol (PCP^-), 2,4,6-trichlorophenol (TCP^-), and 2,6-dichlorophenol (DCP^-) at 23 °C. The isotherm tests were carried out at pH 8.5, where all of the chlorophenols exist predominantly as anions. All three isotherms conformed to Langmuir-type behavior and the corresponding best-fit equations in accordance with Eq. (4.29) are given as:

$$y_{\text{PCP}} = \frac{403 * x_{\text{PCP}}}{1 + 403 * x_{\text{PCP}}} \tag{4.45}$$

$$y_{\text{TCP}} = \frac{36.2 * x_{\text{TCP}}}{1 + 36.2 * x_{\text{TCP}}} \tag{4.46}$$

$$y_{\text{DCP}} = \frac{12.8 * x_{\text{DCP}}}{1 + 12.8 * x_{\text{DCP}}} \tag{4.47}$$

and $\alpha_{\text{PCP/Cl}} = 403$, $\alpha_{\text{TCP/Cl}} = 36.2$, and $\alpha_{\text{DCP/Cl}} = 12.8$.

The sequence of sorption affinity, that is, $PCP^- > TCP^- > DCP^-$ is in accordance with increased chlorine substitution of the parent phenol resulting in increased hydrophobicity or octanol–water partition coefficient of its non-polar moiety. Three separate kinetic studies were carried out with DCP^-, TCP^-, and PCP^-. The concentration of competing chloride ion (50 meq/L) in every case was the same, but much

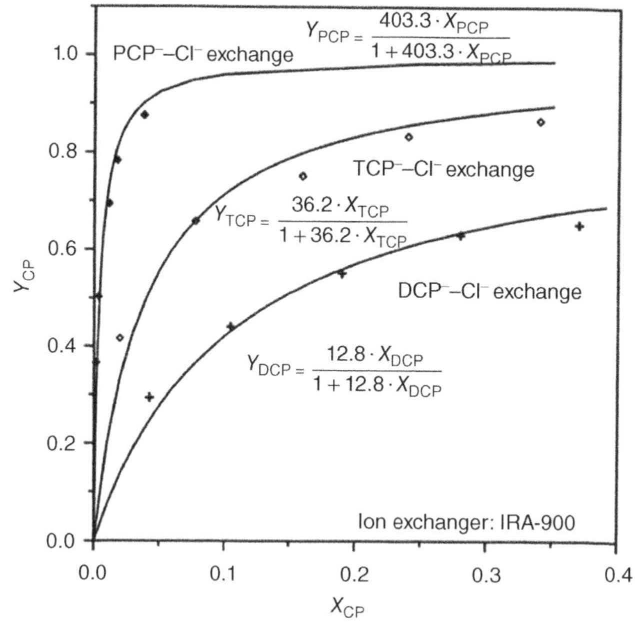

Figure 4.24 Binary isotherms for chlorophenol–chloride for three different chlorophenols at pH = 8.5 where they all exist as monovalent anions. *Source:* Li and SenGupta 2000 [2]. Reproduced with permission of Elsevier.

greater (2000×) than individual chlorophenate anions, that is, chlorophenates were the trace species. Fractional uptake rates of the three chlorophenates are plotted against time in Figure 4.25; the figure also includes the computed best-fit values of effective intraparticle diffusivities ($\overline{D}_{\text{eff}}$) of each chlorophenol. $\overline{D}_{\text{eff}}$ values are inversely related to their relative sorption affinities. Each of the three anionic chlorophenols used in the study is monovalent with one aromatic ring and their liquid-phase self-diffusion coefficients (i.e., D_{P}^{0}) are nearly identical. For a given ion exchanger particle (i.e., IRA-900), $\epsilon_{\text{P}}, \rho_{\text{P}}, Q,$ and τ are essentially constant; total aqueous-phase concentration, C_{T}, was also constant at 50 meq/L for all kinetic experiments. Thus, from Eq. (4.44)

$$\overline{D}_{\text{eff}} = \frac{\text{constant}}{\alpha_{\text{CP}}} \tag{4.48}$$

$$\ln \overline{D}_{\text{eff}} = \text{constant} - \ln \alpha_{\text{CP}}. \tag{4.49}$$

Figure 4.26 shows the log–log plot of experimentally determined $\overline{D}_{\text{eff}}$ versus α_{CP} for three chlorophenols. The linearity of the $\log \overline{D}_{\text{eff}}$ versus $\log \alpha_{\text{CP}}$ plot with a negative slope close to unity is in agreement with Eq. (4.49). The octanol–water partition coefficient (K_{OW}) of an un-dissociated chlorophenol is, also, a representative measure of the non-polar moiety's hydrophobicity and is correlated to the sorption affinity of the chlorophenol

$$\frac{\log \alpha_{\text{CP}}}{\log K_{\text{OW}}} = \text{constant} \tag{4.50}$$

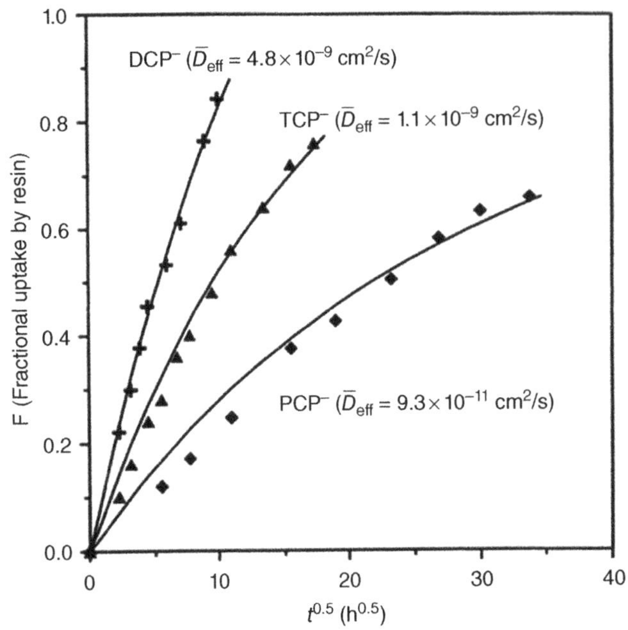

Figure 4.25 Fractional uptake of three different chlorophenates versus the square root of time for anion exchanger IRA-900. *Source:* Li and SenGupta 2000 [2]. Reproduced with permission of Elsevier.

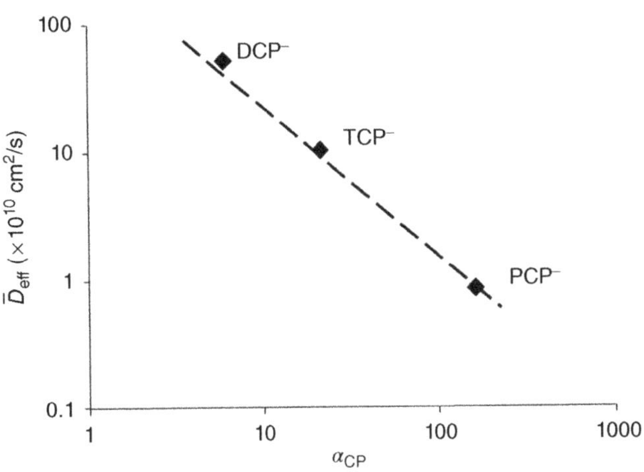

Figure 4.26 Computed effective intraparticle diffusivities plotted against chlorophenate–chloride separation factors (α_{CP}). *Source:* Li and SenGupta 2000 [2]. Reproduced with permission of Elsevier.

Figure 4.16 (presented earlier) shows the plot of log $\overline{D}_{eff}$ versus log K_{OW} values for three chlorophenols. The linearity of the plot validates the dependence of $\overline{D}_{eff}$ on K_{OW}, as predicted by Eq. (4.50). From a phenomenological viewpoint this observation implies that a target aromatic anion with a greater K_{OW} will result in lower $\overline{D}_{eff}$ value and will create a longer mass transfer zone (MTZ) during a fixed-bed column run. The column breakthrough history of that solute will be more gradual.

Without any loss of generality, similar inferences can also be drawn for selective cation exchange. For a chelating exchanger with an iminodiacetate functional group, the selectivity sequence for three divalent cations stand as follows: $Cu^{2+} > Ni^{2+} \gg Ca^{2+}$. The sequence of intraparticle diffusivity during trace ion exchange is:

$$\overline{D}_{Cu} < \overline{D}_{Ni} \ll \overline{D}_{Ca}$$

Equation (4.44) should be revisited to provide a further quantitative validation of the proposed intraparticle pore diffusion model

$$\overline{D}_{eff} = \frac{D_P^0 \epsilon_P C_T}{\tau \rho_P Q} \frac{1}{\alpha_{PCP}} \tag{4.51}$$

Or,

$$D_{PCP}^0 = \overline{D}_{eff} \frac{\tau \rho_P \, Q \, \alpha_{PCP}}{\epsilon_P \, C_T} \tag{4.52}$$

From the batch kinetic test and equilibrium isotherm data: $\overline{D}_{eff} = 9.3 \times 10^{-11}$ cm^2/s, $\alpha_{PCP} = 403$, $C_T = 50$ meq/L, $Q_{wet} = 1.1$ meq/g, and $\rho_P = 1100$ g/L. For exchange of monovalent ions, Yoshida *et al.* [27] experimentally determined ϵ_P and τ values for IRC-200, which is macroporous and biphasic, like IRA-900: $\epsilon_P = 0.29$ and $\tau = 3$.

The liquid-phase self-diffusion coefficient of PCP^- computed from Eq. (4.52) is

$$D_{PCP}^0 = 6.7 \times 10^{-6} \, \text{cm}^2/\text{s} \tag{4.53}$$

For comparison, self-diffusion coefficient of PCP^- in water is now estimated independently using modified Wilke–Chang correlation for organic electrolytes as follows [44]:

$$D_{PCP}^0 = 0.9 \times 7.4 \times 10^{-12} \frac{T(2.6 M_w)^{1/2}}{\mu V_B^{0.6}} \tag{4.54}$$

where M_w is molar weight of water, T is absolute temperature (K), μ is the viscosity of water in centipoise, and V_B is molar volume at the normal boiling point. The value of V_B is estimated using LeBas method [45] and equal to 227.5 cm^3/mol. Thus,

$$D_{PCP}^0(\text{Wilke–Chang}) = 4.7 \times 10^{-6} \, \text{cm}^2/\text{s} \tag{4.55}$$

The self-diffusion coefficient of PCP^- calculated independently from the Wilke–Chang correlation compares quite satisfactorily (same order of magnitude) with the diffusivity computed from the batch kinetic data using the intraparticle pore diffusion model.

4.8.4 Solute Concentration Effect

With biphasic physical morphology of a macroporous ion exchanger, pore diffusion is often, if not always, the rate-limiting step for a trace target solute (A) in selective ion exchange. Under such condition, the intraparticle transport of A can be written as:

$$\epsilon_\mathrm{P}\frac{\partial C_\mathrm{A}}{\partial t} + \rho_\mathrm{P}\frac{\partial q_\mathrm{A}}{\partial t} = \epsilon_\mathrm{P} D_\mathrm{A}\left(\frac{\partial^2 C_\mathrm{A}}{\partial r^2} + \frac{2}{r}\frac{\partial C_\mathrm{A}}{\partial r}\right) \tag{4.56}$$

where D_A is the pore diffusivity of counterion "A." For a general isotherm including Langmuir type, it can be written:

$$\frac{\partial q_\mathrm{A}}{\partial t} = \frac{\partial q_\mathrm{A}}{\partial C_\mathrm{A}}\frac{\partial C_\mathrm{A}}{\partial t} \tag{4.57}$$

Inserting (4.57) into (4.56), it is obtained:

$$\frac{\partial C_\mathrm{A}}{\partial t} = \frac{\epsilon_\mathrm{P} D_\mathrm{A}}{\epsilon_\mathrm{P} + \rho_\mathrm{P}\frac{\partial q_\mathrm{A}}{\partial C_\mathrm{A}}}\left(\frac{\partial^2 C_\mathrm{A}}{\partial r^2} + \frac{2}{r}\frac{\partial C_\mathrm{A}}{\partial r}\right) = \overline{D}_\mathrm{eff}\left(\frac{\partial^2 C_\mathrm{A}}{\partial r^2} + \frac{2}{r}\frac{\partial C_\mathrm{A}}{\partial r}\right) \tag{4.58}$$

$\overline{D}_\mathrm{eff}$ in Eq. (4.58) is the effective intraparticle diffusivity, which is expressed as:

$$\overline{D}_\mathrm{eff} = \frac{\epsilon_\mathrm{P} D_\mathrm{A}}{\epsilon_\mathrm{P} + \rho_\mathrm{P}\frac{\partial q_\mathrm{A}}{\partial C_\mathrm{A}}} \tag{4.59}$$

Note that the effective intraparticle diffusivity, $\overline{D}_\mathrm{eff}$, is strongly influenced by the slope, $\frac{\partial q_\mathrm{A}}{\partial C_\mathrm{A}}$, of the isotherm. Thus, $\overline{D}_\mathrm{eff}$ is minimized at very low concentration of C_A in the linear range of the Langmuir isotherm, where slope is maximized. As the slope gradually decreases in the non-linear range with an increase in concentration, $\overline{D}_\mathrm{eff}$ increases. At the top asymptotic end of Langmuir isotherm, $\frac{\partial q_\mathrm{A}}{\partial C_\mathrm{A}} \approx 0$ and $\overline{D}_\mathrm{eff}$ is equal to the pore diffusivity, D_A.

4.9 Rectangular Isotherms and Shell Progressive Kinetics

The occurrence of rectangular isotherms for a counterion can be viewed as an equilibrium condition where the counterion has extraordinarily high sorption affinity and occupies nearly all the binding sites in the exchanger irrespective of aqueous phase concentration. Figure 4.27 includes both the rectangular isotherm and a typical Langmuir isotherm. Note that at any aqueous-phase concentration greater than zero, the slope of the rectangular isotherm is zero, that is, $\frac{\partial q_\mathrm{A}}{\partial C_\mathrm{A}} = 0$.

To comprehend the physical reality of shell progressive kinetics in conjunction with the underlying mathematical implications, a spherical ion exchange resin bead within which the pore diffusion is the rate-limiting step, will be considered.

From Eq. (4.59),

$$\overline{D}_\mathrm{eff} = \frac{\epsilon_\mathrm{P} D_\mathrm{A}}{\epsilon_\mathrm{P} + \rho_\mathrm{P}\frac{\partial q_\mathrm{A}}{\partial C_\mathrm{A}}} \tag{4.60}$$

Note that for rectangular isotherm $\frac{\partial q_\mathrm{A}}{\partial C_\mathrm{A}} = 0$ and thus $\overline{D}_\mathrm{eff} = D_\mathrm{A} = \frac{D_\mathrm{A}^0}{\tau}$, where D_A^0 is the liquid-phase diffusivity of A and τ is the tortuosity of the ion exchanger.

Thus, within every bead, no concentration gradient exists for "A" in the exchanger phase. The pore phase concentration, however, gradually drops from the periphery

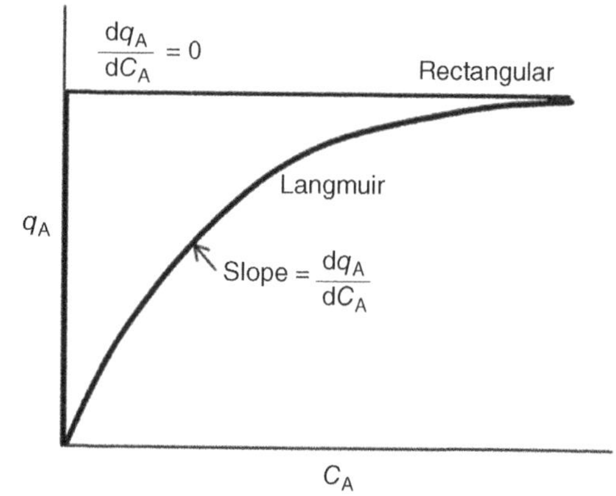

Figure 4.27 Illustrative plots of rectangular and Langmuir isotherms.

to the interior with the progress of the ion exchange reaction with a sharp boundary. This behavior-intraparticle diffusion with a sharp change in boundary due to fast ion exchange process – forms the basis of "shrinking core" or "shell progressive" kinetics corresponding to rectangular isotherm. In reality, this represents a scenario where the incoming counterion "A" has an affinity far greater than "B" with which the ion exchanger is presaturated. Figure 4.28 depicts the change in concentration of A and B within a spherical macroporous ion exchanger bead for shell-progressive kinetics. Note that if R–A and R–B have different colors in the resin phase, the contrast would be distinctly visible for "shrinking core" situations. The "rule of thumb" guideline is that the dimensionless parameter in the Langmuir equation for the incoming ion, $b_A C_A$, should be greater than 20 to exhibit shrinking core phenomenon [46]. Shrinking core phenomenon is also prevalent in adsorption process and a detailed review has been provided by Suzuki [47].

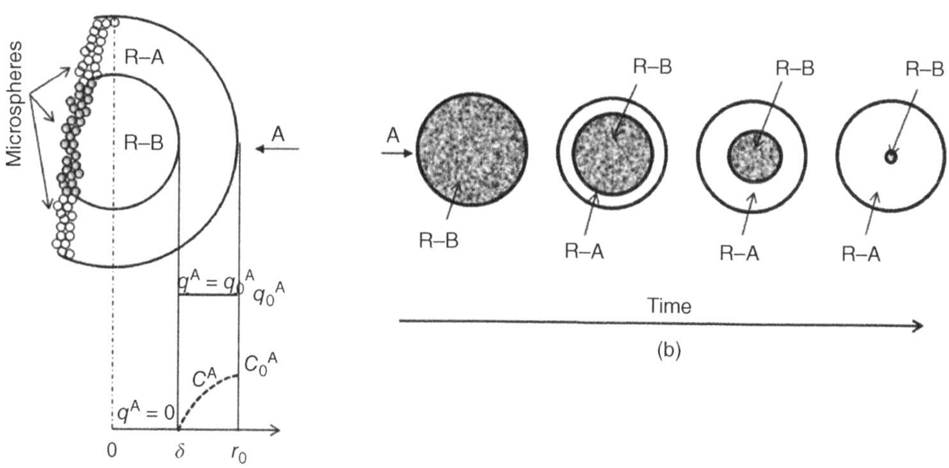

Figure 4.28 Schematic illustrations of concentration changes for A and B (counterions) within a spherical macroporous ion exchanger bead, in accordance with shrinking core or shell-progressive kinetics: affinity of A is far greater than affinity of B.

4.9.1 Anomalies in Arrival Sequence of Solutes

It is imperative that here we discuss one distinctive difference in intraparticle solute transport phenomena that occurs in conventional Langmuir isotherms versus rectangular isotherms. Let us consider the following two scenarios for the intraparticle transport of "A" inside an ion exchange resin bead presaturated with B:

1. First come, first occupy (FCFO or orderly)
2. Last come, first occupy (LCFO or disorderly).

In order to attain equilibrium, counterions enter the ion exchanger bead and move from one site to the next toward the center. For intraparticle diffusion-controlled kinetics, counterions in the exchanger phase proceed in an ordered sequence, that is, a counterion (say A_1) entering earlier inside the exchanger always remains ahead in its path toward the center versus others (say A_2, A_3, A_4, etc.) entering later and simultaneously displacing B ions. Figure 4.29a illustrates the phenomenon where counterions respectfully obey the law of seniority and follow "first come, first occupy" transport to ion exchange sites. In principle, FCFO or the sequential progression of counterions is strictly adhered to during intraparticle diffusion.

In contrast, for shrinking core, or shell progressive kinetics, FCFO is not an accurate representation of the phenomenon. As soon as the first counterion (i.e., A_1) transports to the first ion exchange site at the outermost periphery of the ion exchanger bead, A_1 is irreversibly sorbed following exchange with B. Since the exchanger-phase concentration gradient is zero, in accordance with the property of rectangular isotherms, the next counterion in line, A_2, is unable to desorb A_1. Consequently, A_2 goes past A_1 through the pores filled with solvent/water present in the exchanger and binds to the

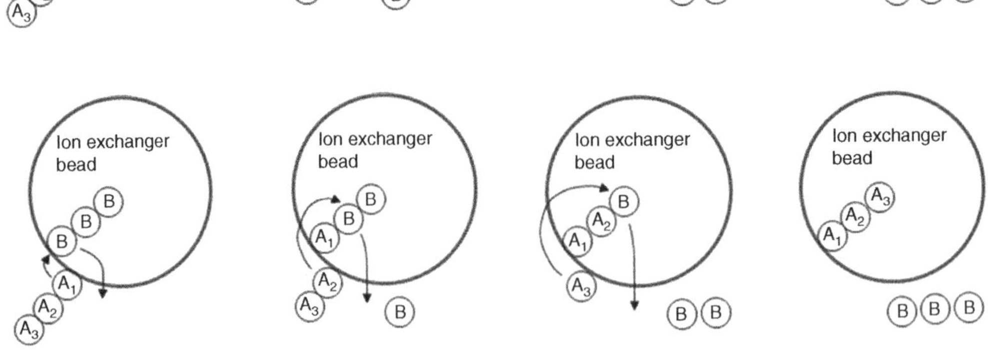

Figure 4.29 Schematic representation of intraparticle transport of counterion A inside a spherical ion exchanger bead presented with exchangeable ion in B-form based on the concept of (1) first come, first occupy (A-top); (2) last come, first arrive (B-bottom).

next ion exchange site in B-form closer to the center. Similarly, the counterion A_3 goes past both A_1 and A_2 and binds irreversibly with the next available ion exchange site. Thus, A_3, which enters the exchanger after A_1 and A_2, is rewarded with a permanent seat at an ion exchange site closer to the center of the bead than A_1 and A_2, in accordance with LCFO (or disorderly) scenario. In fact, the ion exchange site at the center of the bead is presented to the last counterion, after which the resin bead is essentially saturated. Figure 4.29b illustrates the situation emphasizing the following two distinctive features of rectangular isotherms:

1. Last come, first occupy (LCFO) transport of counterions "A" from the bulk solution to the ion exchange sites;
2. Sharp boundary in the exchanger between the exchanging counterions and the pre-saturated ions leading to shrinking core or shell progressive phenomenon.

It is noteworthy that ion exchange accompanied by fast chemical reactions often tends to exhibit shell progressive kinetics and are similar to rectangular isotherms. Such chemical reactions often involve favorable thermodynamic equilibrium with high equilibrium constant values. Some examples include acid-base neutralization, weak-acid/weak-base association and complexation; a few are presented below:

(a) Neutralization of H^+ and OH^-:

$$\overline{\text{R-SO}_3^-\text{H}^+} + \text{Na}^+ + \text{OH}^- \rightarrow \overline{\text{R-SO}_3^-\text{Na}^+} + \text{H}_2\text{O} \tag{4.61}$$

$$\overline{\text{R-N(CH}_3)_3{}^+\text{OH}^-} + \text{H}^+ + \text{Cl}^- \rightarrow \overline{\text{R-N(CH}_3)_3^+\text{Cl}^-} + \text{H}_2\text{O} \tag{4.62}$$

(b) Association of protons on weak ion exchangers:

$$\overline{\text{R-COO}^-\text{Na}^+} + \text{H}^+ + \text{Cl}^- \rightarrow \overline{\text{R-COOH}} + \text{Na}^+ + \text{Cl}^- \tag{4.63}$$

$$\overline{\text{R-N(CH}_2\text{-COO}^-)_2\text{Ni}^{2+}} + 2\text{H}^+ + 2\text{Cl}^- \rightarrow \overline{\text{R-N(CH}_2\text{COOH)}_2} + \text{Ni}^{2+} + 2\text{Cl}^- \tag{4.64}$$

(c) Formation of chelates in the exchanger phase:

$$\overline{\text{R-N(CH}_2\text{-COO}^-)_2\text{Ca}^{2+}} + \text{Cu}^{2+} + 2\text{Cl}^- \rightarrow \overline{\text{R-N(CH}_2\text{COO}^-)_2\text{Cu}^{2+}} + \text{Ca}^{2+} + 2\text{Cl}^- \tag{4.65}$$

Most importantly, all such ion exchange reactions are essentially irreversible unless pH is significantly changed.

4.9.2 Quantitative Interpretation

In accordance with the premise of shrinking core kinetics, the reaction at the periphery of the core is so fast (compared with diffusion) that the counterions (A) are consumed instantly at the ion exchange site before they can proceed. The exchanger phase concentration of the counterion is, therefore, uniform throughout the shell and equal to q_A^o which is essentially equal to the total exchange capacity, Q, when the isotherm is completely rectangular, as illustrated earlier in Figure 4.27. Thus, at the concentration front in the macropore ($r = \delta$), the counterion A is completely sorbed. Under these conditions, consider the pseudo-steady state approximation, where the LHS of Eq. (4.56) is zero, with the boundaries of interest. For this situation, Yoshida and Ruthven have provided the following algebraic solution for appropriate initial and boundary conditions [48]:

$$\frac{6\epsilon_P D_P C_o}{\rho q_o r_0^2} t = 1 + 2 * (1 - F) - 3 * (1 - F)^{2/3} \quad (4.66)$$

where F is the fractional conversion at time t. The half time, $t_{1/2}$, to attain 50% uptake can be computed by putting $F = 0.5$ and is given by

$$t_{1/2} = 0.11 \frac{q_o}{6C_o \rho \epsilon_P} \frac{r_0^2}{D_P} = \text{constant} * \frac{r_0^2}{D_P} \quad (4.67)$$

It is worth noting that for a rectangular isotherm with shell progressive kinetics, which is a special case of intraparticle diffusion, $t_{1/2}$ varies with r_0^2, or the size of the ion exchanger bead, in agreement with the conclusion of Example S4.4.

4.10 Responses to Observations in Section 4.6

Theories of ion exchange kinetics and related quantitative expressions were developed in the last three sections with particular emphasis on intraparticle diffusion. Scientific explanations are now provided for each observation in Section 4.6 in identical sequence aided by the quantitative models presented.

4.10.1 Effect of Concentration on Half-time ($t_{1/2}$)

Note Eq. (4.59) where

$$\overline{D}_{\text{eff}} = \frac{\epsilon_P D_A}{\epsilon_P + \rho_P \frac{\partial q_A}{\partial C_A}} \quad (4.68)$$

Considering the Langmuir isotherm, as the concentration of target contaminant Ni decreases in the presence of other competing ions, the slope or $\partial q_A/\partial C_A$ increases and, hence, $\overline{D}_{\text{eff}}$ decreases. Thus, solute uptake will be kinetically slower at lower concentrations, especially in the lower range of the Langmuir isotherm. $\partial q_A/\partial C_A$ decreases with higher concentrations of the target solute (e.g., Ni) and $\overline{D}_{\text{eff}}$ increases. Consequently, all other conditions remaining the same, the rate of Ni uptake will decrease (i.e., $t_{1/2}$ will increase) with lower nickel concentrations, as evidenced from the nickel uptake graph in Figure 4.15. At high concentrations, the Langmuir isotherm becomes asymptotic, that is, the slope ($\partial q_A/\partial C_A$) tends to be zero, and $\overline{D}_{\text{eff}}$ increases to near D_A.

The above analysis can also be extended to understand why desorption of final traces of solute from the exchanger phase during regeneration is so difficult and often unattainable. During the initial part of the regeneration process, $\overline{D}_{\text{eff}}$ is very high because the solute uptake or exchanger-phase concentration is high. Conversely, $\overline{D}_{\text{eff}}$ decreases to its minimum (i.e., highest slope or $\partial q_A/\partial C_A$) during the last stage of regeneration, when the exchanger has trace concentrations of the target solute. The regeneration during this stage is, thus, very inefficient; for many commercial operations, regeneration is discontinued before complete desorption of the target solute.

4.10.2 Slow Kinetics of Weak-Acid Resin

First, consider the bottom three reactions from Table 4.2 after separating coions and counterions in the aqueous phase:

$$\overline{\text{R-SO}_3^-\text{H}^+} + \text{K}^+ + \text{OH}^- \rightarrow \overline{\text{R-SO}_3^-\text{K}^+} + \text{H}_2\text{O} \tag{4.69}$$

$$2\overline{\text{R-COO}^-\text{Na}^+} + \text{Ca}^{2+} + 2\text{Cl}^- \rightarrow \overline{(\text{R-COO}^-)_2\text{Ca}^{2+}} + 2\text{Na}^+ + 2\text{Cl}^- \tag{4.70}$$

$$2\overline{\text{R-SO}_3^-\text{Na}^+} + \text{Ca}^{2+} + 2\text{Cl}^- \rightarrow \overline{(\text{R-SO}_3^-)_2\text{Ca}^{2+}} + 2\text{Na}^+ + 2\text{Cl}^- \tag{4.71}$$

Note that the progression of the foregoing ion exchange reactions does not depend on the invasion of coions (e.g., OH^- and Cl^-) inside the cation exchanger: the reactions involve exchange of counterions only.

In contrast, for the first ion exchange reaction to proceed, the weak-acid functional groups need to be continuously deprotonated in accordance with the following:

$$\overline{\text{R-COOH}} + \text{K}^+ + \text{OH}^- \rightarrow \overline{\text{R-COO}^-\text{K}^+} + \text{H}_2\text{O} \tag{4.72}$$

However, the same cannot happen for the two following reasons:

1. Unlike SAC exchange resin, WAC resins have orders of magnitude higher affinity for H^+ over K^+. Thus, K^+ cannot desorb H^+ on its own, unless H^+ is removed through neutralization reaction. So, the presence of OH^- near the cation exchange sites is required.
2. For neutralization to occur, OH^- has to enter the cation exchanger phase with fixed ionized carboxylate (R-COO^-) groups balanced by K^+. But due to the Donnan exclusion effect, OH^- is rejected inside the exchanger phase and, hence, H^+ cannot be neutralized.

For a WAC, desorption of H^+ by K^+ is an extremely slow and unfavorable process caused by the rejection of OH^- in the exchanger phase by the Donnan coion exclusion effect exerted by $RCOO^-$. Figure 4.30 illustrates the phenomenon of the slow kinetics as the reaction progresses. An increase in KOH concentration in the aqueous phase will increase the rate of conversion. It is worth noting that the reverse reaction, that is, desorption of K^+ from WAC by H^+, is relatively very fast because it does not require the presence of an anion coion inside the exchanger to proceed.

4.10.3 Chemically Similar Counterions: Drastic Difference in Intraparticle Diffusivity

For trace ion exchange, as deduced in Eq. (4.44), $\overline{D}_{\text{eff}}$ is inversely proportional to α, or the separation factor. As the chlorine substitution increases on phenol, the non-polar moiety of the resulting chlorophenol becomes more hydrophobic. The relative affinity of chlorophenol toward the anion exchanger increases with greater NPM-matrix interactions, as discussed in the previous chapter. Thus, the sorption affinity, or separation factor, follows the sequence $PCP^- > TCP^- > DCP^-$, which influences their

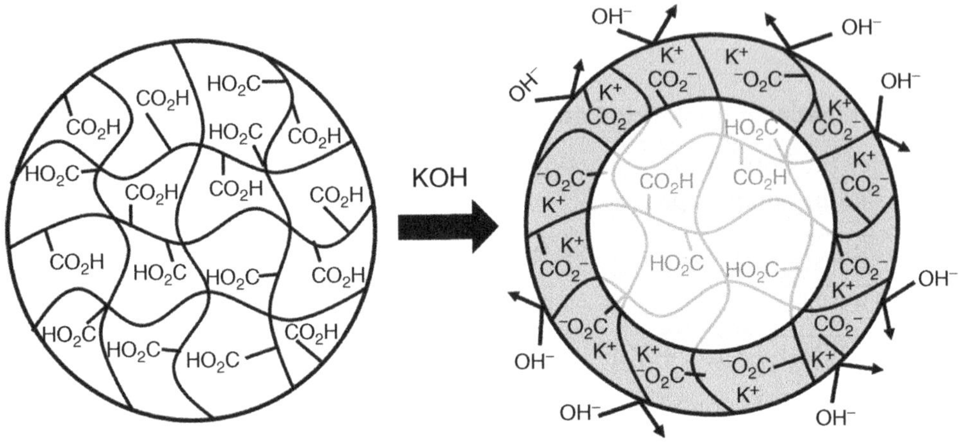

Figure 4.30 Schematic explanation of slow kinetics caused by rejection of OH^- ions due to the Donnan coion exclusion effect.

intraparticle diffusivity, $\overline{D}_{eff}$ as discussed in Section 4.8.3. Thus, $\overline{D}_{eff}$ values are in the reverse order of separation factor, as delineated in Eqs (4.48), (4.50). Self-diffusion coefficients of these chlorophenols in the aqueous phase are not influenced by the foregoing non-polar phenomenon and the aqueous self-diffusion coefficients stay nearly the same.

4.10.4 Gel versus Macroporous

Equations (4.43) and (4.44) show that for a gel-type ion exchanger, $\overline{D}_{eff}$ is independent of C_T or total electrolyte concentration, but $\overline{D}_{eff}$ is directly proportional to C_T for macroporous ion exchangers. $\overline{D}_{eff}$ increases for macroporous anion exchangers with an increase in C_T, resulting in faster uptake, as presented in Figure 4.18, but an identical increase in C_T showed no change in uptake for the gel-type anion exchange resin in Figure 4.17, all other conditions remaining identical. One meaningful outcome of this finding is that by noting the effect of C_T on the uptake rate, it is possible to gather information about the pore structure (gel versus macroporous) of the ion exchanger. From a phenomenological perspective, macroscale kinetic data provide microscopic information about the structure of the adsorbent.

4.10.5 Intraparticle Diffusion during Regeneration

In Figure 4.19, during cation exchange between a gel-type cation exchanger and sodium counterions, at different concentrations, the reason for the observed decrease in the intraparticle self-diffusion coefficient of Na^+ cannot be due to relatively low sorption affinity. An increase in external solution concentration causes the gel ion exchange resin to shrink; as discussed in Section 4.2, the pore water content (ϵ) in the gel phase is reduced. So, the intraparticle self-diffusion coefficient is also reduced as

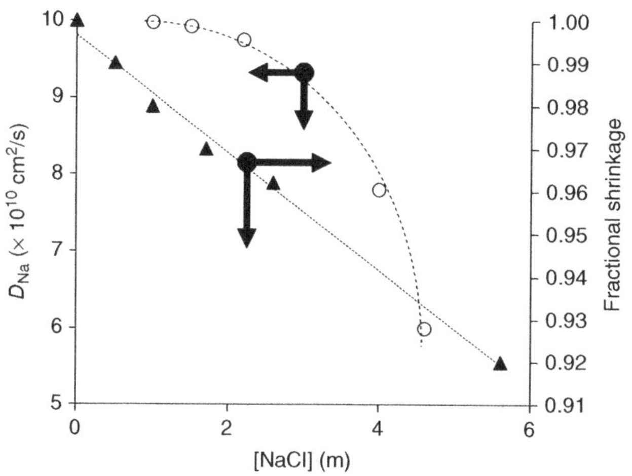

Figure 4.31 Plots of fractional shrinkage of a strong-acid cation exchanger in sodium form and intraparticle diffusion coefficient of sodium (D_{Na}) versus concentration of sodium chloride solution. *Source:* Slater 1991 [40]. Reproduced with permission of Elsevier.

per Mackie's equation:

$$\overline{D}_i = D_i \left(\frac{\epsilon}{2-\epsilon}\right)^2 \tag{4.73}$$

Figure 4.31 includes the plot of intraparticle diffusivity versus external NaCl concentration as presented in Figure 4.19 and the change in fractional volume of the resin is superimposed [40]. The trend in reduction of the diffusivity (i.e., slower intraparticle diffusion) with shrinkage of the resin caused by increased external solution concentration can be readily noted.

4.10.6 Shrinking Core or Shell Progressive Kinetics

As elaborated in Section 4.9, the genesis of shrinking core or shell-progressive kinetics lies in equilibrium conditions leading to a rectangular isotherm. The protonation of WAC resin during Ca^{2+}–H^+ exchange, with a self-sharpening boundary, as illustrated in Figure 4.20, attests to the theory discussed in Section 4.9. Note: WAC resins prefer H^+ well over Ca^{2+} ($\alpha \gg 1$) and the H^+:Ca^{2+} isotherm is truly rectangular.

The observations made with kinetics of plutonium(IV) nitrate ($Pu(NO_3)_5^-$) in Figure 4.21, that is, sharp shell-progressive behavior gradually becoming blurred, defies scientific explanation by rectangular isotherm behavior alone. According to Streat [43], a significant difference in the diffusivity (or mobility) of exchanging counterions accompanied by a gradual change in the ionic composition of the exchanger is the underlying reason for the observed autoradiographs. The weak-base anion exchange resin was presaturated with nitrate (MW = 62), which has orders of magnitude greater diffusivity than plutonium nitrate (MW = 554), a bulky monovalent anion: $D_{NO_3} \cong 10^{-6}\,\text{cm}^2/\text{s} \gg D_{PuN} \cong 5 \times 10^{-10}\,\text{cm}^2/\text{s}$.

In this case, the ion initially present in the bead, that is, nitrate, is much faster than the counterion in the bulk solution. At the start of uptake, the concentration of the faster nitrate ion is lower in the outer shell and higher at the center of the bead. So, a sharp boundary moves from the periphery at the outset, in agreement with Figure 4.21, but the sharpness is blurred as the front moves toward the center of the bead. This is a consequence of the dependence of the interdiffusion coefficient on ionic composition as presented in Eqs (4.20)–(4.23). The observation is open to explanation by classical Nernst–Planck diffusion (Eq. (4.20)) considering interdiffusion of nitrate and plutonyl nitrate. At the outset, when nitrate is a trace species in the periphery, the interdiffusion coefficient is equal to that of nitrate and is a maximum. Thus, the plutonium nitrate uptake is thermodynamically very favorable and kinetically fast in the beginning with a sharp transition.

With the progression of plutonyl nitrate uptake, the interdiffusion coefficient value decreases because nitrate is no longer a trace species. Toward the center of the resin, the decrease in interdiffusion coefficient causes the sharp boundary at the front to become diffuse and blurred, in line with the Nernst–Planck equation and as observed in the autoradiographs of Figure 4.21.

The preceding six examples, although seemingly counterintuitive, are not unique and similar observations abound in many real-life scenarios. An in-depth understanding, even without full comprehension of mathematical models, is likely to be found useful.

4.11 Rate-Limiting Step: Dimensionless Numbers

Earlier discussion in this chapter and several previous studies have validated that the activation energy required for ion exchange processes is very low and reaction kinetics are rarely the rate-limiting step. Often, if not always, transport through diffusion is the rate-limiting step and it primarily takes place in two steps in series: diffusion from the bulk liquid phase to the ion exchanger interface and diffusion from the interface toward the interior of the exchanger. While the first step is known as the external liquid-film diffusion, the second step is called intraparticle diffusion. In a previous section of this chapter, sufficient details have been provided for an experimental protocol called an "interruption test" to identify the rate limiting step under specific operating conditions. In this section, a theoretical protocol using a dimensionless group, referred to as *Biot number*, will be discussed to determine the rate-limiting step.

The *Biot number* is the ratio of the maximum possible solute flux by external liquid-phase film diffusion and the maximum possible flux from intraparticle diffusion. Considering an ion exchanger particle of radius "*r*" similar to Figure S4.5, the maximum possible solute flux resulting from the external liquid-phase film diffusion is

$$J_{\mathrm{L}}^{\max} = \frac{D_{\mathrm{L}}}{\delta}(C_{\mathrm{b}} - C_{\mathrm{s}}) = k_{\mathrm{f}} C_{\mathrm{b}} \tag{4.74}$$

where D_{L} is the liquid phase diffusivity of the solute or counterion, δ is the thickness of the static liquid film or boundary layer, C_{b} is the concentration of the solute in the bulk liquid and C_{s} is the concentration at the ion exchanger-water interface. The maximum liquid-phase flux corresponds to a situation where C_{s} is zero.

The maximum possible solute flux in the ion exchanger or solid phase by intraparticle diffusion is

$$J_{\rm S}^{\max} = \frac{\overline{D}_{\rm eff} q_{\rm b} \rho_{\rm b}}{r} \tag{4.75}$$

Note that $q_{\rm b}$ is the uptake of the solute or counterion in mass units at the exchanger interface in equilibrium with the bulk concentration $C_{\rm b,}$ where $\rho_{\rm b}$ is the bulk density of the adsorbent.

Dimensionless Biot number or Bi is essentially a ratio of the two maximum fluxes. Thus,

$$\text{Bi} = \frac{J_{\rm L}^{\max}}{J_{\rm S}^{\max}} = \frac{k_{\rm f} C_{\rm b}\, r}{\overline{D}_{\rm eff} q_{\rm b} \rho_{\rm P}} \tag{4.76}$$

Also, partition coefficient of the solute is:

$$\lambda = \frac{q_{\rm b} \rho_{\rm b}}{C_{\rm b}} \tag{4.77}$$

Thus,

$$\text{Bi} = \frac{k_{\rm f}\, r}{\overline{D}_{\rm eff} \lambda} \tag{4.78}$$

4.11.1 Implications of Biot Number: Trace Ion Exchange

The parameters $k_{\rm f}, r, \overline{D}_{\rm eff},$ and λ in Bi are known or can be determined independently. The following are the two distinct cases:

Case I. $\text{Bi} \gg 1, J_{\rm L}^{\max} \gg J_{\rm S}^{\max}$.

Intraparticle diffusion is significantly slower and hence the rate-limiting step.

Case II. $\text{Bi} \ll 1, J_{\rm S}^{\max} \gg J_{\rm L}^{\max}$.

External liquid-film diffusion is the rate-limiting step.

When $\text{Bi} > 30$, intraparticle diffusion is the predominant rate-limiting step and any effect of liquid film diffusion can be ignored. Similarly, when $\text{Bi} < 0.5$, external liquid film diffusion is the sole rate-limiting step for the solute uptake. Equation (4.78) delivers an immediate impression that an increase in the value of the partition coefficient, λ, reduces Bi and, thus, renders the external film diffusion as the more rate-limiting step. Many discussions on *Biot number* in the open literature also tend to provide this anomalous conclusion. Such analyses frequently treat $\overline{D}_{\rm eff}$ as a stand-alone self-diffusion coefficient in the exchanger phase and thus ignore the effect of selectivity on $\overline{D}_{\rm eff}$. For a trace species, in a macroporous exchanger, as already demonstrated in Section 4.8.4

$$\overline{D}_{\rm eff} = \frac{\epsilon_{\rm P} D_{\rm P}}{\epsilon_{\rm P} + \rho_{\rm P} \frac{\partial q}{\partial C}} \tag{4.79}$$

where $D_{\rm P}$ is the intraparticle pore diffusivity. For Langmuir isotherm, at trace concentration, $\rho_{\rm P} \frac{\partial q}{\partial C} = \text{constant} = \lambda$.

Thus,

$$\overline{D}_{\text{eff}} = \frac{\epsilon_{\text{P}} D_{\text{P}}}{\epsilon_{\text{P}} + \lambda} \tag{4.80}$$

Considering $\lambda \gg \epsilon_{\text{P}}$

$$\overline{D}_{\text{eff}} = \frac{\epsilon_{\text{P}} D_{\text{P}}}{\lambda} \tag{4.81}$$

Using Eq. (4.81) in Eq. (4.78)

$$\text{Bi} = \frac{k_{\text{f}}\, r}{\epsilon_{\text{P}} D_{\text{P}}} \tag{4.82}$$

Again,

$$D_{\text{P}} = \frac{D_{\text{L}}^{0}}{\tau} \tag{4.83}$$

where D_{L}^{0} is the diffusivity of the counterion in the bulk liquid phase and τ is the dimensionless tortuosity.

Thus,

$$\text{Bi} = \frac{k_{\text{f}} r \tau}{\epsilon_{\text{P}} D_{\text{L}}^{0}} \tag{4.84}$$

In summary, hydrodynamic conditions in the liquid phase (i.e., embedded in the term k_{f}), particle size (i.e., r), intraparticle pathway or tortuosity inside the exchanger (i.e., τ), pore water content (i.e., ϵ_{p}) and the liquid phase diffusion coefficient of the counterion (i.e., D_{L}^{0}) determine the magnitude of the Biot number for a trace species during selective ion exchange. Interestingly enough, the effect of sorption affinity is compensated between the effective intraparticle diffusivity ($\overline{D}_{\text{eff}}$) and the partition coefficient for trace target contaminants.

Example 4.3 *Consider the interruption test results for an anion exchanger column run as shown in Figure 1. Pentachlorophenate (or PCP⁻) is the trace anion in the presence of other major competing anions, namely, chloride, sulfate, and bicarbonate. The kinetics seem to be intraparticle diffusion-controlled. Validate the rate-limiting step using Biot number, Bi.*

For obtaining external liquid-film mass transfer coefficients (k_{f}) in a fixed-bed process, the following empirical correlation [5] is employed:

$$\frac{2k_{\text{f}} r}{D_{\text{L}}} = 2 + 1.58\, Re^{0.4}\, Sc^{1/3} \tag{1}$$

where D_{L} is the liquid phase diffusivity of the solute, Re is the Reynolds number, and Sc is the Schmidt number. The correlation is valid for $0.001 < Re < 5.8$ and $500 < Sc < 70{,}600$. The Reynolds number and Schmidt number are defined by the following relations:

$$Re = \frac{2rv}{\nu} \tag{2}$$

$$Sc = \frac{\nu}{D_L} \quad (3)$$

where ν is the kinematic viscosity and v is the superficial linear velocity. The intraparticle pore diffusivity (D_P) is further related to liquid phase diffusivity (D_L) by:

$$D_P = \frac{D_L}{\tau} \quad (4)$$

where τ is tortuosity. Tortuosity of resins usually ranges from 2 to 6 with an average value of, say, 3. Combining Eqs (1)–(4) and relating them with expressions of the Biot number,

$$N_{Bi} = \frac{\tau}{\epsilon_P}\left(1 + 0.79 Re^{0.4} Sc^{\frac{1}{3}}\right) \quad (5)$$

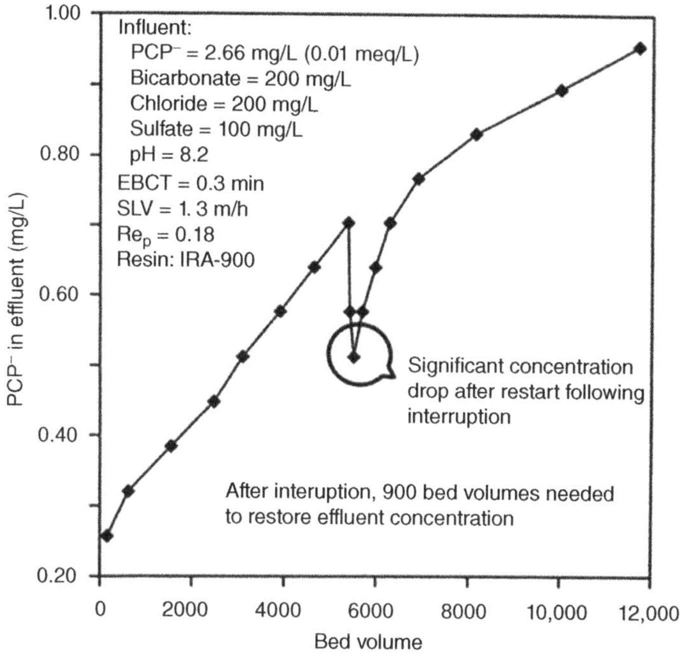

Figure 1. Plot of PCP^- concentration vs bed volumes. There is a significant drop in PCP^- concentration immediately after restart following a 24-h column interruption. This concentration profile is demonstrative of intraparticle diffusion being the rate-limiting step (same as Figure 4.23). *Source:* Li 2000 [37]. Reproduced with permission of American Chemical Society.

The diffusivity D_L of PCP is estimated using the correlation obtained from Wilke and Chang equation [44]

$$D_L = 7.4 \times 10^{-12} \frac{T\,(2.6M_W)^{1/2}}{\mu V_B^{0.6}} \quad (6)$$

where M_W is molar weight of water, T is temperature (K), μ is viscosity of water (cP), and V_B is molar volume of organic compounds at the normal boiling point. The value of V_B for PCP is estimated to be 227.5 cm³/mol. So, D_L for PCP is calculated as:

$$D_L = 7.4 \times 10^{-12} \frac{293\,\text{K}\,(2.6 \times 18)^{1/2}}{1.0 \times 227.5^{0.6}} = 5.3 \times 10^{-6}\,\text{cm}^2/\text{s} \tag{7}$$

For organic electrolytes, the real D_L value is usually 80–90% of the calculated value. Considering an average of 85%,

$$D_L = 0.85 \times 5.3 \times 10^{-6}\,\text{cm}^2/\text{s} = 4.5\ \times 10^{-6}\,\text{cm}^2/\text{s} \tag{8}$$

Given an average particle radius 0.25 mm (i.e., 0.025 cm), superficial linear velocity 1.3 m/h (0.035 cm/s) and kinematic viscosity (ν) 0.01 cm²/s, the calculated Reynolds number,

$$\text{Re} = \frac{2 \times 0.025\,\text{cm} \times 0.035\,\text{cm/s}}{0.01\,\text{cm}^2/\text{s}} = 0.18 \tag{9}$$

The Schmidt number is calculated as:

$$\text{Sc} = \frac{10^{-2}\,\text{cm}^2/\text{s}}{4.5 \times 10^{-6}\,\text{cm}^2/\text{s}} = 2.2 \times 10^3 \tag{10}$$

With $\epsilon_P = 0.3$, $\tau = 3$, the Biot number obtained from Eq. (5) is

$$\text{Bi} = \frac{3}{0.3}(1 + 0.79(0.18)^{0.4}(2.2 \times 10^3)^{\frac{1}{3}} = 62 \tag{11}$$

Thus, the Biot number under the condition of the experiment is greater than 30. This parameter validates that the intraparticle macropore diffusion is the rate-limiting step and the diffusional resistance in the external liquid film is relatively negligible.

Example 4.4 *Given the same conditions, what would the radius be if the Biot number = 15?*

Substituting known parameters into Eq. (5),

$$15 = \frac{3}{0.3}\left(1 + 0.79 \cdot \left(\frac{2 \cdot r \cdot 0.035\,\text{cm/s}}{0.01\,\text{cm}^2/\text{s}}\right)^{0.4} \cdot \left(\frac{0.01\,\text{cm}^2/\text{s}}{4.5\ \times 10^{-6}\,\text{cm}^2/\text{s}}\right)^{1/3}\right) \tag{12}$$

Solving for r,

$$r = 7.4 \times 10^{-5}\,\text{cm} = 0.074\,\text{mm} \tag{13}$$

4.12 Intraparticle Diffusion: From Theory to Practice

By and large, the equilibrium properties, namely, selectivity and regenerability dictate the appropriateness of an ion exchanger or selection of an ion exchange process for a specific application. In this section, three examples will be presented where kinetics have as much importance as, if not more than, equilibrium in deciding the success of the final system. Most importantly, all the examples take advantage of the fundamentals presented in the chapter pertaining to intraparticle diffusion.

4.12.1 Reducing Diffusion Path Length: Short-Bed Process and Shell–Core Resins

Like other heterogeneous processes using small (<1 mm) particles, various reactor configurations and operations have evolved for ion exchange, viz, packed-bed, continuous counter-current, continuous stirred-tank and fluidized bed processes. Of them, fixed- or packed-bed processes are, by far, the most popular , where the mobile liquid-phase passes through the stationary-phase ion exchange beads in a column. This method of unit operation is routinely used for contaminant removal, water softening and water demineralization.

Poor kinetics, often controlled by intraparticle diffusion, is one of the limitations of selective ion exchange processes. During the exhaustion cycle, there are three specific zones for the target solute in the fixed-bed: saturated, mass transfer and unused as shown in Figure 4.32. As the cycle advances, the MTZ moves from the column inlet toward the outlet. For a favorable isotherm, the length of the MTZ, $L_{\text{MTZ},}$ does not change during its passage inside the column, but depends on the particle size of the sorbent. For intraparticle diffusion, L_{MTZ} is proportional to the square of the particle diameter. Improving kinetics in a fixed-bed process also means reducing the L_{MTZ}. However, the pressure drop in the fixed-bed may be a limiting factor, as it's influenced by the particle diameter, that is, sizes of the sorbent particles.

Under laminar conditions, the Kozeny–Carman equation for pressure drop for particle size, d_{P}, can be presented as follows:

$$P = \frac{\mu v L}{K_{\text{P}} d_{\text{P}}^2} \tag{4.85}$$

where K_{P} is the permeability, μ is the viscosity of the liquid stream, L is the length of the packed bed column, and v is the superficial liquid-phase velocity. Assuming K_{P}, μ and v to be remaining constant, pressure drop, ΔP, increases inversely with the square of the particle diameter, d_{P}; small diameter particles in packed-bed columns render excessive pressure drops and are not recommended. But the pressure drop remains the same if L/d_{P}^2 is kept constant.

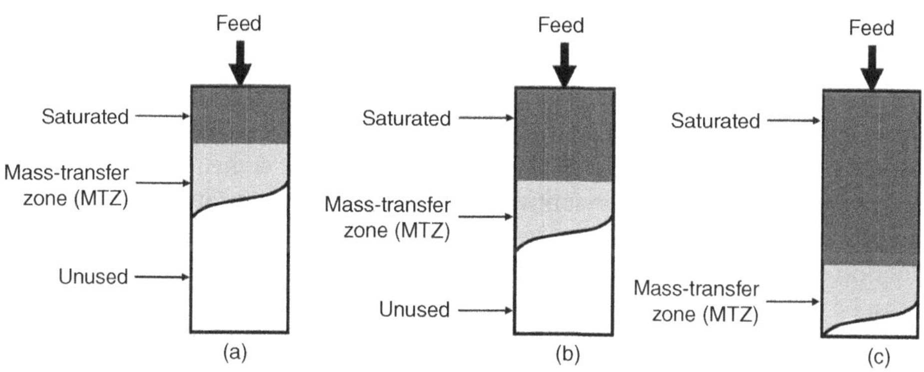

Figure 4.32 Schematic illustration of movement of mass transfer zone (MTZ) from inlet to exit along the bed (a) initial stage when MTZ near the inlet and major part of the column unused; (b) middle of the cycle when MTZ is in the middle with almost even distribution of used and unused zones; (c) before exhaustion when MTZ is near the exit and a major part of the column is used (saturated).

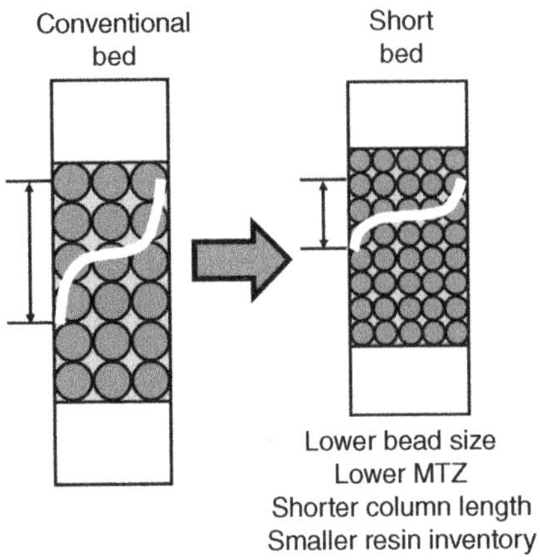

Figure 4.33 Schematic illustration showing benefit of reduction of MTZ.

By reducing d_P, (i) the length of the MTZ in the ion exchange column can be reduced; (ii) the pressure drop can remain unchanged by reducing L so that L/d_P^2 is constant. One concomitant benefit is a significant reduction in resin inventory, total cycle time and equipment size. Illustrations in Figure 4.33 attempt to capture the underlying scientific details and the associated benefits.

The above theoretical principles form the basis of "Recoflo Short-bed IX Process" of Eco-Tec, Ltd., (Ontario, Canada). Recoflo uses much finer particle sizes (0.075–0.15 mm) than normally used in industrial ion exchange processes (0.4–1.0 mm). Depending on the application, total cycle time can vary from as low as 2 min to 1 h. Upon exhaustion, the short bed is regenerated in counter-current mode for maximum efficiency. The entire Recoflo system is assembled in a compact, skid-mounted unit as shown in Figure 4.34, significantly smaller than a conventional packed-bed ion exchange system.

Diffusional path length may also be reduced by using shell–core ion exchange resins, where a spherical bead has been only partially functionalized. For high speed analytical chromatography, shell–core ion exchange resins have been used for long in pellicular form with ion exchange sites practically on the surface only. For large-scale application in sorption processes, shell–core ion exchange resins made from deeply functionalized homogenous copolymer beads with a shell/radius (S/R) ratio greater than 0.4 may offer a significant kinetic advantage with a minor reduction in total capacity. Figure 4.35 shows the relationship between S/R ratio and the volume for a shell–core resin and a standard resin. Note that S/R is essentially the measure of the relative intraparticle diffusion path length between the two resins. Note that for S/R ratios of 0.5, that is, for a reduction of half the diffusion path length, 87.5% of the volume or ion exchange capacity can be retained.

Shell–core resins are now commercially available from Purolite Co. (www.purolite.com) for nearly all types of commercial ion exchange resins. One of the demonstrated benefits of shell–core resins is in the significant reduction of calcium leakage during softening processes with high TDS (total dissolved solids) water at identical regeneration levels. Figure 4.36 shows a comparison of calcium leakage during a softening cycle for two Purolite cation exchange resins in sodium form-C104 and SST104, all

Figure 4.34 RECOFLO short bed IX process. *Source:* Reproduced with kind permission of Eco-Tec. 2014 [49].

Shell radius *s/r*	0.4	0.5	0.6	0.7	0.8	0.9	1.0
Volume ratio	78.4%	87.5%	93.6%	97.3%	98.7%	99.9%	100%

Figure 4.35 The change in volume or ion exchange capacity with *S/R* ratio between a shell–core resin and a standard resin.

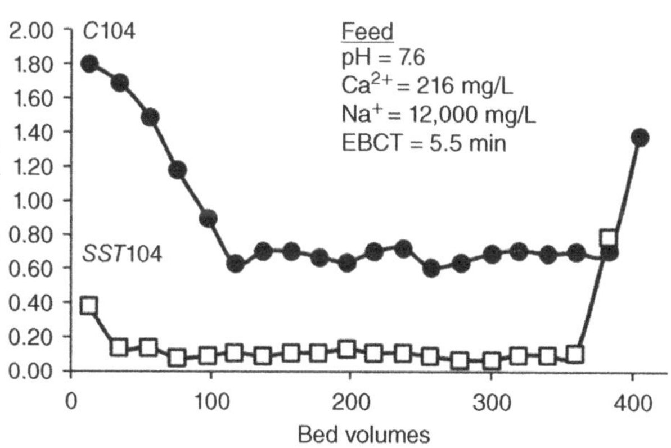

Figure 4.36 Comparison of calcium leakage during cation exchange softening in Na-cycle between Purolite C104 and SST104. *Source:* Adapted from Downey 2006 [50].

other conditions remaining identical [50]. Feed composition and EBCT are also provided in the figure. SST104 is a shallow shell version of C104 with *S*/*R* ratio equal to 0.7.

Shell–core resins, compared to short-bed technology, are similar to standard resins in particle sizes. Thus, the existing vessels can be retrofitted with shell–core resins without any difficulty and the size of the ion exchange plant or system remains essentially unchanged.

4.12.2 Development of Bifunctional Diphonix® Resin

1-Hydroxyethane-1,1-diphosphonic acid (HEDPA) is known to be an effective complexant for a variety of metal ions, even in highly acidic solutions. Diphonix® chelating ion exchange resin with HEDPA functional groups are now commercially available and it outperforms other products in removing U(VI), Pu(IV), and Am(III) in highly acidic medium. Diphonix also strongly coordinates Fe(III) over the range of 0.1–5 M HNO_3. The primary challenge toward the development and commercialization of the Diphonix resin was in overcoming extremely slow ion exchange kinetics, that is, very low intraparticle diffusivity at acidic pH values. The introduction of a secondary non-selective functional group (e.g., sulfonic acid) greatly improved ion exchange kinetics of Diphonix resin and made the new material commercially viable [51].

To fully grasp the gradual progression of development leading to the commercialization of Diphonix, let us consider monofunctional phosphonic acid resin cross-linked with divinylbenzene as shown in Figure 4.37a. At highly acidic pH, phosphonic acid remains fully undissociated (protonated) and, so, has a decreased diameter and decreased water content. Thus, the intraparticle diffusivity of the weak-acid resin, according to Mackie's equation (4.5), is very low and the target metal uptake is an extremely slow process.

In contrast, when the same resin is sulfonated, to yield a bifunctional sulfonic–phosphonic acid group, as illustrated in Figure 4.37b, metal uptake rate increases significantly. The completely ionized sulfonic acid functional groups imbibe water molecules within the resin by osmosis and, thus, prevent the resin from collapsing under highly acidic conditions. Covalently attached sulfonic acid groups do not offer any specific affinity toward the target metal ions, but they greatly enhance the intraparticle diffusion rate in two ways: increased accessibility to selective ion exchange sites and increased free water molecules within the gel phase of the exchanger.

Alexandratos and coworkers made intelligent use of the foregoing phenomena to synthesize the bifunctional Diphonix resin containing both HEDPA and sulfonic acid functional groups, as illustrated in Figure 4.38 [10].

Figure 4.39 shows a comparison of Am(III) uptake rate between sulfonated and unsulfonated Diphonix resin at acidic pH, all other conditions remaining identical. A remarkable enhancement in the uptake rate of Am(III) can be readily noted. Improved kinetics have also been observed for Fe(III) and other transition metal ions.

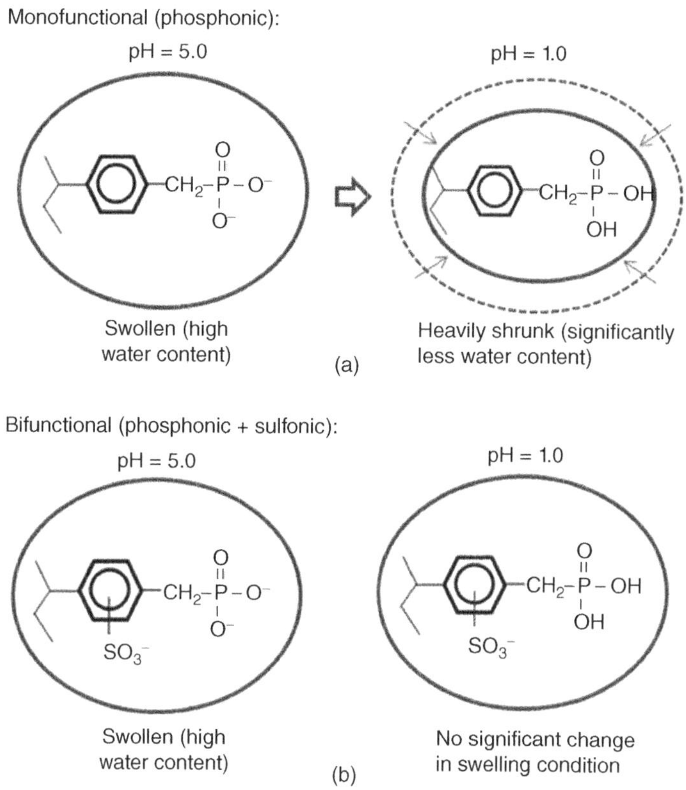

Figure 4.37 Illustration of differences in the degree of swelling under different pH conditions between (a) a monofunctional (phosphonic acid group) and (b) a bifunctional (phosphonic acid and sulfonic acid groups) ion exchanger with a polystyrene matrix resin.

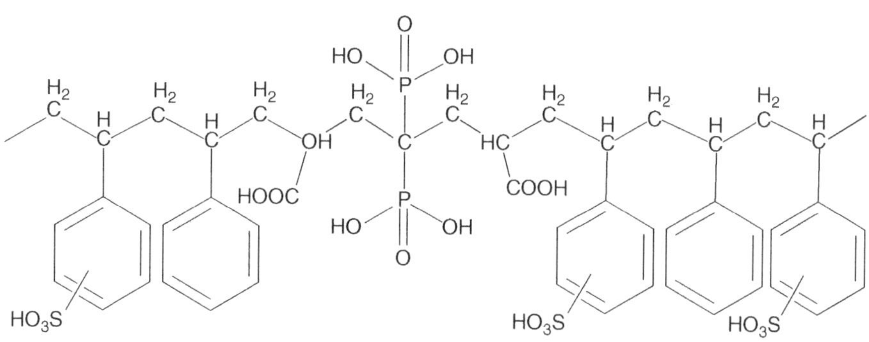

Figure 4.38 Structure of Diphonix resin. $pK_1 = 1.5$, $pK_2 = 2.5$, $pK_3 = 7.2$, $pK_4 = 10.5$ [10].

4.12.3 Ion Exchanger as a Host for Enhanced Kinetics

Metal oxide particles – namely, oxides of Fe(III), Zr(IV), Ti(IV), and Al(III) – are environmentally benign and exhibit excellent sorption behavior at neutral pH toward many anionic ligands, such as arsenate, phosphate and fluoride. Since sorption sites reside predominantly on the surface, metal oxides offer very high sorption capacity at

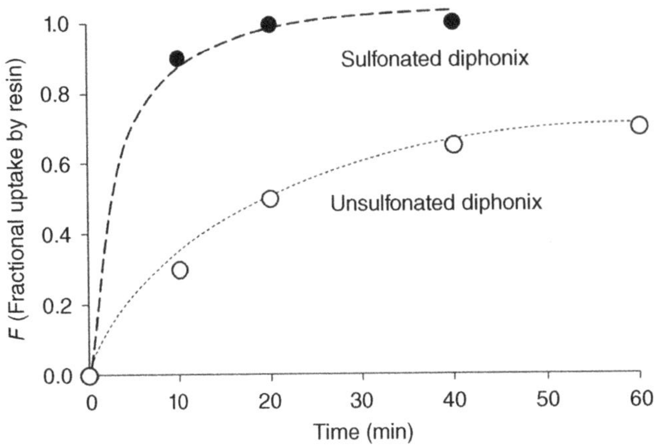

Figure 4.39 A comparison of Am(III) uptake rate between sulfonated and unsulfonated Diphonix resins at acidic pH (~2). *Source:* Adapted from Chiarizia 1994 [51].

nanoscale sizes on a per mass basis because of the high surface area-to-volume ratio. However, metal oxide nanoparticles are almost impermeable in a fixed-bed columnar configuration or any flow-through system. That is why attempts have been made by many to dope porous host materials with such nanoparticles as hydrated Fe(III) oxide (HFO) [52–54]. Such hybrid sorbent materials integrate superior hydraulic behavior of the host material in a fixed-bed column with high sorption capacity of the HFO nanoparticles. Here we will limit our discussion primarily to kinetics of As(V) or $H_2AsO_4^-$ sorption and the comparison between two host materials, namely, (i) macroporous polymeric adsorbent (Amberlite XAD-10) without any functional groups and (ii) gel-type anion exchange resin (Amberlite A400).

Batch kinetic studies were carried out using these two hybrid materials for arsenic sorption in a setup similar to the one in Figure S4.1 with a stirrer speed over 1200 rpm to make intraparticle diffusion the rate-limiting step. Besides the use of two different host materials for HFO nanoparticles, all other conditions for the experiments, including solution composition, were identical. Figure 4.40 shows the fractional kinetic uptake rate curves for the two materials and the computed effective intraparticle diffusivity from the experimental data [55]. Note that Amberlite A400, the gel-type anion exchanger, exhibited more than an order of magnitude greater intraparticle diffusivity value than Amberlite XAD-10, an unfunctionalized polymer resin. In fact, anion exchangers, when used as host materials, always provided faster kinetics than other porous host materials. It is important to note that in the presence of much greater sulfate concentration, arsenate sorption onto anion exchanger host material is negligible and has no impact on the rate.

The following provides a mechanistic explanation of how Amberlite A400 offers faster kinetics and significantly greater intraparticle diffusivity:

Dispersed HFO nanoparticles within Amberlite XAD 10 are separated from one another through pores filled with stagnant water, that is, HFO nanoparticles are discontinuous. So, the passage or progression of arsenate from one HFO nanoparticle to the next, according to the concentration gradient, must overcome pore diffusion

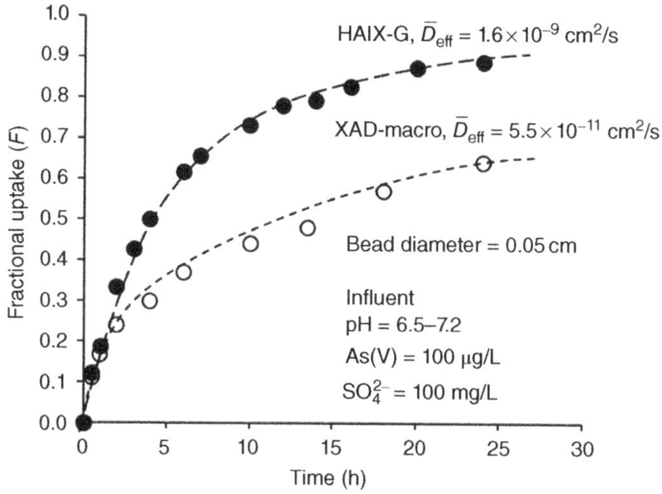

Figure 4.40 Comparison of As(V) fractional uptake during batch kinetic tests between a hybrid anion exchanger (HAIX) and a hybrid polymeric sorbent, both containing iron oxide nanoparticles. *Source:* Cumbal 2004 [55]. Reproduced with permission of Elsevier.

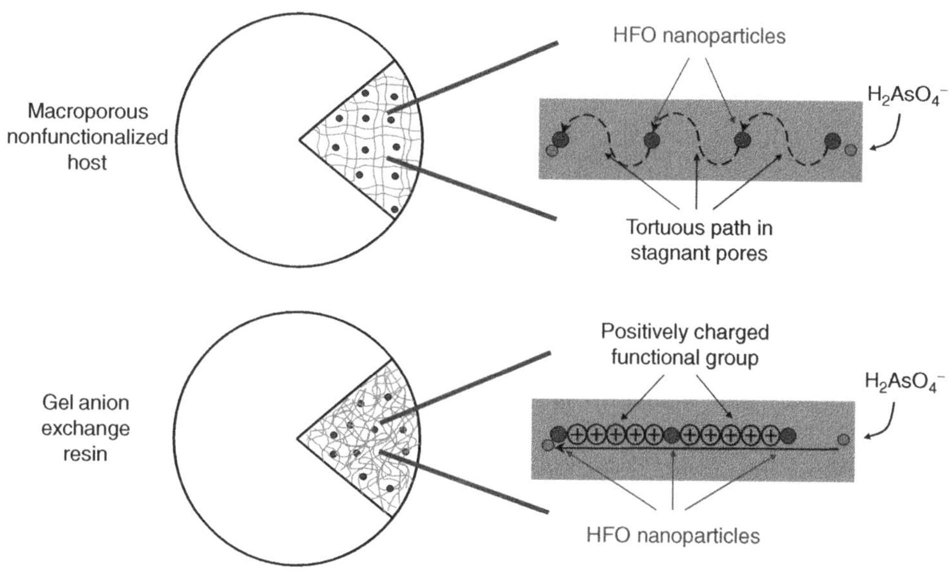

Figure 4.41 Postulated intraparticle transport mechanisms of As(V) inside the two host materials with HFO nanoparticles. *Source:* Cumbal 2004 [55]. Reproduced with permission of Elsevier.

within each polymer bead, and such diffusional resistance for a stagnant liquid phase is always high. For Amberlite A400, however, HFO nanoparticles are dispersed within the gel phase of the anion exchanger with positively charged quaternary ammonium functional groups. These positively charged sites are essentially interconnected with one another in a seamless fashion. Thus, arsenate anion can move from one HFO nanoparticle to the next without encountering major diffusional resistance. The presence of millions of positively charged fixed sites in a single anion exchanger bead (see Example 4.2) is the underlying reason for its much greater intraparticle diffusivity, all other conditions remaining identical. Figure 4.41 provides illustrations in support of the diffusion mechanisms as discussed. It is noteworthy that commercially available, regenerable arsenic-selective sorbents are all essentially HAIXs where HFO nanoparticles have been successfully dispersed within the gel phase of the anion exchange resins [56,57].

Summary

- Ion exchange is not a chemical reaction-controlled process. Almost with no exception, it is a diffusion-controlled process with an activation energy requirement well below 100 kJ/mol.
- For fixed-bed ion exchange processes of general interest, intraparticle diffusion is often the rate-limiting step. A simple-to-run interruption test may confirm this limitation.
- Reduced size of ion exchanger beads or use of ion exchange fibers enhances the rate of intraparticle diffusion by reducing the diffusion path length. The effective intraparticle diffusivity, however, remains unchanged.
- The effective intraparticle diffusivity of trace counterions strongly depends on water content of the exchanger and the counterion's sorption affinity. Intraparticle diffusivity is increased by higher water content and lower sorption affinity.
- Ion exchange is a coupled transport process: sorption of one counterion is always accompanied by desorption of equivalent amounts of other counterions. The effective intraparticle diffusivity is mostly decided by the trace counterion.
- Desorbing the final traces of a counterion is often very difficult due to low intraparticle diffusivity and long diffusion path length. Use of shallow shell ion exchangers (i.e., similar to pellicular) improves desorption efficiency.
- Rectangular isotherms represent extraordinarily high affinity of the counterions. In such cases, intraparticle diffusion occurs with a sharp boundary change due to fast ion exchange process. This phenomenon forms the heart of "shrinking core" or "shell progressive" kinetics.
- The water content of a weak-acid metal selective ion exchanger can be greatly increased, even at very low pH, by introducing strongly ionized sulfonic acid functional group, thus greatly enhancing intraparticle diffusion kinetics. This principle formed the development basis for bifunctional Diphonix resins.
- By reducing particle sizes, resin inventory can be reduced with shorter cycle times. The RECOFLO short-bed ion exchange process is based on that principle.
- The Donnan membrane principle, as it concerns coion exclusion, may help explain unusual observations concerning ion exchange kinetics. For example, deprotonation or dissociation of WAC exchange resins in dilute alkali solution is a very slow process due to the rejection of OH^-, a coion, from diffusing into the cation exchanger caused by the Donnan exclusion effect.

References

1 Liberti, L. (1983) Planning and interpreting kinetic investigations, in *Mass Transfer and Kinetics of Ion Exchange Hague* (eds L. Liberti and F. Helfferich), Martinus and Nijhoff Publishers, Netherland, pp. 181–206.

2 Li, P. and SenGupta, A.K. (2000) Intraparticle diffusion during selective ion exchange with a macroporous exchanger. *Reactive and Functional Polymers*, **44** (3), 273–287.

3 Li, P. (1999) *Sorption of Synthetic Aromatic Anions onto Polymeric Ion Exchangers: Genesis of Selectivity and Effects of Equilibrium Process Variables on Sorption Kinetics*, PhD dissertation, Lehigh University, Bethlehem, PA.

4 Kressman, T. and Kitchener, J. (1949) Cation exchange with a synthetic phenolsulphonate resin. Part V. Kinetics. *Discussions of the Faraday Society*, **7**, 90–104.

5 Crank, J. (1979) *The Mathematics of Diffusion*, 2nd edn, Oxford University Press, London.

6 Chatterjee, P.K. (2011) *Sensing and Detection of Toxic Metals in Water with Innovative Sorption-based Techniques*, PhD dissertation, Lehigh University, Bethlehem, PA.

7 Wheeler, A. (1951) Reaction rates and selectivity in catalyst pores. *Advances in Catalysis*, **3** (5), 433–439.

8 Mackie, J. and Meares, P. (1955) The diffusion of electrolytes in a cation-exchange resin membrane. I. Theoretical. *Proceedings of the Royal Society of London A: Mathematical, Physical and Engineering Sciences*, **232** (1191), 498–509.

9 Chiariza, R., Horwitz, E., Alexandratos, S., and Gula, M. (1997) Diphonix® resin: a review of its properties and applications. *Separation Science and Technology*, **32** (1-4), 1–35.

10 Horwitz, E.P., Alexandratos, S.D., Gatrone, R.C., and Chiarizia, R. (inventors) (1994). Arch Development Corp (assignee). Phosphonic acid based ion exchange resins. Chicago, IL. US patent 5,281,631 A. January 25, 1994.

11 Chatterjee, P.K. and SenGupta, A.K. (2009) Sensing of toxic metals through pH changes using a hybrid sorbent material: concept and experimental validation. *AIChE Journal*, **55** (11), 2997–3004.

12 Chatterjee, P.K. and SenGupta, A.K. (2011) Interference-free detection of trace copper in the presence of EDTA and other metals using two complementary chelating polymers. *Colloids and Surfaces A: Physicochemical and Engineering Aspects*, **384** (1), 432–441.

13 Sarkar, S., Chatterjee, P.K., Cumbal, L.H., and SenGupta, A.K. (2011) Hybrid ion exchanger supported nanocomposites: sorption and sensing for environmental applications. *Chemical Engineering Journal*, **166** (3), 923–931.

14 Selke, W.A. (1956) Mass transfer and equilibria, in *Ion Exchange Technology* (eds F.C. Nachod and J. Schubert), Academic Press, Inc., New York, pp. 52–94.

15 Boyd, G., Adamson, A., and Myers, L. Jr., (1947) The exchange adsorption of ions from aqueous solutions by organic zeolites. II. Kinetics. *Journal of the American Chemical Society*, **69** (11), 2836–2848.

16 Petruzzelli, D., Helfferich, F., Liberti, L. *et al.* (1987) Kinetics of ion exchange with intraparticle rate control: models accounting for interactions in the solid phase. *Reactive Polymers, Ion Exchangers, Sorbents*, **7** (1), 1–13.

17 Chanda, M. and Rempel, G. (1995) Polyethyleneimine gel-coat on silica. High uranium capacity and fast kinetics of gel-coated resin. *Reactive Polymers*, **25** (1), 25–36.
18 Bond, G. (1987) *Heterogeneous Catalysis: Principles and Applications*, 2nd edn, Clarendon, Oxford University Press, Oxford.
19 Price, S., Hebditch, D., and Streat, M. (1988) Diffusion or chemical kinetic control in a chelating ion exchange resin system, in *Ion Exchange for Industry: Society for Industry* (ed. M. Streat), Ellis Horwood Limited, Chichester, pp. 275–285.
20 Price, S.G. (1988) *Kinetics of Ion Exchange in a Chelating Resin*, Open University & Council for National Academic Awards (CNAA), London.
21 Albright, R.L. (1986) Porous polymers as an anchor for catalysis. *Reactive Polymers, Ion Exchangers, Sorbents*, **4** (2), 155–174.
22 Kunin, R., Meitzner, E., and Bortnick, N. (1962) Macroreticular ion exchange resins. *Journal of the American Chemical Society*, **84** (2), 305–306.
23 Abrams, I.M. and Millar, J.R. (1997) A history of the origin and development of macroporous ion-exchange resins. *Reactive and Functional Polymers*, **35** (1), 7–22.
24 Harland, C.E. (1994) *Ion Exchange: Theory and Practice*, 2nd edn, Royal Society of Chemistry, London.
25 Helfferich, F.G. (1990) Models and physical reality in ion-exchange kinetics. *Reactive Polymers*, **13** (1), 191–194.
26 Helfferich, F.G. (1962) *Ion Exchange*, McGraw-Hill Book Company, Inc, New York.
27 Yoshida, H., Kataoka, T., and Ikeda, S. (1985) Intraparticle mass transfer in bidispersed porous ion exchanger. Part I: Isotopic ion exchange. *The Canadian Journal of Chemical Engineering*, **63** (3), 422–429.
28 Yoshida, H. and Kataoka, T. (1985) Intraparticle mass transfer in bidispersed porous ion exchanger. Part II: Mutual ion exchange. *The Canadian Journal of Chemical Engineering*, **63** (3), 430–435.
29 Ruckenstein, E., Vaidyanathan, A.S., and Youngquist, G.R. (1971) Sorption by solids with bidisperse pore structures. *Chemical Engineering Science*, **26** (9), 1305–1318.
30 Greenleaf, J.E., Lin, J., and SenGupta, A.K. (2006) Two novel applications of ion exchange fibers: arsenic removal and chemical-free softening of hard water. *Environmental Progress*, **25** (4), 300–311.
31 Soldatov, V.S. and Bychkova, V.A. (1988) *Ionoobmennye ravnovesiia v mnogokomponentnykh sistemakh (Ion Exchange Equilibrium in Multicomponent Systems)*, Nauka i Tekhnika, Minsk.
32 Dominguez, L., Benak, K.R., and Economy, J. (2001) Design of high efficiency polymeric cation exchange fibers. *Polymers for Advanced Technologies*, **12** (3–4), 197–205.
33 Awual, M.R., Jyo, A., Ihara, T. *et al.* (2011) Enhanced trace phosphate removal from water by zirconium(IV) loaded fibrous adsorbent. *Water Research*, **45** (15), 4592–4600.
34 Lin, J.C. and SenGupta, A.K. (2009) Hybrid anion exchange fibers with dual binding sites: simultaneous and reversible sorption of perchlorate and arsenate. *Environmental Engineering Science*, **26** (11), 1673–1683.
35 Soldatov, V.S. (2008) Syntheses and the main properties of Fiban fibrous ion exchangers. *Solvent Extraction and Ion Exchange*, **26** (5), 457–513.

36 Kunin, R. (1958) *Ion Exchange Resins*, 2nd edn, John Wiley & Sons, Inc., New York.
37 Li, P. and SenGupta, A.K. (2000) Intraparticle diffusion during selective sorption of trace contaminants: the effect of gel versus macroporous morphology. *Environmental Science & Technology*, **34** (24), 5193–5200.
38 Kunin, R. and Barry, R.E. (1949) Carboxylic, weak acid type, cation exchange resin. *Industrial & Engineering Chemistry*, **41** (6), 1269–1272.
39 Mackay, D., Shiu, W.Y., and Ma, K. (1997) *Illustrated Handbook of Physical–Chemical Properties of Environmental Fate for Organic Chemicals*, CRC Press, Boca Raton.
40 Slater, M.J. (1991) *Principles of Ion Exchange Technology*, Butterworth-Heinemann, Oxford.
41 Höll, W. (1984) Optical verification of ion exchange mechanisms in weak electrolyte resins. *Reactive Polymers, Ion Exchangers, Sorbents*, **2** (1-2), 93–101.
42 Phelps, D.S. and Ruthven, D.M. (2001) The kinetics of uptake of Cu ions in ionac SR-5 cation exchange resin. *Adsorption*, **7** (3), 221–229.
43 Streat, M. (1984) Kinetics of slow diffusing species in ion exchangers. *Reactive Polymers, Ion Exchangers, Sorbents*, **2** (1), 79–91.
44 Wilke, C. and Chang, P. (1955) Correlation of diffusion coefficients in dilute solutions. *AIChE Journal*, **1** (2), 264–270.
45 Le Bas, G. (1915) *The Molecular Volumes of Liquid Chemical Compounds, from the Point of View of Kopp*, Longmans, Green and Co., London.
46 Weaver, L.E. and Carta, G. (1996) Protein adsorption on cation exchangers: comparison of macroporous and gel-composite media. *Biotechnology Progress*, **12** (3), 342–355.
47 Suzuki, M. (1990) *Adsorption Engineering (Chemical Engineering Monographs Book 25)*, Kodansha Elsevier, Tokyo.
48 Yoshida, H. and Ruthven, D.M. (1989) Adsorption of gaseous ethylamine on H-form strong-acid ion exchangers. *AIChE Journal*, **35** (11), 1869–1875.
49 Eco-Tec. Ion Exchange Demineralization. (2014) Available at: http://eco-tec.com/products/demineralization/ion-exchange-demineralization/ (accessed 12 January 2017).
50 Downey, D. (2006) High total dissolved solids (HTDS) produced water softening with PUROLITE shallow shell technology resins. *PUROLITE: Application Guide*; AG_SSTEng report_12-12-06:1-15.
51 Chiarizia, R., Horwitz, E., and Alexandratos, S. (1994) Uptake of metal ions by a new chelating ion-exchange resin. Part 4: Kinetics. *Solvent Extraction and Ion Exchange*, **12** (1), 211–237.
52 Hering, J.G., Chen, P., Wilkie, J.A., and Elimelech, M. (1997) Arsenic removal from drinking water during coagulation. *Journal of Environmental Engineering*, **123** (8), 800–807.
53 Miller, S.M. and Zimmerman, J.B. (2010) Novel, bio-based, photoactive arsenic sorbent: TiO_2-impregnated chitosan bead. *Water Research*, **44** (19), 5722–5729.
54 Cumbal, L. and SenGupta, A.K. (2005) Arsenic removal using polymer-supported hydrated iron(III) oxide nanoparticles: role of Donnan membrane effect. *Environmental Science & Technology*, **39** (17), 6508–6515.

55 Cumbal, L. (2004) *Polymer-Supported Hydrated Fe Oxide (HFO) Nanoparticles: Characterization and Environmental Applications*, PhD dissertation, Lehigh University, Bethlehem, PA.
56 Layne Christensen. (2016) *LayneRT*, http://www.layne.com/en/technologies/laynert.aspx.
57 Purolite (2016). *FerrIX^{TM} A33E,* http://www.purolite.com/RelId/619885/isvars/default/potable_water.htm.

5

Solid- and Gas-Phase Ion Exchange

Ion exchange processes described so far in previous chapters have two distinct phases: the ion exchanger and the polar solvent (water). In solid- or gas-phase ion exchange, in addition to ion exchanger and water, solids and/or gases are also present contributing toward ion exchange reactions. In solid-phase ion exchange, the additional solid phase, (or phases) is often a poorly soluble solid with a low solubility product (K_{sp}) value that dissociates into cations and anions in the aqueous phase only to a limited extent. The goal is often, if not always, to dissolve, separate or recover the solid phase fully or partly. During the ion exchange process, cations and/or anions migrate from the poorly soluble solid phase to the ion exchanger through the liquid phase. Similarly, for the gas-phase ion exchange process, the target gases, upon dissolution in water, hydrolyze into cations and/or anions influencing the process of ion exchange. Solid-phase ion exchange may also be controlled, accelerated or retarded, through dissolution of an appropriate gas of choice, that is, solid- and gas-phase ion exchange may proceed simultaneously. In the subsequent sections of this chapter, solid- and gas-phase ion exchange will be discussed separately.

5.1 Solid-Phase Ion Exchange

The discussion in various subsections will address and include different types of solid phases and their role in ion exchange processes.

5.1.1 Poorly Soluble Solids

Dissolution of poorly soluble solids can be enhanced when mediated through ion exchange resins. To illustrate, let us consider a solid phase ($CaSO_4$) where the dissociated cations and anions (Ca^{2+} and SO_4^{2-}) have identical charges and their solubility is not pH-dependent. The solid is in equilibrium with the dissolved ions in water:

$$CaSO_4(s) \rightleftarrows Ca^{2+} + SO_4^{2-} \tag{5.1}$$

Assuming ideality, the solubility product is expressed as

$$K_{sp} = [Ca^{2+}][SO_4^{2-}] \tag{5.2}$$

Ion Exchange in Environmental Processes: Fundamentals, Applications and Sustainable Technology,
First Edition. Arup K. SenGupta.

When the solid phase is brought in contact with a cation exchanger in Na^+ form, the cation exchanger gradually takes up calcium ions,

$$2\overline{R-Na} + Ca^{2+} \rightleftarrows \overline{R_2Ca} + 2Na^+ \tag{5.3}$$

And K_{Na}^{Ca} is the selectivity coefficient (i.e., equilibrium constant under ideal conditions) of the exchange reaction

$$K_{Na}^{Ca} = \frac{\overline{[R_2Ca]}}{\overline{[R-Na]}^2}\frac{[Na^+]^2}{[Ca^{2+}]} \tag{5.4}$$

As Ca^{2+} is removed from the solution by the ion exchanger, dissolution of the solid phase is promoted in accordance with Le Châtelier's principle. The overall reaction, including both dissolution and ion exchange, is

$$CaSO_4(s) + 2\overline{R-Na} \rightleftarrows \overline{R_2Ca} + 2Na^+ + SO_4{}^{2-} \tag{5.5}$$

Thus, the equilibrium constant for the overall reaction is

$$K_{overall} = K_{sp}K_{Na}^{Ca} \tag{5.6}$$

The larger the constant, $K_{overall}$, the greater the amount of solid dissolved by a given amount of ion exchanger. Hence, for a specific poorly soluble solid, the dissolution is favored by high selectivity of the cation exchanger, for example, Ca^{2+} over Na^+ in this case. However, during the dissolution, the sulfate concentration increases and calcium in the aqueous phase decreases. Thus, the dissolution process becomes less and less favorable. Example 5.1 shows how calcium and sulfate concentrations in the aqueous phase change with gradual addition of cation exchange resin.

Understandably, when both cation (in sodium form) and anion exchangers (say, in chloride form) are present, solid phase dissolution takes place more rapidly. The removal of anions (namely, sulfate for solid calcium sulfate) provides an additional driving force for the dissolution of the solid. The situation is further enhanced when the cation and anion exchangers exist in H^+ form and OH^- form respectively. The hydroxyl anion (OH^-) released by the anion exchanger promotes the cation exchange by combining with H^+ to form H_2O in accordance with the following:

$$CaSO_4 \rightarrow Ca^{2+} + SO_4^{2-} \tag{5.7}$$

$$2\overline{RH} + Ca^{2+} \leftrightarrow \overline{(R^-)_2Ca^{2+}} + 2H^+ \tag{5.8}$$

$$2\overline{ROH} + SO_4^{2-} \rightleftarrows \overline{(R^+)_2SO_4^{2-}} + 2OH^- \tag{5.9}$$

$$2H^+ + 2OH^- \rightarrow 2H_2O \tag{5.10}$$

Overall:

$$CaSO_4(s) + 2\overline{RH} + 2\overline{ROH} \rightleftarrows \overline{(R^-)_2Ca^{2+}} + \overline{(R^+)_2SO_4^{2-}} + 2H_2O \tag{5.11}$$

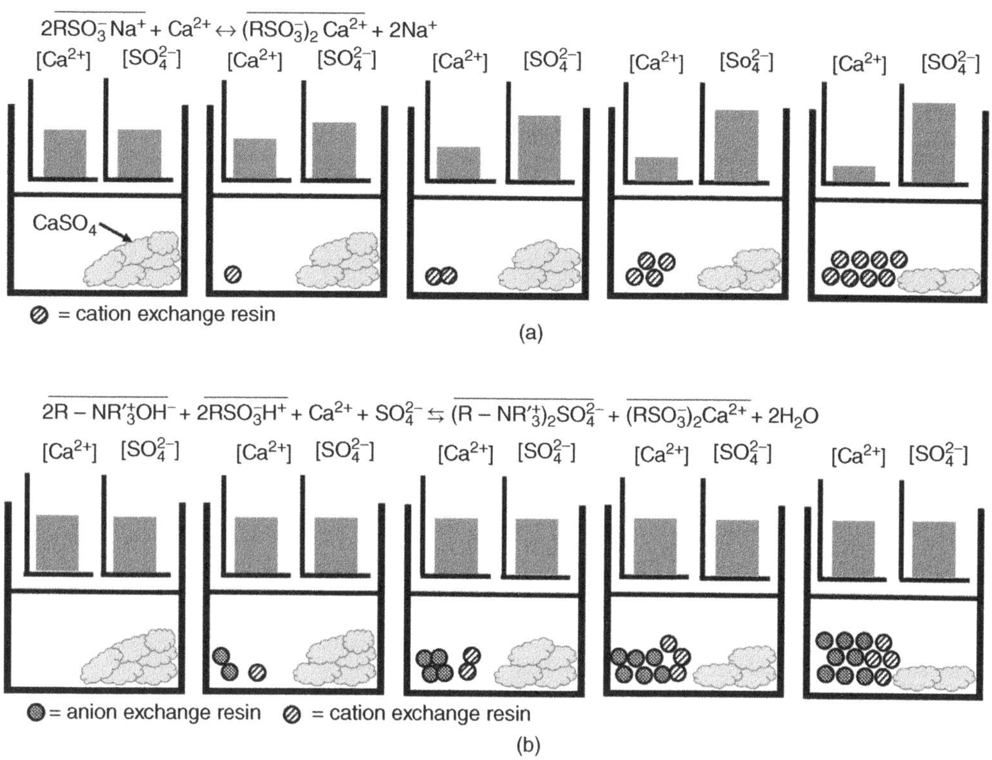

Figure 5.1 (a) Calcium sulfate dissolution and removal with gradual addition of cation exchange resins in Na-form. (b) Calcium sulfate dissolution and removal with gradual addition of cation and anion exchange resins in H- and OH-form, respectively.

Figure 5.1a and b illustrates the two scenarios of calcium sulfate dissolution and removal with gradual addition of ion exchange resins. Note that when only cation exchanger in Na-form is added, calcium in the aqueous phase progressively decreases while both sulfate and sodium increase. On the contrary, when both cation and anion exchange resins in H-form and OH-form, respectively, are added, calcium and sulfate concentrations in the aqueous phase remain essentially the same.

Example 5.1 *In 200 mL of distilled water, 1 g solid $CaSO_4(s)$ is added.*

(i) *Find the equilibrium calcium (Ca^{2+}) and sulfate concentration.*

(ii) *1.0 g of a strong-acid cation exchange resin in Na^+ form is added. Find the new calcium and sulfate concentration at equilibrium.*

(iii) *Altogether 5.0 g of cation exchange resin is added, through five 1.0 g additions. Plot the equilibrium calcium and sulfate concentration in water after each addition.*
Given: Solubility product for $CaSO_4(s)$ is $K_{sp} = 4.9 \times 10^{-5}$;
Cation exchange capacity is 4 meq/g.

(iv) *State assumptions and make comments about the process, if any.*

Solution:

(i) *Solubility or equilibrium calcium concentration may be obtained from the K_{sp} value of $CaSO_4(s)$*

$$CaSO_4(s) \rightleftharpoons Ca^{2+} + SO_4^{2-} \tag{1}$$

In pure water,

$$K_{sp} = [Ca^{2+}][SO_4^{2-}] = x^2 \tag{2}$$

where x is the molar concentration of Ca^{2+}.

Thus, $4.9 \times 10^{-5} = x^2$

$$x = 7 \times 10^{-3}\,M \tag{3}$$

(ii) *Addition of 1.0 g of cation exchange resin will cause the following dissolution – ion exchange reaction:*

$$CaSO_4(s) \rightleftharpoons Ca^{2+} + SO_4^{2-}$$

$$2\overline{(R - SO_3^-)Na^+} + Ca^{2+} \rightleftharpoons \overline{(R - SO_3^-)_2Ca^{2+}} + 2Na^+ \tag{4}$$

Overall:

$$CaSO_4(s) + 2\overline{(R - SO_3^-)Na^+} \rightleftharpoons \overline{(R - SO_3^-)_2Ca^{2+}} + 2Na^+ + SO_4^{2-} \tag{5}$$

Note that sulfate and sodium concentration increases in solution. As sulfate concentration increases with the addition of cation exchange resin, calcium concentration decreases. Assume that strong-acid cation prefers Ca^{2+} well over Na^+, that is, cation exchange sites are all converted to Ca^{2+}. Thus, Ca^{2+} uptake by the cation resin is $1.0g \times 4\frac{meq}{g} = 4\,meq$ or 2 mmol.

Consequently, 4 meq of Na^+ is released in the aqueous phase. Thus, sodium concentration in the aqueous phase becomes

$$[Na^+] = \frac{4\,meq}{0.2L} = 20\,meq/L \tag{6}$$

From electroneutrality,

$$[Na^+] + 2[Ca^{2+}] = 2[SO_4^{2-}] \tag{7}$$

Again,

$$[Ca^{2+}][SO_4^{2-}] = K_{sp} \tag{8}$$

$$[Na^+] + \frac{2K_{sp}}{[SO_4^{2-}]} = 2[SO_4^{2-}] \tag{9}$$

In Eq. (9), $[SO_4{}^{2-}]$ is the only unknown and can be computed.

$$[SO_4^{2-}] = 10\,mmol/L$$

$$[Ca^{2+}] = \frac{K_{sp}}{[SO_4^{2-}]} = 4.9\,mmol/L$$

(iii) *Assuming sodium in the cation exchange gets completely displaced by Ca^{2+}, sodium concentration in the aqueous phase changes with the addition of the cation exchange resin.*

Thus, from the law of electroneutrality (9),

$$0.02\,\mathrm{m} + \frac{2K_{sp}}{[SO_4^{2-}]} = 2[SO_4^{2-}] \tag{10}$$

where "m" is the mass of cation exchange resin added in grams.

For addition of each gram of resin, we may compute $[SO_4{}^{2-}]$ and then $[Ca^{2+}]$ and $[Na^+]$.

Figure 1 shows changes in Ca^{2+}, Na^+, and $SO_4{}^{2-}$ concentration in the aqueous phase with the addition of cation exchange resin when undissolved $CaSO_4$ is always present as a solid.

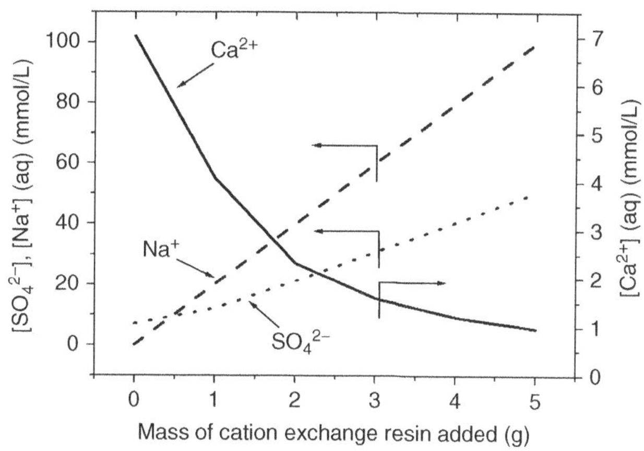

Figure 1. Change in the sodium, calcium and sulfate concentrations in the aqueous phase with gradual addition of cation exchange resins in Na form to a slurry containing $CaSO_4$ solid.

(iv) *Note that calcium concentration decreases while sodium and sulfate increase. Thus, for the process to be viable, the cation exchanger must be very selective toward calcium in preference to Na^+. Also, the process will gradually be kinetically slower as Ca^{2+} concentration is progressively decreased.*

Example 5.2 *Everything remains the same as the previous example, except that in each stage 1.0 g anion exchange resin (OH^- form) and 0.5 g cation exchange resin (H^+ form) are added. While the cation exchange capacity is 4 meq/g, anion exchanger is 2 meq/g. Discuss the difference and relative advantages.*

Solution:

Use of both cation and anion exchange resins in H^+ and OH^- form, respectively, disallow any increase in sulfate concentration while removing calcium from calcium sulfate in the solid phase. The reactions are:

$$CaSO_4(s) \rightleftharpoons Ca^{2+} + SO_4^{2-} \tag{1}$$

$$\overline{2(R-SO_3^-)H^+} + Ca^{2+} \rightleftharpoons \overline{(R-SO_3^-)_2Ca^{2+}} + 2H^+ \quad (2)$$

$$\overline{2(R_4N^+)OH^-} + SO_4^{2-} \rightleftharpoons \overline{(R_4N^+)_2SO_4^{2-}} + 2OH^- \quad (3)$$

$$2H^+ + 2OH^- \rightleftharpoons 2H_2O \quad (4)$$

Overall:

$$CaSO_4(s) + \overline{2(R-SO_3^-)H^+} + \overline{2(R_4N^+)OH^-}$$
$$\rightleftharpoons \overline{(R-SO_3^-)_2Ca^{2+}} + \overline{(R_4N^+)_2SO_4^{2-}} + 2H_2O \quad (5)$$

Since sulfate is also removed, calcium concentration does not change with increased addition of ion exchange resins. In summary, at any time during the process,

$$[Ca^{2+}] = \sqrt{K_{sp}} = [SO_4^{2-}] \quad (6)$$

Also, there is no other cation competing with Ca^{2+} for cation exchange sites.

Example 5.3 *Consider Example 5.2, except for the removal of barium, when*

(i) *$BaSO_4(s)$ alone is the solid phase*
(ii) *$BaSO_4(s)$ is present along with $CaSO_4(s)$*

For $BaSO_4$, $K_{sp} = 1 \times 10^{-10}$

Solution:

(i)

$$BaSO_4(s) \rightleftharpoons Ba^{2+} + SO_4^{2-} \quad (7)$$

$$[Ba^{2+}] = \sqrt{K_{sp}} = 10^{-5} M \quad (8)$$

Due to lower solubility product compared to $CaSO_4(s)$, Ba^{2+} concentration at equilibrium is significantly lower than Ca^{2+}. But the process is otherwise the same. In fact, on a molar basis, the mass of cation and anion exchange resins remains the same for dissolution and separation of $BaSO_4(s)$ from the solid phase as it is for $CaSO_4(s)$.

(ii) Selective separation of $BaSO_4(s)$ changes drastically in the presence of $CaSO_4(s)$, which has a much greater solubility product. From electroneutrality, we get

$$[Ca^{2+}] + [Ba^{2+}] = [SO_4^{2-}] \quad (9)$$

Since the K_{sp} of $BaSO_4(s)$ is nearly five orders of magnitude lower than $CaSO_4(s)$,

$$[Ca^{2+}] \gg [Ba^{2+}]$$

Hence,

$$[Ca^{2+}] = [SO_4^{2-}] = 7 \times 10^{-3} M \quad (10)$$

Thus,

$$[Ba^{2+}] = \frac{K_{sp}}{[SO_4^{2-}]} = \frac{1 \times 10^{-10} M^2}{7 \times 10^{-3} M} = 1.4 \times 10^{-8} M \tag{11}$$

Two things are to be noted:

- Ba^{2+} concentration is nearly three orders of magnitude lower in the presence of $CaSO_4$;
- Calcium concentration is five orders of magnitude greater than barium.

Thus, selective dissolution of barium and subsequent removal are an extremely inefficient separation process in the presence of $CaSO_4$.

Unlike calcium sulfate, anions of many poorly soluble solid phases result from weak acids; such calcium salts are $CaCO_3(s)$, $Ca_3(PO_4)_2(s)$, and $CaF_2(s)$ and the corresponding weak acids are H_2CO_3, H_3PO_4, and HF. Let us take the example of marble or calcite, that is, $CaCO_3(s)$. Using the cation exchanger in H^+ form is particularly useful in such cases due to protonation of the anion ($CO_3{}^{2-}$) and consequent dissolution of the solid phase as presented below:

$$CaCO_3(s) \leftrightarrow Ca^{2+} + CO_3^{2-} \tag{5.12}$$

$$2\overline{RH} + Ca^{2+} \leftrightarrow \overline{(R^-)_2Ca^{2+}} + 2H^+ \tag{5.13}$$

$$2H^+ + CO_3^{2-} \leftrightarrow H_2O + CO_2(g) \tag{5.14}$$

Overall:

$$CaCO_3(s) + 2\overline{RH} \leftrightarrow \overline{(R^-)_2Ca^{2+}} + H_2O + CO_2(g) \tag{5.15}$$

The reaction suggests that limestone or marble can be dissolved under mild chemical conditions using a hydrogen-form cation exchange resin in the absence of any acid. Equally important, there is no build-up of $CO_3{}^{2-}$ with the progress of $CaCO_3(s)$ dissolution and the process is thermodynamically favorable due to evolution and escape of CO_2 gas.

Example 5.4 *Solve Example 5.1 after considering a constant calcium/sodium separation factor of 5.0 (i.e., $\alpha_{Ca/Na} = 5.$) during the entire process of calcium removal.*

5.1.2 Desalting by Ion Exchange Induced Precipitation

Because of the high electrolyte content of seawater, the use of ion exchange resins in a mixed bed or in any conventional manner to desalinate seawater is not possible. For emergency situations, although economically unattractive, ion exchange followed by precipitation may be a relatively simple technique to produce adequate amounts of drinking water. Such a practice was in place during World War II [1,2]. For a typical seawater, besides very high concentrations of Na^+ and Cl^-, Mg^{2+} and SO_4^{2-} are also significantly present. An appropriate mixture of high capacity cation exchange resins preloaded with Ag^+ and Ba^{2+} ions may significantly reduce the salinity of seawater and make it drinkable. The exchange reactions are followed by precipitation of AgCl(s) and $BaSO_4$(s) as shown below

$$\overline{(R^-)Ag^+} + Na^+ + Cl^- \rightarrow \overline{(R^-)Na^+} + AgCl(s)\downarrow \quad (5.16)$$

$$\overline{(R^-)_2Ba^{2+}} + Mg^{2+} + SO_4^{2-} \rightarrow \overline{(R^-)_2Mg^{2+}} + BaSO_4(s)\downarrow \quad (5.17)$$

Very low solubility of silver chloride and barium sulfate essentially drive the reaction to the right-hand side. The sodium, chloride, magnesium, and sulfate ions, the primary constituents of seawater, are filtered off as insoluble $\overline{RNa}$, $\overline{R_2Mg}$, AgCl(s), and $BaSO_4$(s).

Example 5.5 *175 g of cation exchange resins in Ag^+ form and 21 g of cation exchange resins in Ba^{2+} form were added in one liter of typical sea water (35,000 mg/L TDS). Calculate the reduction in salinity or TDS in mg/L. State assumptions.*

Solution:

Assumptions:

- Following ion exchange, all of the silver precipitates as AgCl(s) and all of the barium as $BaSO_4$(s).
- $Q = 3.0\frac{\text{meq}}{\text{g resin}}$

Ag-form resin:

$$175\text{g Ag} - \text{resin} \times 3.0\frac{\text{meq Ag}}{\text{g resin}} \times 1.0\frac{\text{meq NaCl}}{\text{meq Ag}} = 525\,\text{meq NaCl removal capacity}$$

$$525\,\text{meq NaCl removed} \times 58.5\frac{\text{mg NaCl}}{\text{meq NaCl}} = 30{,}710\,\text{mg NaCl removed}$$

Ba-form resin:

$$21\text{g Ba} - \text{resin} \times 3.0\frac{\text{meq Ba}}{\text{g resin}} \times 1.0\frac{\text{meq MgSO}_4}{\text{meq Ba}} = 63\,\text{meq MgSO}_4\,\text{removal capacity}$$

$$63\,\text{meq MgSO}_4\,\text{removed} \times 60.2\frac{\text{mg MgSO}_4}{\text{meq MgSO}_4} = 3790\,\text{mg MgSO}_4\,\text{removed}$$

Overall desalination:

$$\text{TDS reduction} = 30{,}710\,\text{mg NaCl} + 3790\,\text{mg MgSO}_4 = 34{,}500\,\text{mg}$$

$$\text{Final salinity} = 35{,}000\frac{\text{mg}}{L} - 34{,}500\frac{\text{mg}}{L} = 500\frac{\text{mg}}{L}$$

$$\%\text{TDS reduction} = \frac{34{,}500\,\text{mg}/L}{35{,}000\,\text{mg}/L} = 98.5\%\,\text{reduction}$$

5.1.3 Separation of Competing Solid Phases

In the previous sections, we dealt with mostly one solid phase. Many real-life separation problems warrant selective dissolution and subsequent removal of one target solid phase in the presence of others. In this regard, we will particularly concentrate on precipitates of toxic or heavy metals due to its practical significance without loss of generality.

Widespread environmental issues pertain to the disposal of sludge or treatment of soil contaminated with a minor fraction (often <1% by mass) of environmentally regulated heavy metals in the solid phase in an otherwise innocuous background. So, the entire sludge or solid phase may be designated as "hazardous," thus greatly increasing the cost of disposal. Selective and targeted removal of the toxic metals from the background solid phase would constitute an efficient treatment process, as it would render the bulk of the sludge nonhazardous. In addition, such an approach offers an opportunity for the toxic metals to be concentrated and recovered.

It is true that the accompanying non-toxic materials present in the sludge are unimportant from a regulatory viewpoint, but their physical-chemical properties may strongly influence the selective separation of heavy metals. Conceptually, different scenarios are possible and Table 5.1 provides a schematic illustrating various possibilities [3].

While the first one is trivial, the latter two present challenges:

(i) Toxic metal cations bound to the ion-exchange sites of soil;
(ii) Toxic metals present amid a background of high buffer capacity.

The two scenarios need to be discussed separately to distinguish their underlying uniqueness.

Table 5.1 Toxic Metal (TM) contaminated sludge: various scenarios for separation.

Scenario	Schematic	Remarks
Chemically non-interacting solid-phase species; TM ppt. in sand; no Buffer Capacity (BC)	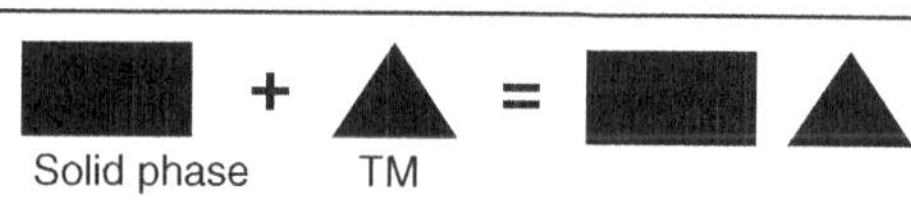	TM dissolution independent of accompanying phases
Ion-exchanging solid phase; dissolved TM bound to I-X sites of soil	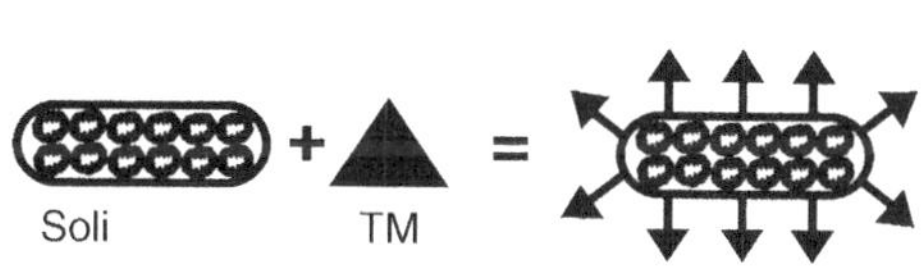	TM dissolution dependent on sorption/desorption phenomenon
Chemically interacting solid phase species; TM ppt. with calcite; high BC	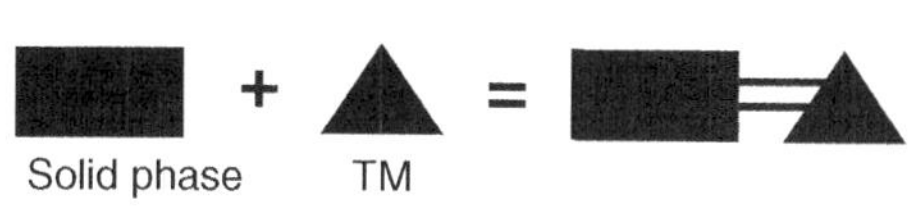	TM dissolution dependent on accompanying solid phase

Source: Sengupta and SenGupta 1996 [3]. Reproduced with permission of Mary Ann Liebert, Inc.

5.1.4 Recovery from Ion Exchange Sites of Soil

Clayey soils, bentonite or illite, in particular, derive their cation exchange capacity primarily from the isomorphous substitution within the lattice of silicon by aluminum. Divalent metal ions are strongly held onto these ion exchange sites primarily through electrostatic interactions. Removal of toxic or heavy metals from such contaminated soils essentially involves the following consecutive steps:

(i) Desorption from the ion exchange sites of the soil into the aqueous phase aided by an innocuous counter ion;
(ii) Selective sorption of toxic metals from the liquid phase into a metal-selective chelating exchanger.

The overall process is characteristically similar to facilitated transport. Non-toxic sodium or calcium ions in the aqueous phase are used to desorb toxic metal ions, (M^{2+}), from the ion exchange sites of the soil and subsequently, M^{2+} ions are sorbed onto metal-selective chelating exchangers releasing an equivalent amount of Na^+ or Ca^{2+}. It is the difference in metal-ion selectivity between soil and the chelating ion exchangers that allows passage of toxic metals from soil to the chelating exchanger quite favorably. Chelating exchangers may further be regenerated to recover metal and reused in the recovery process. Here we consider a chelating cation exchanger with an iminodiacetate functional group.

For the case of Cu(II) desorption from ion exchange sites of bentonite by addition of Ca^{2+} in the aqueous phase, that is, driver cation, the exchange reactions involved can be summarized as follows:

$$\overline{(Z^-)_2Cu^{2+}} + Ca^{2+}(aq) \leftrightarrow \overline{(Z^-)_2Ca^{2+}} + Cu^{2+}(aq) \tag{5.18}$$

$$\begin{aligned} &\overline{R-N-(CH_2COO^-)_2Ca^{2+}} + Cu^{2+}(aq) \\ &\quad \leftrightarrow 2\overline{R-N-(CH_2COO^-)_2Cu^{2+}} + Ca^{2+}(aq) \end{aligned} \tag{5.19}$$

Overall:

$$\begin{aligned} &\overline{(Z^-)_2Cu^{2+}} + \overline{R-N-(CH_2COO^-)_2Ca^{2+}} \\ &\quad \leftrightarrow 2\overline{R-N-(CH_2COO^-)_2Cu^{2+}} + \overline{(Z^-)_2Ca^{2+}} \end{aligned} \tag{5.20}$$

where Z and R represent the lattice of bentonite clay and the matrix of the chelating ion exchanger, respectively.

For the proposed process to succeed, the overall reaction must be thermodynamically favorable. Considering ideality, the equilibrium constant for such a reaction is given as follows:

$$K_{overall} = \frac{q_{Ca}^{Z} q_{Cu}^{R}}{q_{Cu}^{Z} q_{Ca}^{R}} \tag{5.21}$$

Superscripts Z and R denote the soil phase and the ion exchanger phase, respectively, while q_{Ca} and q_{Cu} represent the calcium and copper concentrations in the corresponding solid phase. Multiplying both numerator and denominator of the equation by C_{Ca}/C_{Cu} (C_i denotes the aqueous phase concentration of species i), we obtain:

$$K_{overall} = \left[\frac{q_{Cu}^{R}/C_{Cu}}{q_{Ca}^{R}/C_{Ca}}\right]\left[\frac{q_{Ca}^{Z}/C_{Ca}}{q_{Cu}^{Z}/C_{Cu}}\right] = \frac{\alpha_{Cu/Ca}^{R}}{\alpha_{Cu/Ca}^{Z}} \tag{5.22}$$

Thus, $K_{overall}$ is the ratio of the Cu/Ca separation factor between the ion exchanger and the bentonite clay. As discussed earlier, due to the presence of the chelating functional group (iminodiacetate moiety), the dimensionless Cu/Ca separation factor for the ion exchanger is often two orders of magnitude greater compared to bentonite. As a result, the overall process is quite selective for decontamination of clays with high ion exchange capacity. Alkaline or alkaline-earth metal cations, such as Na^+ or Ca^{2+} can act as shuttles in the facilitated transport of heavy metals from the soil to the chelating ion exchanger. Figure 5.2 illustrates how heavy metal cations move from the ion exchange sites of soil into metal-selective chelating exchanger while the extracting solution consisting of Na^+ and/or Ca^{2+} can be recycled [4].

5.1.5 Composite or Cloth-like Ion Exchanger (CIX)

Conventional fixed-bed sorption processes and membrane separation processes are unable to handle sludge/slurry with high-suspended solids (1–10%) content. Extensive pre-treatment to remove suspended solids is an essential requirement for such cases. Composite ion exchanger or CIX with cloth-like physical configuration is an appropriate candidate for treating slurry or sludge, for it is not susceptible to fouling by suspended solids with an appropriate reactor configuration. CIX is essentially fine spherical chelating ion exchanger beads entrapped in thin sheets (about 0.5 mm thick) of porous poly(tetrafluoroethylene) (PTFE) [5–9]. When dry, these composite sheets consist of >80% particles (polymeric ion exchanger) and <20% PTFE by weight. They

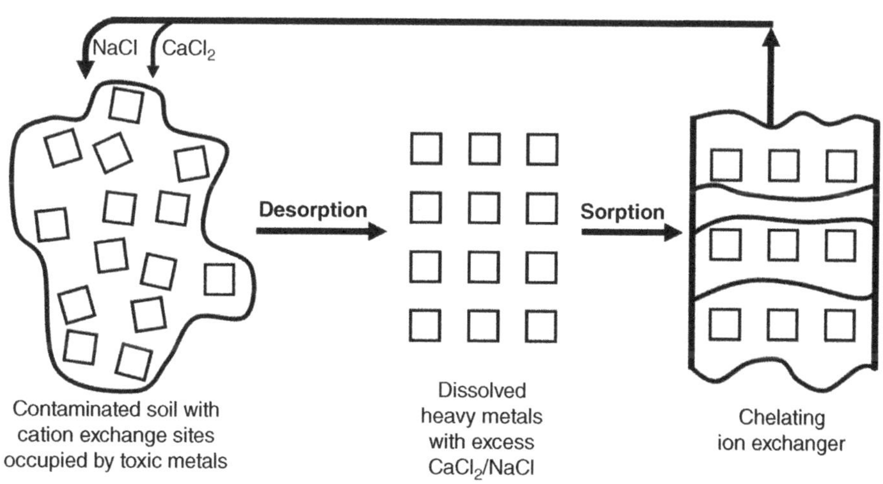

Figure 5.2 Conceptualized two-step process illustrating the sequential desorption of toxic metals from contaminated soils followed by uptake onto the chelating ion exchanger, releasing $CaCl_2$/NaCl for reuse. *Source:* Reprinted with permission from Sengupta and SenGupta 2000 [4].

are porous (usually >40% voids) with pore size distributions that are uniformly below 0.5 μm. The ion-exchange microspheres are usually <100 μm in diameter and have a total thickness ≈0.5 mm. Thus, they are effective filters that remove suspended solids >0.5 μm from permeating fluids. Because o f such sheet-like configuration, this material can be easily introduced into or withdrawn from a reactor with high concentration of suspended solids, with the target solutes being adsorbed onto or desorbed from the microadsorbents. In this section, the chelating functionality of the microspheres chosen is iminodiacetate (IDA). Figure 5.3a shows the electron microphotograph of the composite IDA exchanger. Figure 5.3b provides a schematic depicting how the microbeads are trapped within the fibrous network of PTFE and Figure 5.3c shows the photograph confirming cloth-like morphology of CIX [8]. Table 5.2 provides the salient properties of the CIX; note that the chelating microbeads constitute 90% of the composite membrane by mass. This feature allows the membrane to achieve the same level of performance as the parent chelating beads used in a fixed-bed operation. More details about characterization of the membrane are available in the open literature [10–12].

The subject CIX material differs fundamentally from traditional ion exchange membranes used in industrial process like Donnan dialysis (DD) and electrodialysis (ED) because of its high porosity. DO and ED membranes have very low porosity and are strongly influenced by the Donnan Coion Exclusion principle, which does not allow anions to pass through cation-exchange membrane and vice versa. However, in the case of CIX, large gaps between ion exchangers allow anions to pass through freely even though it is a cation-exchange membrane. The suspended solids that are >0.5 μm are not able to penetrate across the skin of the membrane because of the pore size of the material. However, water molecules and ions can easily move in and out of the thickness of the sheet, thus allowing unimpeded ion exchange between target ions in solutions (toxic metals in this case) and the counter ions of the CIX, as shown schematically in Figure 5.2. After a designated time, the CIX can be withdrawn and chemically regenerated with dilute (3–5%) mineral acid solution.

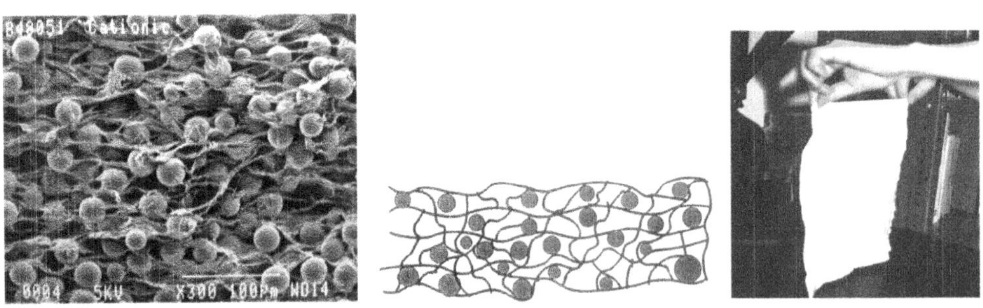

(a) (b) (c)

Figure 5.3 (a) Electron micrograph of the composite IDA membrane. (b) Schematic of microbeads in a fibrous network of PTFE. (c) Cloth-like configuration. *Source:* Sengupta and SenGupta 1993 [8]. Reproduced with permission of American Chemical Society.

Table 5.2 Properties of the composite chelex material.

Composition	90% chelating resin beads Distributed in 10% PTFE (by mass)
Pore size (nominal)	0.4 μm
Nominal capacity	3 meq/g membrane
Membrane thickness	0.2–0.5 mm
Ionic form	Na^+
Resin matrix	Styrene-divinylbenzene
Functional group	Iminodiacetate
Average bead size	100 μm
pH stability	1–14
Temperature range	0–75 °C
Chemical stability	Methanol; 1 N NaOH
Commercial availability	Bio-Rad Inc., CA

Source: Sengupta and SenGupta 1996 [3]. Reproduced with permission of Mary Ann Liebert, Inc.

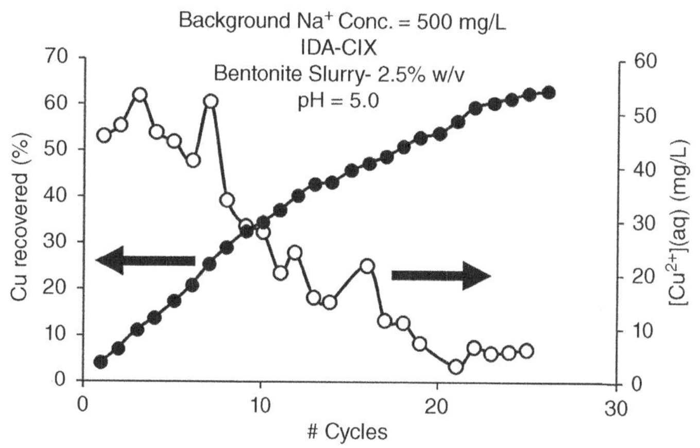

Figure 5.4 Copper(II) recovery from the ion-exchange sites of bentonite clay during the cyclic process. *Source:* Sengupta and SenGupta 2001 [13]. Reproduced with permission of Elsevier.

Using copper-loaded bentonite slurry, laboratory experiments were simulated in agreement with the conceptual process illustrated in Figure 5.2. Figure 5.4 demonstrates the plot of percentage recovery of Cu(II) and the aqueous-phase Cu(II) concentration versus the number of cycles for the case of Cu(II) loaded bentonite. Note that 60% copper recovery was achieved in less than 30 cycles. Additional details are available in the open literature [13,14].

5.1.6 Heavy Metals (Me^{2+}) with Solids Possessing High Buffer Capacity

The accompanying background solid phase, although innocuous, may influence the dissolution of heavy metals (Me^{2+}), thus affecting its selective separation from the bulk solid phase. For example, if a heavy metal precipitate, say $Me(OH)_2(s)$ or $MeCO_3(s)$, is present as a minor contaminant in the background of bulk calcite or $CaCO_3(s)$, the pH of the aqueous-phase will always be slightly alkaline. So, heavy metal concentration will be low and orders of magnitude lower than that of Ca^{2+}. Any selective separation of the heavy metal from the background solid phase would be inefficient in such cases.

For quantitative description of a representative scenario, let us consider a sludge containing a minor amount of $PbCO_3(s)$ in the background of bulk $CaCO_3(s)$ open to atmosphere. For an insightful understanding of the effect of $CaCO_3(s)$ as an accompanying phase on dissolved Pb^{2+} concentration for an open system, the Supplementary Reading S.5.1 is advised [12]. However, the reader can skip the Section S.5.1, if the following two outcomes are duly noted:

(i) Ratio of Ca^{2+}/Pb^{2+} in the aqueous phase is proportional to the ratio of their carbonate solubility products. Thus, Ca^{2+} concentration is in the order of magnitude greater than Pb^{2+} making selective separation inefficient.
(ii) An increase in carbon dioxide pressure increases aqueous phase Pb^{2+} concentration, but Ca^{2+}/Pb^{2+} ratio remains unchanged.

Supplementary Reading S5.1

Case 1

Let us take the generic case of a heavy metal precipitate, say $MeCO_3(s)$. If it is the only solid phase, and $CO_2(g)$ is introduced to lower the pH and concomitantly increase the aqueous phase $[Me^{2+}]$, relevant equations and commonly accepted equilibrium constants are

$$MeCO_3(s) \leftrightharpoons Me^{2+} + CO_3^{2-};\ K_{sp} \tag{S5.1}$$

$$CO_3^{2-} + H^+ \leftrightharpoons HCO_3^-;\ 1/K_{a,2} \tag{S5.2}$$

$$H_2CO_3^* \leftrightharpoons H^+ + HCO_3^-;\ K_{a,1} \tag{S5.3}$$

$$CO_2(g) + H_2O \leftrightharpoons H_2CO_3^*;\ K_H \tag{S5.4}$$

Adding Eqs (S5.1) through (S5.4), we get:

$$MeCO_3(s) + CO_2(g) + H_2O \leftrightharpoons Me^{2+} + 2HCO_3^- \tag{S5.5}$$

$$K_{eq,1} = \frac{K_{sp}K_{a,1}K_H}{K_{a,2}} = \frac{[Me^{2+}][HCO_3^-]^2}{pCO_2(g)} \tag{S5.6}$$

The electroneutrality equation is:

$$[H^+] + 2[Me^{2+}] = [OH^-] + [HCO_3^-] + 2[CO_3^{2-}] \tag{S5.7}$$

For pH ϵ (4.3, 8.3), $[H^+]$, $[OH^-]$, and $[CO_3^{2-}]$ can be ignored, and the simplified electroneutrality equation may be written as:

$$2[Me^{2+}] = [HCO_3^-] \tag{S5.8}$$

Equation ((S5.6)) may be rewritten as:

$$K_{eq,1} = \frac{K_{sp}K_{a,1}K_H}{K_{a,2}} = \frac{[Me^{2+}]\{2[Me^{2+}]\}^2}{pCO_2(g)}$$

$$4[Me^{2+}]^3 = \frac{Ksp\,Ka,1\,K_H}{Ka,2} pCO_2(g) \tag{S5.9}$$

$$[Me^{2+}] = constant * K_{sp}^{1/3} * pCO_2(g)^{1/3}$$

Thus, a plot of log(Me^{2+}) versus log[$pCO_2(g)$] would have a slope of 1/3. Also, from Eq. ((S5.3))

$$K_{a,1} = \frac{[H^+][HCO_3^-]}{[H_2CO3^*]} = \frac{[H^+][HCO_3^-]}{K_H pCO_2(g)} \tag{S5.10}$$

Thus,

$$[H^+] = \frac{pCO_{2(g)}\,K_{a,1}\,K_H}{[HCO_3^-]} \tag{S5.11}$$

Substituting $[HCO_3^-]$ from Eq. (S5.8) to Eq. (S5.11),

$$[H^+] = \frac{pCO_{2(g)}\,K_{a,1}\,K_H}{2[Me^{2+}]} \tag{S5.12}$$

Substituting [Me^{2+}] from Eq. (S5.9),

$$[H^+] = \frac{pCO_2(g)\,K_{a,1}\,K_H}{2\left(\frac{pCO_2 K_{sp}\,K_{a,1}\,K_H}{4K_{a,2}}\right)^{\frac{1}{3}}}$$

$$[H^+] = constant * pCO_2(g)^{2/3} \tag{S5.13}$$

Thus, a plot of pH versus log[$pCO_2(g)$] would have a slope of −2/3.

Let us substitute $MeCO_3$ with $PbCO_3$ with a solubility product of 1.46×10^{-13} M.

Figure S5.1 shows the profile of log(Pb^{2+}) and pH as a function of partial pressure of $CO_2(g)$ when $PbCO_3(s)$ alone in water is in equilibrium with carbon dioxide.

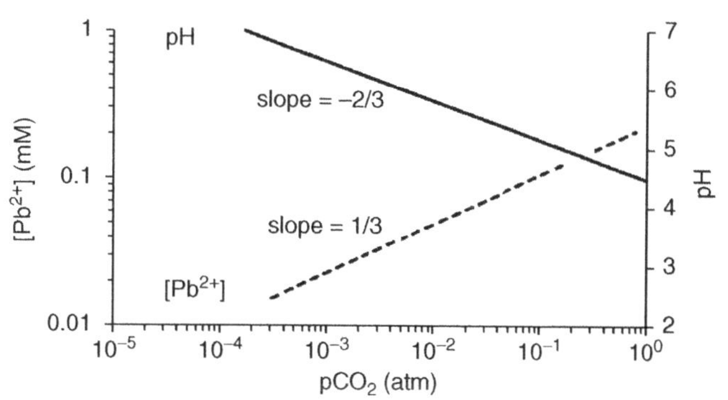

Figure S5.1. *Change in pH and Pb^{2+} concentration of a $PbCO_3(s)$ slurry as a function of $pCO_2(g)$.*

Case 2

Heterogeneous equilibria- more than one solid phase present, for example, $PbCO_3(s)$ and $CaCO_3(s)$ are present together.

(Continued)

Supplementary Reading S5.1 (Continued)

Relevant equations:

$$PbCO_3(s) \leftrightharpoons Pb^{2+} + CO_3^{2-};\ K_{sp1} \quad (S5.14)$$

$$CaCO_3(s) \leftrightharpoons Ca^{2+} + CO_3^{2-};\ K_{sp2} \quad (S5.15)$$

$$2CO_3^{2-} + 2H^+ \leftrightharpoons 2HCO_3^-;\ (1/K_{a,2})^2 \quad (S5.16)$$

$$2H_2CO_3^* \leftrightharpoons 2H^+ + 2HCO_3^-;\ K_{a,1}{}^2 \quad (S5.17)$$

$$2CO_2(g) + 2H_2O \leftrightharpoons 2H_2CO_3^*;\ (K_H)^2 \quad (S5.18)$$

Adding Eq. (S5.14) through (S5.18),

$$PbCO_3(s) + CaCO_3(s) + 2CO_2(g) + 2H_2O \leftrightharpoons Pb^{2+} + Ca^{2+} + 4HCO_3^- \quad (S5.19)$$

$$K_{eq,2} = \frac{K_{sp1}K_{sp2}(K_{a,1})^2(K_H)^2}{(K_{a,2})^2} = [Pb^{2+}][Ca^{2+}]\frac{[HCO_3^-]^4}{[pCO_2(g)]^2} \quad (S5.20)$$

The electroneutrality equation is:

$$[H^+] + 2[Pb^{2+}] + 2[Ca^{2+}] = [OH^-] + [HCO_3^-] + 2[CO_3^{2-}] \quad (S5.21)$$

If pH $\in$ *(4.3, 8.3),* $[H^+]$, $[OH^-]$, *and* $[CO_3{}^{2-}]$ *can be ignored.* $K_{sp1} \ll K_{sp2}$, *thus, at a given pH* $[Ca^{2+}] \gg [Pb^{2+}]$, $[Pb^{2+}]$ *can be ignored, and Eq. (S5.21) may be rewritten as:*

$$2[Ca^{2+}] = [HCO_3^-] \quad (S5.22)$$

Also,

$$K_{sp,1} = [Pb^{2+}][CO_3^{2-}] \therefore [Pb^{2+}] = \frac{K_{sp,1}}{[CO_3^{2-}]}$$

$$K_{sp,2} = [Ca^{2+}][CO_3^{2-}] \therefore [Ca^{2+}] = \frac{K_{sp,2}}{[CO_3^{2-}]}$$

$$\frac{K_{sp,1}}{K_{sp,2}} = \frac{[Pb^{2+}]}{[Ca^{2+}]}$$

$$\therefore [Ca^{2+}] = [Pb^{2+}]\frac{K_{sp,2}}{K_{sp,1}} \quad (S5.23)$$

Substituting $[Ca^{2+}]$ *from Eq. (S5.23) to Eq. (S5.22),*

$$2[Pb^{2+}]\left[\frac{K_{sp,2}}{K_{sp,1}}\right] = [HCO_3^-] \quad (S5.24)$$

Substituting $[HCO_3^-]$ *from Eq. (S5.24) and* $[Ca^{2+}]$ *from Eq. (S5.23) into Eq. (S5.20)*

$$K_{eq,2} = \frac{[Pb^{2+}]}{[pCO_2(g)]^2}[Pb^{2+}]\frac{K_{sp,2}}{K_{sp,1}}\left[2\,[Pb^{2+}]\left[\frac{K_{sp,2}}{K_{sp,1}}\right]\right]^4$$

$$K_{eq,2} = 16\frac{[Pb^{2+}]^6}{[pCO_2(g)]^2}\left[\frac{K_{sp,2}}{K_{sp,1}}\right]^5 \quad (S5.25)$$

Again,

$$K_{eq,2} = K_{sp1}K_{sp2}\left(\frac{K_{a,1}K_H}{K_{a,2}}\right)^2 \tag{S5.26}$$

Thus,

$$[Pb^{2+}]^6 = \frac{[pCO_2(g)]^2}{16}\left[\frac{K_{sp,1}}{K_{sp,2}}\right]^5 K_{eq,2}$$

$$[Pb^{2+}]^6 = \frac{[pCO_2(g)]^2}{16}\left[\frac{K_{sp,1}}{K_{sp,2}}\right]^5 K_{sp1}K_{sp2}\left(\frac{K_{a,1}K_H}{K_{a,2}}\right)^2$$

$$[Pb^{2+}]^6 = constant \cdot [pCO_2(g)]^2 \frac{K_{sp,1}{}^6}{K_{sp,2}{}^4} \tag{S5.27}$$

For a constant pCO_2 (g),

$$[Pb^{2+}] \propto K_{sp,1} \tag{S5.28}$$

$$[Pb^{2+}] \propto K_{sp,2}^{-2/3} \tag{S5.29}$$

Thus, we see in Case 1,

$$[Pb^{2+}] \propto K_{sp,1}^{1/3} \tag{S5.30}$$

Whereas in Case 2,

$$[Pb^{2+}] \propto K_{sp,1} \tag{S5.31}$$

Moreover, we also note that in Case 2,

$$[Pb^{2+}] \propto K_{sp,2}^{-2/3} \tag{S5.32}$$

that is, solubility of $PbCO_3$ is also dependent on the solubility product of $CaCO_3$. Further, from Eq. (S5.22),

$$2[Ca^{2+}] = [HCO_3^-] \tag{S5.33}$$

Substituting $[HCO_3^-]$ from Eq. (S5.22) into Eq. (S5.20)

$$K_{eq,2} = K_{sp1}K_{sp2}\left(\frac{K_{a,1}K_H}{K_{a,2}}\right)^2 = [Ca^{2+}]\frac{K_{sp,1}}{K_{sp,2}}[Ca^{2+}]\frac{[2[Ca^{2+}]]^4}{[pCO_2(g)]^2} \tag{S5.34}$$

$$K_{eq,2} = 16\frac{[Ca^{2+}]^6}{[pCO_2(g)]^2}\frac{K_{sp,1}}{K_{sp,2}} \tag{S5.35}$$

Thus,

$$[Ca^{2+}]^6 = \frac{[pCO_2(g)]^2}{16}[K_{sp2}]^2\left(\frac{K_{a,1}K_H}{K_{a,2}}\right)^2 \tag{S5.36}$$

$$[Ca^{2+}] \propto [pCO_2(g)]^{1/3} \tag{S5.37}$$

(Continued)

Supplementary Reading S5.1 (Continued)

$$[Ca^{2+}] \propto [K_{sp2}]^{1/3} \tag{S5.38}$$

Thus, the solubility of Ca^{2+} is not dependent on solubility product of $PbCO_3$, but the opposite does not hold.

Also, since

$$[Pb^{2+}] = [Ca^{2+}]\frac{K_{sp,1}}{K_{sp,2}} \tag{S5.39}$$

Thus, $[Ca^{2+}] \gg [Pb^{2+}]$ at all pCO_2. Furthermore,

$$K_{a,1} = \frac{[H^+][HCO_3^-]}{[H_2CO3*]} = \frac{[H^+][HCO_3^-]}{K_H pCO_2(g)} \tag{S5.40}$$

And,

$$[Ca^{2+}] \propto [pCO_2(g)]^{1/3} \tag{S5.41}$$

$$[HCO_3^-] \propto [pCO_2(g)]^{1/3} \tag{S5.42}$$

$$[H^+] = \frac{K_{a,1}K_H pCO_2(g)}{[HCO_3^-]} \tag{S5.43}$$

$$[H^+] \propto \frac{[pCO_2(g)]}{[pCO_2(g)]^{1/3}} \tag{S5.44}$$

$$[H^+] \propto [pCO_2(g)]^{2/3} \tag{S5.45}$$

Figure S5.2 depicts the dissolution behaviors in accordance with the foregoing prediction, that is, plots $\log(Pb^{2+})$, $\log(Ca^{2+})$ and pH versus $\log[pCO_2(g)]$.

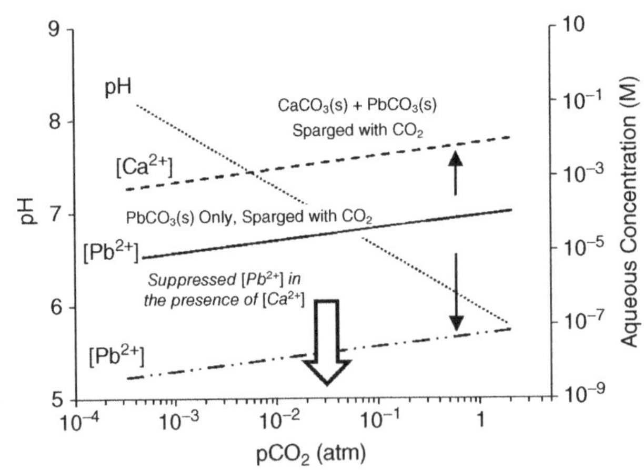

***Figure S5.2** Theoretically computed dissolved lead and calcium concentration profiles upon addition of $CO_2(g)$ to a sludge containing (i) only $PbCO_3(s)$ and (ii) $CaCO_3(s)$ and $PbCO_3(s)$. Source: Sengupta [3]. Reproduced with permission of Mary Ann Liebert, Inc.*

The following observations in Figure S5.2 are noteworthy: (i) in the presence of accompanying $CaCO_3$(s), dissolved lead concentration is greatly suppressed; (ii) dissolved calcium concentration is independent of the presence of $PbCO_3$(s); (iii) slope of dissolved lead concentration on partial pressure of CO_2 is independent of the accompanying solid phase, that is, $CaCO_3$(s).

5.1.7 Ligand-Induced Metal Recovery with a Chelating Exchanger

The primary challenge lies in recovering or separating the toxic metals with minimum disturbance to the accompanying bulk solid phase, namely, $CaCO_3$(s). To achieve that goal, aqueous-phase metal concentration must be increased relative to calcium. Addition of a mineral acid is undesirable because that will cause major dissolution of $CaCO_3$(s). Use of a moderately strong ligand that is environmentally benign may achieve an efficient separation provided the following criteria are satisfied:

(i) The ligand should have significantly higher stability constant for the target metal than that for calcium;
(ii) The chelating exchanger should possess high metal ion affinity to break the labile metal-ligand complex, sorb the metal ions into the exchanger and release the ligand back into the aqueous phase for continuation of the cyclic process.

Important reactions with L^{n-} as the added ligand can be presented as follows:

$$CaCO_3(s) \leftrightarrow Ca^{2+} + CO_3^{2-} \tag{5.23}$$

$$MeCO_3(s) \leftrightarrow Me^{2+} + CO_3^{2-} \tag{5.24}$$

$$Ca^{2+} + L^{n-} \leftrightarrow \text{Minimum Conversion} \tag{5.25}$$

$$Me^{2+} + L^{n-} \leftrightarrow (MeL)^{2-n} \tag{5.26}$$

$$2RH + (MeL)^{2-n} \leftrightarrow \overline{(R^-)_2Me^{2+}} + 2H^+ + L^{n-} \tag{5.27}$$

Again, this is essentially a cyclic process where L^{n-} acts as a shuttle between the two solid phases, namely, precipitates and the ion exchanger.

Laboratory investigations were carried out to simulate a cyclic process using oxalate as the facilitating ligand [7,8,13]. Slurry was prepared by mixing 5 g of $CaCO_3$(s), 45 g of fine sand, 13.4 g of sodium oxalate and 0.38 g of CuO(s) in one liter of water and pH was maintained at 9.0. Aqueous phase oxalate concentration was 4000 mg/L, and <1% CuO(s) was present in the solid phase of the sludge.

The CIX sheet with metal-selective iminodiacetate functional group was used for selective separation of copper from sludge. Figure 5.5 presents a conceptual schematic of the process configuration where the copper-contaminated sludge gradually becomes free of copper [3]. Figure 5.6 validates that the dissolved sludge phase concentrations of Cu and Ca remained fairly constant with the number of cycles, suggesting that they are controlled by solubility products of the solid phases. Obviously, at pH of 9.0, free copper ion, Cu^{2+}, is practically absent; most of the dissolved copper exists primarily as neutral or anionic Cu-oxalate complex. Predicted total dissolved copper, as computed

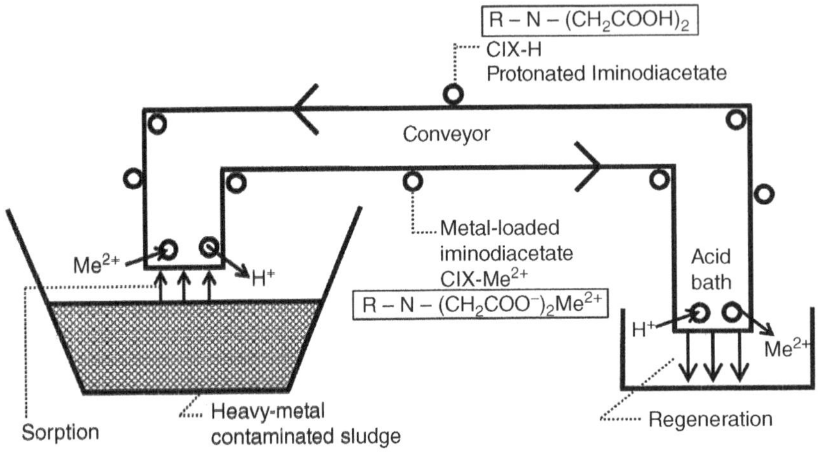

Figure 5.5 Conceptualized heavy metal decontamination process using a CIX sheet, where heavy metals are continuously separated from sludge and concentrated in the regeneration tank. *Source:* Sengupta and SenGupta 1996 [3]. Reproduced with permission of Mary Ann Liebert, Inc.

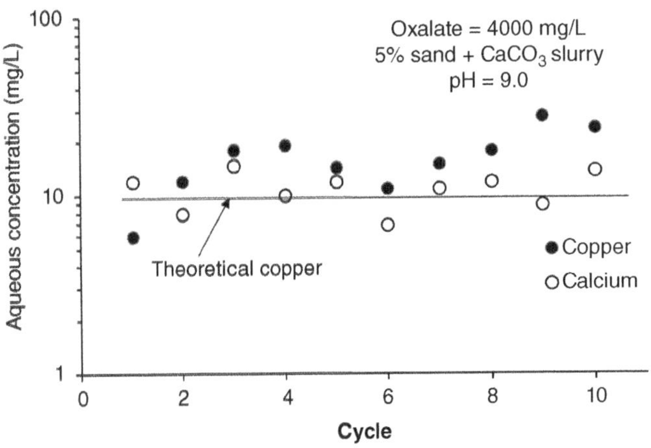

Figure 5.6 Dissolved copper and calcium concentration during the recovery process at pH = 9.0 for oxalate concentration of 4000 mg/L. *Source:* Sengupta and SenGupta 1993 [8]. Reproduced with permission of American Chemical Society.

from the stability constant values available in the open literature, is well in agreement with the experimental values. Figure 5.7 shows the recovery of Cu and Ca in the regenerant solution. Although copper is present primarily as copper–oxalate complex in solution, copper recovery is significant and increases steadily with every cycle. In comparison, calcium recovery was much lower and tended to approach an asymptotic concentration in the regenerant with an increase in number of cycles, confirming selective separation or removal of copper from the bulk solid phase containing mostly calcium carbonate and sand. Results of the study provided convincing evidence that relatively small amount of toxic metal precipitates can be selectively separated from other bulk solid phases using ligand-assisted chelating ion exchanger.

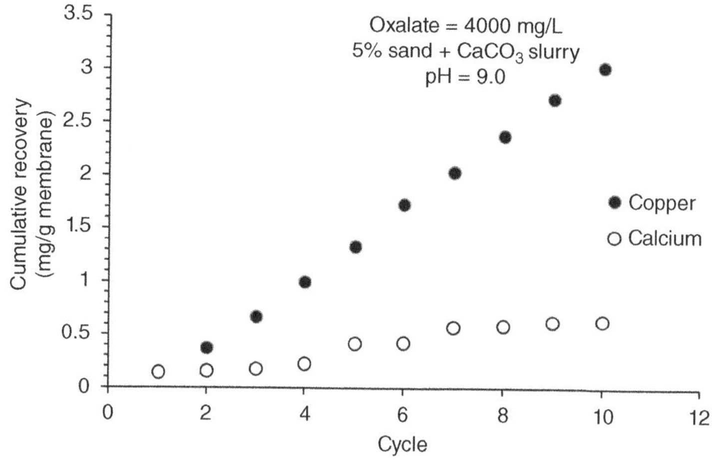

Figure 5.7 Cumulative copper and calcium recovery with increase in number of cycles at alkaline pH. *Source:* Sengupta and SenGupta 1993 [8]. Reproduced with permission of American Chemical Society.

5.2 Coagulant Recovery from Water Treatment Sludge

Possible recovery of alum from water treatment plant sludge is an example of solid phase separation with global significance. We will discuss it as a pertinent example in this section. In drinking water treatment plants around the world, alum, $Al_2(SO_4)_3{\cdot}14H_2O$, is the most widely used coagulant for efficient removal of particulate solids and colloids from surface water supply. Alum is finally converted during the process into insoluble aluminum hydroxide, a major component (25–60%) of the solids in water treatment plant sludge, called water treatment residuals (WTR). The WTR are essentially bulky, gelatinous slurry composed of suspended inorganic particles, natural organic matters (NOM), trace amounts of heavy metal precipitates and aluminum hydroxide. The WTR disposal into landfills, waterways, or through land application is a concern in both developed and developing countries and it is receiving scrutiny for its high aluminum content [15,16]. Water treatment plants in the United States alone produce over 2 million tons of aluminum-laden WTR every day. Because of the magnitude and pervasive nature of the problem, the prospect of alum recovery from WTR and its reuse has received considerable attention during the last three decades [17–19].

In the past, attempts have been made to recover alum through the acid digestion process [17–19]. In this process, WTR is sufficiently acidified with sulfuric acid, so that insoluble aluminum hydroxide is dissolved in the form of dilute liquid alum. Subsequently, the supernatant liquid, rich in dissolved aluminum, is decanted. The stoichiometry of this reaction can be written as follows for alum-based WTR:

$$2Al(OH)_3 \cdot 3H_2O + 3H_2SO_4 + 2H_2O \leftrightarrow Al_2(SO_4)_3 \cdot 14H_2O \tag{5.28}$$

Although operationally simple, the process is non-selective, that is, along with alum it recovers also all other substances that are soluble under acidic conditions or that exist as colloids. Thus, naturally occurring organic material (humates and fulvates), which are generally removed quite well by alum coagulation, will be present in the recovered alum as dissolved organic matters. Should this recovered alum be reused as a coagulant, trihalomethane formation potential (THMFP) in the treated water upon chlorination would tend to increase significantly. The non-selective dissolution would mean that "toxic metals" are also dissolved in the decanted alum.

If an alkali digestion process is tried, given the amphoteric nature of aluminum oxide, the process will dissolve aluminum at higher pH. However, the simultaneous dissolution of NOMs will remain an issue. Figure 5.8 shows the limitations of both acid and alkali digestion processes for sludge collected from the Allentown Water Treatment Plant (AWTP) in Pennsylvania, USA [18]. It may be noted that the dissolution of aluminum is accompanied by high concentrations of dissolved organic carbon (DOC) at both low and high pH values.

5.2.1 Development of Donnan IX Membrane Process

A simple-to-operate Donnan membrane or Donnan dialysis process has recently been developed allowing selective alum recovery from WTR [18,20]. The key features of the process are summarized below:

- Recovered alum is essentially free of NOM and particulate matters;
- The concentration of aluminum in the recovered alum can be significantly greater than that in the WTR;
- The process works on electrochemical potential gradient across a cation exchange membrane, thus avoiding fouling of the membrane caused by NOM or particulate matters;
- The volume of disposable sludge is greatly reduced and sulfuric acid is the only chemical required for the process.

Figure 5.8a shows a general schematic of the process [20] while (b) illustrates the underlying principles of selective alum recovery and rejection of NOM and particulate matters [18]. Cation exchange membranes form the heart of the process and are available commercially from several manufacturers. More than 75% aluminum was recovered in 20 hours from the residuals of AWTP. Figure 5.9 shows a visual comparison of the clarifier sludge from the AWTP (a), recovered alum coagulant after acid digestion (b) and the recovered alum by the Donnan membrane process (c).

The recovered alum by Donnan dialysis was clear and transparent, with practically no turbidity and NOM, similar to fresh liquid alum. Figure 5.10 shows the distribution of different species present in the recovered alum. Besides aluminum and Fe(III) both of which are desirable for efficient coagulation, other constituents are essentially negligible. When Jar Tests were performed on Lehigh River water, it was found that the recovered alum was equally effective in reducing turbidity as fresh commercial alum.

5.2.2 Alum Recovery: Governing Donnan Equilibrium

Let us consider aluminum sulfate and sulfuric acid solutions divided by a cation-exchange membrane that allows only cations to migrate from one side to

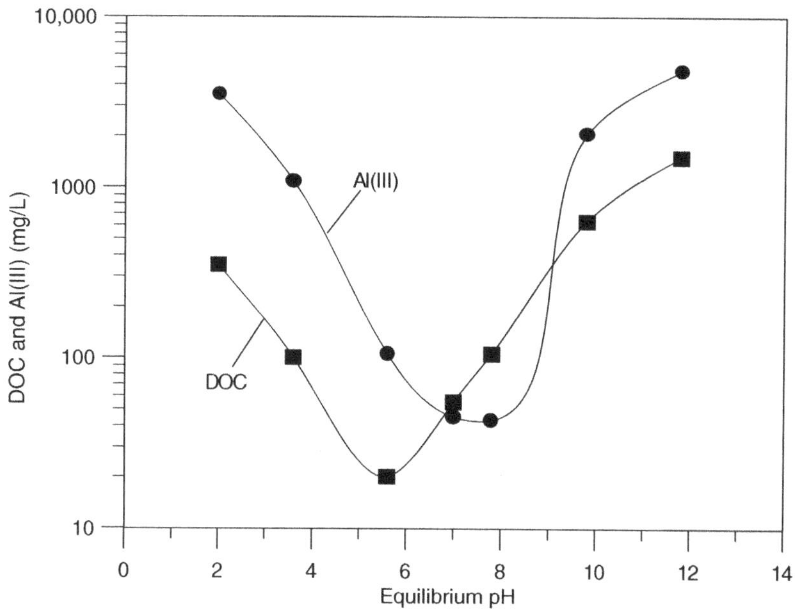

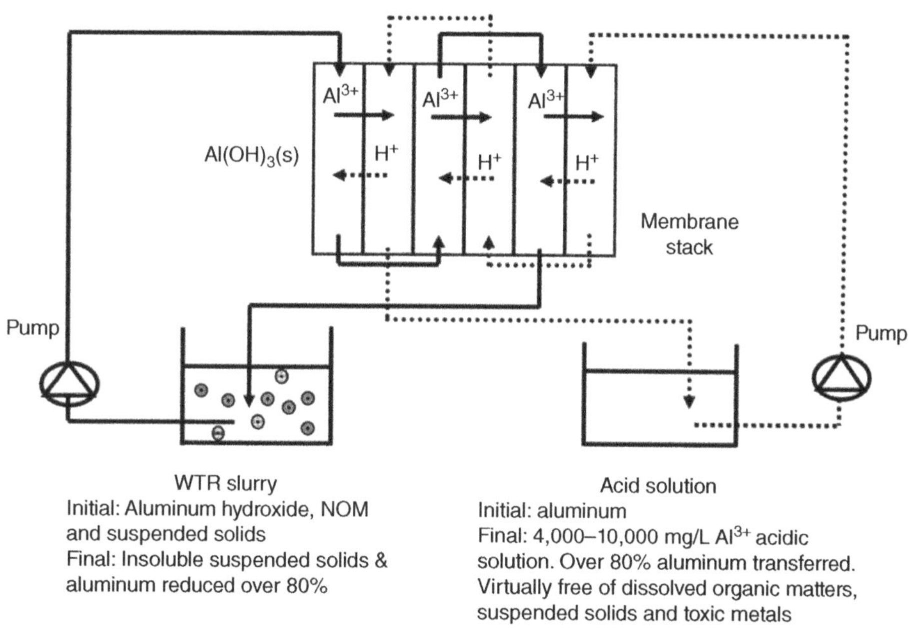

(a)

Figure 5.8 Dissolved Organic Carbon (DOC) and Al (III) in the AWTP sludge at varying pH levels. *Source:* Prakash and SenGupta 2003 [18]. Reproduced with permission of American Chemical Society. (a) A general schematic of Donnan membrane process for alum recovery from water treatment residuals. *Source:* Reprinted with permission from SenGupta and Prakash 2002 [20]. (b) Underlying principles of selective alum recovery using Donnan membrane process with cation exchange membrane. *Source:* Prakash and SenGupta 2003 [18]. Reproduced with permission of American Chemical Society.

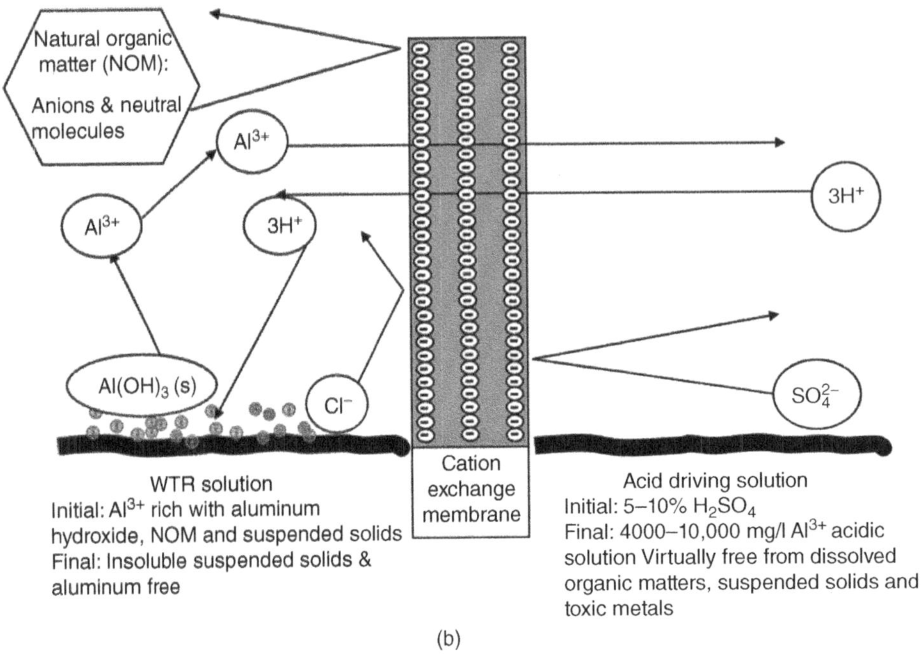

Figure 5.8 (*Continued*)

Figure 5.9 Visual comparison of the clarifier sludge from the AWTP (a), recovered alum coagulant after acid digestion (b) and the recovered alum by the Donnan membrane process (c). *Source:* Prakash and SenGupta 2003 [18]. Reproduced with permission of American Chemical Society.

the other but rejects any passage of anions according to the Donnan coion exclusion principle [21]. At equilibrium, the electrochemical potential of aluminum ion Al^{3+} ion ($\bar{\eta}$) in the electrolyte solution on the left-hand side (LHS) of the membrane will be the same as that in the electrolyte solution on the right-hand side (RHS), that is,

$$\bar{\eta}_{Al}^{L} = \bar{\eta}_{Al}^{R} \tag{5.29}$$

$$\mu_{Al}^{o} + RT\ \ln(a_{Al}^{L}) + zF\varnothing^{L} = \mu_{Al}^{o} + RT\ \ln(a_{Al}^{R}) + zF\varnothing^{R} \tag{5.30}$$

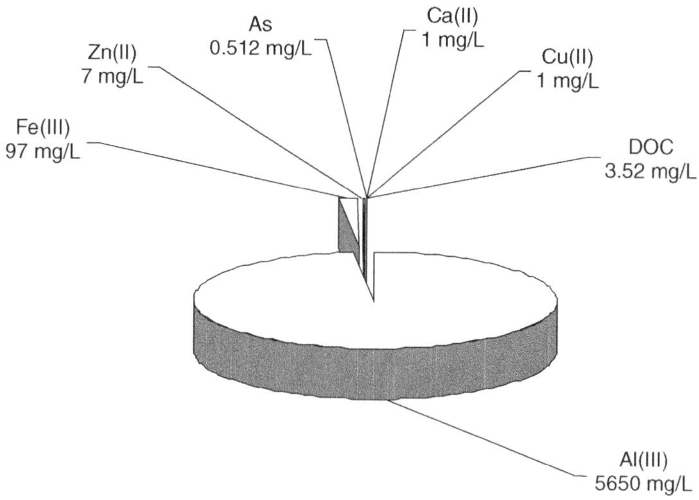

Figure 5.10 Distribution of different species present in the recovered alum by Donnan membrane process. *Source:* Prakash and SenGupta 2003 [18]. Reproduced with permission of American Chemical Society.

where superscripts "0", "L", and "R" refer to standard state, LHS and RHS and μ, a, F and $\varnothing$ denote chemical potential, activity, Faraday Constant and electrical potential, respectively. The 'z' refers to the charge of the diffusing ion, which is 3 for trivalent aluminum ion Al^{3+}. Equation (5.30) yields the following equality for aluminum ions on two sides of the membrane:

$$\frac{F(\varnothing^{L} - \varnothing^{R})}{RT} = \ln\left(\frac{a_{Al}^{R}}{a_{Al}^{L}}\right)^{1/3} \tag{5.31}$$

In a similar way, it can be shown for hydrogen ions that

$$\frac{F(\varnothing^{L} - \varnothing^{R})}{RT} = \ln\left(\frac{a_{H}^{R}}{a_{H}^{L}}\right) \tag{5.32}$$

Assuming non-ideality effects are about the same on both sides of the membrane, activities can be replaced by molar concentrations. Equations (5.29) and (5.30) then yield the following:

$$\left(\frac{C_{Al}^{R}}{C_{Al}^{L}}\right) = \left(\frac{C_{H}^{R}}{C_{H}^{L}}\right)^{3} \tag{5.33}$$

If the ratio $\frac{C_{H}^{R}}{C_{H}^{L}}$ is 10, it means C_{Al}^{R} is 1000 times greater than C_{Al}^{L}. Thus, by maintaining high hydrogen ion concentration on the right-hand side of the membrane, aluminum ions can be driven from the LHS to the RHS even against a positive concentration gradient, that is, from a lower concentration region to a higher concentration one. Figure 5.8, presented earlier, depicts the conceptualized selective alum recovery from WTR, highlighting the following: (i) aluminum hydroxide precipitates can be dissolved and then concentrated on the right-hand side; (ii) negatively charged NOM, sulfate and chloride cannot permeate through the membrane; and (iii) the transmembrane pressure does not influence the aluminum transfer flux.

5.2.3 Process Validation

Experiments were carried out using water treatment plant sludge from the AWTP in Pennsylvania. Figure 5.11 shows the results of the process for a period of 24 h; the percentage aluminum recovery and the concentration of aluminum in the two chambers were plotted against time. Over 70% recovery (72%) was attained in 24 h. The noteworthy observation is that the recovered aluminum concentration was 6650 mg/L as Al and it was significantly greater than the total aluminum concentration (2400 mg/L) present in the parent sludge [18,21].

Ferric salts (chloride or sulfate) are also used as coagulants in water treatment plants and the resulting ferric hydroxide precipitates constitute a major portion of the clarifier sludge or WTR. In principle, the Donnan membrane process is capable of selectively recovering Fe(III) coagulants from these WTR as well. To validate it, the WTR from the Baxter plant (Philadelphia, PA), which utilized $FeCl_3$ as coagulant, was used in several test runs. Figure 5.12 shows percentage Fe(III) recovery and the concentration of Fe(III) in the feed and recovery side with time. Nearly 75% recovery is attained in 24 h. The resulting Fe(III) is essentially free of NOM, particulate matter and other impurities. Figure 5.13 shows the visual comparison of the recovered Fe(III) coagulant between the traditional acid digestion process and the Donnan membrane process. Higher transparency of the coagulant from Baxter Plant, recovered by Donnan membrane process, is readily noticeable due to the absence of turbidity and NOM. It is worth mentioning that intraparticle diffusion of Al^{3+} or Fe^{3+} within the ion exchange membrane is the rate-limiting step and the subject has been extensively discussed in the open literature [21,22]. Before leaving this section, it is only appropriate to mention that with global pursuit of enhanced sustainability, solid phase separation with recovery will only grow. Such separations will warrant novel applications of ion exchange processes.

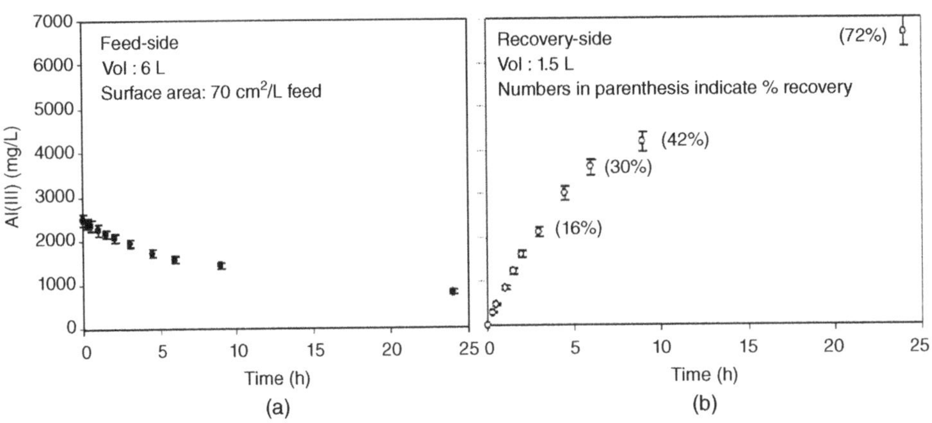

Figure 5.11 Aluminum recovery from AWTP residuals during Donnan membrane process: (a) decrease in Al concentration in feed; (b) percentage recovery and increase in Al concentration in recovery solution. *Source:* Prakash and SenGupta 2003 [18]. Reproduced with permission of American Chemical Society.

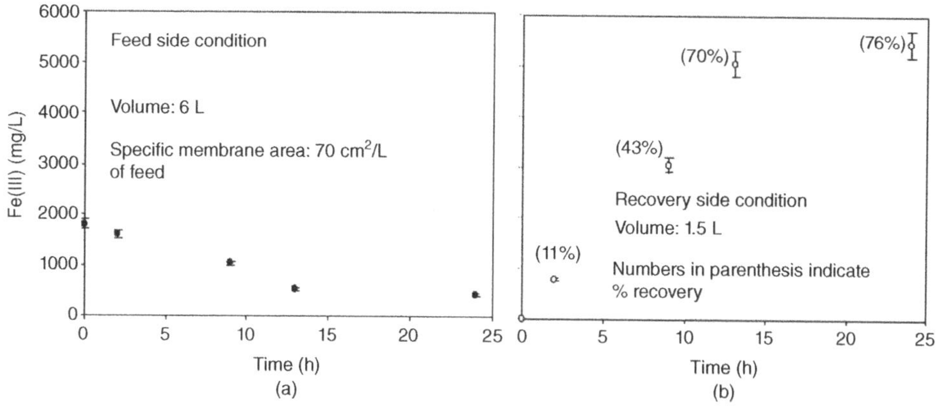

Figure 5.12 Fe(III) recovery from ferric chloride based WTR from Baxter Water Treatment Plant during Donnan membrane process: (a) decrease in Fe concentration in feed; (b) percentage recovery and increase in Fe concentration in recovery solution. *Source:* Prakash and SenGupta 2003 [18]. Reproduced with permission of American Chemical Society.

(a) (b)

Figure 5.13 Visual Comparison of recovered ferric coagulant from Baxter Plant residuals by Acid digestion process (a) and Donnan membrane process (b). *Source:* Prakash and SenGupta 2005 [21]. Reproduced with permission of John Wiley & Sons.

5.3 Gas Phase Ion Exchange

The distinctive feature for sorption of gases onto an ion exchanger is that the ion exchanger may act as a solid reacting phase, be it an acid, a base or a redox active agent. Acid gases are sorbed on the basic anion exchangers, and basic gases are sorbed on the acid cation exchangers by an acid/base neutralization reaction. In normal conditions, the ion exchangers do not sorb neutral atmospheric gases, namely, oxygen and nitrogen. Thus, this property of an ion exchanger offers an opportunity for allowing recovery or separation of acidic or basic gases from contaminated atmosphere. Such gases may include sulfur dioxide, nitric oxides, hydrogen sulfide, carbon dioxide, ammonia and others. Ion exchangers, during sorption of acidic or basic gases, always remain partly hydrated and exhibit high affinity for water vapor. Gas drying through selective removal of water vapor by ion exchanger from the background of other gases is thus a viable proposition and in use for specific applications. Also, relative humidity of atmosphere can be controlled through use of ion exchangers. Other gaseous exchange processes can be classified by appropriately designating gases of interest:

I. Acidic Gases (CO_2, SO_2, HCl, NO_2, H_2S)
II. Basic Gases (NH_3)
III. Redox Active Gases (O_2, Hg, Cl_2)
IV. Ligand Gases (NH_3, H_2S)

In the following sections, we will consider fundamentals governing removal of gases and properties and physical configuration of ion exchangers influencing gas separation processes.

5.3.1 Sorption of Acidic and Basic Gases

During gas separation, ion exchangers remain partially or fully hydrated and the water of hydration is a significant component of the process along with the solid ion exchanger and the gas. If needed, other polar solvents, namely, ethanol or acetone, can replace water of hydration. For general treatment of the gaseous ion exchange, let us consider removal of sulfur dioxide or SO_2 from air (or flue gas) with a hydrated weak-base anion (WBA) exchanger in accordance with the following steps:

Step 1. Sulfur Dioxide Dissolution in the Ion Exchanger

$$SO_2(g) + (H_2O)_{IX} \leftrightarrow (H_2SO_3)_{IX} \tag{5.34}$$

Step 2. Dissociation of Sulfurous Acid

$$(H_2SO_3)_{IX} \leftrightarrow (H^+)_{IX} + (HSO_3^-)_{IX} \tag{5.35}$$

Step 3. Protonation of weak-base Ion Exchanger

$$(R_3N)_{IX} + (H^+)_{IX} + (HSO_3^-)_{IX} \leftrightarrow (R_3NH^+HSO_3^-)_{IX} \tag{5.36}$$

The overall reaction can be presented as

$$SO_2(g) + (H_2O)_{IX} + (R_3N)_{IX} \leftrightarrow (R_3NH^+HSO_3^-)_{IX} \tag{5.37}$$

subscript 'IX' refers to the ion exchanger phase. The overall equilibrium constant for the reaction is

$$K_{\text{overall}} = K_H K_{HA} K_{IX} \tag{5.38}$$

where K_H represents Henry's constant (Eq. 5.34), K_{HA} is the exchanger-phase acid dissociation constant (Eq. 5.35) and K_{IX} is the equilibrium constant of the WBA exchange resin (Eq. 5.36).

The WBA exchanger can be replaced with a strong-base anion exchanger in OH^- form as follows:

$$(R_4N^+(OH^-))_{IX} + (H^+)_{IX} + (HSO_3^-)_{IX} \rightarrow (R_4N^+(HSO_3^-))_{IX} + H_2O \tag{5.39}$$

Since H_2CO_3 is a weaker acid than H_2SO_3, strong-base anion exchanger may also be used in carbonate form to remove SO_2 as follows:

$$((R_4N^+)_2(CO_3^{2-}))_{IX} + (H^+)_{IX} + (HSO_3^-)_{IX} \rightarrow ((R_4N^+)_2SO_3^{2-})_{IX} + H_2O + CO_2\uparrow \tag{5.40}$$

Note that sorption of every mole of SO_2 is accompanied by release of one mole of CO_2 in the atmosphere. When the SO_2 sorption capacity is exhausted, passing a solution of Na_2CO_3 can regenerate the anion exchange resin.

Ammonia, an alkaline gas, can similarly be removed with a strong-acid or weak-acid cation exchanger in H^+ form with the overall reactions presented as follows:

Strong acid:

$$NH_3(g) + (H_2O)_{IX} + (RSO_3^-H^+)_{IX} \leftrightarrow (RSO_3^-NH_4^+)_{IX} + H_2O \tag{5.41}$$

Weak acid:

$$NH_3(g) + (H_2O)_{IX} + (RCOOH)_{IX} \leftrightarrow (RCOO^-NH_4^+)_{IX} + H_2O \tag{5.42}$$

It is apparent that $(H_2O)_{IX}$ or water content of the ion exchanger that is acting as a solid acid or base, influences the uptake of the target gas. In a dynamic operation, the water content of the ion exchanger is dependent on the relative humidity of the gas stream being treated. Figure 5.14 shows the variation of water content of a strong-base anion exchange resin (Dowex-1 from Dow Chemical Co.) with change in relative humidity. Obviously, removal of acidic or basic gases is thus significantly dependent on relative humidity of the gas stream unless the ion exchangers are soaked wet intermittently [23].

5.3.2 CO_2 and SO_2 Capture with Weak-Base Anion (WBA) Exchanger

Carbon dioxide, although far less toxic and acidic compared to sulfur dioxide, has attained notoriety due to its greenhouse characteristics and large emissions. As hydrocarbon-based combustion accounts for over 80% of these emissions, carbon dioxide has been singularly held responsible as a driving force in climatic change. One

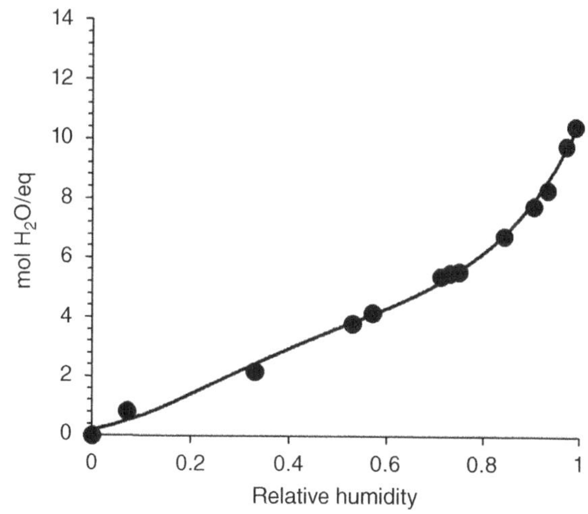

Figure 5.14 Isobaric curves of ion exchanger Dowex-1 × 6 (OH-ionic form) for water content at different relative humidity. *Source:* Boyd and Soldano 1953 [23]. Reproduced with permission of Wiley-VCH.

solution for reduction in CO_2 emissions is through carbon capture and sequestration (CCS), specifically from post-combustion point sources such as coal, oil and natural gas fired power plant facilities. Obviously, weak-base polymeric anion exchange resins are worthy candidates for application in CO_2 capture from fossil energy power plant flue gas. Besides CO_2, flue gas contains primarily nitrogen followed by water vapor and unburnt oxygen. A good sorbent for CO_2 capture must be selective toward CO_2, but at the same time minimize sorption of other species such as water, although some degree of hydration of the ion exchanger resin is essential. A sorbent capable of CO_2 capture from post-combustion flue gases must selectively remove 12–14 vol% CO_2 from the background of an N_2 and O_2 stream with 12–15% H_2O.

A WBA resin, Lewatit VP OC 1065 (Lanxess), which is a primary amine functionalized macroporous polystyrene beads with divinylbenzene cross-linking, was extensively investigated for its CO_2 capture capacity [24]. The water uptake of the dry WBA exchange resin, subsequent CO_2 adsorption and desorption were determined by a thermogravimetric analyzer (TGA) under representative conditions. Figure 5.15 illustrates four major steps of CO_2 adsorption–desorption cycle:

Step 1: Drying the WBA resin under an N_2 stream at 140 °C.
Step 2: Exposing the resin to 9.1 vol% H_2O stream at 50 °C until stabilization of mass uptake.
Step 3: Equilibration with a gas stream containing 9.1% H_2O, 11.1% CO_2 and the balance being N_2 at 50 °C.
Step 4: Desorption at 150 °C with N_2.

Note that the WBA resin provides significant CO_2 capture capacities for the sorption–desorption cycles. Most importantly, CO_2 capture was significant even in

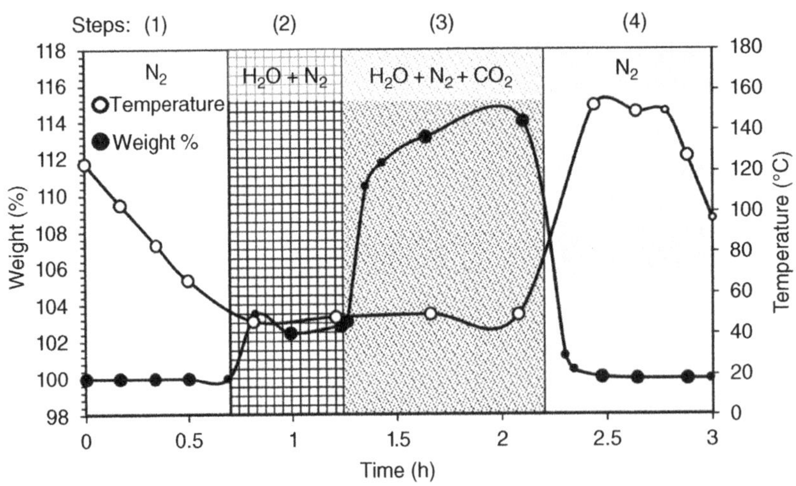

Figure 5.15 Thermogravimetric mass uptake of water, water with CO_2, with regeneration under N_2 at 150 °C. *Source:* Alesi and Kitchin 2012 [24]. Reproduced with permission of American Chemical Society.

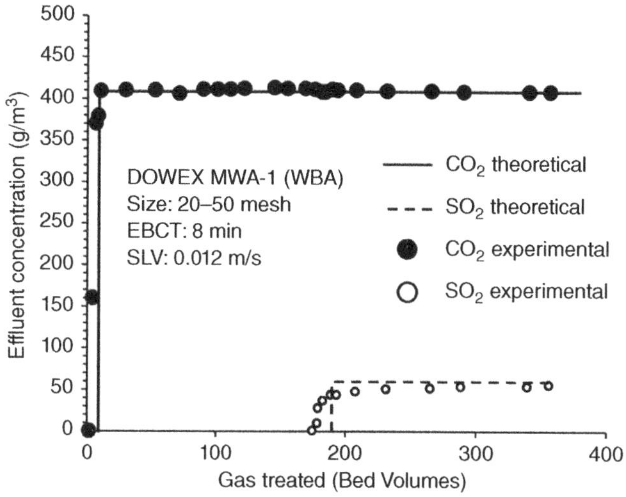

Figure 5.16 Comparison of experimental and predicted concentration histories for a mixture of SO_2 and CO_2. *Source:* Chen and Pinto 1991 [25]. Reproduced with permission of Elsevier.

the presence of 9.1 vol% H_2O during Step 3 when the mass of the resin rose about 10.5% by weight. From an energy consumption viewpoint, the WBA resin offers significant advantage over solvent extraction using monoethanolamine (MEA).

Between SO_2 and CO_2, SO_2 is significantly more acidic.

$$H_2SO_3 \leftrightarrow H^+ + HSO_3^-, \ pK_a = 1.81 \tag{5.43}$$

$$H_2CO_3 \leftrightarrow H^+ + HCO_3^-, \ pK_a = 6.3 \tag{5.44}$$

Thus, SO_2 offers significantly higher sorption capacity than CO_2. Preferential SO_2 sorption by WBA resin in the presence of large excess of CO_2 has the potential of being an effective flue gas desulfurization (FGD) process. Figure 5.16 shows the breakthrough curves for the mixture of SO_2 and CO_2, of which the feed concentrations are 56.8 and 409.2 g/m^3, respectively [25].

The breakthrough time for SO_2 is nearly two orders of magnitude greater than CO_2. Equilibrium models for ideal plug flow reactors are adequate to predict the effluent histories of CO_2 and SO_2 [26].

5.3.3 Effect of Ion Exchanger Morphology

Gel versus Macroporous

It was well known that amine species (e.g., ethylamine) in aqueous solution are sorbed strongly onto strong-acid ion exchangers in H^+ form in accordance with the following reaction:

$$\overline{RSO_3^-H^+} + NH_2 - R^1 \leftrightarrow \overline{RSO_3^-NH_3^+ - R^1} \tag{5.45}$$

where $NH_2 - R^1$ denotes the amine and the right-hand side (RHS) represents the amine-resin complex. Yoshida and Ruthven [27] determined equilibrium isotherms

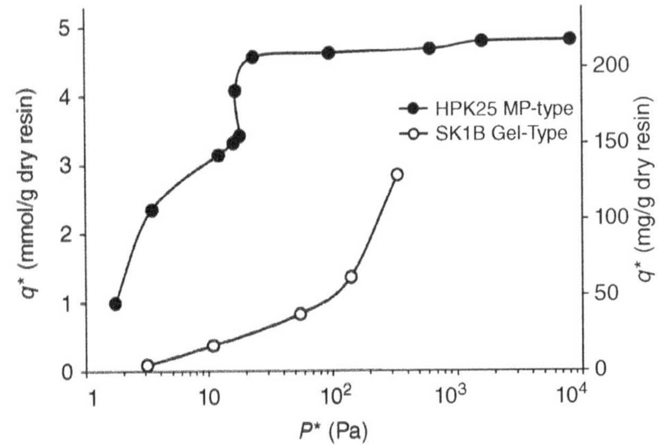

Figure 5.17 Equilibrium isotherms of ethylamine for MP-type and gel-type H-form resins.*Source:* Yoshida and Ruthven 1989 [27]. Reproduced with permission of John Wiley & Sons.

and kinetic data for sorption of gaseous ethylamine on two different H^+ form strong-acid exchangers: a homogeneous DIAION SK1B (gel-type) and a macroporous type DIAION HPK25 (MP). Gaseous amines and ammonia are generated in various chemical processes, including the manufacture of cellophane, rayon, and paper, as well as sewage disposal. Because of their strong smell, it is important to remove these compounds from the waste gases, and this has been traditionally accomplished by acid water washing. The possibility of a sorption process would, however, have the potential advantage of both removing and recovering amines in concentrated form.

Figure 5.17 shows the experimental equilibrium isotherms at 65 °C for the two above mentioned resins in H^+ form. It is evident that the equilibrium for the MP-type resin (HPK25) is very favorable, and, under most practical conditions, the isotherm may be considered as rectangular. Chemisorption is postulated to be the primary binding mechanism in accordance with the reaction between the H^+ form resin (R-H) and the amine gas (R^1-NH_2) as presented in Eq. (5.45). In comparison, adsorption on the gel-type resin (DIAION HPK25) is much weaker.

Yoshida and Kataoka [28,29] investigated sorption of several amines from both aqueous solution and vapor onto H^+ form cation exchange resins. For aqueous solution, both gel- and macroporous-type resins exhibited high uptake but for vapor phase, gel-type performed very poorly. A possible scientific explanation for this difference was provided as follows by Yoshida and Kataoka [27]:

> *The strong base anion exchange resin particles swell in an aqueous solution and shrink in the gaseous system due to reduced hydration. The degree of shrinkage caused by dehydration of the gel-type resin is much higher than for the macroporous-type resin. Because of the shrinkage in the gel-type resin, a relatively large fraction of bound H^+ ions may be inaccessible to incoming amine molecules, thus reducing the energy of interaction between the amine and H^+ ions. Macroporous-type resins shrink significantly less than gel-type and the pores around the fixed H^+ ions may remain open and accessible to amine molecules.*

The foregoing explanation is at best speculative and ignores the much slower intraparticle diffusion- controlled kinetics of the gel-type resin. Isotherms in Figure 5.17 were obtained after a two-day-long equilibration, absent adequate moisture. Note that the sorption capacity of the gel resin is far from reaching a plateau implying the possibility of very slow uptake as the underlying mechanism for the drastic difference in amine uptake between the macroporous and gel-type resins, all other conditions remaining identical. Figure 5.18 shows the experimental uptake rate data of ethylamine for the macroporous resin for different particle sizes.

The half time, $t_{1/2}$, corresponding to 50% uptake, was computed for different particle sizes and $t_{1/2}$ values were subsequently plotted against particle diameters in a log plot (Figure 5.19). Note that the slope is equal to, say, 2, that is, half-time is about square of the particle diameter, suggesting that the sorption rate is controlled by intraparticle

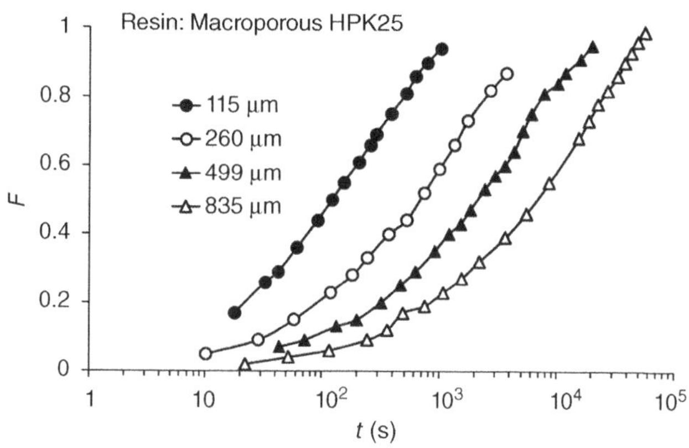

Figure 5.18 Experimental uptake curves (i.e., plot of fractional uptake, *F*, versus time), for ethylamine on macroporous resin (H^+-form) for different particle sizes. *Source:* Yoshida and Ruthven 1989 [27]. Reproduced with permission of John Wiley & Sons.

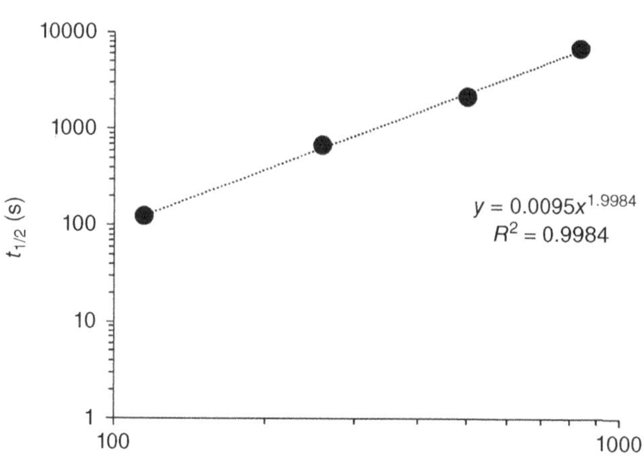

Figure 5.19 Half-time experimental uptake rate data of ethylamine for the macroporous resin for different particle sizes. *Source:* Yoshida and Ruthven 1989 [27]. Reproduced with permission of John Wiley & Sons.

diffusion. This conclusion was further supported by examination of the initial rate data, which demonstrates that, in the initial part, the uptake increases, say,linearly with the square root of time.

Ion Exchange Fibers

Besides gel- and macroporous-type ion exchange resins, ion exchange fibers have the attributes to be effective sorbents for removal of target gases. This is particularly so because sorption/desorption kinetics with fibers is often over an order of magnitude faster than the commercial resin beads. However, one misconception in the open literature about the genesis of fiber's faster kinetics is quite prevalent. The rate of sorption/desorption on fibrous ion exchangers is not faster due to larger specific surfaces. In fact, macroporous ion exchangers often have hundreds of square meters of surface area per gram compared to 1 m^2/g of ion exchange fibers. The underlying reason for fibers' fast kinetics is its significantly shorter intraparticle diffusion path length compared to resin beads. The subject has been adequately discussed in Chapter 4.

Currently, only a few tons of ion exchange fibers are produced annually in the world by several companies, mostly outside the USA. The materials are often formed from staple fibers with filaments of uniform 5–50 μm effective diameter and length of 30–80 mm. Figure 5.20a shows different ion exchange materials including fibers while (b) shows a typical illustration of deploying ion exchange fibers in tandem with existing ventilation or air filtration or flue gas treatment [30].

In Belarus, Soldatov and his associates have carried out extensive research using commercially available ion exchange fibers (Fiban) synthesized through chemical modification of polypropylene or polyacrylonitrile fibers [30]. These ion exchange fibers exhibit different water uptakes at varying relative humidity as shown in Figure 5.21.

For strong-acid ion exchange fibers in H^+ form, both bound and free water molecules are present within the ion exchanger. That is why ammonia removal by Fiban K-1 (strong-acid sulfonic acid functional group covalently attached onto polypropylene fibers) is very good, rapid and independent of relative humidity as shown in Figure 5.22a. In contrast, ammonia sorption onto weak-acid Fiban K-4 (carboxylic acid functional group in H^- form) is strongly dependent on gas-phase relative humidity as shown in Figure 5.22b; the higher the humidity, the higher is the ammonia sorption. Both free and bound water molecules are essentially absent in weak-acid ion exchange fibers and that is postulated to be the primary reason for an increased ammonia uptake with an increase in relative humidity.

5.3.4 Redox Active Gases: Hydrogen Sulfide and Oxygen

Hydrogen sulfide is an extremely odorous and toxic gas even at the parts per billion level. Air purification through H_2S removal is warranted in the working environment of several industries including sewage pipe lines, natural gas extraction and processing, animal skin processing, cattle farming and chemical manufacturing. Hydrogen sulfide

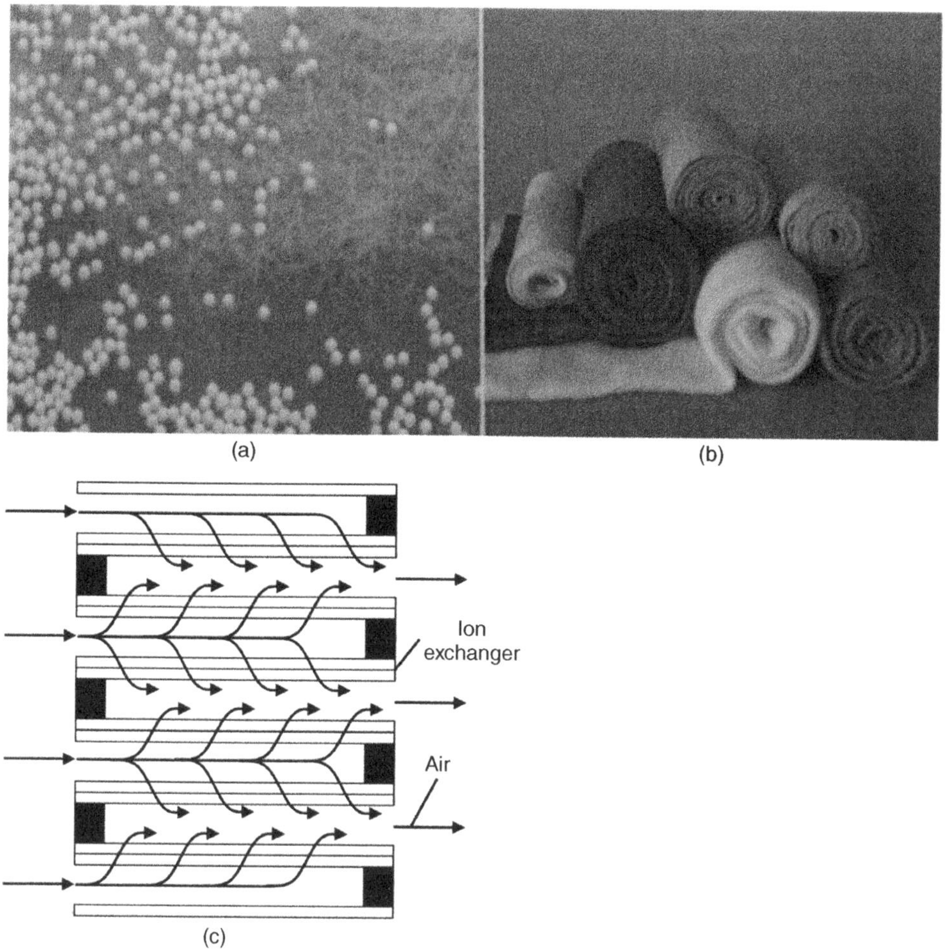

Figure 5.20 Ion exchangers in different physical forms. (a) Granular and fibrous forms (diameter of granular ion exchanger = 0.5 mm). (b) Nonwoven needle punctured Fiban material (A-top). (c) Placing ion exchange canvas in filtering chamber of frame type (RIF) ion exchange filter, top view (B-bottom). *Source:* Soldatov and Kosandrovich 2011 [30]. Reproduced with permission of Taylor & Francis.

is a weak-acid gas and previous studies reported sorption of H_2S by OH^- forms of anion exchangers [31–57]. This approach is, however, not useful for practical air purification because of formidable competition with carbon dioxide, whose concentration in ambient air is about 300 mg/m^3 and greatly exceeds the allowable concentration of H_2S. Further, H_2S is a weaker acid than H_2CO_3.

To overcome the foregoing shortcomings, Soldatov and coworkers investigated the catalytic conversion of H_2S to elemental sulfur using ion exchange fibers as carriers of the catalyst. The Fe(III)–ethylenediamine tetraacetate (EDTA) complex is an efficient catalyst of this process. Experimental results demonstrated removal of H_2S from air via catalytic oxidation of H_2S to elemental sulfur by atmospheric oxygen in the presence

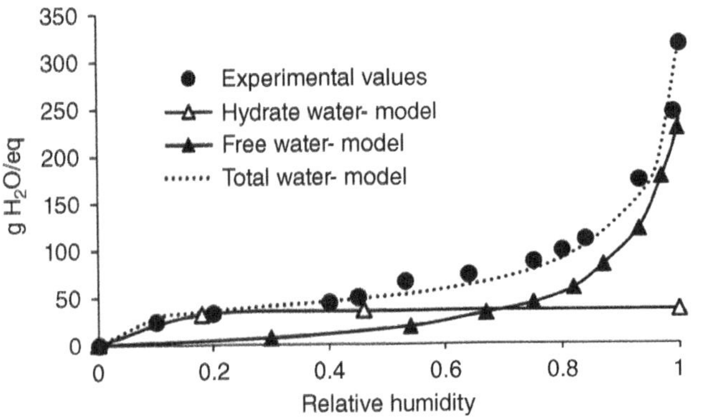

Figure 5.21 Isobaric curves of ion exchanger Fiban K-1 (H^+-form). Experimental values and modeling values. *Source:* Soldatov and Kosandrovich 2011 [30]. Reproduced with permission of Taylor & Francis.

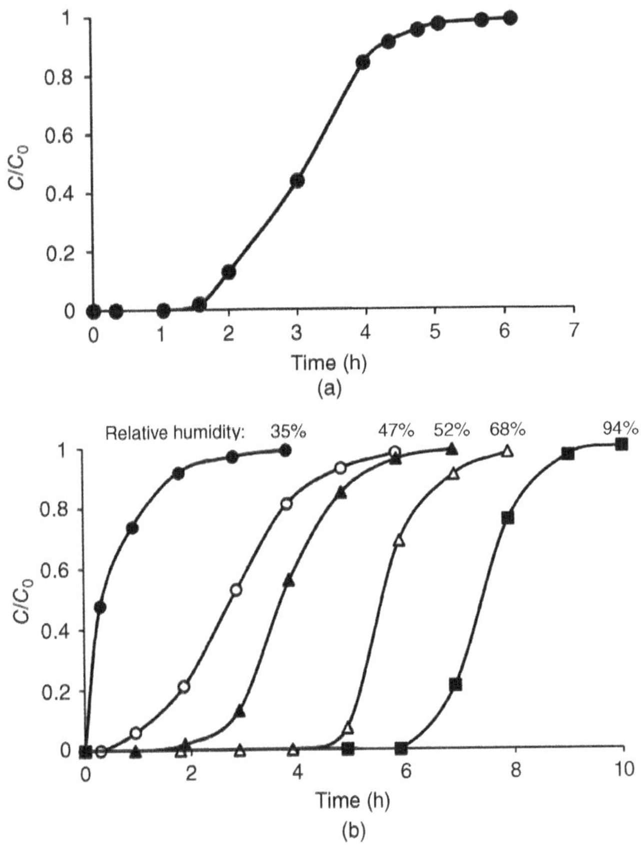

Figure 5.22 (a) Breakthrough and sorption curve of ammonia on strong-acid cation exchanger (sulfonic acid functionality) Fiban K-1 (H^+-form) at relative humidity of 7.5–85%. $T = 25\,°C$, $v = 0.09$ m/s, $[NH_3] = 17$ mg/m^3, thickness of filtering layer = 3 mm (top); (b) Breakthrough and sorption curves of ammonia on carboxylic weak-acid cation exchanger Fiban K-4 (H^+-form) at various relative air humidity levels. $T = 25\,°C$, $v = 0.08$ m/s, $[NH_3] = 18$ mg/m^3, thickness of filtering layer = 6 mm (B-bottom). *Source:* Soldatov and Kosandrovich 2011 [30]. Reproduced with permission of Taylor & Francis.

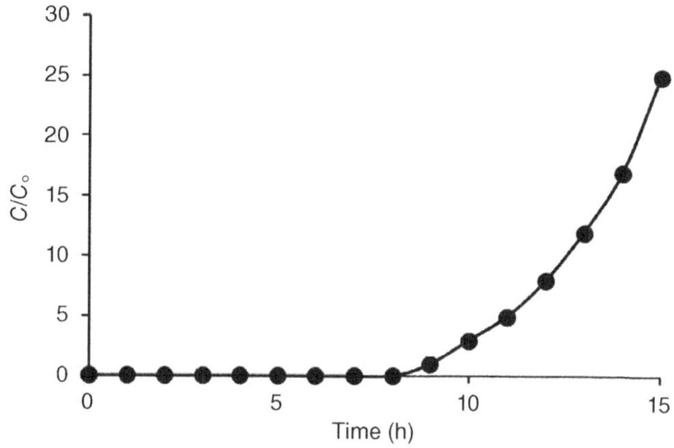

Figure 5.23 Breakthrough curve of H_2S on fibrous catalyst (Fiban AK-22 parent material). $C_0(H_2S) = 60\ mg/m^3$; thickness of single catalyst layer = 3 mm; catalyst contained 0.18 mmol Fe/g. *Source:* Soldatov and Kosandrovich 2011 [30]. Reproduced with permission of Taylor & Francis.

of Fe(III)–EDTA complex [58]. Specific steps of the process in the exchanger phase are as follows:

$$H_2S + OH^- \leftrightarrow HS^- + H_2O \tag{5.46}$$

$$2Fe^{3+} + EDTA + HS^- \leftrightarrow S^0 + H^+ + 2Fe^{2+} + EDTA \tag{5.47}$$

$$2Fe^{2+} + EDTA + H_2O + {}^1\!/_2O_2 \leftrightarrow 2Fe^{3+} + EDTA + 2OH^- \tag{5.48}$$

Figure 5.23 presents the results of H_2S removal from air by fibrous catalyst at 50% relative humidity. As the need for domestic and industrial air purification receives stricter scrutiny regarding contaminating gases, ion exchangers may find increased application opportunities due to its small footprint, operational simplicity and no time lag for start-up and shutdown.

Removal of trace amounts of mercury (Hg) from the flue gases of coal-fired boilers has lately surfaced as a major environmental challenge [59]. Elemental mercury (Hg^0) is significantly present in the flue gas: it is volatile and not amenable to removal by adsorption or scrubbing. Mercury vapor can be removed from flue gas or air by chelating ion exchange fibers dispersed with MnO_2(s) nanoparticles. The removal process involves oxidation of Hg^0 to Hg^{2+} followed by selective binding onto fibers containing iminodiacetate functional groups in accordance with the following two consecutive reactions:

$$Hg^0 + MnO_2(s) + 4H^+ \rightarrow Hg^{2+} + Mn^{2+} + 2H_2O \tag{5.49}$$

$$\begin{aligned} &4\overline{(RNCH_2COO^-)Na^+} + Hg^{2+} + Mn^{2+} \\ &\quad \rightarrow \overline{(RNCH_2COO^-)_2Hg^{2+}} + \overline{(RNCH_2COO^-)_2Mn^{2+}} + 4Na^+ \end{aligned} \tag{5.50}$$

Note that both Hg^{2+} and Mn^{2+} are retained within the hybrid fiber.

5.4 CO_2 Gas as a Regenerant for IX Softening Processes: A Case Study

Sequestration of carbon dioxide, especially from the exhaust of the chemical industries and electric power utilities, has gained special significance in the wake of escalating efforts to curb global warming caused by carbon dioxide emission. Carbon dioxide is a weakly acidic gas and thus has the potential to replace acids as regenerants, especially for weak-acid cation exchange resins. Figure 5.24 provides a general schematic illustrating the possible use of CO_2 rich stack gas as a potential regenerant. Note that besides 12–17% CO_2, the stack gas mostly contains nitrogen, oxygen and water vapor. Earlier, Figure 5.15 presented an engineered process to capture carbon dioxide from flue gases and desorb it with nitrogen.

The concept of using CO_2 as a substitute for strong acids is nothing new, but the efficiency of regeneration has been rather poor [61]. Removal of hardness (i.e., calcium or magnesium ions) from water is a universally practiced application called softening. Current practice includes use of strong-acid cation exchange resin and regeneration with concentrated brine (e.g., 10–12% NaCl). Disposal of spent regenerant with high concentration of NaCl is being increasingly disallowed and discouraged in many urban and arid areas due to its adverse environmental impact. Use of weak-acid cation exchange resins and possible regeneration with carbon dioxide has been proposed as a remedy to the situation. Three important steps of carbon dioxide regeneration for calcium- loaded weak-acid resins are:

$$2CO_2(g) + 2H_2O \Leftrightarrow 2H_2CO_3\,(aq) \tag{5.51}$$

$$2H_2CO_3 \Leftrightarrow 2H^+ + 2HCO_3^- \tag{5.52}$$

$$\overline{(R-COO)_2Ca} + 2H^+ \Leftrightarrow 2\overline{R-COOH} + Ca^{2+} \tag{5.53}$$

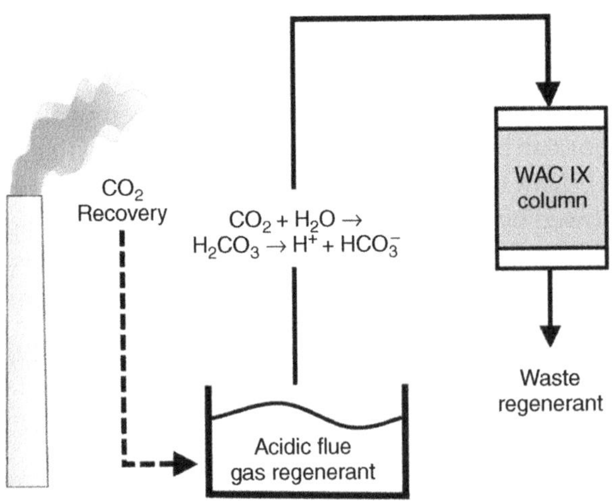

Figure 5.24 A schematic illustrating the possible use of CO_2-rich stack gas as a regenerant. *Source:* Greenleaf and SenGupta 2009 [60]. Reproduced with permission of American Society of Civil Engineers.

Overall,

$$\overline{(R-COO)_2Ca} + 2CO_2(g) + 2H_2O \Leftrightarrow 2\overline{R-COOH} + Ca^{2+} + 2HCO_3^- \quad (5.54)$$

$$K_{overall} = \frac{[\overline{R-COOH}]^2[Ca^{2+}][HCO_3^-]^2}{[\overline{(R-COO)_2Ca}]P_{CO_2}^2} \quad (5.55)$$

By combining equilibrium constants for Eqs (5.51)–(5.53) (i.e., Henry's constant for carbon dioxide, first acid dissociation constant of carbonic acid and the selectivity coefficient of hydrogen ion over calcium), the overall equilibrium constant ($K_{overall}$) may be written as:

$$K_{overall} = (K_H)^2(K_{a1})^2K_{IX} \quad (5.56)$$

in which K_H = Henry's constant for CO_2 dissolution in water; K_{a1} = first acid dissociation constant of carbonic acid, and K_{IX} = equilibrium constant of ion exchange between H^+ and Ca^{2+}. Considering ideality and rearranging:

$$[Ca^{2+}][HCO_3^-]^2 = (K_H)^2(K_{a1})^2K_{IX}\frac{[\overline{(R-COO)_2Ca}]}{[\overline{R-COOH}]^2}(P_{CO_2}^2) \quad (5.57)$$

From electroneutrality,

$$2[Ca^{2+}] \approx [HCO_3^-] \quad (5.58)$$

For any given percentage of regeneration efficiency, the calcium and the hydrogen loading of the ion exchange fibers are constant. With this condition and the equality of Eq. (5.58), the Eq. (5.57) simplifies into:

$$[Ca^{2+}] = \text{Constant}(P_{CO_2})^{2/3} \quad (5.59)$$

An increase in carbon dioxide partial pressure enhances regeneration efficiency resulting in higher calcium concentration in the eluent. An experimental regeneration study was carried out for both weak-acid ion-exchange resins and fibers under otherwise identical conditions as shown in Figure 5.25.

Besides difference in morphology between spherical beads and thin cylindrical fibers, the two materials had similar chemical makeup, that is, weak-acid carboxylate functional groups covalently attached to a polymer substrate as presented in Table 5.3. Strictly from an equilibrium consideration as represented by Eq. (5.59), the efficiency of regeneration should remain identical for both spherical beads and cylindrical fibers.

Figure 5.26 presents a comparison of two fixed-bed column runs under identical conditions; one run used the weak-acid cation exchange resin (C-104, Purolite Co.), while the other used Fiban K-4 weak-acid IX fiber. The influent composition and hydrodynamic conditions (i.e., EBCT and SLV) were the same as provided in Figure 5.26. Although the overall calcium removal capacity was greater for the spherical resin beads, the calcium breakthrough from the IX-fiber column was much sharper. The 10% of influent calcium breakthrough occurred at less than 3000 mL of solution throughput for C-104 resin. In comparison, with the IX-fiber column, the same breakthrough occurred at nearly 4000 mL. Also, the 24-h interruption test results, shown in the inset, demonstrate that the intraparticle diffusion is more

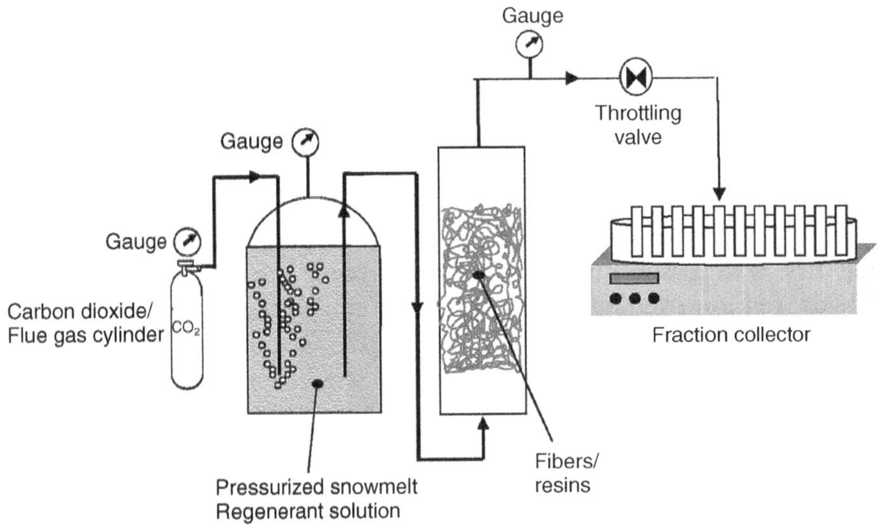

Figure 5.25 Laboratory setup depicting the CO_2-sparged snowmelt system used in the regeneration of both fiber and resin ion-exchange materials. *Source:* Greenleaf and SenGupta 2006 [62]. Reproduced with permission of American Chemical Society.

Table 5.3 Salient properties of weak-acid ion exchange fiber and weak-acid ion exchange resin beads.

Description	Weak-acid ion exchange fiber (Fiban K-4)	Weak-acid ion exchange resin beads
Diameter	10–50 μm	500–1200 μm
Physical shape	Cylindrical	Spherical
Functionality	Carboxylate (COO^-)	Carboxylate (COO^-)
Capacity (air-dried)	4–5 meq/g	5–8 meq/g
Equipment configuration	Fixed-bed	Fixed-bed

Source: Greenleaf and SenGupta 2006 [62]. Reproduced with permission of American Chemical Society.

pronounced with larger ion exchange resin beads as evidenced from higher drop in calcium concentration right after restart.

Carbon dioxide regeneration results for both fibers and resin beads are presented in Figure 5.27a and b, respectively, at different partial pressures. One of the striking findings is that while ion exchange fibers exhibited high regeneration efficiency and that again increased with an increase in CO_2 partial pressure, weak-acid spherical resin beads performed poorly and was not amenable to efficient regeneration even at high partial pressures of carbon dioxide. Figure 5.28 provides bar charts showing a comparison of percentage recoveries of calcium between IX-fibers and weak-acid cation resin beads. Even at high CO_2 partial pressure, for example, 6.8 atm, calcium recovery from resin beads is less than 10% while it is over 90% for IX-fibers.

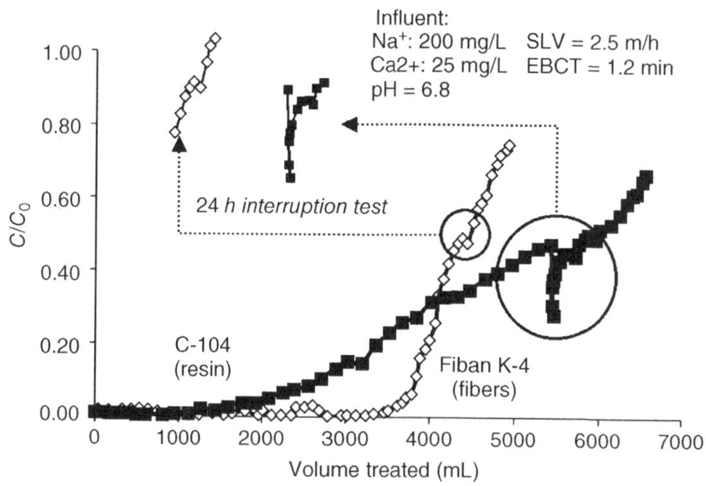

Figure 5.26 Fixed-bed column runs for hardness removal using two different ion-exchange materials under otherwise identical conditions: (EBCT – empty bed contact time; SLV – superficial liquid velocity). *Source:* Greenleaf and SenGupta 2006 [62]. Reproduced with permission of American Chemical Society.

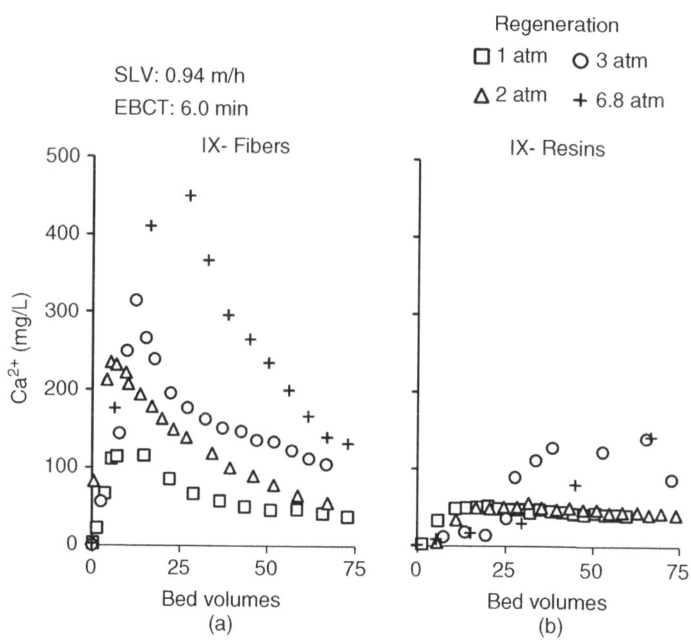

Figure 5.27 Effluent calcium concentration profiles for (a) IX-fibers and (b) resins during CO_2-sparged snowmelt regeneration at different carbon dioxide partial pressures. *Source:* Greenleaf and SenGupta 2006 [62]. Reproduced with permission of American Chemical Society.

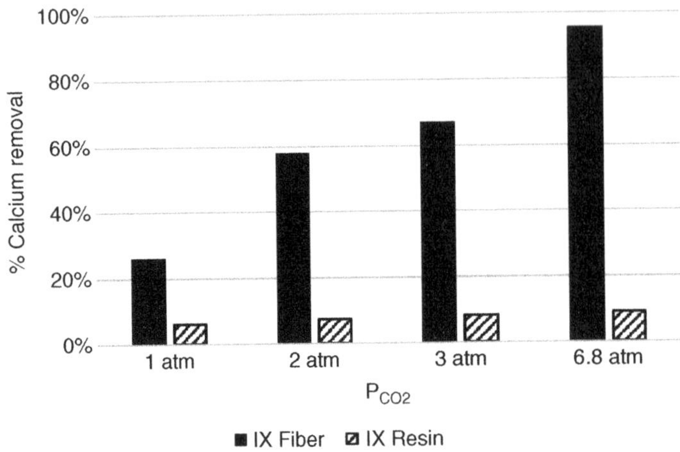

Figure 5.28 Effluent calcium concentration profiles for IX-fibers and IX resins during CO_2-sparged snowmelt regeneration at different carbon dioxide partial pressures. *Source:* Greenleaf and SenGupta 2006 [62]. Reproduced with permission of American Chemical Society.

To develop a mechanistic understanding of the poor regenerability of calcium-loaded spherical resin beads with carbon dioxide, let us consider a single bead as a cross-linked polyelectrolyte gel with carboxylate functional groups. The affinity sequence for weak-acid carboxylate functional groups stands as follows: $H^+ \gg Ca^{2+} > Mg^{2+} > Na^+$. Uptake of H^+ during regeneration by a weak-acid carboxylate group is essentially an association reaction leading to a major decrease in its osmotic pressure, thus causing expulsion of water from the gel phase. A spherical ion-exchange resin bead, therefore, gradually shrinks with the progress of regeneration through the uptake of hydrogen ions that involves counter transport of H^+ and Ca^{2+}. At the onset, hydrogen ions would initially displace the outermost (i.e., peripheral) calcium ions. Such an exchange would, however, dramatically decrease the water content of the regenerated portion, thus decreasing the effective intraparticle diffusivity near the outer periphery of the resin bead. The progress of the regeneration process increases the depth of the relatively impervious skin, thus further slowing down the counter-transport of H^+ and Ca^{2+}. Scientifically, this hypothesis is in agreement with the premise of the ion exchange kinetics accompanied by very favorable chemical reactions [63]. Previous studies with weak-acid cation-exchange resins also provided optical confirmation of shrunk periphery during acid regeneration [11]. For carbon dioxide regeneration, hydrogen ion concentration in the bulk liquid phase cannot be as high as it is normally with mineral acid regeneration. Thus, the concentration gradient across the shrunk periphery is too small to overcome the diffusional resistance. The poor regenerability of resin beads with carbon dioxide is thus attributed to enhanced diffusional resistance offered by the shrunk peripheral layers with very low water content.

In contrast, the ion-exchange sites for fibers reside primarily on the surface, and the phenomenon of intraparticle diffusion, as demonstrated earlier, is of less significance.

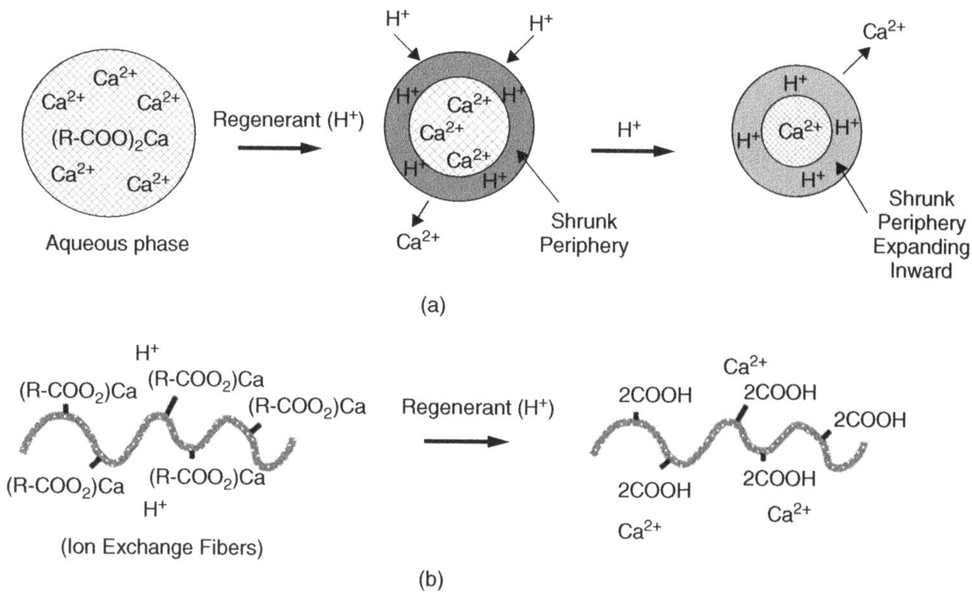

Figure 5.29 Schematic illustrating the difference in desorption mechanisms between (a) resin beads and (b) IX-fibers. *Source:* Greenleaf *et al.* 2006 [64]. Reproduced with permission of John Wiley & Sons.

Protonation of weak-acid functional groups has only marginal impact on diffusional resistance, and hence, the carbon dioxide regeneration is efficient for IX-fibers. Figure 5.29 provides a schematic illustrating the difference in desorption mechanism between resin beads and fibers. Although not produced on a large scale yet, IX-fibers demonstrate a unique kinetic advantage over widely used polymeric ion exchange resins for efficient regeneration with carbon dioxide. Many application opportunities are likely to emerge.

Summary

- Ion exchange processes may include solid and/or gas phases for specific separation goals, in addition to water and ion exchanging materials.
- Insoluble solid phases (i.e., $K_{sp} < 10^{-5}$) can be gently dissolved and separated with ion exchange resins, while avoiding the use of aggressive chemicals.
- Removal of solid phases involves sorption onto ion exchangers followed by gradual dissolution. Type and combination of ion exchange resins significantly influence the efficiency of dissolution and the removal of solid phases.
- Effective separation of a small amount of toxic metal precipitates from a background of large innocuous solids is possible with intelligent use of selective ion exchange processes.
- Buffer capacity and ion exchange property of the accompanying bulk solid phases are the two most significant challenges for solid-phase separation of toxic metal precipitates.

- By using the Donnan membrane principle, cation exchange membranes can selectively recover high purity alum from water treatment plant sludge.
- Acidic, basic and redox active gases can be removed through the choice of appropriate ion exchange materials.
- Efficiency of gas removal often depends on the degree of acidity or basicity of the gases. For example, due to its higher acidity, SO_2 can be removed in preference to CO_2 by using a WBA exchanger.
- Sorption rates of gases onto ion exchange resins are controlled by intraparticle diffusion; macroporous ion exchangers are more accessible and offer faster kinetics than gel-type ion exchangers.
- By virtue of its weak-acid properties, CO_2 can be a viable substitute for mineral acid regenerant for weak-acid cation exchange fibers with relatively short intraparticle diffusion-path length.

References

1 Tiger, H., Sussman, S., Lane, M., and Calise, V. (1946) Desalting sea water. *Industrial & Engineering Chemistry*, **38** (11), 1130–1137.
2 Glueck, A. (1968) Desalination by an ion exchange-precipitation-complex process. *Desalination*, **4** (1), 32–37.
3 Sengupta, S. and SenGupta, A.K. (1996) Solid phase heavy metal separation using composite ion-exchange membranes. *Hazardous Waste and Hazardous Materials*, **13** (2), 245–263.
4 Sengupta, S. and SenGupta, A.K. (2000) Decontamination of heavy-metal-laden sludges and soils using a new ion-exchange process. *Environmental Science and Pollution Control Series*, **23**, 541–572.
5 Ree, B.R., Errede, L.A., Jefson, G.B. and Langager, B.A. inventors (1979). Minnesota Mining and Manufacturing Company, assignee. Method of making polytetrafluoroethylene composite sheet. US 4,153,661 A. 1979 May 8.
6 Hagen, D.F., Mary, S.J.S., Errede, L.A. and Carr, P.W. inventors (1989). Minnesota Mining and Manufacturing Company, assignee. Composite chromatographic article. US4,810,381A. 1989 Mar 7.
7 Sengupta, S. (1993) A new separation and decontamination technique for heavy-metal-laden sludges using sorptive/desorptive ion-exchange membranes. PhD dissertation. Lehigh University.
8 Sengupta, S. and SenGupta, A.K. (1993) Characterizing a new class of sorptive/desorptive ion exchange membranes for decontamination of heavy-metal-laden sludges. *Environmental Science & Technology*, **27** (10), 2133–2140.
9 Sengupta, S. and SenGupta, A.K. (1997) Heavy-metal separation from sludge using chelating ion exchangers with nontraditional morphology. *Reactive and Functional Polymers*, **35** (1), 111–134.
10 Garcia, A.A. and King, C.J. (1989) The use of basic polymer sorbents for the recovery of acetic acid from dilute aqueous solution. *Industrial and Engineering Chemistry Research*, **28** (2), 204–212.

11 Höll, W. (1984) Optical verification of ion exchange mechanisms in weak electrolyte resins. *Reactive Polymers, Ion Exchangers, Sorbents,* **2** (1–2), 93–101.
12 Helfferich, F. (1965) Ion-exchange kinetics. V. Ion exchange accompanied by reactions. *The Journal of Physical Chemistry,* **69** (4), 1178–1187.
13 Sengupta, S. and SenGupta, A.K. (2001) Chelating ion-exchangers embedded in PTFE for decontamination of heavy-metal-laden sludges and soils. *Colloids and Surfaces A: Physicochemical and Engineering Aspects* , **191** (1), 79–95.
14 Sengupta, S. and Nawaz, T. (2016) Solid phase heavy metals separation with selective ion exchangers: two novel morphologies, in *Ion Exchange and Solvent Extraction: A Series of Advances,* vol. **22** (ed. A.K. SenGupta), CRC Press, Boca Raton, FL, pp. 99–146.
15 Driscoll, C.T., Baker, J.P., Bisogni, J.J., and Schofield, C.L. (1980) Effect of aluminium speciation on fish in dilute acidified waters. *Nature,* **284**, 161–164.
16 Lamb, D.S. and Bailey, G.C. (1981) Acute and chronic effects of alum to midge larva (Diptera: Chironomidae). *Bulletin of Environmental Contamination and Toxicology,* **27** (1), 59–67.
17 Cornwell, D.A. and Westerhoff, G.P. (1981) Management of water treatment plant sludges, in *Sludge and Its Ultimate Disposal Ann Arbor* (ed. J.A. Borchardt), Ann Arbor Science, MI, pp. 31–62.
18 Prakash, P. and SenGupta, A.K. (2003) Selective coagulant recovery from water treatment plant residuals using Donnan membrane process. *Environmental Science & Technology,* **37** (19), 4468–4474.
19 Saunders, M.F., Roeder, M.L., Rivers, R.B. and Magara, Y. (1991) Coagulant recovery: A critical assessment. Denver, CO: American Water Works Association Research Foundation (AWWARF); No. 1P-5C-90565-7/91-OM.
20 SenGupta, A.K. and Prakash, P. inventors (2002). SenGupta AK, assignee. Process for selective coagulant recovery from water treatment plant sludge. US patent 6,495,047 B1. 2002 December 17.
21 Prakash, P. and SenGupta, A.K. (2005) Modeling Al3/H ion transport in Donnan membrane process for coagulant recovery. *AIChE Journal,* **51** (1), 333–344.
22 Prakash, P., Hoskins, D., and SenGupta, A.K. (2004) Application of homogeneous and heterogeneous cation-exchange membranes in coagulant recovery from water treatment plant residuals using Donnan membrane process. *Journal of Membrane Science,* **237** (1), 131–144.
23 Boyd, G.E. and Soldano, B.A. (1953) Osmotic free energies of ion exchangers. *Zeitschrift für Elektrochemie, Berichte der Bunsengesellschaft für physikalische Chemie,* **57** (3), 162–172.
24 Alesi, W.R. Jr., and Kitchin, J.R. (2012) Evaluation of a primary amine-functionalized ion-exchange resin for CO2 capture. *Industrial and Engineering Chemistry Research,* **51** (19), 6907–6915.
25 Chen, T. and Pinto, N.G. (1991) Fixed-bed adsorption of acid gases on a macroreticular ion-exchange resin. *Reactive Polymers,* **14** (2), 151–168.
26 Yoshida, H. (1995) Chapter 8: Sorption of gaseous pollutants on ion exchangers, in *Ion Exchange Technology: Advances in Pollution Control Lancaster* (ed. A.K. SenGupta), Technomic Publishing Co., Inc., PA, pp. 315–352.

27 Yoshida, H. and Ruthven, D.M. (1989) Adsorption of gaseous ethylamine on H-form strong-acid ion exchangers. *AIChE Journal*, **35** (11), 1869–1875.
28 Yoshida, H. and Kataoka, T. (1987) Adsorption of amines and ammonia on H^+-form ion exchanger. *Chemical Engineering Science*, **42** (7), 1805–1814.
29 Yoshida, H. and Kataoka, T. (1986) Recovery of amine and ammonia by ion exchange method: comparison of ligand sorption and ion exchange accompanied by neutralization reaction. *Solvent Extraction and Ion Exchange*, **4** (6), 1171–1191.
30 Soldatov, V.S. and Kosandrovich, E.H. (2011) Chapter 2: Ion exchangers for air purification, in *Ion Exchange and Solvent Extraction: A Series of Advances*, vol. **20** (ed. A.K. SenGupta), CRC Press, Boca Raton, FL, pp. 45–116.
31 Ashirov, A. (1983) *Ion Exchange Purification of Wastewaters, Solutions and Gases*, vol. **295**, Khimiya, Leningrad.
32 Chikin, G.A. and Miagkoj, O.N. (1984) Ion exchangers in gas sorption technologies. *Ion Exchange Methods of Substance Purification* Voronezh University. p. 326.
33 Vulikh, A.I., Bogatyriov, V.A. and Aloviajnikov, A.A. (1970) Application of ion exchange resins for sorption and purification of gases. *Zhurnal Vsesoyuznogo khimicheskogo obschestva Mendeleeva*, **15** (4), 425–429.
34 Shramban, B.I. (1972) Investigation of the process of sorption of hydrogen fluoride from the gas phase by anion exchanger AV-17. MITCT Moscow.
35 Soldatov, V. (2008) Syntheses and the main properties of fiban fibrous ion exchangers. *Solvent Extraction and Ion Exchange*, **26** (5), 457–513.
36 Soldatov, V., Pawłowski, L., Shunkevich, A., and Wasąg, H. (2004) New materials and technologies for environmental engineering, Part I. Syntheses and structure of ion exchange fibres. *Monografie Komitetu Inżynierii Środowiska PAN*, **21**, 1–127.
37 Zagorodni, A.A. (2006) *Ion Exchange Materials: Properties and Applications*, Elsevier, Amsterdam, Netherlands.
38 Soldatov, V., Tsigankov, V., and Elinson, I. (1990) Sorption of water vapor by salt forms of fibrous anion exchanger FIBAN A-1. *Russian Journal of Applied Chemistry (Zhurnal prikladnoy khimii (in Russian))*, **63** (10), 2285–2291.
39 Crabb, C.R. and Mcdonald, L.S. inventors (1966). Dow Chemical Co, assignee. Desiccant regeneration. US3,275,549 A. 1966 Sep 27.
40 Radl, V. and Krejkar, E. (1962) Cation exchangers as drying agents for gases and liquids. *Chemicky Prumysl (Poll)*, **12** (10), 579–582.
41 Wymore, C. (1962) Sulfonic-type cation-exchange resins as desiccants. *Industrial & Engineering Chemistry Product Research and Development*, **1** (3), 173–178.
42 Shamilov, T., Zhirova, L., and Kadyrova, M. (1980) The gas drying by fibrous anion exchangers. *Chemistry and Industry (Khimicheskaya promyshlennost (in Russian))*, **3**, 181.
43 Miagkoj, O., Krutskikh, A. and Astakhova, E. (1981) The gas drying by fibrous anion exchangers. Application of ion exchange materials (Primenenie ionoobmennykh materialov (in Russian)). Voronezh University, Voronezh, 50.
44 Glueckauf, E. and Kitt, G.P. (1955) A theoretical treatment of cation exchangers. III. The hydration of cations in polystyrene sulphonates. *Proceedings of the Royal Society of London A: Mathematical, Physical and Engineering Sciences*, **228** (1174), 322–341.

45 Gregor, H.P., Sundheim, B.R., Held, K.M., and Waxman, M.H. (1952) Studies on ion-exchange resins. V. Water vapor sorption. *Journal of Colloid Science*, **7** (5), 511–534.

46 Gregor, H.P. and Frederick, M. (1953) Thermodynamic properties of ion exchange resins; free energy of swelling as related to ion selectivities. *Annals of the New York Academy of Sciences*, **57** (3), 87–104.

47 Van Koevelen, D. (1972) *Properties of Polymers—Correlations with Chemical Structure*, Khimia, p. 416.

48 Chalych, A. (1987) *Diffusion in Polymeric Systems*, Khimia, Moscow, p. 364.

49 Soldatov, V.S., Tsigankov, V., and Elinson, I. (1990) Quantitative description of water sorption by salts forms of fibrous anion exchanger FIBAN A-1. *Russ J Appl Chem (Zhurnal prikladnoy khimii (in Russian))*, **63** (10), 2291–2296.

50 Gantman, A. and Veshev, S. (1985) Water uptake and swelling of ion exchangers. *Russian Journal of Physical Chemistry (Zhurnal fizicheskoy khimii (in Russian))*, **59** (10), 2615–2618.

51 Sosinovich, Z., Hogfeldt, E., and Novitskaya, L. (1978) Investigation of the hydration of strong anion exchangers using the model of stepwise hydration. *Proceedings of the National Academy of Sciences Belarussian SSR (Doklady Akademii Nauk BSSR (in Russian))*, **22** (10), 920–923.

52 Kats, B., Kutarov, V., and Kutovaya, L. (1991) Kinetics of water-vapor sorption by anion-exchange fibers based on cellulose or polyacrylonitrile. *Journal of Applied Chemistry of the USSR*, **64** (8), 1568–1571.

53 Brunauer S. *Adsorption of Gases and Vapors*, vol. **1**. London: Princeton University Press; H. Milford, Oxford University Press; 1943.

54 Jovanović, D. (1969) Physical adsorption of gases. *Kolloid-Zeitschrift und Zeitschrift für Polymere*, **235** (1), 1214–1225.

55 White, H.J. and Eyring, H. (1947) The adsorption of water by swelling high polymeric materials. *Textile Research Journal*, **17** (10), 523–553.

56 Sosinovich, Z.I., Novitskaya, L.V., Soldatov, V.S., and Hoegfeld, T. (1985) Thermodynamics of water sorption on Downex 1 of different crosslinking and ionic form. *Ion Exchange and Solvent Extraction*, **9**, 303.

57 Pollio, F. and Kunin, R. (1968) Macroreticular Ion Exchange Resins as H2S Sorbents. *Industrial & Engineering Chemistry Product Research and Development*, 7 (1), 62–65.

58 Potapova, L., Yegiazarov, Y., and Soldatov, V. (1998) Oxidation of hydrogen sulfide on fibrous anion exchanger Fiban A-1 with coordination saturated complex Fe-EDTA. *Doklady Akademii Nauk Belarusi*, **42**, 54.

59 Pavlish, J.H., Sondreal, E.A., Mann, M.D. *et al.* (2003) Status review of mercury control options for coal-fired power plants. *Fuel Processing Technology*, **82** (2), 89–165.

60 Greenleaf, J.E. and SenGupta, A.K. (2009) Flue gas carbon dioxide sequestration during water softening with ion-exchange fibers. *Journal of Environmental Engineering*, **135** (6), 386–396.

61 Kunin, R. and Myers, R.J. (1950) *Ion Exchange Resins*, John Wiley & Sons, Inc., New York.

62 Greenleaf, J.E. and SenGupta, A.K. (2006) Environmentally benign hardness removal using ion-exchange fibers and snowmelt. *Environmental Science & Technology*, **40** (1), 370–376.
63 Höll, W. and Sontheimer, H. (1977) Kinetics of the protonation of weak acid ion exchange resins. *Chemical Engineering Science*, **32** (7), 755–762.
64 Greenleaf, J.E., Lin, J., and Sengupta, A.K. (2006) Two novel applications of ion exchange fibers: arsenic removal and chemical-free softening of hard water. *Environmental Progress*, **25** (4), 300–311.

6

Hybrid Ion Exchange Nanotechnology (HIX-Nanotech)

From a physical–chemical viewpoint, every polymeric ion exchange resin bead is essentially a cross-linked polyelectrolyte that is insoluble in water. Today, there remains tens of ion exchange resin manufacturers around the world with hundreds of different products to serve a wide variety of applications. As different as they may appear, these ion exchangers can be well defined and the specific interactions can be well characterized by five independent composition variables, namely, matrix, functionality, cross-linking, pore structure, and physical configuration. In addition to the diverse groups of ion exchangers characterized by these five composition variables, a new class of polymeric–inorganic ion exchangers has been synthesized. Here, a dispersed phase of metal nanoparticles (MNPs) or metal oxide nanoparticles (MONPs) is embedded in the ion exchanger phase; together they offer a synergy not available otherwise. The resulting new materials are heterogeneous, even at the sub-10 nm scale, and termed "hybrid ion exchangers" or HIX. Figure 6.1a illustrates the fundamental composition variables of a typical spherical ion exchanger bead, namely, functional groups, matrix, crosslinking, and pore structure. Figure 6.1b represents HIX or hybrid ion exchanger where zirconium oxide nanoparticles have been dispersed primarily in the gel phase of the ion exchanger (scanning electron microphotograph or SEM and tunneling electron microphotograph or TEM are included). Thus, the HIX has two distinctive sites: ion exchange functional groups and ZrO_2 surfaces.

Due to their extremely high surface area-to-volume ratio, nanoparticles of many metals and metal oxides offer fast kinetics and enhanced sorption capacity for many reactions of environmental significance. For example, (i) hydrated Fe(III) oxides or HFO particles can selectively sorb dissolved heavy metals, for example, zinc and copper, or metalloids, for example, arsenic oxyacids/oxyanions; (ii) Mn(IV) oxides are fairly strong solid-phase oxidizing agents; (iii) magnetite (Fe_3O_4) crystals are capable of imparting magnetic activity; (iv) elemental Zn^0 or Fe^0 are excellent reducing agents for both inorganic and organic contaminants [1–10]. Figure 6.2 depicts the properties of several MNPs or MONPs [11]. The synthesis of these nanoparticles and their aggregates is environmentally safe, operationally simple and inexpensive. However, the nanoparticles are not suitable for direct application in fixed-bed columns, reactive barriers or any flow-through processes due to poor durability and excessive pressure drop across their path. On the contrary, many commercially available porous polymeric beads are very durable, have excellent hydraulic properties and have low

Ion Exchange in Environmental Processes: Fundamentals, Applications and Sustainable Technology, First Edition. Arup K. SenGupta.

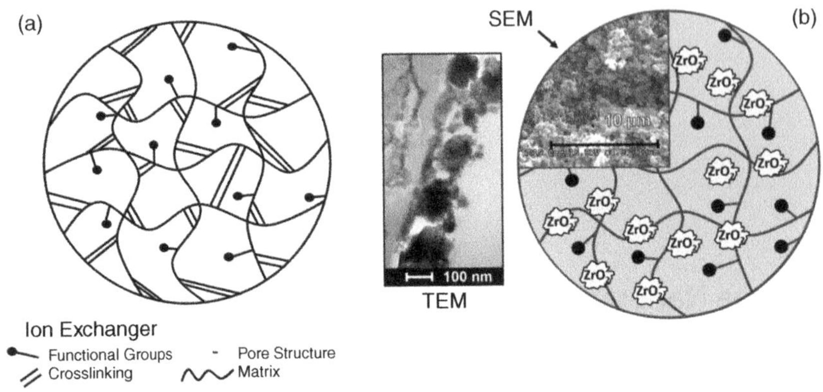

Figure 6.1 (a) A spherical ion exchanger bead with conventional composition variables; (b) A hybrid ion exchanger (HIX) bead with ZrO_2 nanoparticles dispersed within the gel phase. *Source:* Sarkar *et al.* 2011 [1]. Reproduced with permission of Taylor & Francis.

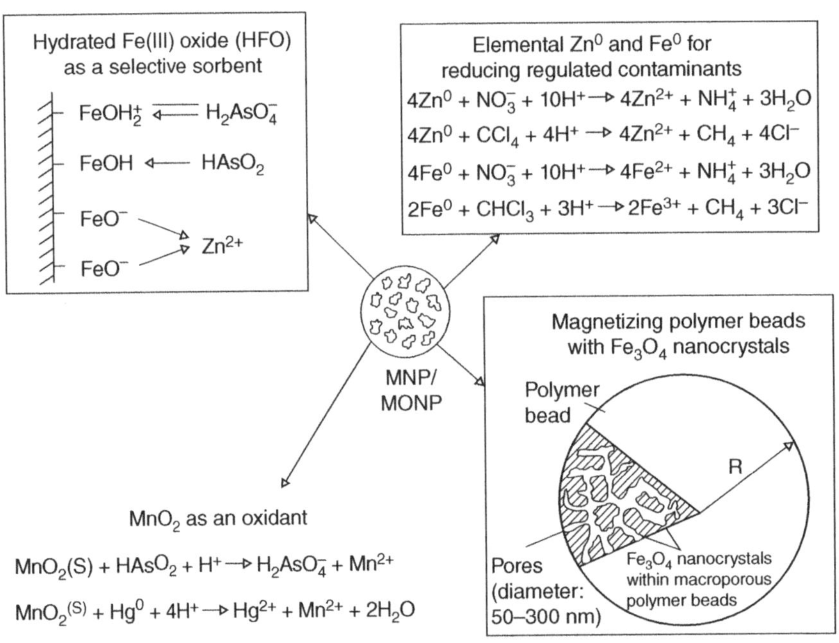

Figure 6.2 Favorable properties of some metal and metal oxide nanoparticles. *Source:* Cumbal *et al.* 2003 [11]. Reproduced with permission of Elsevier.

pressure drop in fixed-bed columns. Conceptually, it is worth developing a new class of hybrid polymeric inorganic materials that combines the excellent hydraulic characteristics of spherical polymer beads with favorable sorption, redox and/or magnetic properties of inorganic nanoparticles. In such a material, the host (i.e., polymer beads) improves the hydraulic permeability in the flow-through systems with no apparent influence on the behavior of the MNPs and MONPs. This chapter mostly emphasizes how the choice of the functional groups of the polymeric host materials can be harnessed to alter (enhance or diminish) the intrinsic properties of the nanomaterials.

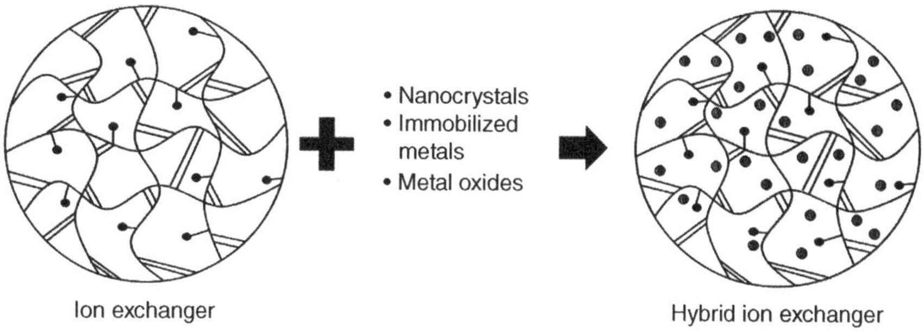

Figure 6.3 A schematic of a hybrid ion exchanger. *Source:* Sarkar *et al.* 2011 [1]. Reproduced with permission of Taylor & Francis.

An HIX essentially contains two phases: (i) functionalized polymeric host, that is, ion exchanger; and (ii) metal or MONPs dispersed within the polymer phase. Although the nanoparticles can be incorporated within the polymeric phase during the pre-polymerization steps, it is often difficult to control and retain the desired properties of the inorganic nanoparticles in the final product [12–17]. Moreover, hybrid materials prepared in this way, for example, magnetic polymers, tend to be overly expensive due to the proprietary nature of their manufacturing processes. In this chapter, we shall denote only those materials as HIXs for which the inorganic nanoparticles have been incorporated inside the polymeric ion exchanger phase in the post-polymerization stage. Figure 6.3 is a schematic diagram representing the generic composition of HIXs.

The primary focus of the chapter pertains to preparation, characterization and application of the following three classes of HIXs:

(i) Magnetically active polymeric particles (MAPPs);
(ii) HIX-NanoFe: Containing HFO nanoparticles or for selective removal of ligands; and,
(iii) HIX-NanoZr: Containing zirconium oxide nanoparticles for concurrent defluoridation and desalination.

For the second and the third class of HIXs, the synergistic role of the Donnan membrane principle needs special recognition and will be discussed.

6.1 Magnetically Active Polymer Particles (MAPPs)

Magnetic polymers have long been pursued by biomedical, electronics, and materials science professionals for their unique properties and their potential to be incorporated in novel processes, for example, protein and biomolecule separation, water and wastewater treatment, color imaging and information storage. In many cases of biological importance, magnetic polymers are applied as carriers for cells and biomolecules, for example, nucleic acids and proteins. In these applications, magnetic polymers offer the advantages of being easily manipulated, automated and/or miniaturized. Also, fast and cost-efficient separation of the magnetic carriers from a biological mixture without filtration or centrifugation makes magnetic polymers particularly attractive.

In the field of environmental separation and control of contaminants, polymeric sorbent materials with specific affinities for heavy metals, metalloids, (in)organic ligands, chlorophenols and pesticides are extensively used for remediation purposes [18–24]. In environmental forensics, the origin of any particular pollutant of interest in natural waters can be tracked down using highly selective polymeric adsorbents. Despite their excellent sorption properties for targeted contaminants, it often becomes impractical to use polymeric sorbents in complex matrices because of difficulties in retrieving the saturated sorbents from the matrix. The problem can be overcome if the non-magnetic (i.e., diamagnetic) polymeric sorbents are imparted with magnetic activity. Then, by applying a magnetic field, these sorbents can be recovered from complex matrices. When magnetized, the polymeric sorbent materials can be very effectively used to sequester target contaminants and can easily be recovered from complex environmental matrices, such as slurries (high suspended solids content), high viscosity liquids, radioactive liquids or a medium with high concentrations of biomass. Thus, introduction of magnetic activity into a wide array of commercially available or custom tailored polymeric materials can make them fit for application in complex environmental systems which are not possible otherwise.

MAPPs of different morphologies can be prepared to satisfy specific needs. MAPPs can have either a superparamagnetic core embedded inside a polymeric shell or a magnetic material heterogeneously dispersed within the polymeric sorbent. For the core–shell type of MAPP, magnetic core materials are encapsulated by a polymer coating which is separately applied to the magnetic core by either phase inversion or solvent evaporation [25–28]. Morphological properties of these magnetic polymers, such as particle size and particle size distribution, depend on the nature of the polymer and the polymer coating method. When monomers and magnetic particles are mixed together and the monomer is polymerized using different polymerization techniques, the resultant magnetic polymeric material has evenly distributed magnetic material within its polymeric matrix [29–32]. The preparation of these two types of magnetically active polymers are proprietary in nature and the products are overly expensive. Besides, because of the rigorous reactions involved in the polymerization or polymer deposition stages, there is very little flexibility in developing a wide variety of magnetically active polymers with specific sorption affinity for different types of environmental contaminants.

For diverse environmental application purposes, MAPPs should be prepared following in situ methods, so that the following can occur: (i) the magnetization process is universally applicable for a wide range of reusable polymeric sorbent particles; (ii) the imparted magnetic activity, or the process of magnetization, does not interfere with the sorption properties (equilibrium or kinetics); (iii) MAPPs retain their magnetic activity over many cycles of operation. In one in situ method, preformed magnetite nanocrystals are deposited inside the polymer-phase through swelling of the polymer, followed by intraparticle diffusion of the magnetite and adhesion within the polymer. Micron-size polystyrene (PS) particles were swollen in an aqueous solution of *N*-methyl-2-pyrrolidone (NMP) and then mixed with superparamagnetic iron oxide nanoparticles [33]. The magnetic nanoparticles then diffused into and were entrapped within the polymer microspheres. Challenges associated with the process are related to proper solvent use: the polymeric support tends to dissolve and be lost within the solution.

Of all the ferromagnetic materials, magnetite is environmentally benign, inexpensive and chemically stable once formed. The magnetic activity of a polymeric sorbent can be greatly enhanced by irreversibly dispersing nanoscale magnetite particles within the polymeric phase of the ion exchanger. The challenge in magnetization of polymeric beads lies in controlling the process conditions so that the formation of Fe_3O_4(s) is preferred to nonmagnetic $Fe(OH)_3$(s) and $Fe(OH)_2$(s). Under a reducing environment (i.e., absence of oxygen), $Fe(OH)_2$(s) is the predominant solid phase, while under a moderate to highly oxidizing environment, $Fe(OH)_3$(s) predominates. The environmental conditions under which magnetite crystals grow, are very sensitive to pH and redox conditions. There is a very small sliver of their pe–pH diagram in which the ferromagnetic material is preferably formed between the conditions for non-magnetic hydrated Fe(II) or Fe(III) oxides [34]; hence it is a challenging task. Figure 6.4 shows the stability or predominance diagram at 25 °C (E_h–pH) depicting the pertinent solid phases. Considering oxygen to be the sole electron acceptor, the solid-phase transition between Fe_3O_4(s) and $Fe(OH)_3$(s) can be presented as follows:

$$4Fe_3O_4 + O_2 + 18H_2O \rightarrow 12Fe(OH)_3 \quad (6.1)$$

The magnetization process demands the presence of an extremely low concentration of dissolved oxygen which will oxidize Fe^{2+} to Fe_3O_4 without forming nonmagnetic $Fe(OH)_3$(s). All these factors are to be taken into consideration while designing the process of forming magnetite crystals within the polymer beads.

Figure 6.5 elaborates a scheme for dispersing magnetite nanocrystals within a polymeric cation exchange resin with sulfonic acid groups. The procedure can be applied with necessary adjustments to other types of functionalized polymers with both macroporous type and gel-type morphology [11,35].

Sulfonic acid groups have very low affinity toward hydrogen ions [36]. When a cation exchange resin with sulfonic acid functional groups in hydrogen form is contacted

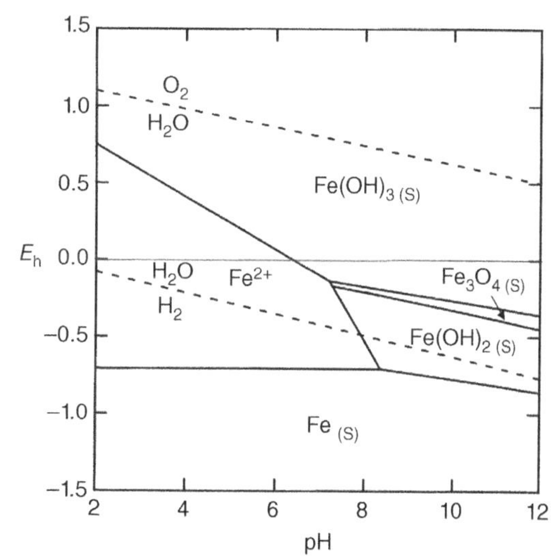

Figure 6.4 Predominance diagram of different species of Fe and Fe oxides. *Source:* Cumbal *et al.* 2003 [11]. Reproduced with permission of Elsevier.

Step I: Loading of Fe^{2+} at acidic pH

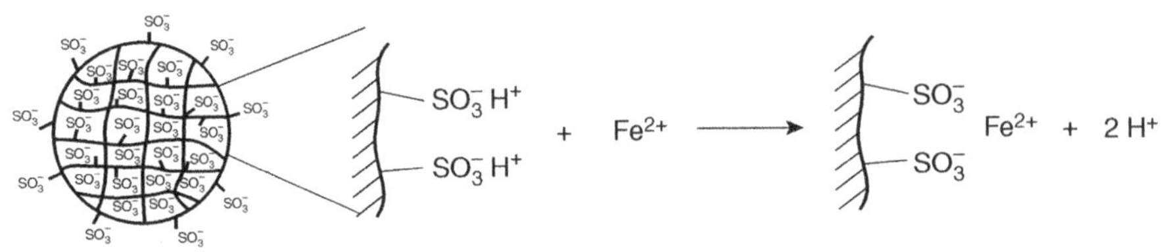

Step II: Magnetite formation mechanisms

$$3\,Fe^{2+} + \tfrac{1}{2}\,O_2 + 3H_2O \rightarrow Fe_3O_4\,(s) + 6H^+$$

$$Fe^{2+} + 2\,OH^- \rightarrow Fe(OH)_2\,(S)$$

$$3\,Fe(OH)_2\,(s) + \tfrac{1}{2}\,O_2 \rightarrow Fe_3O_4\,(s) + 3H_2O$$

$$4Fe^{2+} + O_2 + 10H^+ \rightarrow 4Fe(OH)_3\,(s) + 8H^+$$

Step III: Washing with an alcohol or a solvent with low dielectric constant

Figure 6.5 Steps involved in the preparation of MAPPs with a cation exchanger. *Source:* Cumbal *et al.* 2003 [11]. Reproduced with permission of Elsevier.

with an acidic solution containing Fe^{2+} ions, the Fe^{2+} ions are immediately taken up by the cation exchanger in exchange for H^+ ions. The concentrations of Fe(II) species within the ion exchanger is very high, in the range of 2 N (eq/L) or more. In the next step, as the ion exchangers in Fe(II) form is contacted with a high concentration of Na^+ ions from an alkaline solution of NaCl and NaOH, Na^+ ions replace the Fe^{2+} ions from the ion exchange sites. The unbound Fe(II) concentration inside the polymer phase is very high and the pH inside is also alkaline. If trace concentrations of oxygen are allowed to be present within the polymer phase, the following three different iron oxides can form: $Fe(OH)_2(s)$ (ferrous hydroxide), $Fe(OH)_3(s)$ (ferric hydroxide) and $Fe_3O_4(s)$ (magnetite). Both the Fe(II) and Fe(III) hydroxides are non-magnetic while the magnetite crystals are ferromagnetic.

In the laboratory, the process of magnetization for different polymeric sorbent particles was carried out in a batch reactor as shown in Figure 6.6. Specific steps followed in the laboratory were as follows: (i) Prepare 500 mL of ferrous chloride solution (500 mg/L as Fe) inside the reactor vessel at acidic conditions (pH 2.5–3.5) under nitrogen. Heat up the solution to, say, 60–70 °C and introduce 10 g of polymeric sorbent particles inside the reactor in a porous nylon pouch. (ii) After 15–20 min, slowly raise the pH value to 10 by adding 5% NaCl + 0.5% NaOH solution and concurrently bubble 0.1–1% oxygen (v/v) with carrier nitrogen. Stir for 60 min. Formation of magnetite can be noticed by a dark black color. (iii) Stop stirring, switch off the heating plate, and remove the pouch containing polymeric particles. Rinse with

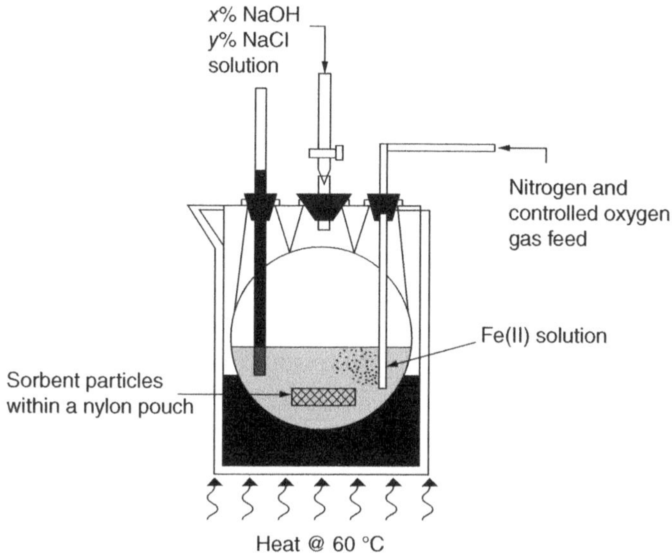

Figure 6.6 A schematic of the batch reactor used for the synthesis of MAPPs. *Source:* Leun and SenGupta 2000 [35]. Reproduced with permission of American Chemical Society.

distilled water and dry the particles in a vacuum furnace for 1 h. *Note:* Rinsing with ethanol helps produce relatively dry particles. (iv) Repeat the process for the second cycle as necessary.

6.1.1 Characterization of MAPPs

The procedure for dispersing magnetite nanoparticles inside the polymers is non-invasive; the chemical reactions involved in dispersal of magnetite nanocrystals within the bead did not interfere with or alter the polymeric phase. The physical morphology of the polymeric phase remained unaltered, except the color, which turned black. Figure 6.7 is a photograph showing an enlarged view of a magnetized Purolite C145 cation exchange resin bead. The resin looked the same as the parent resin except

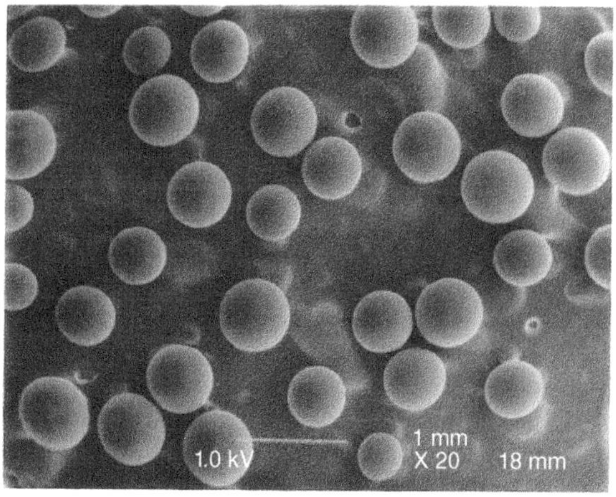

Figure 6.7 Enlarged view of magnetized Purolite C145 polymer beads (20 × magnification). Physical configurations of the particles were unchanged following magnetization. *Source:* Leun and SenGupta 2000 [35]. Reproduced with permission of American Chemical Society.

for a change in color. When both the parent and magnetized resins were cut for a visual observation at the laboratory, both samples had similar-looking surfaces, but a pitch dark magnetic coating was observed only within the MAPP. Figure 6.8 shows X-ray diffractograms performed on a sliced MAPP and a pure magnetite crystal used as a reference standard. The similarity in the peaks between the two samples proves the existence of magnetite crystals inside the bead. The degree of magnetization can be visually found by comparing responses to a laboratory magnet used on four different kinds of materials, Figure 6.9. Only the ion exchanger dispersed with magnetite showed significant magnetic activity, whereas the following did not show any magnetic activity: (i) the parent ion exchanger; (ii) ion exchanger dispersed with ferric (Fe(III)) hydroxide; and (iii) ion exchanger dispersed with ferrous (Fe(II)) hydroxide.

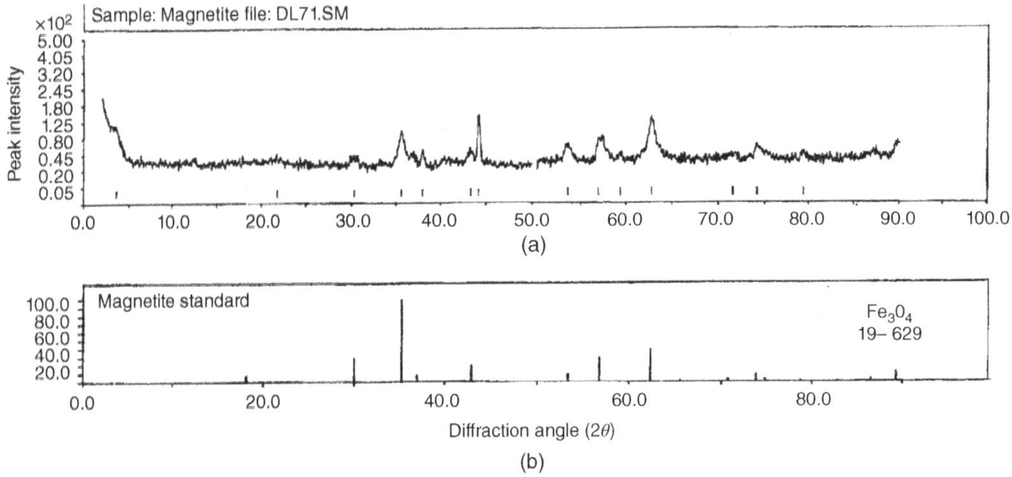

Figure 6.8 X-ray diffractograms and characteristic peaks of (a) sliced magnetized polymeric particle (Purolite C145) and (b) magnetite standard. *Source:* Leun and SenGupta 2000 [35]. Reproduced with permission of American Chemical Society.

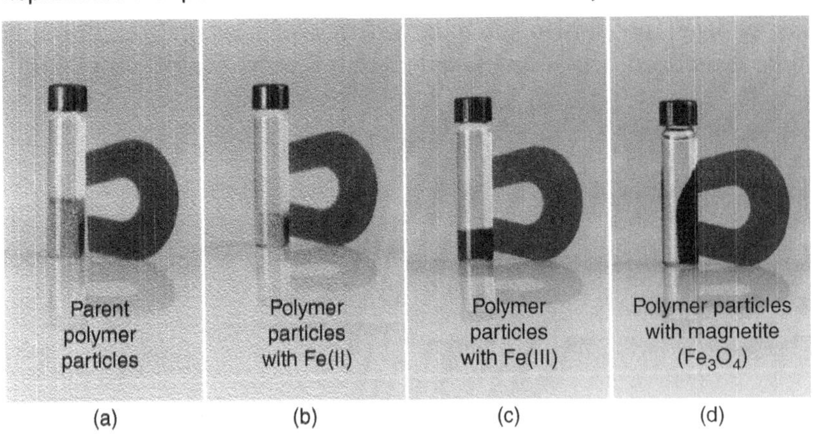

Figure 6.9 Comparison of responses to a laboratory magnet for four types of Diphonix Polymer beads: (a) no treatment, and dispersed with (b) Fe(II) hydroxide, (c) Fe(III) hydroxide, and (d) magnetite. *Source:* Leun and SenGupta 2000 [35]. Reproduced with permission of American Chemical Society.

6.1.2 Factors Affecting Acquired Magnetic Activity

A dimensionless parameter termed magnetic susceptibility (χ_m) is used to determine the magnetic behavior of a material. Magnetic susceptibility is defined as the degree of magnetization induced in a material under the influence of one unit of magnetic field. For nonmagnetic particles, the value of magnetic susceptibility is negative. Water is diamagnetic with magnetic susceptibility of $\gamma_m = -9 \times 10^{-6}$, whereas pure magnetite has a magnetic susceptibility of $\gamma_m = 1 - 5.7$ [37]. The magnetic susceptibility of the MAPPs prepared in the laboratory at Lehigh University was measured using a susceptibility meter (Sapphire Instruments SI-2). The sample particles were placed within a copper coil of the instrument to which a magnetic field was applied externally and carried through the coil. The core logging device measured the magnetic susceptibility of the sample in terms of the dimensionless volume susceptibility.

Figure 6.10 shows the experimentally-determined specific magnetic susceptibilities of different MAPPs prepared following the method detailed earlier. Although the degree of magnetization was different, all forms of hybrid sorbents acquired magnetic activity, regardless of the physical and chemical nature of the parent polymeric sorbents. The magnetic activity of all the hybrid sorbents increased when the loading of magnetite was continued for multiple cycles. A closer look at the acquired magnetic susceptibility values in Figure 6.10, and an examination of the chemical nature of the functional groups of the different polymeric sorbents detailed in Table 6.1,

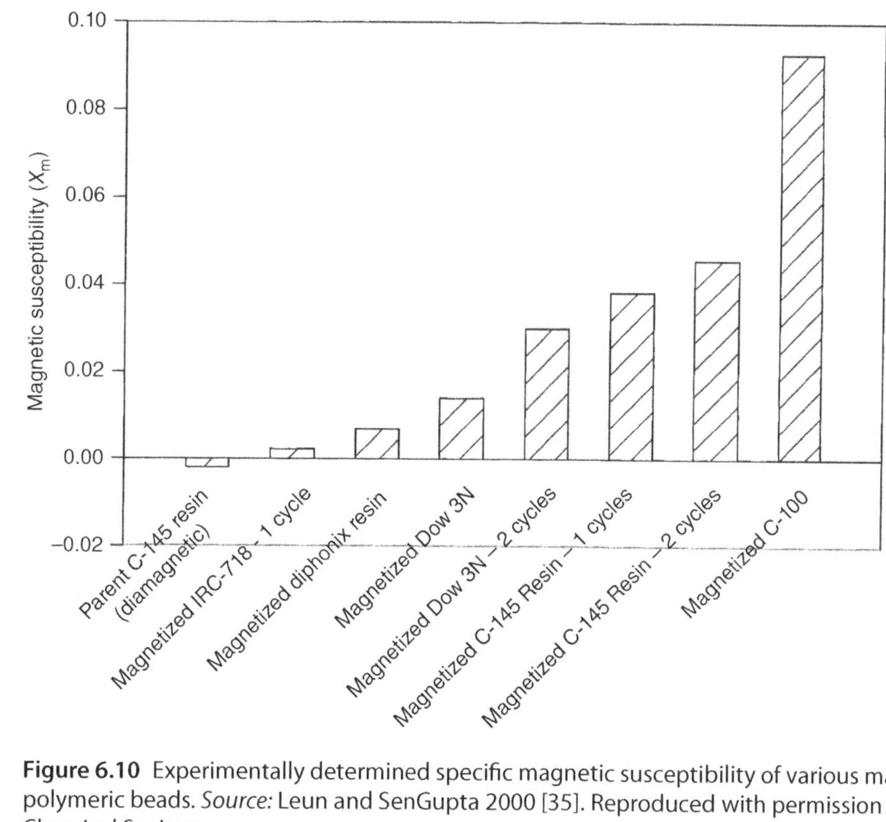

Figure 6.10 Experimentally determined specific magnetic susceptibility of various magnetized polymeric beads. *Source:* Leun and SenGupta 2000 [35]. Reproduced with permission of American Chemical Society.

Table 6.1 Magnetic activity imparted in the polymer phase containing different types of functional groups.

Characteristic	Composition of the functional group	Manufacturer trade name
High metal-ion affinity	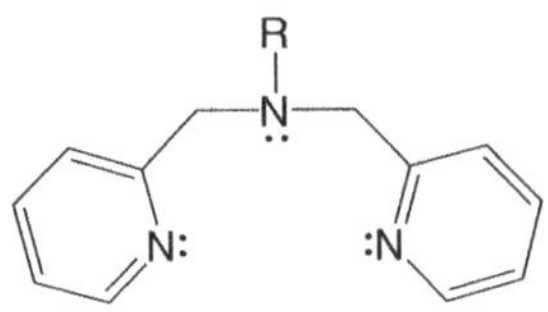	Dow 3N or XFS 4195
High affinity for chromate, benzene sulfonate, pentachlorophenate	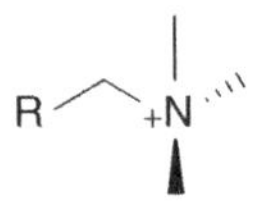	Rohm and Haas IRA-900 Purolite A500
High arsenic selectivity	Polymeric–inorganic hybrid sorbent	LayneRT, Purolite FerrIX A33E
Cation exchange resin	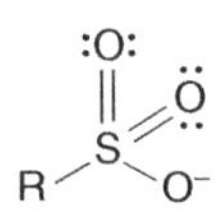	Purolite C145
Metal-selective multi-functional cation exchange	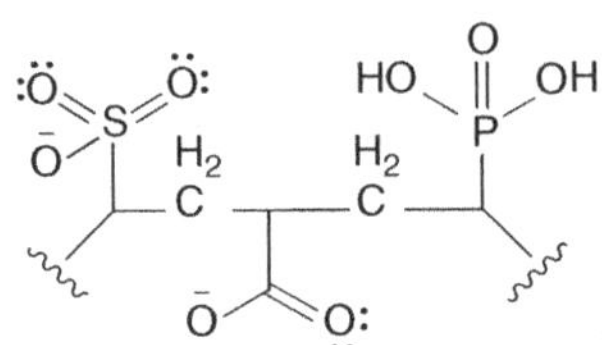	Eichrome industries, Diphonix
High metal-ion affinity	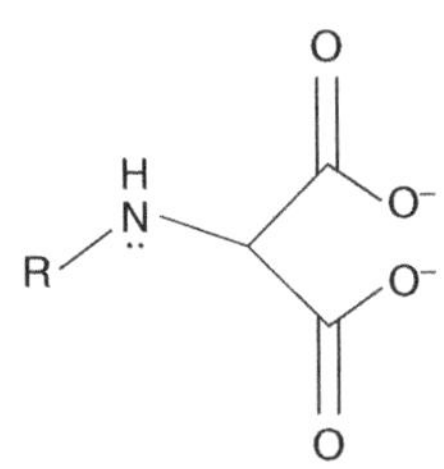	Rohm and Haas IRC-718

Source: Leun and SenGupta 2000 [35]. Reproduced with permission of American Chemical Society.

reveal that the degree of magnetic susceptibility acquired depends on the nature of the functional groups.

For the magnetized Purolite C100, the magnetic susceptibility was significantly higher than the other resins, after only one cycle. The magnetic susceptibility of magnetized Purolite C145 and Dow 3N resins were the next in the series of decreasing magnetic susceptibility. The following is to be noted: (i) C100 is a gel-type strong-acid cation exchange resin with sulfonic acid functional groups; (ii) C145 is a macroporous cation exchange resin with sulfonic acid functional group; (iii) Dow 3N has weakly basic characteristics, but has a high affinity for metals like copper; and (iv) Diphonix is a metal-selective bifunctional ion exchanger. It was found that the acidity of the functional groups and the acquired magnetic activity are correlated. Strong-acid cation exchangers have sulfonic acid functional groups, which have a very low affinity toward hydrogen ions. But, the affinity toward hydrogen ion increases as the acidity of the functional groups decreases. During the first step of magnetization, the ion exchanger in hydrogen form is contacted with a solution containing Fe^{2+} ions, so that Fe^{2+} ions exchange with the hydrogen ions. The uptake of Fe^{2+} ions is the highest for strong-acid cation exchangers compared to other resins with low acidity characteristics. It is estimated that the concentration of Fe^{2+} ions inside the strong-acid cation exchanger reaches a concentration of 2 N during this step. Thus, when magnetite nanoparticles are formed in the next step, via controlled oxidation of the Fe(II) counterions on the functional groups of the resin, the strong-acid cation exchange resin acquires greater magnetization compared to other resins. Although other functional groups did not provide conditions as favorable for acquiring magnetic activity, the degree of magnetic activity of the ion exchangers can always be enhanced through repetition of the loading procedure and the loading of more magnetite nanoparticles. All other conditions remaining identical, higher magnetic susceptibility of a material allows one to have more effective magnetic separations at lower magnetic fields. Therefore, partial functionalization of any polymeric sorbent with sulfonic acid functional groups enables attaining greater magnetization and widens the opportunities for using magnetic separation techniques in complex environmental conditions.

6.1.3 Retention of Magnetic Activity and Sorption Behavior

To be viable in complex environmental separation processes, it is important for MAPPs to simultaneously retain both the magnetic activity as well as the specific sorption capacity. To assess this capability of the MAPPs, magnetized DOW 3N ion exchangers were utilized. Previous studies have shown that DOW 3N and XFS 4195 have high selectivity and high capacity for copper ions, even at low pH [38,39]. Once exhausted, these resins can be effectively regenerated by a solution of ammonia [19,40,41]. Magnetized DOW 3N particles were subjected to 15 adsorption and desorption cycles to find out the effects of magnetization on the sorption–desorption characteristics of the resin, as well as the retention of magnetic susceptibility over cycles of adsorption–desorption. Each cycle consisted of equilibration with 20 mg/L copper solution at pH 4.0 followed by desorption with 5% ammonia solution. Figure 6.11 shows the experimentally determined copper sorption capacities of

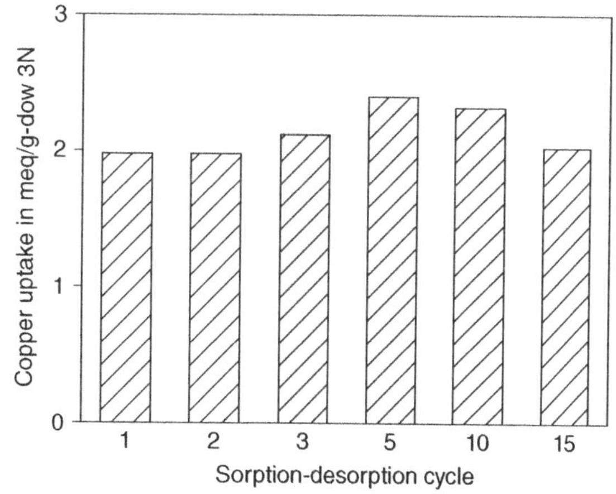

Figure 6.11 Copper sorption capacities of magnetized metal-selective DOW 3N polymer beads over 15 consecutive sorption–desorption cycles. *Source:* Leun and SenGupta 2000 [35]. Reproduced with permission of American Chemical Society.

magnetic DOW 3N resin, 2 meq Cu/g, which were comparable to that of the parent resin. Figure 6.12 shows the specific magnetic susceptibility of the MAPP sorbent is constant over 15 cycles of use, although it was exposed to many cycles of harsh chemical conditions; adsorption at low pH and desorption at high pH. Since magnetic activity is imparted by the magnetite nanoparticles dispersed within the polymer matrix, retention of magnetic activity implies that there was no loss of magnetite from inside the MAPP beads even after several adsorption–desorption cycles.

Figure 6.13 shows the results of the kinetic tests carried out for zinc removal using both magnetized and parent Diphonix resin beads. The presence of magnetite did not

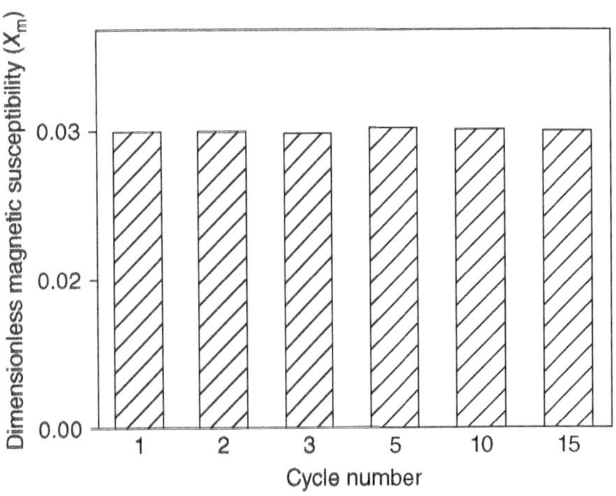

Figure 6.12 Specific magnetic susceptibilities of the same DOW 3N polymer beads in Figure 6.11 over 15 cycles. *Source:* Leun and SenGupta 2000 [35]. Reproduced with permission of American Chemical Society.

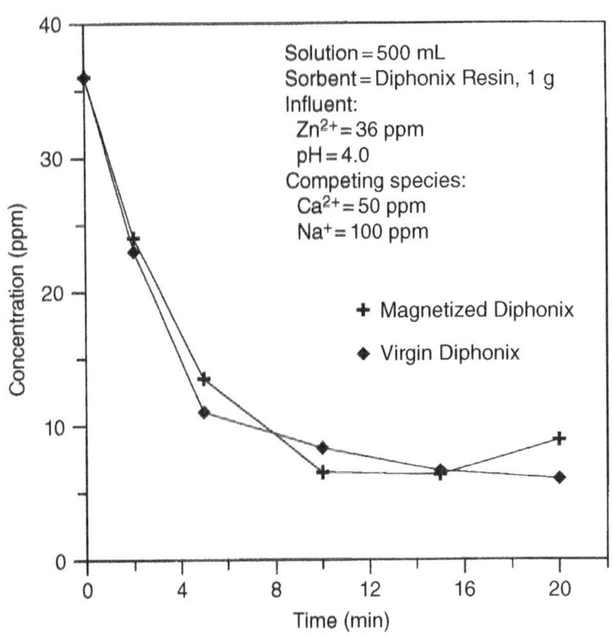

Figure 6.13 Comparison of batch kinetic test results for zinc removal between parent and magnetized Diphonix resins. *Source:* Leun and SenGupta 2000 [35]. Reproduced with permission of American Chemical Society.

have any effect on the uptake rate of zinc ions. Under batch conditions, intraparticle diffusion is usually the rate-limiting step. Thus, it may be concluded that the presence of fine magnetite particles, or the process of forming the magnetite particles inside the beads, did not influence the effective intraparticle diffusivity of the polymer particle.

6.2 Hybrid Nanosorbents for Selective Sorption of Ligands (e.g., HIX-NanoFe)

Oxides of some polyvalent metals, such as Al(III), Fe(III), Ti(IV), Zr(IV), etc. are environmentally benign and have amphoteric sorption behavior around neutral pH. As these metal oxides are inexpensive, readily available and chemically stable over a wide pH range, they have been considered for use as selective sorbents. At a pH value lower than their pH of zero-point charge pH_{zpc} or isoelectric point (p*I*), the metal oxides behave like Lewis acids, that is, electron pair acceptors. At $pH < pH_{zpc}$, surfaces of these oxides can selectively bind dissolved ligands, that is, Lewis bases. Lewis bases donate electron pairs to a central metal atom of the metal oxide to form or coordinate inner sphere complexes. Examples of environmentally significant ligands are inorganic metal oxyanions/oxyacids (e.g., arsenic, phosphorus, selenium, etc.) and deprotonated organic acids with carboxylate functional groups [42–52]. Investigations involving extended X-ray absorption fine structure (EXAFS) spectroscopy has confirmed that both As(III) and As(V) are selectively bound to hydrated ferric oxide (HFO) surfaces through coordinate bonding [53]. Table 6.2 shows the ligand characteristics or electron

Table 6.2 Ligand characteristics or electron pair donating ability for As(III) and As(V) species, as well as for phosphorus oxyanions.

Parent oxyacids	pK_a values	Predominant dissolved species at pH 6.0	Predominant dissolved species at pH 8.0	Sorption interaction
As(V): H_3AsO_4	$pK_{a1} = 2.2$ $pK_{a2} = 6.98$ $pK_{a3} = 11.6$	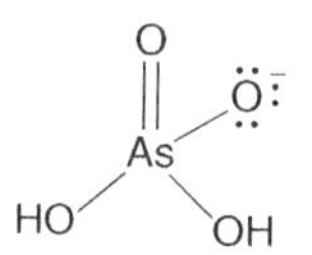 $H_2AsO_4^-$	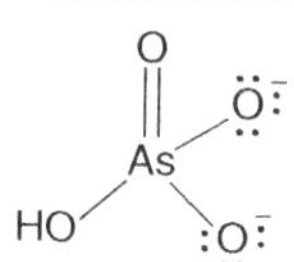 $HAsO_4^{2-}$	As(V) species or arsenate can undergo coulombic (ion exchange) and Lewis acid–base interactions
As(III): $HAsO_2$	$pK_{a1} = 9.2$	O=As—OH $HAsO_2$	O=As—OH $HAsO_2$	As(III) species or arsenite can undergo only Lewis acid–base interactions
P(V): H_3PO_4	$pK_{a1} = 2.12$ $pK_{a2} = 7.21$ $pK_{a3} = 12.7$	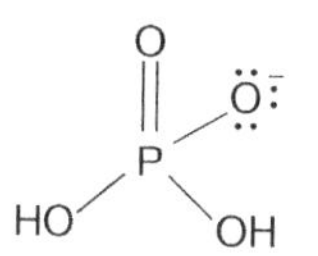 $H_2PO_4^-$	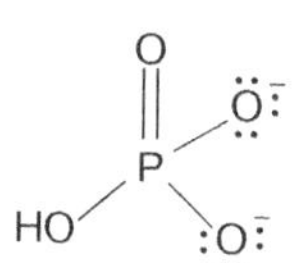 HPO_4^{2-}	P(V) species or arsenates can undergo coulombic (ion exchange) as well as Lewis acid–base interaction

Source: Cumbal and SenGupta 2005 [54]. Reproduced with permission of American Chemical Society.

pair donating ability for As(III) and As(V) species as well as for phosphorus oxyanions [54]. The metal oxides can selectively sorb trace ligands from the background of other commonly occurring anions, for example, chloride, sulfate, nitrate, etc. These common anions are capable of forming outer sphere complexes only with the metal oxide surfaces through Coulombic interactions. Compared to their crystalline forms, amorphous metal oxides have higher surface area per unit mass or volume. Since sorption sites primarily remain at the surface, amorphous metal oxides show higher sorption capacity than other forms.

HFO particulates are popular adsorbents for arsenic and phosphorus species. The sizes of freshly precipitated amorphous particles are in the range of 20–100 nm. Although the small size helps to achieve a very high specific surface area, it is nearly impossible to use these nanoscale particles in the field to remove trace contaminants from the contaminated ground water or wastewater. A fixed bed adsorption process is universally acceptable for water and wastewater treatment because it is simple, requires virtually no start-up time and is forgiving toward any fluctuation of the influent concentrations. The nanoscale adsorbents are not suitable for application in fixed beds because of high pressure drop and poor mechanical strength. There have been efforts to develop granulated HFO or ferric hydroxide particles and they have been used in fixed bed columns for selective removal of arsenic from contaminated water. However, the mechanical properties of granular Fe(III) particles remain weaker compared to polymeric and other inorganic sorbents normally used in fixed beds. Another drawback of the weaker Fe(III) materialsis that the sorbents are not regenerable and produce contaminated waste needing disposal after one cycle of operation.

To overcome the foregoing problems associated with the isolated use of nanosorbents, it is necessary to disperse the nanosorbents irreversibly within a host material which can provide mechanical strength and sorption enhancement for their use in fixed beds. There have been attempts to use naturally occurring or synthetic material (polymeric or inorganic) as hosts to contain the nanosorbents, for example, alginate, zeolites, activated carbon, chitosan, cellulose, polymeric sorbents, and polymeric cation exchangers [55–65]. All these support materials improved the permeability and durability of the adsorbent in fixed bed columns. But, it is necessary to understand how and to what extent the chemical and physical nature of the host material influences the sorption behavior of the hybrid nanosorbent. The morphology of the polymeric host materials, such as pore size and its distributions, influence the size and nature of the nanoparticles dispersed within the pores. Also, for host materials with charged surface functional groups, such as polymeric cation exchangers, the nature and charge density of the groups influences the process and extent of nanoparticle dispersion. The effect of the charged functional groups on the sorption behavior of the hybrid material is another important consideration. Depending on the type of functional group on the polymeric host, three types of hybrid nanosorbents are possible: (i) hybrid anion exchange resins (HAIX) with positively charged functional groups; (ii) hybrid cation exchange resins (HCIX) with negatively charged functional groups; and (iii) hybrid resins with no ion exchange capability (HNIX). We shall focus on the first two types of HIXs with HFO nanoparticles (HIX-NanoFe): synthesis, characterization, use and the role of the Donnan membrane effect.

Table 6.3 Salient features of two different polymeric supports (gel-type) containing cation exchange/anion exchange functional groups.

	Purolite A400/IRA-900	**Purolite C100**
Structure	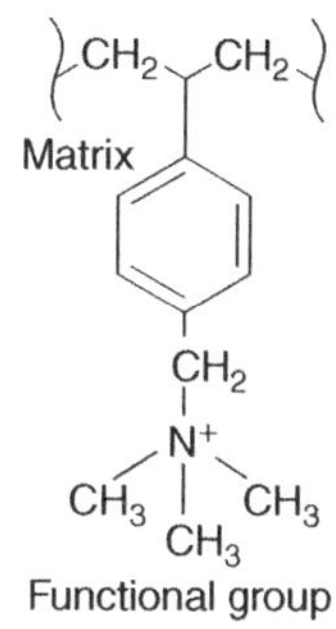	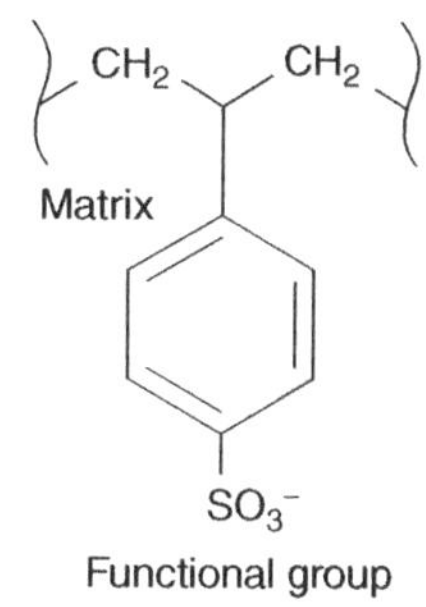
Functional groups	Type I quaternary ammonium	Sulfonic acid
Matrix	Gel (IRA-900)/Macroreticular (A400) on PS-DVB	Gel PS-DVB
Capacity (eq/L)	1.3 (A400)/1.0 (IRA-900)	2.0

These supports were used for synthesis of hybrid sorbent materials (PS-DVB = polystyrene crosslinked with divinylbenzene).
Source: Cumbal and SenGupta 2005 [54]. Reproduced with permission of American Chemical Society.

6.2.1 Synthesis of Hybrid Ion Exchange Nanomaterials

Table 6.3 has the salient features of polymeric supports (gel type) that were used for synthesis of the hybrid sorbent materials: (i) Purolite A400 (anion exchange groups); (ii) Purolite C100 (cation exchange groups). Similar exchangers with both gel and macroporous types of morphology from other manufacturers can also be used.

The preparation of the hybrid cation exchanger or HCIX-NanoFe(III) consisted of the following three steps:

Step 1. Loading of Fe(III) onto the sulfonic acid sites of the cation exchanger by passing 4% (w/v) $FeCl_3$ solution at pH 2.

Step 2. Simultaneous desorption of Fe(III) and precipitation of Fe(III) hydroxides within the gel phase of the exchanger through passage of a solution containing both NaCl and NaOH, each at 5% (w/v).

Step 3. Rinsing and washing of the hybrid resin with a 50/50 ethanol–water solution and mild thermal treatment at 50–60 °C for 60 min.

Figure 6.14 depicts the major steps involved in the process. Step 2 was repeated twice to achieve greater loading of Fe(III) nanoparticles. Experimental observations suggest that after Step 3, both amorphous and crystalline phases were present. A fraction of the nanoscale HFO particles coalesced to form agglomerates. During preparation, turbulence and mechanical stirring did not result in any loss of HFO particles. High concentration of sulfonic acid functional groups allowed high and uniform loading of HFO particles within the polymeric beads, say, 9–12% of Fe by mass.

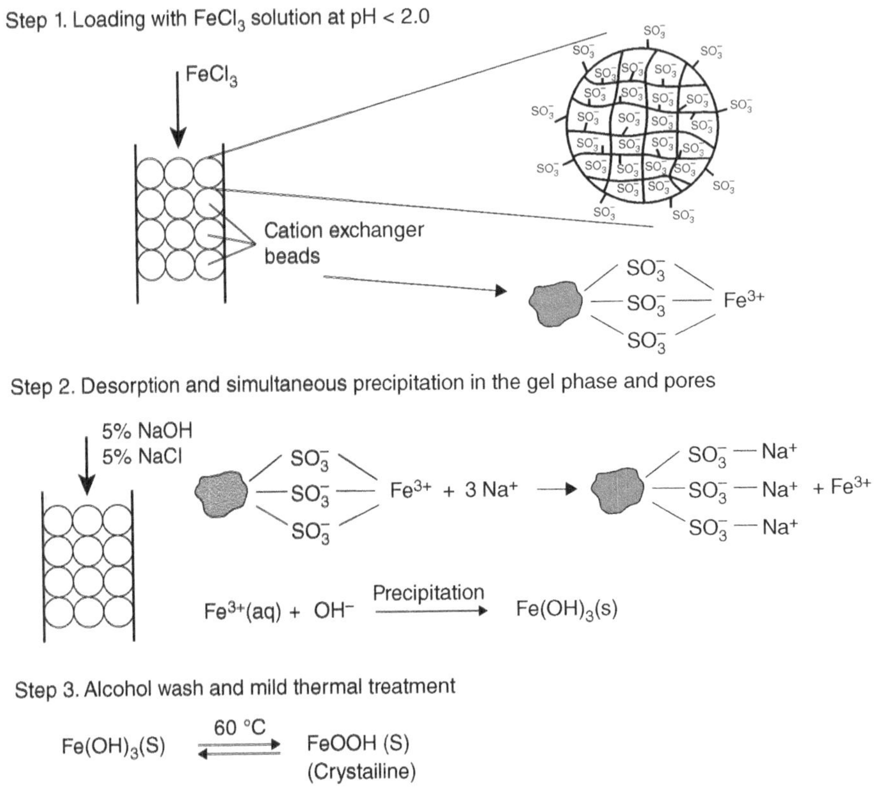

Figure 6.14 Illustration of a three-step procedure to disperse both crystalline and amorphous HFO nanoparticles inside polymeric cation exchange beads to create HCIX-NanoFe(III). *Source:* DeMarco *et al.* 2003 [63]. Reproduced with permission of Elsevier.

Dispersing HFO nanoparticles inside a strong-base anion exchanger is scientifically challenging because both ferric ions (Fe^{3+}) and the functional groups (quaternary ammonium, (R_4N^+)) are positively charged. Owing to high Donnan exclusion potential caused by the presence of non-diffusible quaternary ammonium functional groups inside the resin, it is difficult for the Fe^{3+} ions to diffuse inside the anion exchanger. However, an economically and technically viable process for preparation of the hybrid anion exchanger was developed at Lehigh University [66] and the process has since been commercialized by Layne Christensen and Purolite Company in the US. It is known that the Donnan exclusion potential increases with the increase in the valence of the coion. Thus, all other conditions remaining identical, equilibrium concentration of ferrous ions (Fe^{2+}) inside the anion exchanger shall be higher compared to ferric (Fe^{3+}) ions. Subsequently, oxidation of the ferrous ion within the ion exchanger will convert it to ferric ion which can then initiate and form HFO nanoparticles. The salient steps followed for the method of dispersal of HFO inside anion exchange resin are as follows: First, the anion exchanger is contacted with a solution of potassium permanganate ($KMnO_4$) or sodium hypochlorite (NaOCl), transforming the anion exchanger to MnO_4^- or OCl^- form. Second, the anion exchange resin is contacted with 5% solution of ferrous chloride. The chloride ion replaces the MnO_4^- or OCl^- ion from the ion

6.2.4 Support of Hybrid Ion Exchangers: Cation versus Anion

Two separate column runs were carried out with similar influent solutions and under identical hydrodynamic conditions. In both cases, As(V) or arsenate was a trace species versus other competing electrolytes (namely, sulfate, chloride, and bicarbonate anions). Gel-type hybrid cation exchangers (HCIX-G) and gel-type hybrid anion exchangers (HAIX-G) loaded with HFO nanoparticles were the sorbents used for the fixed-bed column runs. Their general characteristics were summarized in Table 6.3. Figure 6.23 shows the comparison of As(V) effluent histories between the two column runs; the marked difference in the performance of the two sorbents is quite apparent. Note that despite greater HFO content, HCIX-G was essentially unable to remove As(V), which broke through almost immediately after the start of the column run. On the contrary, HAIX-G with HFO nanoparticles dispersed within an anion exchanger showed excellent arsenic removal capacity. Only 10% of arsenic breakthrough occurred after nearly 10,000 bed volumes.

The results shown in Figure 6.23 are intriguing because the performance of HAIX-NanoFe(III) with lower HFO content was better than HCIX-NanoFe(III) by at least 3 orders of magnitude. The results of the earlier column runs with anion exchange resins have already proved that the anion exchange functional groups do not contribute to the sorption of arsenic to a significant extent in the presence of competing sulfate and chloride anions. Such a drastic difference in arsenic removal capacity results from the Donnan membrane effect exerted by the ion exchanger support. The following section explains the reason for the enhanced arsenic adsorption capacity of the anion exchanger in accordance with the Donnan membrane principle.

Donnan Membrane Equilibrium and Coion Exclusion Effect

The Donnan membrane equilibrium principle essentially deals with completely ionized electrolytes in a heterogeneous system where certain ions are unable to permeate,

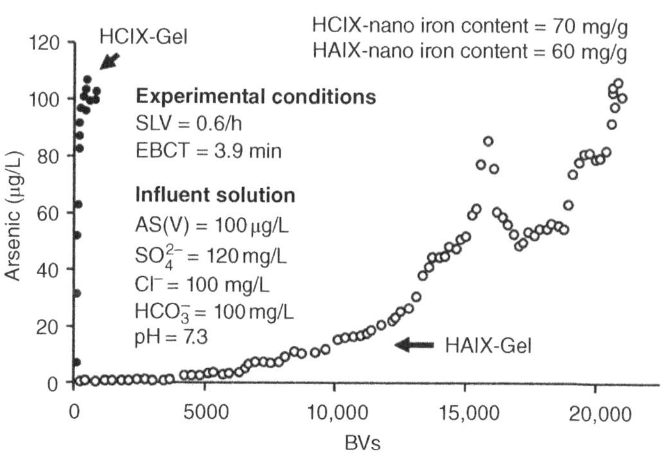

Figure 6.23 Comparison of As(V) effluent histories between HCIX-G-NanoFe(III) and HAIX-G-NanoFe(III) for two separate column runs under otherwise identical conditions. *Source:* Cumbal and SenGupta 2005 [54]. Reproduced with permission of American Chemical Society.

that is, non-diffusible from one phase to the other through the interface [78, 79]. The gel phase of an ion exchanger can be viewed as a cross-linked polyelectrolyte where functional groups (such as quaternary ammonium groups for anion exchanger and sulfonic acid groups for cation exchanger) are covalently attached to the matrix and hence are nondiffusible. In contrast, both the coions and counterions in the bulk liquid phase in contact with the ion exchanger are mobile and can freely move under chemical or electrical potential gradient. So, the fixed nature of the functional group of the ion exchanger acts like a virtual semi-permeable membrane which restricts the movement of one particular type of ion across the phase boundary.

For a membrane that is completely permeable to both Na^+ and $H_2AsO_4^-$, the Donnan membrane principle provides the following equality at equilibrium, assuming ideality:

$$[Na^+]_L[H_2AsO_4^-]_L = [Na^+]_R[H_2AsO_4^-]_R \tag{6.2}$$

where subscripts "L" and "R" refer to the solution at the left-hand and right-hand sides of the membrane, respectively, and [ ] represents molar concentration or activity under ideal conditions. If sodium arsenate is the only electrolyte present in the solution phase and the volume on either side of the membrane is the same (say 1.0 L), the constraint from electroneutrality requires that

$$[Na^+]_L = [H_2AsO_4^-]_L \tag{6.3}$$

$$[Na^+]_R = [H_2AsO_4^-]_R \tag{6.4}$$

Thus,

$$\frac{[H_2AsO_4^-]_L}{[H_2AsO_4^-]_R} = \frac{[Na^+]_R}{[Na^+]_L} = 1 \tag{6.5}$$

The above equality is understandably trivial, as shown in Figure 6.24 (Case I). But the distribution of $H_2AsO_4^-$ on both sides of the membrane is greatly altered in the presence of a semi-permeable membrane. Consider the illustration in Case II where the salt NaR is initially present on the left-hand side of the membrane at 1.0 M concentration. But the resulting anion (R^-) cannot permeate through the membrane. All other conditions are essentially the same as in Case I (i.e., the membrane is completely permeable to both Na^+ and $H_2AsO_4^-$ and the initial concentration of NaH_2AsO_4 on the right-hand side is 0.01 M). At equilibrium, the equality stated in Eq. (6.2) will hold even in the presence of non-permeating R^-. Hence, Na^+ and $H_2AsO_4^-$ will redistribute to arrive at the following equilibrium condition:

$$\frac{[H_2AsO_4^-]_L}{[H_2AsO_4^-]_R} = \frac{[Na^+]_R}{[Na^+]_L} = \frac{1}{99} \tag{6.6}$$

Note that the monovalent arsenate concentration on the left-hand side of the membrane, $[H_2AsO_4^-]_L$, is nearly 2 orders of magnitude lower than $[H_2AsO_4^-]_R$. Although the membrane is permeable to Na^+ and $H_2AsO_4^-$, the presence of electrolytically dissociated NaR at high concentration suppresses the permeability of $H_2AsO_4^-$ in one direction. This phenomenon is an outcome of the Donnan coion exclusion effect and does not result from Coulombic or electrostatic interaction. The derivation of Eq. (6.6) ((and, also, Eq. (6.7) later)) can be readily followed by consulting the recently published English translation of Donnan's original paper [78]. It is assumed that the volume on

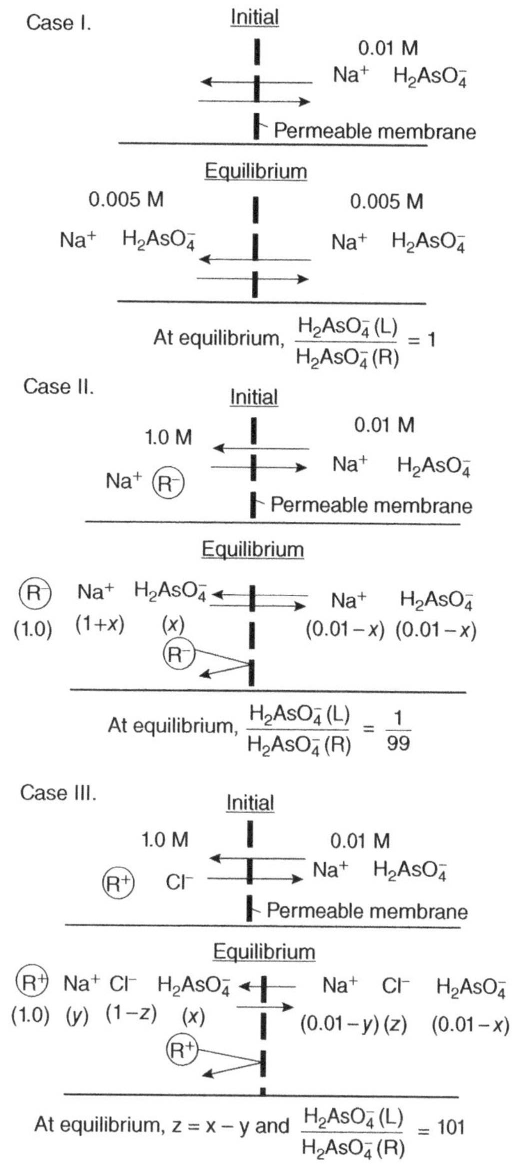

Figure 6.24 Depiction of three specific cases presenting the Donnan distribution of arsenate ($H_2AsO_4^-$) when the membrane is permeable to (Case I) all the ions; (Case II) all the ions except R^-; and (Case III) all the ions except R^+. *Source:* Cumbal and SenGupta 2005 [54]. Reproduced with permission of American Chemical Society.

each side of the membrane is not altered by osmosis. Should HFO particles be now added to the solutions on both sides of the membrane at equilibrium, arsenic sorption capacity on the left-hand side would be relatively low due to significantly lower aqueous-phase arsenic concentration caused by the presence of non-permeating R^-.

The illustration in Case III (Figure 6.24) is, in principle, similar to Case II except that the non-permeating ion (R^+) is a cation. The relative distributions of Na^+, Cl^-, and $H_2AsO_4^-$ on both sides of the membrane, after necessary calculations, stand at equilibrium as follows:

$$\frac{[Na^+]_R}{[Na^+]_L} = \frac{[H_2AsO_4^-]_L}{[H_2AsO_4^-]_R} = \frac{[Cl^-]_L}{[Cl^-]_R} = \frac{101}{1} \tag{6.7}$$

Note that contrary to Case II, arsenic concentration on the left-hand side, $[H_2AsO_4^-]_L$, is nearly 2 orders of magnitude greater than $[H_2AsO_4^-]_R$. The addition of HFO particles in the LHS of the membrane will therefore offer very high sorption capacity due to an enhanced arsenic concentration at equilibrium. The chloride concentration is also enhanced on the left-hand side of the membrane, but HFO particles have negligible sorption affinity for chloride anions. The fixed nature of charged functional groups residing on the ion exchanger makes the phase boundary between the ion exchanger and bulk solution to act like a virtual semi-permeable membrane illustrated in the aforementioned discussion.

To further elucidate the lack of arsenic removal capacity of the hybrid cation exchanger or HCIX-NanoFe(III), let us consider a typical cation exchanger bead of 0.5 mm or 500 μm diameter (d_p).

Considering a bead density (ρ_b) of 1100 kg/m^3, the mass (m) of a single bead is 7.2×10^{-5} g.

The estimated capacity of C100 cation-exchange resin is $q = 4.0$ eq/kg. Then, the number of fixed negative charges (N_e) in a single bead is

$$N_e = \frac{q_m}{1000} \times 6.02 \times 10^{23} = 1.73 \times 10^{17} \text{ charges}$$

where 6.02×10^{23} is Avogadro's number.

Thus, inside a tiny 500 μm cation exchange bead, there are 1.73×10^{17} number of covalently attached sulfonic acid groups with negative charges that cannot permeate from the exchanger phase to the aqueous phase. Comparing this situation with Case II scenario of Figure 6.24, it may be readily inferred that the monovalent and divalent arsenate species will be completely prevented from permeating into the polymer phase due to the Donnan exclusion effect. HFO nanoparticles dispersed inside the cation exchanger are thus inaccessible to arsenate. That is why HCIX-NanoFe(III) did not show any arsenic removal capacity for the column run presented in Figure 6.23. It is worth mentioning that activated carbon, zeolite, alginate, etc. also contain significant concentrations of negatively charged functional groups, namely, carboxylate and aluminosilicate groups. These substrates may be easily dispersed with HFO nanoparticles, but arsenic removal capacity will not be fully attained due to the Donnan exclusion effect. In a previous study, alginate loaded with HFO showed poor arsenic removal capacity during fixed-bed column runs [55].

To the contrary, a polymeric anion exchanger is an excellent substrate because it allows enhanced permeation of anions within the polymer phase due to its high concentrations of fixed positive charges. Figure 6.25 provides a schematic illustrating the difference between cation and anion exchangers as substrate materials. Note that both cation and anion exchanger beads are electrically neutral. Electrostatic repulsion/attraction is not the underlying reason for the difference in permeation of arsenate into the polymer phase. The presence of high concentration of non-diffusing fixed charges (R^+ or R^-) in the polymer phase act as highly permeable or impermeable interface for arsenate, thus influencing its sorption onto the HFO particles embedded in the polymer phase. The polymeric substrate, therefore, acts not only as a robust and hydraulically suitable support material, but also influences the adsorption capacity of the hybrid ion exchange resin. Judicious design of the hybrid polymeric sorbent

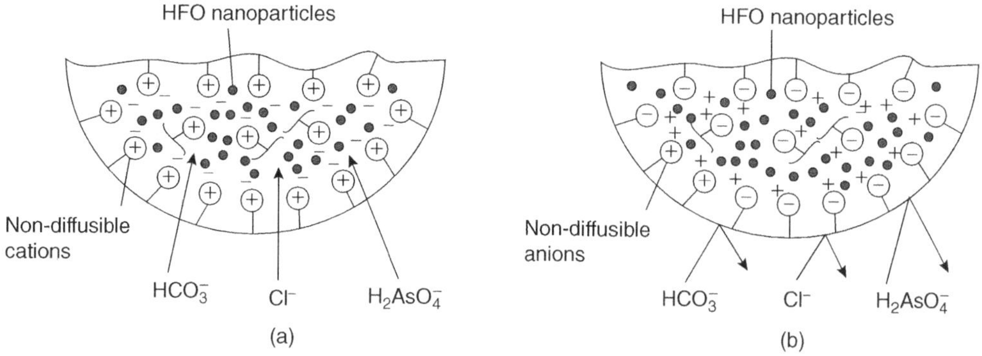

Figure 6.25 Schematic illustrating (a) enhanced permeation of anions into the hybrid sorbent in the presence of non-diffusible cations (anion exchanger) and (b) exclusion of anions from the hybrid sorbent in the presence of non-diffusible anions (cation exchanger). *Source:* Cumbal and SenGupta 2005 [54]. Reproduced with permission of American Chemical Society.

with the appropriate nature of functional group helps offer a synergy leading to the enhancement of sorption capacity of the MONPs, not otherwise achievable. The Donnan membrane principle can thus be successfully used to design and develop novel materials and engineered processes [80].

6.2.5 Efficiency of Regeneration and Field Application

Due to its chemical stability and durable physical structure, the hybrid polymeric sorbent is amenable to regeneration. Regeneration and subsequent containment of the treatment residuals make the removal process cost-effective and environmentally sustainable. Since HIX-Nano can be regenerated and reused over many cycles, the cost of the treated water is low and the sorbent is economically attractive. Also, it is possible to concentrate the arsenic (or phosphate) removed in a small volume of spent regenerant which can subsequently be transformed to a small mass of solids that do not leach arsenic (or used as fertilizer for phosphate) [81,82].

Regeneration of the exhausted resin used for arsenic removal was successfully accomplished using a solution containing a 2% (w/w) of NaCl and 2% (w/w) of NaOH. Figure 6.26 shows the eluent concentration profile for the HAIX-NanoFe(III) used for arsenic removal. It is observed that nearly 95% of the adsorbed arsenic comes out of the HIX phase within 15 bed volumes.

Although unknown nearly 25 years ago, natural arsenic contamination of groundwater has emerged as a major global crisis affecting over 50 countries [83–85]. In the United States, nearly 10,000 communities are now required to introduce additional treatment to reduce the arsenic level in groundwater to be in compliance with the current Safe Drinking Water Act (SDWA) promulgated by the USEPA in 2006. However, the adverse health effects and loss of human lives resulting from drinking arsenic-contaminated groundwater are most apparent in south and southeast Asia, namely, Cambodia, Bangladesh, Laos, Nepal, Vietnam and the eastern region of India. The November 10, 1998 circulation of the New York Times provided front page coverage reporting arsenic-related calamities and loss of human lives in Bangladesh and India, calling it the worst natural calamity in recent times.

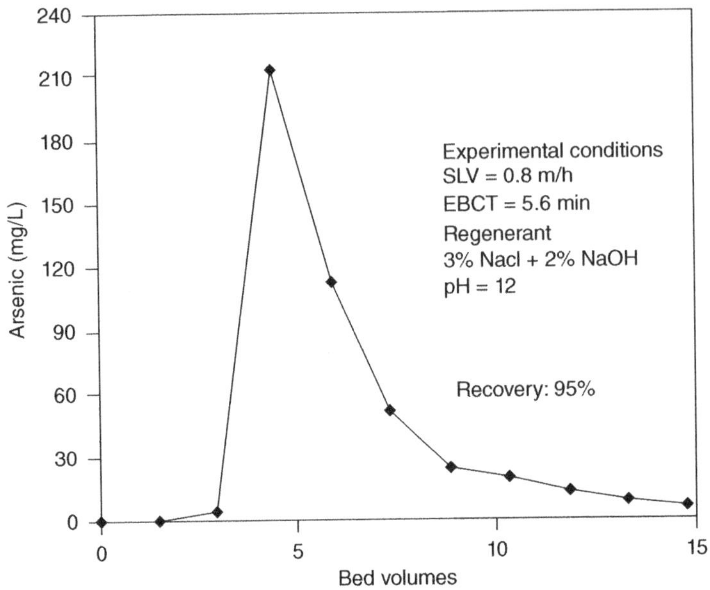

Figure 6.26 Dissolved arsenic concentration profile during desorption of HAIX-NanoFe(III) using 2% NaOH and 3% NaCl as the regenerant. *Source:* Cumbal and SenGupta 2005 [54]. Reproduced with permission of American Chemical Society.

HAIX-NanoFe(III) has been used for treatment of arsenic contaminated water in many places around the world including the remote villages of Cambodia, Nepal, Bangladesh and India, besides the USA. Figure 6.27 shows a representative breakthrough profile of arsenic for an arsenic removal unit located at Ashoknagar in the N. 24 Parganas district of West Bengal, India. Phosphate and silica broke through at the initial stage of the column run. It was also observed that for identical conditions, the arsenic removal units using the hybrid anion exchanger has a superior performance versus those using activated alumina as adsorbent media. Detailed data on the performance of the arsenic removal units and the sustainability of the process of arsenic removal in terms of regeneration and reuse of the media, ecologically sound management of treatment residuals, and so forth can be found in the open literature [81,82,86].

In the USA, HAIX-NanoFe(III) is now commercially available under the trade names Layne^{RT} (Layne Christensen Co., Arizona) and FerrIX^{TM} A33E (Purolite Co., PA). To date, more than 1 million people around the world drink arsenic-safe water through use of HAIX-NanoFe. Figure 6.28a shows a photograph of a plant in Sahuarita, Arizona that uses Layne^{RT}. Figure 6.28b represents the effluent histories of a pilot scale run at the same location where the HAIX-NanoFe(III) was observed to perform better than GFO, a commercially available iron- oxide-based media. The Donnan membrane effect and nanoscale sizes of HFO particles are considered to be the underlying reasons for the superior performance of HAIX-NanoFe with only 18% Fe content.

6.2.6 Hybrid Ion Exchange Fibers: Simultaneous Perchlorate and Arsenic Removal

Like arsenic, perchlorate is also viewed as a trace contaminant of major environmental concern. It has adverse health effects even at extremely low concentrations. Both

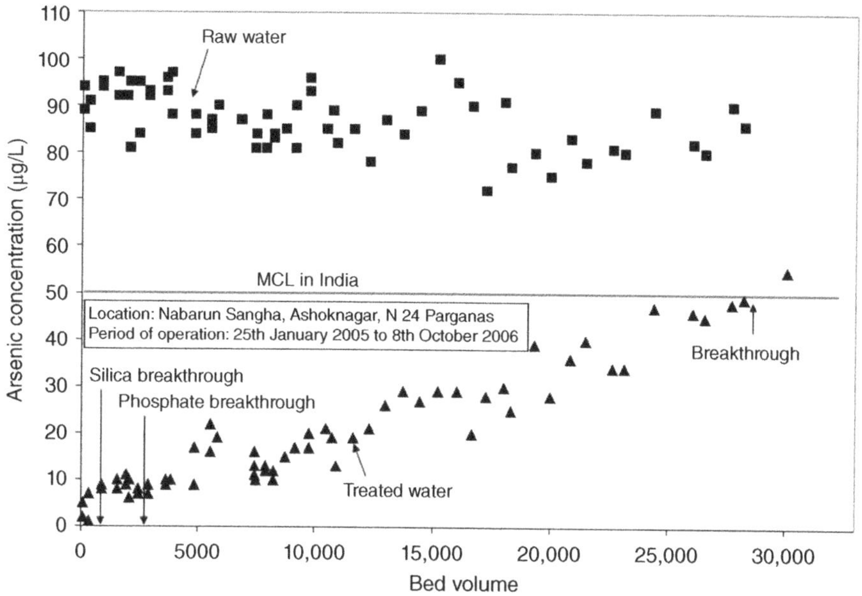

Figure 6.27 Arsenic breakthrough history of an arsenic removal unit at Ashoknagar in West Bengal, India using HAIX-NanoFe(III). *Source:* Sarkar *et al.* 2007 [67]. Reproduced with permission of Elsevier.

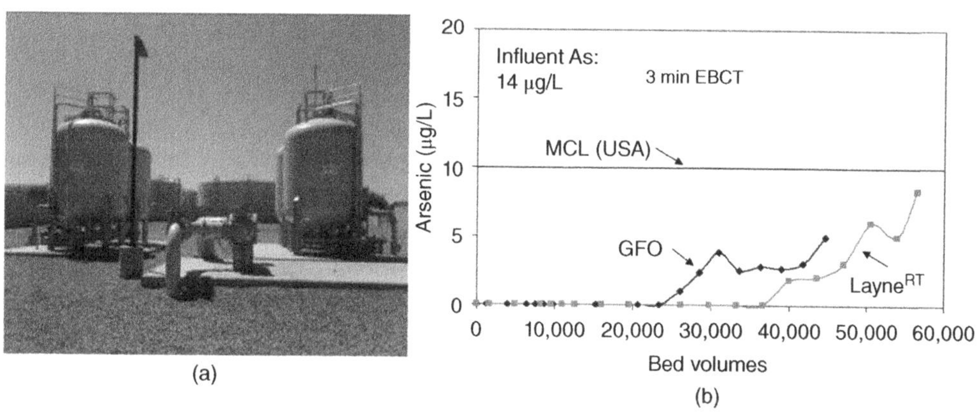

Figure 6.28 (a) Photograph of a plant in Sahuarita, Arizona using LayneRT for arsenic removal and (b) arsenic breakthrough profiles during a pilot run prior to installation (GFO: granulated ferric oxide, MCL: maximum contaminant limit). *Source:* Sarkar *et al.* 2011 [86]. Reproduced with permission of Elsevier.

arsenic and perchlorate exist in water as oxyanions, but their chemistry and genesis in groundwater are different. Unlike arsenic, perchlorate is a poorly hydrated inert anion and more mobile in subsurface water compared to arsenic [87]. In several arsenic-contaminated groundwater aquifers in the Western United States and elsewhere, perchlorate has been reported [88]. Perchlorate is not a natural contaminant in groundwater, but indiscriminate use of perchlorate as an oxidant for solid rocket fuel by aerospace and military industries has led to groundwater contamination over a long period. Contrary to arsenate, perchlorate has no sorption affinity toward

HFO surface binding sites. However, anion exchanger host material with quaternary ammonium functional groups exhibits high sorption affinity for perchlorate [88–91]. HAIX-NanoFe(III) has two characteristically different binding sites: HFO surfaces with high affinity toward anionic ligands and quaternary ammonium functional groups with preference toward hydrophobic anions. That is why HAIX-NanoFe(III) resin is efficient in removing both arsenate and perchlorate anions simultaneously from contaminated water; the observation has been confirmed by others [88,91].

The selective sorption of arsenate ligand onto the HFO surfaces strongly depends on pH. Hence, ligands like arsenic can be easily and rapidly desorbed from the binding sites by raising the pH with an innocuous regenerant like 2% NaOH. On the other hand, both sorption and desorption of perchlorate are kinetically controlled by intraparticle diffusion and is independent of pH. In previous efforts, perchlorate could not be efficiently regenerated from exhausted HAIX-NanoFe(III) resins with brine solution. The possible reason behind such inefficient regeneration was a significantly long diffusion path length for spherical ion exchanger beads, which had an average size of 500–600 μm [88,89]. Reducing the size of the ion exchange resins might prove to be effective, but the major disadvantage is that smaller size of resins would cause drastic increase in fixed bed pressure drop or head loss for identical liquid flow rate. An efficient regeneration technique for perchlorate has been demonstrated recentlythrough the use of $FeCl_4^-$ at very low pH conditions [90,92]. However, such a regeneration technique is not suitable for HAIX-NanoFe(III) because highly acidic pH would promote dissolution of HFO nanoparticles from the HAIX beads.

Cylindrical ion exchange fibers with diameters in the range of 20–50 μm have short diffusion path length and are amenable to use in fixed bed columns. Their properties, both similarities and dissimilarities, have been adequately addressed in Section 4.4.3. Due to very high void fractions, there is no increase in the head loss across the bed. Fibers with polystyrene–divinylbenzene matrix and quaternary ammonium functionality, if impregnated with HFO nanoparticles, should exhibit high selectivity toward both arsenic and perchlorate just like their resin counterpart/version, but at the same time should deliver high regeneration capability with NaCl solutions. A superior regenerability is expected due to a shorter diffusion path length in ion exchange fibers. The fibers were obtained from the Institute of Physical Organic Chemistry of the National Academy of Science in Belarus. Fiban A-1 fibers were used for the laboratory study; the properties and structure of the fibers can be found elsewhere in open literature [91,93,94]. Dispersal of HFO nanoparticles within the fiber was accomplished following a two-step process similar to the one followed for preparation of HAIX-NanoFe(III) resin beads. HFO content was about 110 mg/g as Fe.

Figure 6.29a shows an enlarged view of the hybrid anion exchange fiber (HAIX-F-NanoFe(III)), (b) shows the SEM image of the surface of the parent fiber at 2000× magnification, and (c) shows the surface of the hybrid fiber at 2000 × magnification [93]. The change in the roughness of the surface due to the HFO loading can be readily observed. Further, Figure 6.30a shows a cross-sectional SEM image of a sliced HAIX-F-NanoFe(III) and (b) shows the energy dispersive X-ray mapping of iron across the cross-section. A greater concentration of HFO nanoparticles can be observed at the periphery of the fibers.

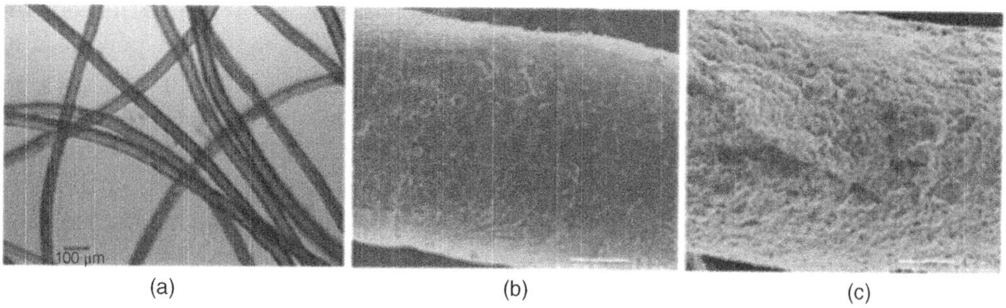

Figure 6.29 (a) Photograph of HAIX-F-NanoFe(III) at 10× magnification, (b) SEM image of the surface of the parent fiber at 2000× magnification, and (c) SEM image of the surface of the hybrid fiber at 2000× magnification. *Source:* Greenleaf and SenGupta 2006 [93]. Reproduced with permission of American Chemical Society.

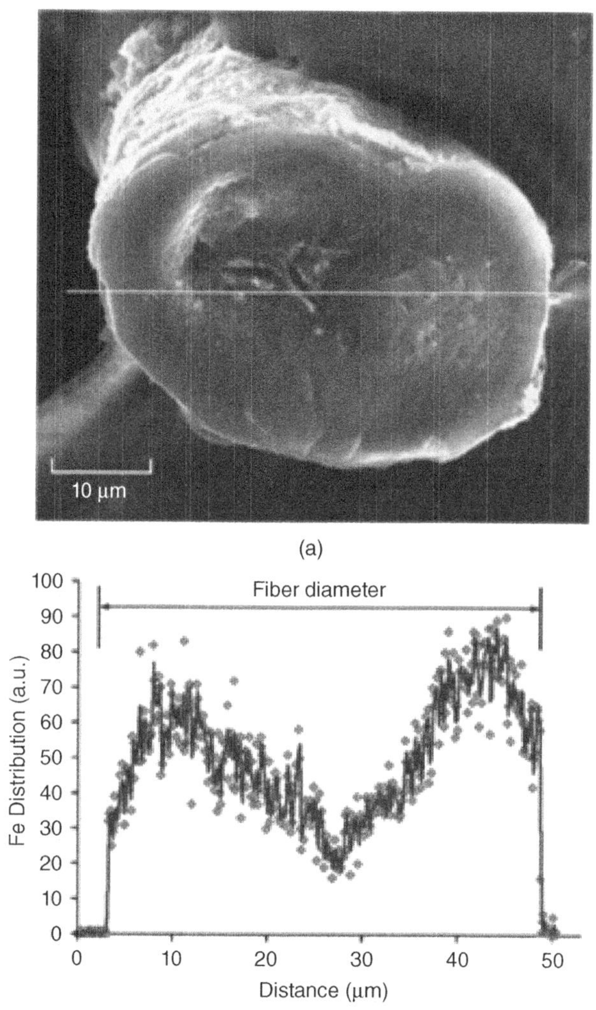

Figure 6.30 (a) SEM image of cross-section of a hybrid anion exchange fiber at 2000× magnification; and (b) energy dispersive X-ray (EDX) mapping of iron along the diameter of the cross-section of the hybrid fiber. *Source:* Lin and SenGupta 2009 [91]. Reproduced with permission of Mary Ann Liebert, Inc.

Fixed bed column runs with 1 g of hybrid ion exchange fibers were carried out in the laboratory with synthetic feed water containing both arsenate and perchlorate ions along with other commonly occurring anions present in the natural water. Upon exhaustion, the exhausted HAIX-F-NanoFe(III) column was regenerated in two consecutive steps using 2% NaOH followed by 10% NaCl. Prior to resuming the column run for the next cycle, the bed was protonated using carbon dioxide sparged water. No excess pressure drop or head loss was observed during the lengthy column runs.

Figure 6.31a and b shows the effluent histories for arsenic and perchlorate for three consecutive column runs. The bed containing the fibers was regenerated in between the column runs following the regeneration protocol mentioned above. The results indicate that there was only a minor change in the perchlorate and arsenic effluent histories during consecutive runs. Figure 6.32 shows the overall uptake capacity of

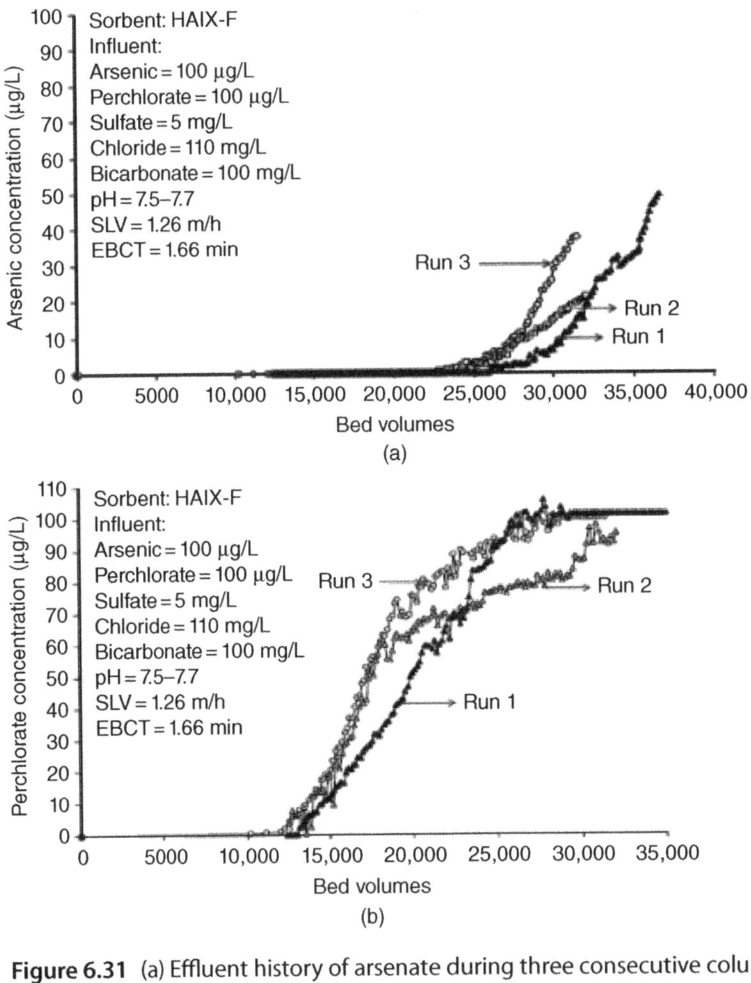

Figure 6.31 (a) Effluent history of arsenate during three consecutive column runs with HAIX-F-NanoFe(III) and (b) Effluent history of perchlorate during three consecutive runs with HAIX-F-NanoFe(III). *Source:* Lin and SenGupta 2009 [91]. Reproduced with permission of Mary Ann Liebert, Inc.

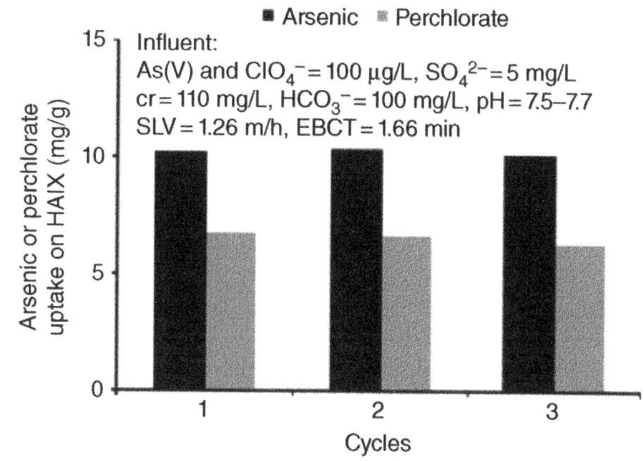

Figure 6.32 Arsenate and perchlorate uptake during multiple cycles of sorption–desorption using HAIX-F-NanoFe(III). *Source:* Lin and SenGupta 2009 [91]. Reproduced with permission of Mary Ann Liebert, Inc.

the hybrid fibers for both arsenate and perchlorate for each column run. The overall sorption uptakes remained practically unchanged even after two regenerations. This confirms that the HFO nanoparticles are irreversibly contained within the gel phase of the ion exchange fiber and the regeneration process did not have any adverse effect on the concentration or activity of the hybrid ion exchange fiber. For thousands of bed volumes, breakthrough was undetectable for both arsenate and perchlorate even for the third run.

Figure 6.33 shows the elution concentration profile for arsenic and perchlorate during the two-stage regeneration process after the first and second cycle. In each case nearly 95% recoveries of arsenate and perchlorate were recorded. The first step of the regeneration process using 2% NaOH solution was effective in regenerating the adsorbed arsenic from the exhausted hybrid fiber. The rise in pH causes the surface functional groups of HFO to change their polarity. As a result of such polarity reversal, the arsenic is rejected by the adsorbent, that is, reversal of ligand exchange.

It is interesting to note that the regeneration process allowed near complete separation of arsenic-rich effluent from that of perchlorate. The strong-base anion exchangers are least selective toward hydroxyl ions. Hence, the first step of regeneration involving 2% NaOH solution could not desorb any significant concentration of perchlorate from the anion exchange sites, though it was effective in desorbing arsenate ions from the HFO nanoparticles. Cl^- ions at high concentration (10%) were chosen as an effective regenerant because at such high ionic concentrations, the selectivity for the chloride ion is higher than sulfate ions due to the selectivity reversal effect described in Chapter 2. However, for common strong-base anion exchange resins, very low intraparticle diffusivity of perchlorate in the presence of chloride is a major obstacle toward their efficient regeneration. So, reducing the intraparticle diffusion path length by replacing relatively large spherical ion exchange resin beads with ion exchange fibers is a clever strategy for improving the regeneration efficiency. All other conditions remaining identical, the regeneration of commercially available spherical ion exchangers should require 50× more volume of regenerant compared to that for the ion exchange fibers for perchlorate desorption.

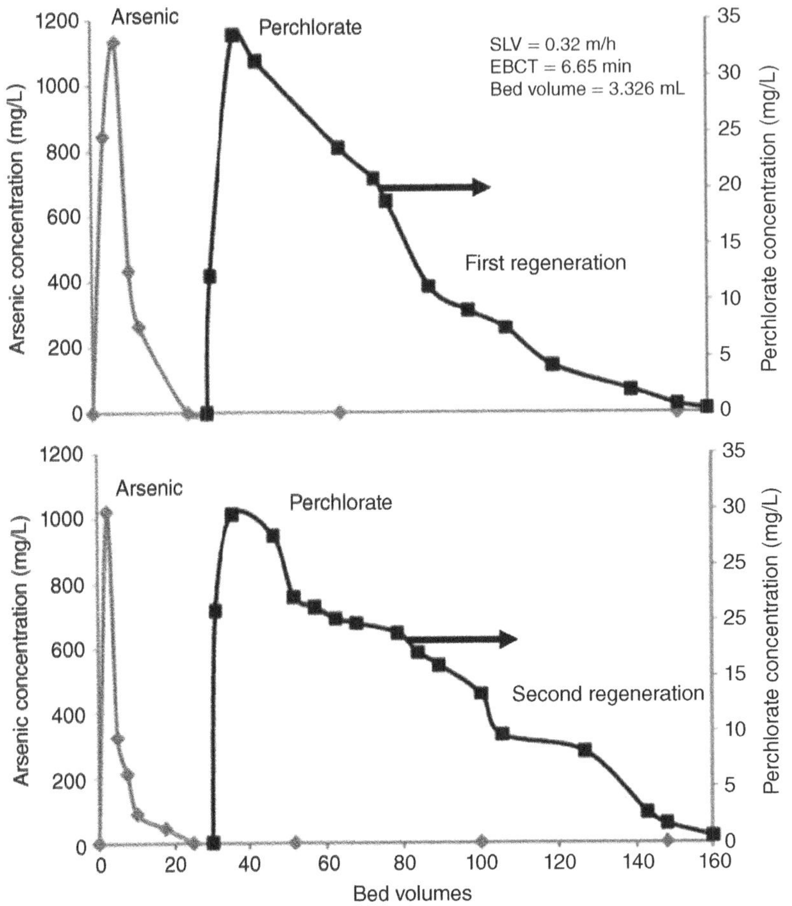

Figure 6.33 Elution concentration profiles of arsenate and perchlorate during two-stage regeneration after the first and second column run. *Source:* Lin and SenGupta 2009 [91]. Reproduced with permission of Mary Ann Liebert, Inc.

6.3 HAIX-NanoZr(IV): Simultaneous Defluoridation and Desalination

Well over 200 million people across the world, mostly in Asia and Eastern Africa are currently at risk for dental and skeletal fluorosis due to drinking fluoride-contaminated groundwater. Current treatment solutions are not adequate in a number of ways: short adsorbent lifespan, high waste generation, poorly treated water aesthetics, continuous dependence on electricity, poor recovery of extracted groundwater and economic unsustainability. Fe(III) oxide nanoparticles do not exhibit any sorption affinity for fluoride (F^-) which is a hard anion. On the contrary, Zr(IV) oxide nanoparticles or ZrO_2 are quite selective toward F^- in the presence of other commonly encountered anions, namely, sulfate and chloride [95–98]. However, fluoride sorption onto ZrO_2 is inversely dependent on pH; as pH decreases, zeta potential of ZrO_2 particles becomes increasingly positive and more selective toward fluoride. Note that Figure 6.34 shows

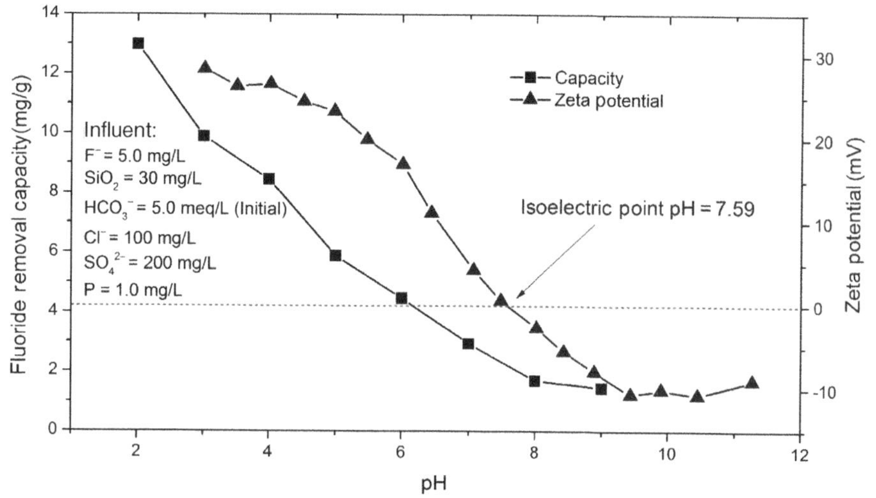

Figure 6.34 Zeta potential and fluoride sorption capacity of zirconium(IV) oxide as a function of pH. *Source:* Unpublished data from Lehigh University.

a plot of experimentally determined fluoride sorption capacity and zeta potential of ZrO_2 nanoparticles as a function of pH.

It is evident that under slightly acidic conditions, around pH 4.0, fluoride removal capacity of ZrO_2 is greatly enhanced while the zeta potential is significantly positive. Many fluoride-contaminated groundwater sources in Asia have total dissolved solid (TDS) levels greater than 500 mg/L, the permissible secondary standard of the World Health Organization (WHO) for drinking water. Very often, if not always, alkalinity or HCO_3^- contributes to a significant portion of the TDS. A process scheme was subsequently conceptualized and implemented to reduce TDS contributed by alkalinity, while simultaneously enhancing fluoride removal capacity [97]. Figure 6.35 illustrates how introduction of a weak-acid cation exchanger ahead of HAIX-NanoZr(IV) helps achieve: first, reduction of pH and removal of $NaHCO_3$ and $Ca(HCO_3)_2$ while reducing the pH; and second, attaining significantly higher fluoride removal at the reduced pH. Very easily the treated water can be pH-adjusted as needed.

6.3.1 Field-Scale Validation

At Piraya village in the state of Maharashtra, India, many groundwater wells are naturally contaminated with fluoride levels higher than 2 mg/L. Villagers, especially children, bear clear signs of dental fluorosis. TDS in this location slightly exceeds 500 mg/L. So, any adsorption technology alone that addresses fluoride removal is inappropriate. Reverse osmosis (RO) is routinely used under such circumstances, but it suffers from the following shortcomings:

1. Over 50% of the water extracted from underground is routinely wasted as reject;
2. Supply of electricity is needed for its operation;
3. Cost of electricity is a major component of the operating cost.

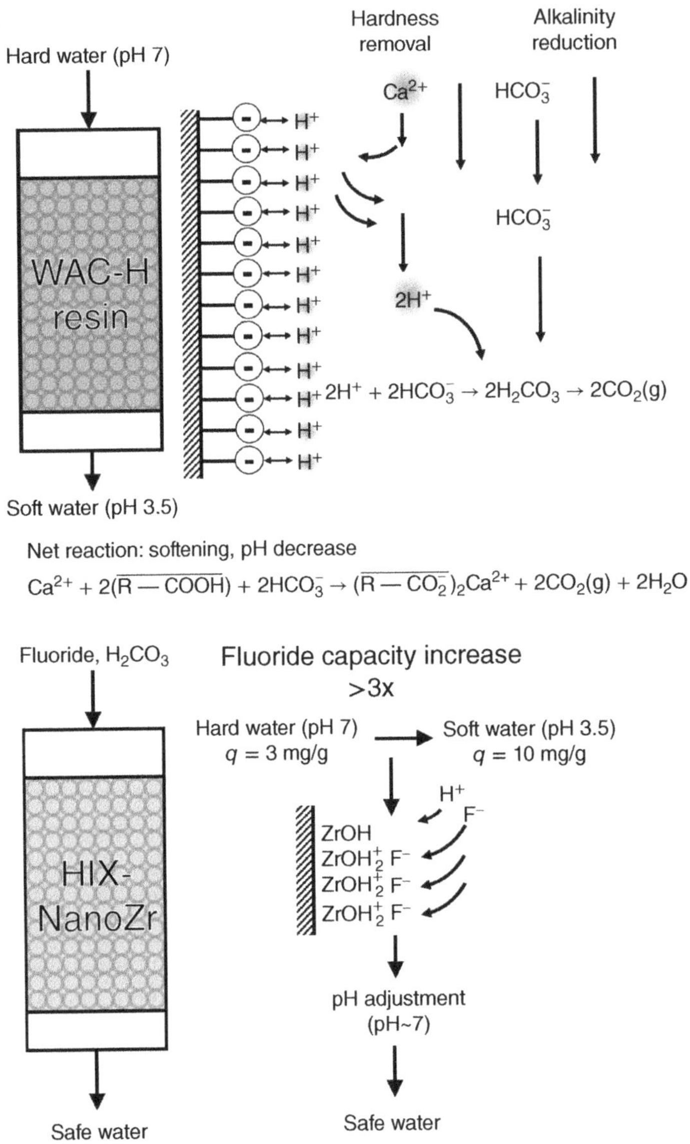

Figure 6.35 Conceptualized mechanism of TDS reduction with simultaneous enhancement in fluoride removal.

Extensive field scale tests were carried out at the Piraya site that consisted of the use of HAIX-NanoZr(IV) sorbent preceded by a weak-acid cation (WAC) exchange resin. Figure 6.36a shows how TDS was consistently brought down from slightly over 500 to less than 300 mg/L. At the same time, fluoride in the treated water was always below the drinking water limit of 1.5 mg/L, as shown in Figure 6.36b.

Four cycles were conducted over 2 years with identical results confirming the reproducibility of the process. Scanning electron microscopy with energy-dispersive X-ray (SEM-EDX) was done for a single HAIX-NanoZr(IV) particle following fluoride removal to map Zr, F, and Cl distribution, Figure 6.37.

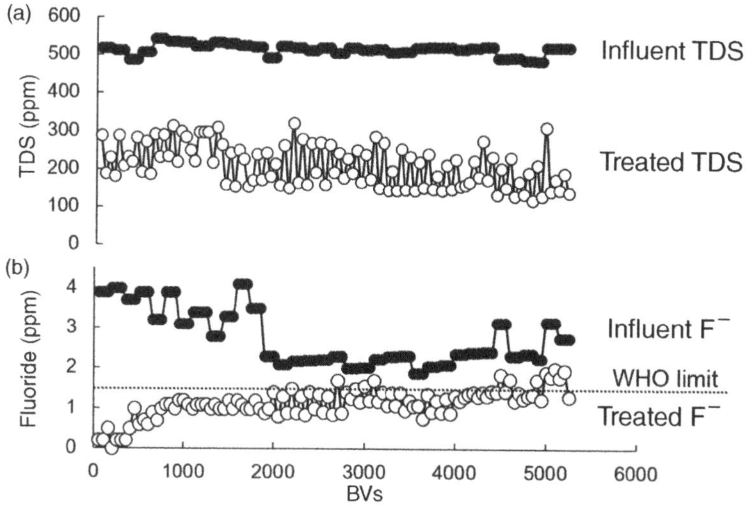

Figure 6.36 Field test results of the HAIX-NanoZr(IV) process at Piraya, Maharashtra, India for (a) TDS reduction, and (b) fluoride reduction. *Source:* Unpublished data from Lehigh University.

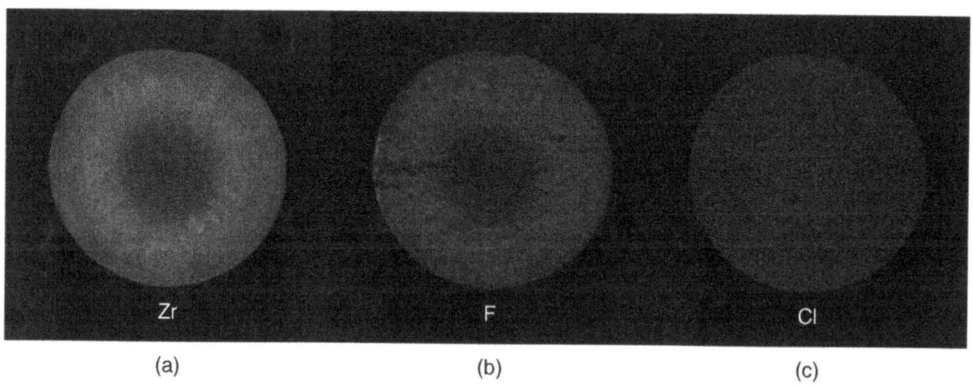

Figure 6.37 SEM-EDX elemental scans of an HAIX-NanoZr(IV) resin bead for (a) zirconium, (b) fluoride, and (c) chloride. *Source:* Unpublished data from Lehigh University.

Note that Zr and F distribution are identical implying fluoride is sorbed only at the ZrO_2 sorption sites. On the contrary, Cl mapping does not coincide with Zr and is evenly distributed over the entire particle. This observation confirms that chloride is sorbed onto the exchange sites of the parent anion exchanger distributed throughout the particle caused solely by Coulombic interaction. Mechanistically, ZrO_2 nanoparticles do not contribute toward sorption of chloride or sulfate.

The first community-based fluoride removal system with simultaneous desalination was installed in the state of Andhra Pradesh in Moparapalli village; Figure 6.38 shows a photograph of the system. There are altogether three separate columns: (i) WAC exchanger for pH adjustment and TDS reduction through alkalinity removal; (ii) HAIX-NanoZr(IV) for selective fluoride sorption; and (iii) dolomite ($CaMg(CO_3)_2$) for pH adjustment. Note that no chemical dosing is needed during the daily operation. Table 6.4 shows both influent and treated water quality. It is quite apparent that

Figure 6.38 A photograph of the HAIX-NanoZr(IV) system (a) and people collecting water (b) in Nalhati, West Bengal, India. *Source:* Unpublished data from Lehigh University.

Table 6.4 Influent and treated water quality from the HAIX-NanoZr(IV) system in Moparapalli village, Andhra Pradesh, India after operation of 1200 bed volumes.

	Units	Raw water	Treated water
pH		7.58	7.22
Electrical conductivity	μS/cm	826	275
Total dissolved solids	mg/L	**537**	179
Total alkalinity as $CaCO_3$	mg/L	168	56
Total hardness as $CaCO_3$	mg/L	228	76
Calcium as Ca	mg/L	37	13
Magnesium as Mg	mg/L	33	11
Fluoride as F	mg/L	**4.18**	0.09

Note: Bolded values are outside of Indian drinking water standards. Unpublished data from Lehigh University.

HAIX-NanoZr technology has significantly brought down TDS (i.e., nearly 50%) while removing fluoride. Recently, twenty community-based plants were installed in Anantapur district for simultaneous fluoride and TDS reduction in compliance with drinking water standards. It is likely that HAIX-NanoZr(IV) in the second column will require regeneration every six months; the protocol is nearly the same as HAIX-NanoFe(III) and described in the open literature [95–98].

6.4 Promise of HIX-Nanotechnology

Polymer-supported metal and MONPs have grown as a field with a multitude of application opportunities [99]. HIX-Nanotechnology is a specific area of the field where polymeric ion exchange resins are used as the support materials, often aided by the Donnan membrane principle. By aptly choosing the type of functional groups, the sorption and separation processes can be effectively controlled, as demonstrated in earlier sections. Like sorption, redox reactions also may be made quite selective. For example, let us consider chemical reduction of nitrate (NO_3^-) and carbon tetrachloride by zero-valent iron, or Fe^0:

$$4Fe^0 + NO_3^- + 10H^+ \rightarrow 4Fe^{2+} + NH_4^+ + 3H_2O \tag{6.8}$$

$$4Fe^0 + CCl_4 + 4H^+ \rightarrow 4Fe^{2+} + CH_4 + 4Cl^- \tag{6.9}$$

Use of chlorinated organic solvents was quite popular in the 1970s for their chemical stability and non-flammability versus acetone, kerosene and other hydrocarbons. But unregulated discharge was responsible for contaminating many groundwater wells providing potable water. Iron nanoparticles were demonstrated to reduce chlorinated organics into innocuous hydrocarbons [2,100–103]. However, the presence of nitrate adversely interferes with the reduction of target chlorinated hydrocarbons, that is, the bulk of Fe^0 nanoparticles can be consumed to react with and reduce nitrate instead of the target chlorinated hydrocarbons. Techniques have been developed to dope cation exchange resins with iron (Fe^0) nanoparticles through a two-step process: (i) loading with Fe^{2+}; and (ii) reducing Fe^{2+} in situ with hydrazine or other reducing agents.

$$2\overline{(R - SO_3^-)Na^+} + Fe^{2+} \leftrightarrow \overline{(R - SO_3^-)_2Fe^{2+}} + 2Na^+ \tag{6.10}$$

$$\overline{(R - SO_3^-)_2Fe^{2+}} + \frac{1}{2}N_2H_4 + 2Na^+ \leftrightarrow 2\overline{(R - SO_3^-)Na^+} + Fe^0 + \frac{1}{2}N_2 + 2H^+ \tag{6.11}$$

Cation exchanger doped with Fe^0 nanoparticles may reduce carbon tetrachloride or TCE while nitrate will be rejected by the negatively charged functional groups of the cation exchanger through the Donnan exclusion effect, as illustrated in Figure 6.39. Besides iron, other MNPs, for example, Zn^0, Pd^0 and Cu^0 also may be introduced in the ion exchanger matrix.

It is only appropriate to mention that impregnating MNPs within polymeric ion exchange resins was first carried out as early as 1949 by Mills and Dickinson. Through in situ synthesis, these researchers introduced copper MNPs or "colloidal copper" in weak-base anion exchange resins and subsequently used this polymer–metal nanocomposite to remove oxygen from water [104].

$$2Cu^0 + O_2 + 4H^+ \rightarrow 2H_2O + 2Cu^{2+} \tag{6.12}$$

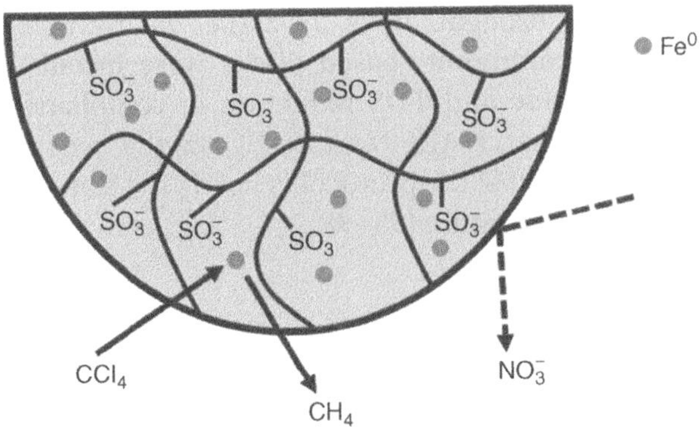

Figure 6.39 A methodology illustrating selective reductive dechlorination of CCl_4 in the presence of nitrate (NO_3^-) through use of the Donnan exclusion effect.

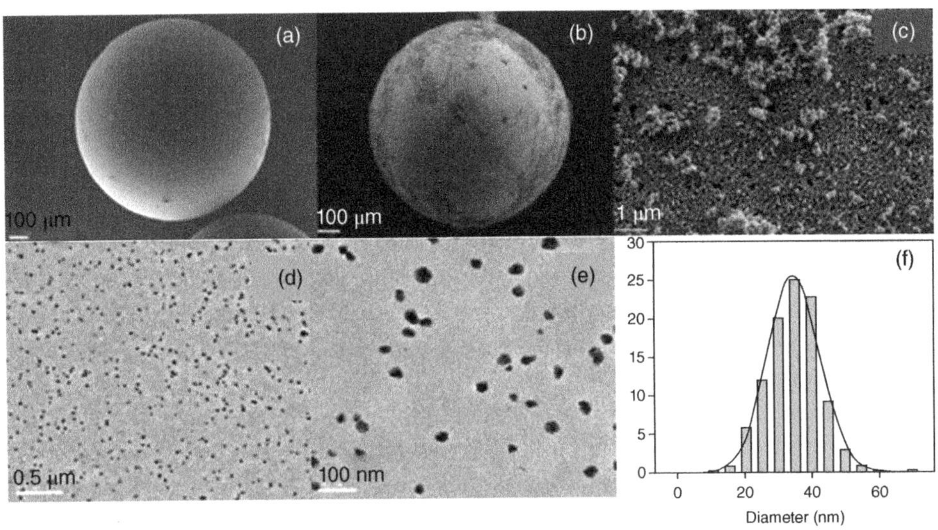

Figure 6.40 SEM images of Purolite A520E resin before (a) and after (b and c) Donnan effect driven intermatrix synthesis of Pd-polymer-stabilized metal nanocatalysts (Pd-PSMNCs). TEM images (d, e) and size distribution histogram (f) of Pd-PSMNC. *Source:* Arrieta *et al.* 2012 [105]. Reproduced with permission of Elsevier.

Techniques have been elaborately presented in the open literature [105,106] pertaining to preparation of MNPs within ion exchange matrices. Both cation and anion exchange resins have been impregnated with various mono- and bi-metallic nanoparticles (e.g., Ag, Pd, Ag/Fe_3O_4, Fe/Pd). Figure 6.40 shows palladium nanoparticles formed within an anion exchange resin through in situ synthesis.

The procedure for forming Pd^0 nanoparticles inside anion exchange resins involved two consecutive steps:

Loading of borohydride reducing agent:

$$\overline{(R^+)Cl^-} + NaBH_4 \rightarrow \overline{(R^+)BH_4^-} + Cl^- + Na^+ \tag{6.13}$$

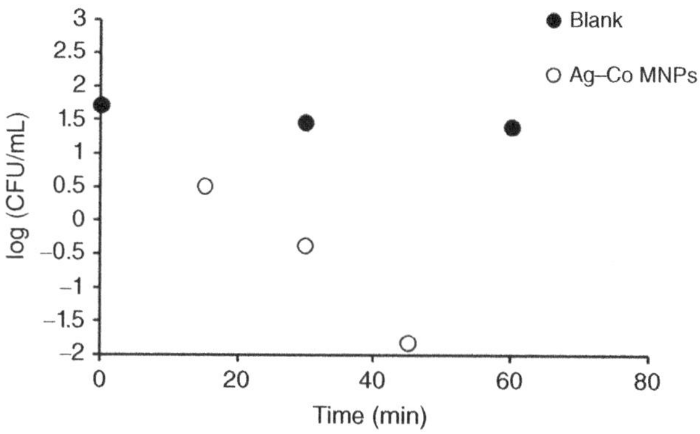

Figure 6.41 Kinetics of bactericidal treatment of *Escherichia coli*-contaminated water with FIBAN-Ag-Co filter. *Source:* Macanás *et al.* 2011 [106]. Reproduced with permission of Elsevier.

Metal reduction:

$$\overline{2(R^+)BH_4^-} + [Pd(NH_3)_4]Cl_2 + 6H_2O \rightarrow \overline{2(R^+)Cl^- + Pd^0} + 2B(OH)_3 + 4NH_3 + 7H_2(g)\uparrow \quad (6.14)$$

Catalytic activity of palladium nanoparticles has been evaluated and is strongly correlated to the total mass of palladium nanoparticles immobilized in the ion exchange resin.

Another potential application of HIX-Nanotechnology related to water treatment pertains to antimicrobial activity. Ion exchange materials are widely used for removal of hardness, toxic metals, metalloids and regulated anions. Further modification with bactericidal properties through immobilization of MNPs empowers ion exchange resins with disinfection to eliminate microbiological contaminants.

One such material is a fibrous cation exchanger (e.g., FIBAN-K1 or FIBAN K-4) with immobilized core shell Ag-Co MNPs distributed primarily at the surface of the fibers so that they are readily accessible for microbial disinfection [106]. Figure 6.41 reveals the high bactericidal activities for water contaminated with *Escherichia coli.*

Before bringing this chapter to a close, it is pertinent to emphasize the two distinct properties of HIX-Nanomaterials in contrast with other polymer–metal nanocomposites. First, the presence of MNPs and/or MONPs does not interfere with the functional groups of the ion exchanger, thus providing an opportunity for a synergy not available otherwise. Second, MNPs or MONPs, once immobilized in the ion exchanger, do not significantly aggregate or coalesce, that is, the surface area available for sorption/reaction does not diminish with prolonged use.

Summary

- Every ion exchanger has five composition variables, namely, matrix, cross-linking, functional groups, pore structure, and physical configuration. By intelligently introducing MNPs and MONPs within the ion exchanger, a sixth composition variable is introduced and a host of new application opportunities are unveiled.

- Polymeric ion exchangers are diamagnetic, that is, they have very poor magnetic susceptibility. By introducing magnetite nanocrystals within the gel phase, ion exchangers can be made magnetically active, that is, they can be separated from water and other solids through application of a magnetic field.
- Sorption capacities and specific affinities of polymeric ion exchangers are not altered due to the introduction of magnetite nanocrystals.
- Capacities of ion exchange resins do not depend on surface area. On the contrary, sorption properties of Zr(IV) and Fe(III) oxide particles involving Lewis acid–base interactions are highly surface dependent. Thus, introducing Fe(III) and Zr(IV) oxide nanoparticles within the ion exchanger materials offers new separation opportunities.
- The Donnan membrane principle offers a unique strategy to use an ion exchanger as a host material to maximize surface sorption properties of Zr(IV) and Fe(III) oxide nanoparticles. Hybrid anion exchangers with Fe(III) and Zr(IV) oxide nanoparticles (HAIX-NanoFe and HAIX-NanoZr, respectively) were synthesized with high sorption affinities toward arsenate, arsenite, phosphate, fluoride and other environmentally significant target ligands. Because of their enhanced mechanical strength and durability, HAIX-nanomaterials are amenable to regeneration and reuse for tens of cycles.
- Amphoteric sorption properties of Zr(IV) and Fe(III) oxide nanomaterials are tunable through use of ion exchanger host materials. Anion exchangers dispersed with ZrO_2 nanoparticles selectively sorb arsenate and phosphate but completely reject Cu(II) and Zn(II). Cation exchangers doped with ZrO_2 nanoparticles exhibit exactly the opposite sorption behaviors.
- HAIX-Nanomaterials have two distinctly different sorption sites: quaternary ammonium functionality and metal oxide surfaces. HAIX-Nanomaterials can thus simultaneously remove perchlorate and arsenate from contaminated groundwater, nitrate and phosphate from the treated wastewater.
- HAIX-NanoZr can be intelligently used along with a WAC exchanger to achieve partial desalination with simultaneous removal of arsenic and/or fluoride.
- Doped with palladium nanoparticles, HIX-Nanotechnology is finding novel applications as a catalyst.
- Over 2 million people in both the developed and developing countries currently drink arsenic-and fluoride-safe water through use of HAIX-Nanotechnology.

References

1 Sarkar, S., Prakash, P., and SenGupta, A.K. (2011) Chapter 7. Polymeric and inorganic hybrid ion exchanger: preparation, characterization and environmental applications, in Ion Exchange and Solvent Extraction: A Series of Advances, vol. 20 (ed. A.K. SenGupta), CRC Press, Boca Raton, FL, pp. 293–342.

2 Wang, C.G. and Zhang, W. (1997) Synthesizing nanoscale iron particles for rapid and complete dechlorination of TCE and PCBs. *Environmental Science & Technology*, **31**, 2154–2156.

3 Matheson, L.J. and Tratnyek, P.G. (1994) Reductive dehalogenation of chlorinated methanes by iron metal. *Environmental Science & Technology*, **28**, 2045–2053.
4 Miehr, R., Tratnyek, P.G., Bandstra, J.Z. *et al.* (2004) The diversity of contaminant reduction reactions by zero-valent iron: role of the reductate. *Environmental Science & Technology*, **38**, 139–147.
5 Cheng, F., Muftikian, R., Fernando, Q., and Korte, N. (1997) Reduction of nitrate to ammonia by zero-valent iron. *Chemosphere*, **35** (11), 2689–2695.
6 Manning, B.A., Fendorf, S.E., and Goldberg, S. (1998) Surface structures and stability of arsenic(III) on goethite: spectroscopic evidence for inner-sphere complexes. *Environmental Science & Technology*, **32** (16), 2383–2388.
7 Cotton, F.A. and Wilkinson, J. (1998) *Advanced Inorganic Chemistry*, Wiley-Interscience, New York.
8 Bajpai, S. and Chaudhury, M. (1999) Removal of arsenic from ground water by manganese dioxide-coated sand. *Journal of Environmental Engineering Division ASCE.*, **125** (8), 782–784.
9 Xiong, Z., Zhao, D., and Pan, G. (2007) Rapid and complete destruction of perchlorate in water and ion-exchange brine using stabilized zero-valent iron nanoparticles. *Water Research*, **41** (15), 3497–3505.
10 Chen, S., Cheng, C., Li, C. *et al.* (2007) Reduction of chromate from electroplating wastewater from pH 1 to 2 using fluidized zero valent iron process. *Journal of Hazardous Materials*, **142** (1–2), 362–367.
11 Cumbal, L., Greenleaf, J.E., Leun, D., and SenGupta, A.K. (2003) Polymer supported inorganic nanoparticles: characterization and environmental applications. *Reactive and Functional Polymers*, **54**, 167–180.
12 Lee, Y., Rho, J., and Jung, B. (2003) Preparation of magnetic ion-exchange resins by the suspension polymerization of styrene with magnetite. *Journal of Applied Polymer Science*, **89** (8), 2058–2067.
13 Eldridge, R.J., Norret, M., Dahlke, T.W., Ballard, M.J. and Adams, B.J. (2009) Complexing resins and method for preparation thereof. US Patent 7,514,500, Apr. 7.
14 Shenhar, R., Norsten, T.B., and Rotello, V.A. (2005) Polymer-mediated nanoparticle assembly: structural control and applications. *Advanced Materials*, **17** (6), 657–669.
15 Schubert, U., Gao, Y., and Kogler, F.R. (2007) Tuning the properties of nanostructured inorganic–organic hybrid polymers obtained from metal oxide clusters as building blocks. *Progress in Solid State Chemistry*, **35** (1), 161–170.
16 Balazs, A.C., Emrick, T., and Russell, T.P. (2006) Nanoparticle polymer composites: where two small worlds meet. *Science*, **314** (5802), 1107–1110.
17 Xu, Z.Z., Wang, C.C., Yang, W.L. *et al.* (2004) Encapsulation of nanosized magnetic iron oxide by polyacrylamide via inverse miniemulsion polymerization. *Journal of Magnetism and Magnetic Materials*, **277** (1–2), 136–143.
18 Li, P. and SenGupta, A.K. (1998) Genesis of selectivity and reversibility for sorption of synthetic aromatic anions onto polymeric sorbents. *Environmental Science & Technology*, **32** (23), 3756–3766.
19 Zhao, D. and SenGupta, A.K. (1998) Ultimate removal of phosphate using a new class of anion exchangers. *Water Research*, **32** (5), 1613–1625.
20 SenGupta, A.K., Zhu, Y., and Hauze, D. (1991) Metal-ion-binding onto chelating exchangers with multiple nitrogen donor atoms. *Environmental Science & Technology*, **25**, 481–488.

21 Browski, A.D., Hubicki, Z., Podkos-cielny, P., and Robens, E. (2004) Selective removal of the heavy metal ions from waters and industrial wastewaters by ion-exchange method. *Chemosphere*, **56** (2), 91–106.
22 Ghurye, G.L., Clifford, D.A., and Tripp, A.R. (1999) Combined arsenic and nitrate removal by ion exchange. *Journal AWWA*, **91** (10), 85–96.
23 Kim, J. and Benjamin, M.M. (2004) Modeling a novel ion exchange process for arsenic and nitrate removal. *Water Research*, **38** (8), 2053–2062.
24 Bolto, B., Dixon, D., Elridge, R. *et al.* (2002) Removal of natural organic matter by ion exchange. *Water Research*, **36** (20), 5057–5065.
25 Furusawa, K., Nagashima, K., and Anzai, C. (1994) Synthetic process to control the total size and component distribution of multilayer magnetic composite particles. *Colloid & Polymer Science*, **272** (9), 1104–1110.
26 Tanyolac, D. and Ozdural, A.R. (2000) Preparation of low-cost magnetic nitrocellulose micro-beads. *Reactive and Functional Polymers*, **45**, 235–242.
27 Tanyolac, D. and Ozdural, A.R. (2000) A new low cost porous magnetic material: magnetic polyvinylbutyral microbeads. *Reactive and Functional Polymers*, **43**, 279–286.
28 Ren, J., Hong, H., Ren, T., and Teng, X. (2006) Preparation and characterization of magnetic PLA–PEG composite nanoparticles for drug targeting. *Reactive and Functional Polymers*, **66**, 944–951.
29 Yanase, N., Noguchi, H., Asakura, H., and Suzuta, T.J. (1993) Preparation of magnetic latex particles by emulsion polymerization of styrene in the presence of a ferrofluid. *Journal of Applied Polymer Science*, **50**, 765–776.
30 Kondo, A., Kamura, H., and Higashitani, K. (1994) Development and application of thermo-sensitive magnetic immunomicrospheres for antibody purification. *Applied Microbiology and Biotechnology*, **41** (1), 99–105.
31 Horak, D. (2001) Magnetic polyglycidylmethacrylate microspheres by dispersion polymerization. *Journal of Polymer Science Part A: Polymer Chemistry*, **39** (21), 3707–3715.
32 Yang, C., Liu, H., Guan, Y. *et al.* (2005) Preparation of magnetic poly(methylmethacrylate–divinylbenzene–glycidylmethacrylate) microspheres by spraying suspension polymerization and their use for protein adsorption. *Journal of Magnetism and Magnetic Materials*, **293** (1), 187–192.
33 Lee, W. and Chung, T. (2008) Preparation of styrene-based, magnetic polymer microspheres by a swelling and penetration process. *Reactive and Functional Polymers*, **68** (10), 1441–1447.
34 Drever, J.I. (1988) *Geochemistry of Natural Water*, Prentice-Hall, Englewood Cliffs, NJ.
35 Leun, D. and SenGupta, A.K. (2000) Preparation and characterization of magnetically active polymeric particles (MAPP) for complex environmental separations. *Environmental Science & Technology*, **34**, 3276–3282.
36 Helfferich, F.G. (1962) Ion Exchange, McGraw-Hill, New York.
37 Blum, P. (1997) “PP Handbook: Chapter 4 Magnetic Susceptibility”, downloaded from www-odp.tamu.edu/publications/tnotes/tn26/CHAP4.PDF on 4/21/2010.
38 Diniz, C.V., Ciminelli, V.S.T., and Doyle, F.M. (2005) The use of the chelating resin Dowex M-4195 in the adsorption of selected heavy metal ions from manganese solutions. *Hydrometallurgy*, **78** (3–4), 147–155.
39 SenGupta, A.K. and Zhu, Y. (1992) Metals sorption by chelating polymers: a unique role of ionic strength. *AICHE Journal*, **38** (1), 153–157.

40 Brown, C.J. and Dejak, M.J. inventors, Eco-Tec Limited, assignee. (1987) Process for removal of copper from solutions of chelating agent and copper. US Patent 4,666,683.
41 Zhu, Y., Millan, E., and SenGupta, A.K. (1990) Toward separation of toxic metal(II) cations by chelating polymers: some noteworthy observations. *Reactive Polymers*, **13** (3), 241–253.
42 Stum, W. and Morgan, J.J. (1996) Aquatic Chemistry: Chemical Equilibria and Rates in Natural Waters, Wiley-Interscience, New York.
43 Ghosh, M.M. and Yuan, J.R. (1987) Adsorption of inorganic arsenic and organicoarsenicals on hydrous oxides. *Environmental Progress*, **3**, 150–157.
44 Dutta, P.K., Ray, A.K., Sharma, V.K., and Millero, F.J. (2004) Adsorption of arsenate and arsenite on titanium dioxide suspensions. *Journal of Colloid and Interface Science*, **278**, 270–275.
45 Pierce, M.L. and Moore, C.B. (1982) Adsorption of As(III) and As(V) on amorphous iron hydroxide. *Water Research*, **6**, 1247.
46 Jang, J.-H. (2004) Surface chemistry of hydrous ferric oxide and hematite as based on their reactions with Fe(II) and U(VI). PhD dissertation. The Pennsylvania State University.
47 Dixit, S. and Hering, J.G. (2003) Comparison of arsenic(V) and arsenic(III) sorption onto iron oxide minerals: implications for arsenic mobility. *Environmental Science & Technology*, **37**, 4182–4189.
48 Driehaus, W., Jekel, M., and Hildebrandt, U. (1998) Granular ferric hydroxide – a new adsorbent for the removal of arsenic from natural water. *Journal of Water Supply: Research and Technology – AQUA*, **47** (1), 30–35.
49 McNeil, L.S. and Edwards, M. (1995) Soluble arsenic removal at water treatment plants. *Journal AWWA*, **87** (4), 105–114.
50 Dzombak, D.A. and Morel, F.M.M. (1990) Surface Complexation Modeling. *Hydrous Ferric Oxide*, Wiley, New York.
51 Kartinen, E.O. and Martin, C.J. (1995) An overview of arsenic removal processes. *Desalination*, **103**, 79–88.
52 Gao, Y., SenGupta, A.K., and Simpson, D. (1995) A new hybrid inorganic sorbent for heavy metals removal. *Water Research*, **29** (9), 2195–2205.
53 Manning, B.A., Fendorf, S.E., and Goldberg, S. (1998) Surface structures and stability of As(III) on goethite: spectroscopic evidence for inner-sphere complexes. *Environmental Science & Technology*, **32** (16), 2383–2388.
54 Cumbal, L. and SenGupta, A.K. (2005) Arsenic removal using polymer-supported hydrated iron(III) oxide nanoparticles: role of Donnan membrane effect. *Environmental Science & Technology*, **39**, 6508–6515.
55 Min, J.M. and Hering, J. (1998) Arsenate sorption by Fe(III)-doped alginate gels. *Water Research*, **32**, 1544–1552.
56 Zouboulis, A.I. and Katsoyiannis, I.A. (2002) Arsenic removal using iron oxide loaded alginate beads. *Industrial and Engineering Chemistry Research*, **41**, 6149–6155.
57 Guo, X. and Chen, F. (2005) Removal of arsenic by bead cellulose loaded with iron oxyhydroxide from groundwater. *Environmental Science & Technology*, **39**, 6808–6818.
58 Xu, Y., Nakajima, T., and Ohki, A. (2002) Adsorption and removal of arsenic(V) from drinking water by aluminum-loaded Shirasu-zeolite. *Journal of Hazardous Materials*, **92** (3), 275–287.

59 Chen, W., Parette, R., Zou, J. *et al.* (2007) Arsenic removal by iron modified activated carbon. *Water Research*, **41** (9), 1851–1858.
60 Vaughan, R.L. Jr., and Reed, B.E. (2005) Modeling As(V) removal by a iron oxide impregnated activated carbon using the surface complexation approach. *Water Research*, **39** (6), 1005–1014.
61 Munoz, J.A., Gonzalo, A., and Valiente, M. (2002) Arsenic adsorption by Fe(III)-loaded open-celled cellulose sponge. Thermodynamic and selectively aspects. *Environmental Science & Technology*, **36** (14), 3405–3411.
62 Onyango, M.S., Kojima, Y., Matsuda, H., and Ochieng, A. (2003) Adsorption kinetics of arsenic removal from groundwater by iron-modified zeolites. *Journal of Chemical Engineering of Japan*, **36** (12), 1516–1522.
63 DeMarco, M.J., SenGupta, A.K., and Greenleaf, J.E. (2003) Arsenic removal using a polymeric/inorganic hybrid sorbent. *Water Research*, **37** (1), 164–176.
64 Katsoyiannis, I.A. and Zouboulis, A.I. (2002) Removal of arsenic from contaminated water sources by sorption onto iron-oxide-coated polymeric materials. *Water Research*, **36**, 5141–5155.
65 Gupta, A., Chauhan, V.S., and Sankararamakrishnan, N. (2009) Preparation and evaluation of iron–chitosan composites for removal of As(III) and As(V) from arsenic contaminated real life groundwater. *Water Research*, **43** (15), 3862–3870.
66 SenGupta, A.K. and Cumbal, L.H. (2007) Hybrid anion exchanger for selective removal of contaminating ligands from fluids and method of manufacture thereof. US Patent 7,291,578.
67 Sarkar, S., Blaney, L.M., Gupta, A. *et al.* (2007) Use of ArsenXnp, a hybrid anion exchanger for arsenic removal in remote villages in the Indian Subcontinent. *Reactive and Functional Polymers*, **67** (12), 1599–1611.
68 Puttamraju, P. and SenGupta, A.K. (2006) Evidence of tunable on–off sorption behaviors of metal oxide nanoparticles: role of ion exchanger support. *Industrial and Engineering Chemistry Research*, **45** (22), 7737–7742.
69 Luis, C. (2004) Polymer-supported hydrated Fe oxide (HFO) nanoparticles: characterization and environmental applications. PhD dissertation. Lehigh University, Bethlehem, PA.
70 Sylvester, P., Westerhoff, P., Möller, T. *et al.* (2007) A hybrid sorbent utilizing nanoparticles of hydrous iron oxide for arsenic removal from drinking water. *Environmental Engineering Science*, **24**, 104–112.
71 Vagliasandi, F.G.A. and Benjamin, M.M. (1998) Arsenic removal in fresh and NOM-preloaded ion exchange packed bed adsorption reactors. *Water Science and Technology*, **38** (6), 337–343.
72 Korngold, E., Belayev, N., and Aronov, L. (2001) Removal of arsenic from drinking water by anion exchangers. *Desalination*, **141**, 8184.
73 Chwirka, J.D., Thomson, B.M., and Stomp, J.M. (2000) Removing arsenic from groundwater. *Journal AWWA*, **92** (3), 79–88.
74 Driehaus, W., Seith, R., and Jekel, M. (1995) Oxidation of arsenic(III) with manganese oxides in water treatment. *Water Research*, **29** (1), 297–305.
75 Frank, P. and Clifford, D. (1986) As(III) oxidation and removal from drinking water. EPA Project Summary, Report No. EPA/600/S2-86/021, Water Engineering Research Laboratory, Environment Protection Agency, Office of Research and Development, Cincinnati, OH.
76 Ficklin, W. (1983) Separation of As(III) and As(V) in groundwaters by ion exchange. *Talanta*, **30** (5), 371.

77 Clifford, D., Ceber, L. and Chow, S. (1983) *As(III)/As(V) Separation by Chloride-Form by Ion-Exchange Resins.* Proceedings of the AWWA WQTC. Norfolk, VA.
78 Donnan, F.G. (1995) Theory of membrane equilibria and membrane potentials in the presence of non-dialysing electrolytes. A contribution to physical–chemical physiology. *Journal of Membrane Science*, **100**, 45–55.
79 Donnan, F.G. (1911) Theorie der Membrangleichgewichte und Membranpotentiale bei Vorhandensein von nicht dialysierenden Elektrolyten. Ein Beitrag zur physikalisch-chemischen Physiologie. *Berichte der Bunsengesellschaft für physikalische Chemie* **17** (14), 572–581.
80 Sarkar, S., Prakash, P., and SenGupta, A.K. (2010) Donnan membrane principle: opportunities for sustainable engineered processes and materials. *Environmental Science & Technology*, **44** (4), 1161–1166.
81 Sarkar, S., Gupta, A., Biswas, R. *et al.* (2005) Well-head arsenic removal units in remote villages of Indian Subcontinent: field results and performance evaluation. *Water Research*, **39** (10), 2196–2206.
82 Sarkar, S., Blaney, L.M., Gupta, A. *et al.* (2008) Arsenic removal from groundwater and its safe containment in a rural environment: validation of a sustainable approach. *Environmental Science & Technology*, **42** (12), 4268–4273.
83 Ravenscroft P, Brammer H, Richards K. Arsenic Pollution: A Global Synthesis : John Wiley & Sons; 2009.
84 Brammer, H. and Ravenscroft, P. (2009) Arsenic in groundwater: a threat to sustainable agriculture in South and South-east Asia. *Environment International*, **35** (3), 647–654.
85 Nickson, R., McArthur, J., Burgess, W. *et al.* (1998) Arsenic poisoning of Bangladesh groundwater. *Nature*, **395** (6700), 338.
86 Sarkar, S., Chatterjee, P.K., Cumbal, L.H., and SenGupta, A.K. (2011) Hybrid ion exchanger supported nanocomposites: sorption and sensing for environmental applications. *Chemical Engineering Journal*, **166** (3), 923–931.
87 Marcus, Y. (1985) Ion Solvation, Wiley-Interscience, London.
88 Hristovski, K., Westerhoff, P., Moller, T. *et al.* (2008) Simultaneous removal of perchlorate and arsenate by ion-exchange media modified with nanostructured iron (hydr)oxide. *Journal of Hazardous Materials*, **152**, 397.
89 Tripp, A.R. and Clifford, D.A. (2004) Selectivity considerations in modeling the treatment of perchlorate using ion-exchange processes, in Ion Exchange and Solvent Extraction (eds A.K. SenGupta and Y. Marcus), Marcel Dekker, New York.
90 Tripp, A.R. and Clifford, D.A. (2006) Ion exchange for the remediation of perchlorate contaminated drinking water. *Journal AWWA*, **98**, 105.
91 Lin, J.C. and SenGupta, A.K. (2009) Hybrid anion exchange fibers with dual binding sites: simultaneous and reversible sorption of perchlorate and arsenate. *Environmental Engineering Science*, **26** (11), 1673–1683.
92 Gu, B., Brown, G.M., Maya, L. *et al.* (2001) Regeneration of perchlorate loaded anion exchange resins by novel tetrachloroferrate displacement technique. *Environmental Science & Technology*, **35**, 3363.
93 Greenleaf, J.E. and SenGupta, A.K. (2006) Environmentally benign hardness removal using ion exchange fibers and snowmelt. *Environmental Science & Technology*, **40**, 370–376.

94 Greenleaf, J.E., Lin, J.C., and SenGupta, A.K. (2006) Two novel applications of ion exchange fibers: arsenic removal and chemical-free softening of hard water. *Environmental Progress*, **25** (4), 300–311.
95 Padungthon, S., German, M., Wiriyathamcharoen, S., and SenGupta, A.K. (2015) Polymeric anion exchanger supported hydrated Zr(IV) oxide nanoparticles: a reusable hybrid sorbent for selective trace arsenic removal. *Reactive and Functional Polymers*, **93**, 84–94.
96 SenGupta, A.K. and Padungthon, S. (2015) inventors; Lehigh University, assignee. Hybrid anion exchanger impregnated with hydrated zirconium oxide for selective removal of contaminating ligand and methods of manufacture and use thereof. US patent 9,120,093. September 1.
97 SenGupta, A.K., Li, J. and German, M. inventors; Lehigh University, assignee. The Process of Defluoridation with Concurrent Desalination (PDCD) through use of Hybrid Anion Exchanger. US patent 62/313,381. 2016 Mar 25.
98 Padungthon, S., Li, J., German, M., and SenGupta, A.K. (2014) Hybrid anion exchanger with dispersed zirconium oxide nanoparticles: a durable and reusable fluoride-selective sorbent. *Environmental Engineering Science*, **31** (7), 360–372.
99 Sarkar, S., Guibal, E., Quignard, F., and SenGupta, A.K. (2012) Polymer-supported metals and metal oxide nanoparticles: synthesis, characterization, and applications. *Journal of Nanoparticle Research*, **14** (2), 1–24.
100 Zhang, W., Wang, C., and Lien, H. (1998) Treatment of chlorinated organic contaminants with nanoscale bimetallic particles. *Catalysis Today*, **40** (4), 387–395.
101 Elliott, D.W. and Zhang, W. (2001) Field assessment of nanoscale bimetallic particles for groundwater treatment. *Environmental Science & Technology*, **35** (24), 4922–4926.
102 Zhang, W. (2003) Nanoscale iron particles for environmental remediation: an overview. *Journal of Nanoparticle Research*, **5** (3–4), 323–332.
103 Li, X., Elliott, D.W., and Zhang, W. (2006) Zero-valent iron nanoparticles for abatement of environmental pollutants: materials and engineering aspects. *Critical Reviews in Solid State and Materials Sciences*, **31** (4), 111–122.
104 Mills, G.F. and Dickinson, B.N. (1949 Dec) Oxygen removal from water by amine exchange resins. *Industrial & Engineering Chemistry*, **41** (12), 2842–4.
105 Bastos-Arrieta, J., Shafir, A., Alonso, A. *et al.* (2012) Donnan exclusion driven intermatrix synthesis of reusable polymer stabilized palladium nanocatalysts. *Catalysis Today*, **193** (1), 207–212.
106 Macanás, J., Ruiz, P., Alonso, A. *et al.* (2011) Chapter 1: Ion-exchange assisted synthesis of polymer-stabilized metal nanoparticles, in Ion Exchange and Solvent Extraction: A Series of Advances, vol. 20, CRC Press, pp. 1–44.

7

Heavy Metal Chelation and Polymeric Ligand Exchange

Since the industrial revolution, the beneficial use of metals has seen a tremendous growth around the world and essentially reshaped our civilization. For a long time, a number of these metals, often referred to as heavy metals for their toxicity, were indiscriminately released into the environment, including the waterways. It required decades for the fate, transport and deleterious effects of these heavy metals to be comprehended. Gradually their impact on the quality of life on our planet surfaced as a matter of grave concern, ushering in a series of environmental regulations during the last 50 years. A specific class of ion exchangers, often called chelating ion exchangers, emerged as robust sorbents to remove and sequester heavy metals from water. As already discussed in previous chapters, phosphate, arsenate, fluoride and organic carboxylates are environmentally significant ligands that exhibit high sorption affinity toward oxides of Fe(III), Zr(IV), Al(III), and Ti(IV). Polymeric sorbents for these anionic ligands do not exist to date.

This chapter is devoted to the underlying science, development and application of ion exchanging polymers concerning heavy metal chelation and ligand sorption.

7.1 Heavy Metals and Chelating Ion Exchangers

7.1.1 Heavy Metals: What are They?

The term "heavy metal," in spite of its widespread use among professionals and laymen, has no rigorous scientific basis or a chemical definition. Although many of the elements listed under "heavy metals" have specific gravities greater than five, major exceptions to this rule remain. In hindsight, this group should preferably have been referred to as "toxic elements" for they are all included in the United States Environmental Protection Agency's (USEPA's) list of priority pollutants. Figure 7.1 shows the periodic table containing the heavy metals that are of significant environmental concern. For comparison, commonly occurring light alkali and alkali-earth metals have also been included in the same figure. Strictly from a chemical viewpoint, heavy metals constitute transition and post-transition elements along with metalloids, namely, arsenic and selenium. They are indeed significantly heavier (i.e., higher specific gravities) than sodium, calcium and other light metals. These heavy metal elements often exist in different oxidation states in soil, water, and air. The reactivities, ionic charges and solubilities of these metals in water vary widely. For their short- and long-term toxic effects,

Ion Exchange in Environmental Processes: Fundamentals, Applications and Sustainable Technology,
First Edition. Arup K. SenGupta.

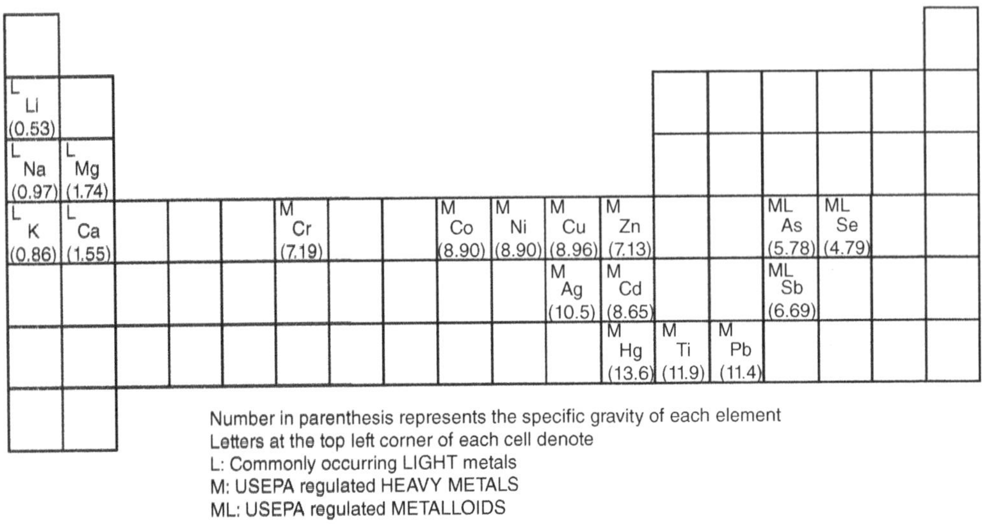

Figure 7.1 A modified periodic table showing common regulated heavy metals, metalloids and unregulated light metals. *Source:* Sengupta 2001 [1]. Reproduced with permission of Taylor & Francis.

the maximum permissible concentrations of these heavy metals in drinking water, as well as in municipal and industrial discharges, are closely regulated through legislation. Yet, barring cadmium, mercury and lead, heavy metals are also required micronutrients, that is, essential ingredients for living cells. Toxicity effects of these elements are, thus, largely a function of concentration. These elements are beneficial and have nutritional values lower than some critical dosages but become inhibitory to toxic with an increase in concentration, as shown in Figure 7.2. The threshold toxic concentrations differ for each heavy metal and are governed primarily by the chemistry of each heavy metal in question and associated physiological effects. On the contrary, nonessential heavy metal elements are inhibitory at all concentrations.

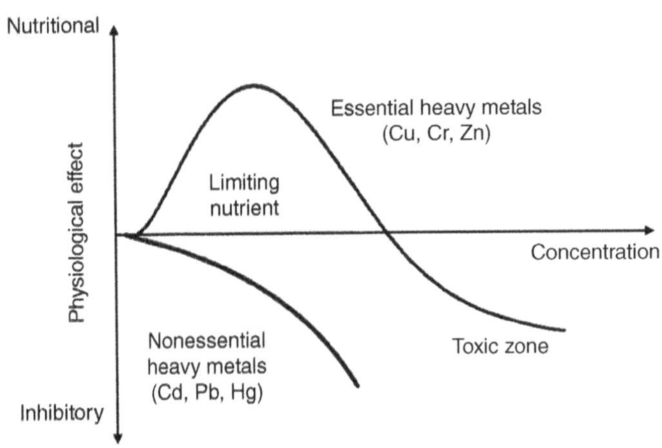

Figure 7.2 Nutritional and inhibitory effects of heavy metal concentrations on living cells/microorganisms. *Source:* Sengupta 2001 [1]. Reproduced with permission of Taylor & Francis.

Metal cycles on a regional and global basis have been profoundly modified by human activity and industrial development during the last 50 years. While industrial mining, metallurgy, electroplating, etc., have greatly boosted the production and usage of heavy metals in our life cycles, the lowering of pH in rain and surface waters, and the increased use of surfactants, have greatly enhanced the mobility of heavy metals in the environment. Understandably, the presence of heavy metals in aquatic, terrestrial and atmospheric environments is of concern. In the aqueous-phase, such heavy metals may exist as cations, anions, nonionized species and complex macromolecules. As most of the heavy metals and their compounds have extremely high boiling points, they are practically absent in the atmosphere under ambient conditions, with the glaring exception of elemental mercury. Flue gases from fossil fuel-fired steam generators and waste incinerators are major industrial sources of mercury emissions into the atmosphere. Higher volatility and relative inertness compared to other heavy metals allows elemental mercury to persist in the environment for a prolonged period of time. Following the phase-out of leaded gasoline in industrial countries, the short- and long-term presence of lead in the atmosphere has greatly subsided. In the soil-phase, heavy metals exist primarily as insoluble precipitates or as bound solutes on the surface sorption sites of microparticles. Mobility and fate of the heavy metals in the soil-phase are often influenced by the chemical composition of the contacting liquid-phase [1]. In recent years, leaching of lead from lead pipes and solder joints, and subsequent contamination of drinking water have been adversely affecting human health and have emerged as a matter of grave social and political concern [2–9].

7.1.2 Properties of Heavy Metals and Separation Strategies

The speciation and fate of metals in the natural environment as well as their separation and/or control by engineered processes are ultimately governed by the electronic structures of the heavy metals. Such electronic structures also dictate the biochemical actions of metals as nutrients or toxicants. To develop an insight, let us consider the electronic configurations of a light metal cation (say Ca^{2+}) and a heavy metal cation (say Cu^{2+}) as shown below:

$$Ca^{2+}:\ 1s^2 2s^2 2p^6 3s^2 3p^6 \tag{7.1}$$

$$Cu^{2+}:\ 1s^2 2s^2 2p^6 3s^2 3p^6 3d^9 \tag{7.2}$$

Note that Ca^{2+} has the noble gas configuration of Krypton, that is, its outermost electron shell is completely filled, and the octet formation is satisfied. Thus, Ca^{2+} is not a good electron acceptor and, hence, a poor Lewis acid. Ions like Ca^{2+} are not readily deformed by electric fields and have low polarizabilities. They are referred to as "hard" cations, and they form only outer sphere complexes with aqueous-phase ligands containing primarily oxygen donor atoms.

In contrast, the transition metal cation, Cu^{2+} or Cu(II), has an incomplete d-orbital and contains electron clouds more readily deformable by electric fields of other species. In general, these ions are fairly strong Lewis acids and tend to form inner sphere complexes with ligands in the aqueous-phase. Electrostatically, Ca^{2+} and Cu^{2+} are identical, that is, both Ca^{2+} and Cu^{2+} have two charges. But, Cu(II) is a stronger Lewis acid or electron acceptor and a relatively "soft" cation. Table 7.1 classifies

Table 7.1 Classification of selected metal cations.

Type	Name	Properties
Hard cations	Na^+, K^+, Mg^{2+}, Ca^{2+}, Al^{3+}, Be^{2+}, etc.	Spherically symmetric and electronic configurations conform to inert gases; form only outer-sphere complexes with hard ligands containing oxygen donor atoms; weak affinity toward ligands with nitrogen and sulfur donor atoms. Besides beryllium, most are non-toxic at low concentrations
Borderline cations	Fe^{3+}, Cu^{2+}, Pb^{2+}, Fe^{2+}, Ni^{2+}, Zn^{2+}, Co^{2+}, Mn^{2+}	Spherically asymmetric, and electronic configurations do not conform to inert gases. Form inner-sphere complexes with O- and N-atom-containing ligands. Excepting iron and manganese, all are toxic beyond a threshold level
Soft cations	Hg^+, Cu^+, Hg^{2+}, Ag^+, Cd^{2+}	Spherically asymmetric, and electronic configurations do not conform to inert gases; exhibit high affinity toward S-atom-containing ligands. They are most toxic from a physiological viewpoint

several metal cations, in three categories, namely, hard, borderline and soft [10,11]. Note that most of the heavy metals of interest fall under "borderline" and "soft." In general, the toxicity of metals increases as one moves from hard cations to borderline and then to soft. Relative affinities vary widely for these metal ions to form complexes with O-, N-, and S-containing ligands. While hard cations prefer oxygen-donating ligands (Lewis bases), borderline and soft cations exhibit higher affinities toward nitrogen- and sulfur-containing ligands. The soft cations thus bind strongly with sulfhydryl groups in proteins of cells. Because sulfhydryl groups form active sites on proteins, their blockages through heavy metal binding result in severe toxic effects [12]. Mercury and lead are particularly notorious in forming very stable complexes with sulfhydryl groups and are classified as neurotoxins.

The foregoing phenomenon prompted Nierboer and Richardson to recommend that toxic metals be classified by their relative complex forming abilities with O-, N-, P-, and S-containing ligands, for such affinities are the primary determinants of physiological toxicity caused by the metals [13]. The fact that many heavy metals bind strongly onto proteins also suggests that the functional groups in proteins, when immobilized onto a solid-phase, may selectively capture dissolved heavy metals from the aqueous-phase.

Ionic charges, Lewis acidity/basicity, surface functional group sorption affinities, aqueous-phase solubilities, metal-ligand complex sizes, redox states, etc., can

Table 7.2 Size of a heavy metal cation (Me^{2+}) in water in different physicochemical states.

Dissolved state	Speciation	Approximate diameter (nm)
Water	H_2O	0.2
Hydrated free metal ion	$[Me(H_2O)_n]^{2+}$	~0.5
Inorganic complexes	$[Me(NH_3)_n]^{2+}$ $[MeOH]^+$ $[Me(OH)_2]^0$ $[MeCO_3]_2^{2-}$	<1.5
Organic complexes	$[Me(COO)_2]^0$ $[Me(NH_3^+)_n]^{2+}$ $[Me(EDTA)]^{2-}$	1–5
Macromolecules/colloids	Me–Humate complex Me–Fulvate complex Me–NOM-coated silica	10–500
Surface binding onto microparticles	$\left\{\begin{matrix} FeO^- \\ FeO^- \end{matrix}\right. (Me^{2+})$	100–10,000
Precipitates	$Me(OH)_2(s)$ $MeCO_3(s)$	>500

be manipulated to achieve efficient control and separation of heavy metals from aqueous-phases and other complex systems. Table 7.2 provides the estimated sizes of divalent heavy metal cations, Me(II), in different physicochemical forms. Figure 7.3 shows a schematic illustrating a wide variety of strategies for heavy metals separation. Understandably, each of them has the potential to be a viable metal separation process under a specific set of conditions. In certain instances, combinations of more than one, that is, a hybrid process, may be the most suitable. But all such applications tend to have one major drawback – they are unable to recover industrial heavy metals with a high degree of purity for reuse. With pollution prevention guidelines and the concept of industrial ecology in place, research and development are underway to separate individual heavy metals and enhance their purities in recovered materials.

7.1.3 Emergence of Chelating Exchangers

Most of the heavy metal cations of interest, such as Cu^{2+}, Hg^{2+}, Pb^{2+}, Ni^{2+}, Cd^{2+}, Zn^{2+}, etc. are transition-metal cations and exhibit Lewis-acid characteristics (electron acceptors). With organic and inorganic ligands (Lewis bases), all these heavy metal cations form fairly strong complexes. Most of the complexes of these metal cations, depending on their coordination number, have regular or slightly distorted tetrahedral, octahedral, or square pyramid structures [14]. Because Ca^{2+}, Mg^{2+}, and Na^+ – the most

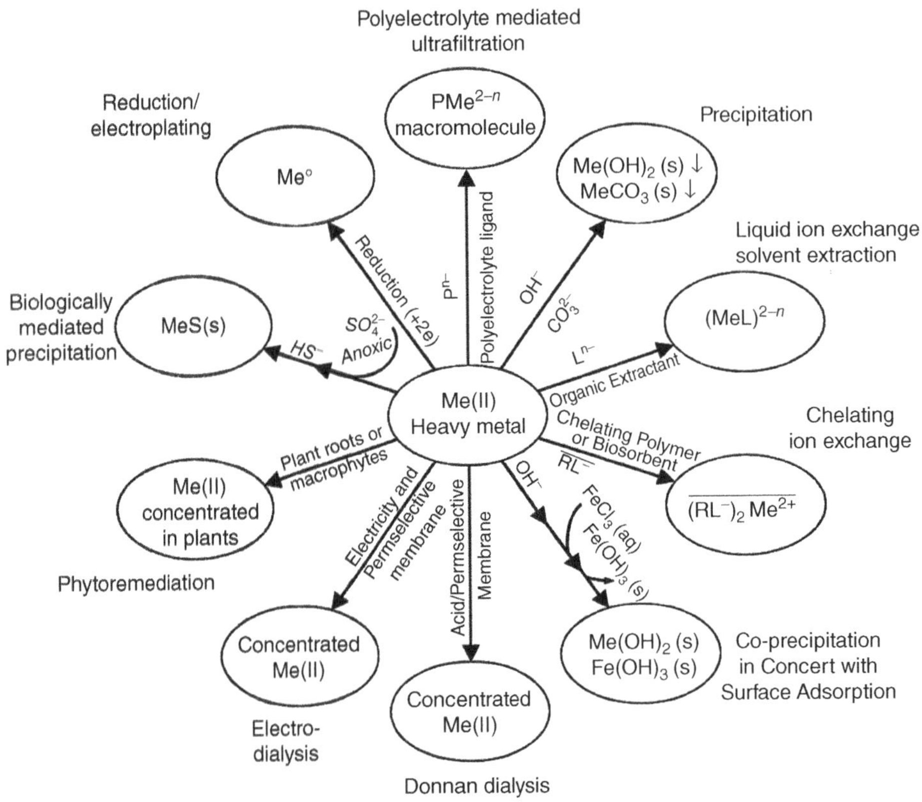

Figure 7.3 A schematic illustrating various engineered processes for heavy metals separation. *Source:* Sengupta 2001 [1]. Reproduced with permission of Taylor & Francis.

commonly encountered competing nontoxic cations in water and wastewater – do not undergo such strong complexation, incorporating organic ligands as functional groups into the polymer matrix of the ion exchanger through covalent bonding was a natural progression of ideas to improve the exchanger's selectivity toward the toxic metal ions. These functionalized polymers are often referred to as chelating polymers, coordinating polymers or metal-selective ion exchange resins.

Tens of polymeric chelating exchangers have been synthesized to date, and are commercially available with various types of covalently attached functional groups. Physically, they are all the same, that is, spherical beads with high mechanical strength and durability. Figure 7.4 illustrates several commercially available chelating exchangers with linear polymer chains, cross-linkings and a variety of covalently attached functional groups. Understandably, it is the Lewis acid–base (LAB) interaction that governs the binding affinity of a heavy metal cation to a chelating exchanger. Such binding affinities (often expressed as separation factor values) are correlated to corresponding aqueous-phase stability constant values between the heavy metal ions and the representative ligands, and they can be modeled by their linear free energy relationships (LFERs) [15]. Figure 7.5 shows the relationship between copper/calcium separation factor values for three commercial chelating exchangers and

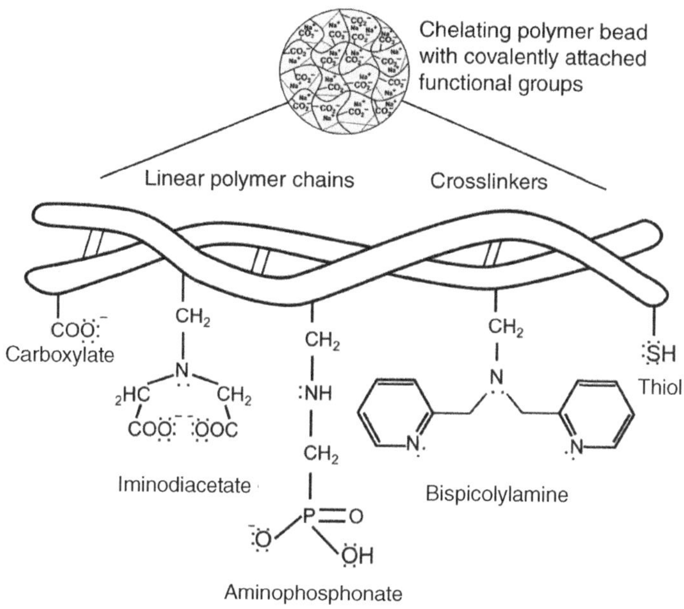

Figure 7.4 An illustration depicting a chelating polymer bead with different covalently attached functional groups. *Source:* Sengupta 2001 [1]. Reproduced with permission of Taylor & Francis.

Figure 7.5 Relationship between copper/calcium separation factors for commercial chelating exchangers and corresponding aqueous-phase stability constant values with representative ligands. *Source:* Sengupta 2001 [1]. Reproduced with permission of Taylor & Francis.

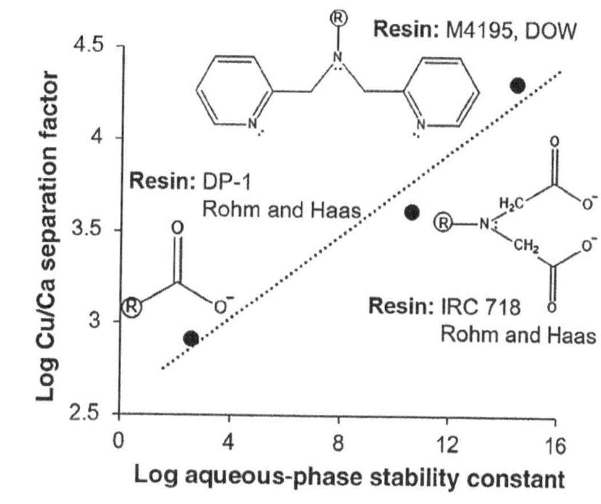

the corresponding aqueous-phase stability constant values for representative ligands [16]. Noteworthy is the fact that as the composition of the functional groups in Figure 7.5 changes from hard oxygen donor atoms (i.e., carboxylate) to relatively soft nitrogen donor atoms (bispicolylamine), the affinity of Cu(II), a borderline Lewis acid, is greatly enhanced over the hard cation, Ca^{2+}. Understandably, the composition of the functional groups in chelating exchangers can be judiciously tailored to improve specific affinities toward target metal ions. Chelating exchangers with S-containing thiol functional groups offer significantly higher selectivity for soft Hg(II) over Cu(II) and Zn(II).

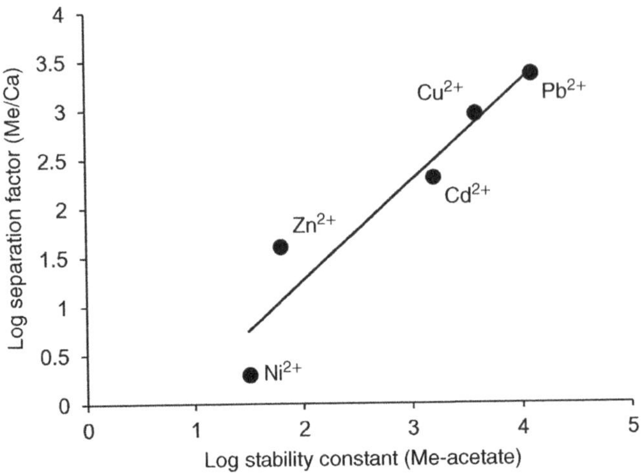

Figure 7.6 Relationship between experimentally determined metal/calcium separation factors for IRC DP-1 and aqueous-phase metal-acetate stability constant values. *Source:* Sengupta 2001 [1]. Reproduced with permission of Taylor & Francis.

Along the same vein, Figure 7.6 shows the separation factor values of five different heavy metal cations for a weak-acid cation exchange resin with carboxylate functional groups (IRC DP-1, Rohm and Haas Co., Philadelphia, PA, USA). Note that the sequence and relative affinity of dissolved heavy metals are strongly correlated to their aqueous-phase metal acetate stability constant values [17].

Barring a few exceptions, LAB interactions, aided by sorption, sieving, precipitation, etc., constitute the primary mechanism for heavy metal removal from the aqueous phase. Many biorenewable materials, such as naturally occurring humus, for example, dead bacteria, fungi and seaweeds, contain surface functional groups (e.g., carboxylate, carbonyl, phenolic) with moderate to high affinity toward heavy metals. Significant progress has been made recently in modifying such materials into chemically stable and mechanically durable sorbents [17,18]. As we lay an increased emphasis on sustainable development, these sorbent materials are likely to be economically competitive, and large-scale commercial production will follow.

7.1.4 Lewis Acid–Base Interactions in Chelating Ion Exchangers

Ion exchangers with iminodiacetate functional groups happen to be the first group of commercial chelating exchangers prepared solely with the purpose of sequestering toxic metal ions in the presence of much higher concentrations of competing calcium and sodium. A typical ion exchange reaction between a metal ion, Me^{2+}, and Na^+ for this resin may be presented in the following way:

$$\overline{R-N(CH_2COO^-Na^+)_2} + Me^{2+} \rightarrow \overline{R-N(CH_2COO^-)_2Me^{2+}} + 2Na^+ \qquad (7.3)$$

Equation (7.3) however, fails to reveal the LAB type interaction between the metal ion and the iminodiacetate functionality. Assuming the metal ion has four coordinated water molecules in the aqueous phase, $[Me(H_2O)_4]^{2+}$, the overall exchange involves the following:

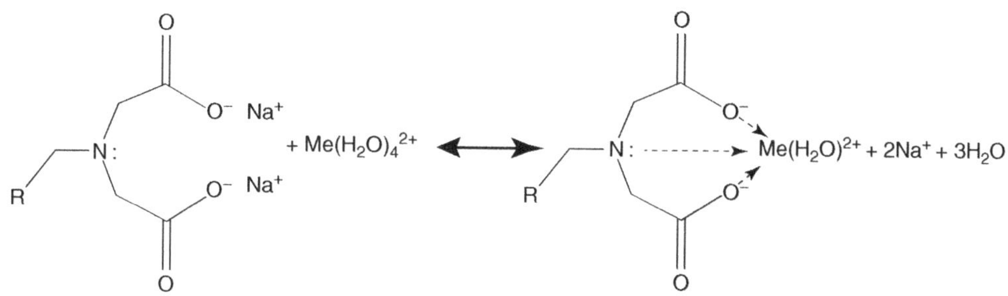

(7.4)

Note that three water molecules (i.e., ligands) from the coordination sphere of the metal ion are replaced by one nitrogen and two oxygen donor atoms in the iminodiacetate functionality. The arrows indicate the metal-ligand or LAB interaction, and the high metal-ion selectivity for this type of functional group is often attributed to the accompanying coordination reaction in conjunction with exchange of ions. Figure 7.7 provides experimentally determined, Me^{2+}/Ca^{2+} separation factors for three commercial ion exchange resins (iminodiacetate functionality – IRC-718, thiol functionality – GT-73, and picolylamine functionality – XFS-4195) for various heavy

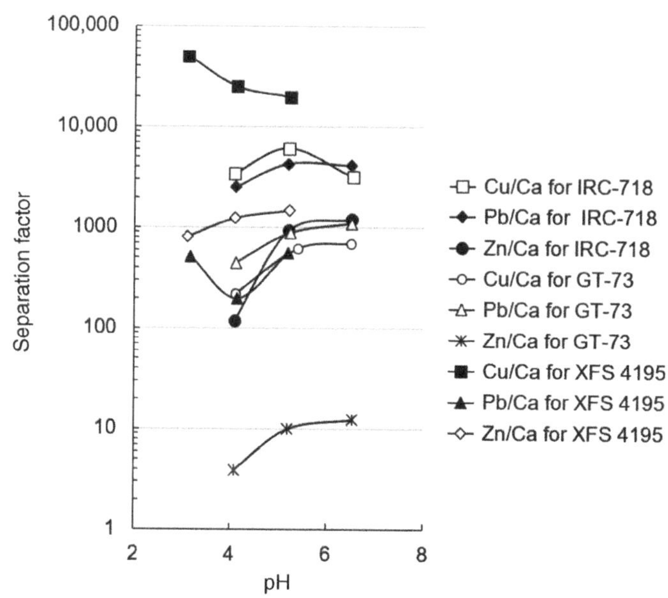

Figure 7.7 Experimentally determined Me(II)-Calcium separation factor values as a function of pH for various resins: IRC-718 = iminodiacetate functionality, GT-73 = thiol functionality and XFS 4195 = bispicolylamine functionality. *Source:* Sengupta and SenGupta 2002 [19]. Reproduced with permission of Taylor & Francis.

metals at varying pH values, and the high selectivity of the metal ions can be readily noted [17], where

$$\text{Separation factor, } \alpha_{Me/Ca} = \frac{\overline{[RMe]}[Ca^{2+}]}{\overline{[RCa]}[Me^{2+}]} \tag{7.5}$$

As discussed in Chapter 2, separation factor is a dimensionless measure of relative selectivity between two competing ions, and in this case, is equal to the ratio of the distribution coefficient of the metal ion concentration between exchanger and aqueous phases to that of calcium ion. Like Ca^{2+}, these metal ions have a charge 2+ and, therefore, only Coulombic/electrostatic interaction cannot be the reason for such a high Me^{2+}/Ca^{2+} separation factor. On the contrary, such a high selectivity is always attributed to the relatively strong Lewis acid characteristic of the toxic-metal cations, favoring their selective uptake through coordination reactions.

To characterize high metal-ion selectivity for chelating ion exchangers from a thermodynamic perspective, the metal ion uptake can be divided into two consecutive steps – ion exchange (IX) followed by LAB interaction, that is,

$$Me^{2+}(aq) \xrightarrow{\text{IX (Step I)}} \overline{RMe} \xrightarrow{\text{LAB (Step II)}} \overline{RMe} \tag{7.6}$$

At the standard state, the overall free energy change at equilibrium between $Me^{2+}(aq)$ and $\overline{RMe}$ is given by the following:

$$\Delta G^0_{overall} = \Delta G^0_{IX} + \Delta G^0_{LAB} \tag{7.7}$$

or

$$-RT\ln K_{overall} = -RT\ln K_{IX} - RT\ln K_{LAB} \tag{7.8}$$

or

$$K_{overall} = K_{IX} \cdot K_{LAB} \tag{7.9}$$

In general, for ion exchangers with chelating functionality, K_{LAB} is very high for most of the heavy-metal ions of interest due to their Lewis acid characteristics. So, the overall equilibrium constants, according to Eq. (7.9), are also very high. For sodium, Na^+, LAB interaction (step II) is practically absent and, hence,

$$K_{overall} = K_{IX} \tag{7.10}$$

For Ca^{2+}, however, LAB interaction is present, but is much weaker compared to most of the heavy metal cations, and thus, as a general rule, the selectivity sequence for chelating exchangers may be written as follows;

$$K_{overall}(\text{heavy metal}) \gg K_{overall}(\text{calcium}) \gg K_{overall}(\text{sodium}) \tag{7.11}$$

Such a high metal-ion selectivity and more stringent environmental regulations have aroused high interest in the application of these chelating polymers for removal, separation and purification of metal ions from heavy-metal contaminated water and wastewater streams [20–28].

In spite of wide variation in the composition of chelating functionalities, nitrogen, oxygen, and sulfur are the donor atoms in almost every chelating exchanger synthesized to date. Identifying the active donor atoms for a given application may provide useful clues to assess metal-ion selectivity and other related properties for a given chelating polymer. These donor atoms form only a part of the complete chelating functionality, which is essentially either weak-acid (e.g., carboxylate, diacetate, thiol, etc.) or weak-base (tertiary amine, pyridine, etc.). Due to the weak-acid or weak-base characteristic, these chelating functionalities exhibit high affinity toward the hydrogen ion. As a result, selective uptake of heavy metal cations by chelating exchangers under highly acidic conditions ($pH < 2.0$) is adversely affected due to strong competition from H^+. On the other hand, at neutral to alkaline pH, heavy metal cations are quite insoluble because of low solubility product values for their hydroxides, carbonates, sulfides, etc. For effective heavy metal removal, the optimum pH range for most of the chelating polymers is often limited to pH 2.0–7.0.

In general, metal(II)-ligand complexes in the aqueous phase have regular or slightly distorted tetrahedral, octahedral or square pyramid structures depending on the metal ion's coordination number that, in turn, is related to the number of non-bonding electrons in its d-orbital [14,29]. It is true that in a chelating polymer with multiple binding sites, the metal ions try to reproduce their aqueous-phase stereochemistry. However, the functionality in a chelating polymer is often rigidly bound to a repeating monomer (e.g., styrene) that, again, is fixed as part of a three-dimensional network cross-linked through divinylbenzene. As a result, the donor atoms (N, O, or S) in the polymer phase will experience considerable strain to orient themselves spatially around the receptor metal ions. This strain, which may be viewed as an extra thermodynamic parameter, may not allow the individual functionality in the polymer phase to reproduce its aqueous-phase metal-ligand configuration.

Experimental results are available for more widely used chelating exchangers with iminodiacetate, amino-phosphonate and carboxylate functionalities. In this context, the consensus is that a metal(II) ion can bind at most with one nitrogen and two oxygen atoms for iminodiacetate or amino-phosphonate exchangers and with two oxygen atoms from two neighboring carboxylate groups for carboxylate-type exchangers [30,31]. Table 7.3 shows a general schematic for these binding mechanisms where carboxylate, iminodiacetate and amino-phosphonate exchangers act as polydentate ligands for a metal(II) ion with a coordination number of 4 or 6. Also note that in all three cases, the 2+ charge of the metal ion is neutralized by the fixed negative charge in the polymer phase. This is the reason why these binding mechanisms are viewed as cation exchange accompanied by chelation, and the anions in the aqueous phase are excluded from the polymeric exchangers due to the Donnan coion exclusion effect [31]. Such a model (cation exchange followed by chelation) can quantitatively explain the equilibrium behavior of these resins quite satisfactorily. For several chelating exchangers with bi- or polydentate functionality, however, individual donor atoms have been reported [32] to be binding metal ions independently on a molar basis.

It is only appropriate to mention the tragic revelation of lead-contaminated drinking water in Flint, MI and its adverse health impact on children who unknowingly ingested the water for months [2–9]. Leaching from lead pipes and joints under corrosive conditions was responsible for contamination. Dissolved lead does not have color, odor or taste and is a neurotoxin at concentrations as low as 15 μg/L. Results in Figure 7.7 demonstrate extremely high lead selectivity for several commercial chelating

Table 7.3 A schematic representation of the nature of metal ion binding (cation exchange followed by chelation) for aminophosphonate, iminodiacetate and carboxylate functional groups.

Functionality	Formula	Donor atoms
Aminophosphonate	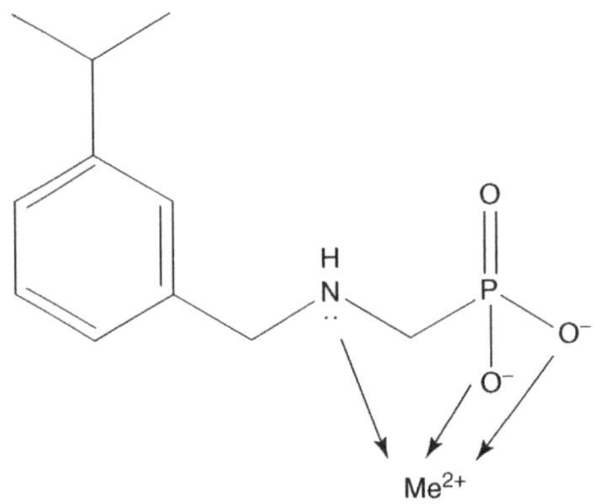	Nitrogen, Oxygen
Iminodiacetate	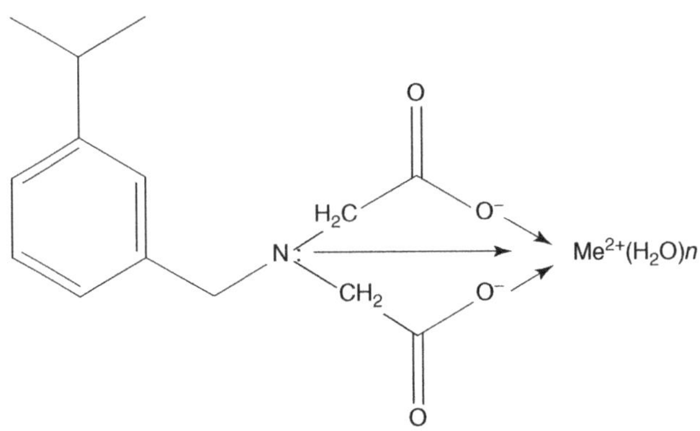	Nitrogen, Oxygen
Carboxylate	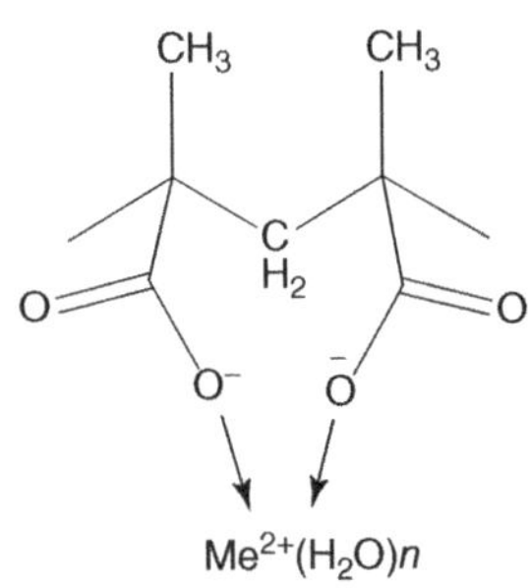	Oxygen

ion exchangers. Their use in point-of-use home filters (e.g., Brita) will ensure elimination of lead from large volumes of contaminated water.

7.1.5 Regeneration, Kinetics and Metals Affinity

Chelating exchangers' high preference for H^+ is often viewed as a shortcoming for heavy metal removal under highly acidic conditions, but it offers an excellent regeneration of metal-loaded chelating polymers with moderately concentrated

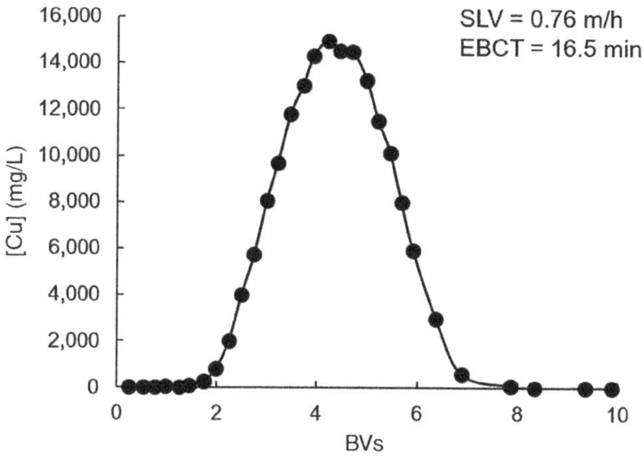

Figure 7.8 Demonstration of high regeneration efficiency of copper-loaded IRC-718 (iminodiacetate functionality) with 2% HCl. *Source:* Zhu *et al.* 1990 [33]. Reproduced with permission of Elsevier.

(2–10%) mineral acid. From a practical viewpoint, high regeneration efficiency of a chelating exchanger is just as desirable as high metal-ion affinity. Figure 7.8 shows the regeneration of copper-loaded IRC-718 (iminodiacetate functionality) with 2% HCl [33]. As expected, copper desorption/elution was very sharp, with copper concentration in the spent regenerant as high as 15,000 mg/L, indicating efficient regeneration in accordance with the principles of displacement chromatography [31]. Incidentally, several toxic metals, namely, Cu, Pb, Hg, Cd, Zn, and Ni, are included in EPA's list of priority pollutants [34], and one of the major challenges is to minimize the volume of these metal-contaminated wastes. High metal-ion selectivity of the chelating exchangers accompanied by excellent acid-regeneration efficiency offers opportunities to concentrate and reduce the volume of metal-laden dilute wastewater streams, often by over 1000× [35].

For chelating ion exchangers, both sorption (i.e., metals uptake) and regeneration (i.e., H^+ elution) are highly favorable from an equilibrium viewpoint. Let us consider the exchange of Cu^{2+} and Ca^{2+} during sorption and regeneration with HCl for chelating exchangers with iminodiacetate functionalities:

Sorption:

$$\overline{R - N(CH_2COO^-)_2Ca^{2+}} + Cu^{2+} \rightarrow \overline{R - N(CH_2COO^-)_2Cu^{2+}} + Ca^{2+} \tag{7.12}$$

Regeneration:

$$\overline{R - N(CH_2COO^-)_2Cu^{2+}} + 2H^+ \rightarrow \overline{R - N(CH_2COOH)_2} + Cu^{2+} \tag{7.13}$$

Both Cu^{2+}–Ca^{2+} and H^+–Cu^{2+} equilibria represent rectangular-type isotherms. As discussed in Chapter 4 (Section 4.9), intraparticle diffusion-controlled kinetics conforms to shell progressive kinetics. The uptake of Cu^{2+} during sorption, and H^+ during regeneration can be represented as shell-progressive or shrinking-core kinetics. Visually, the progress of this special class of diffusion can be presented as shown in Figure 7.9.

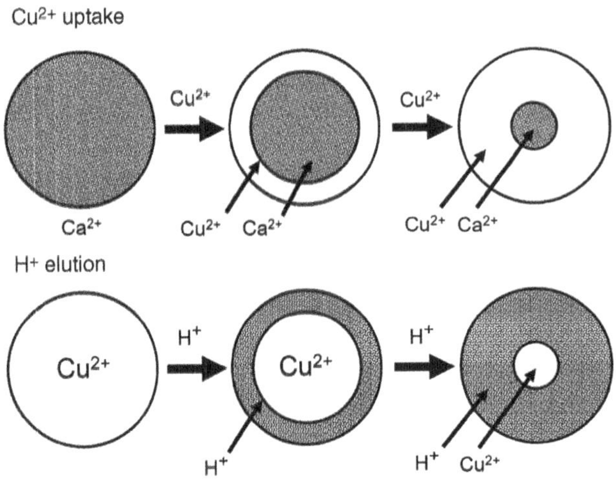

Figure 7.9 Visual representation of shrinking-core kinetics for copper sorption and regeneration with a chelating ion exchanger.

Uptake kinetics of Cu^{2+} ions onto a chelating ion exchanger have been studied and photographs confirm shrinking-core kinetics [36]. For a rectangular isotherm with shell progressive kinetics, the half-time ($t_{1/2}$), or time to attain 50% uptake, varies with r_o^2, as shown earlier in Eq. (4.67), where r_o is the radius of the ion exchanger bead. Obviously, reducing the bead size, or using thin ion exchange fibers will greatly enhance the sorption kinetics.

It is important to recognize that metals sorption onto chelating exchangers is greatly influenced by the presence of ligands in the aqueous phase and the prevailing pH. In the presence of polydentate organic ligands, common in many wastewater streams, the metal-removal capacity of chelating exchangers can be seriously impaired. An experimental study was undertaken to investigate this aspect, using ethylenediamine (EN), a bidentate weakly basic ligand with high affinity toward Me(II) cations [33]. Figure 7.10 presents how Cu/Ca separation factor is influenced by EN/M_T ratio and pH, where M_T corresponds to the concentration of heavy metal cations.

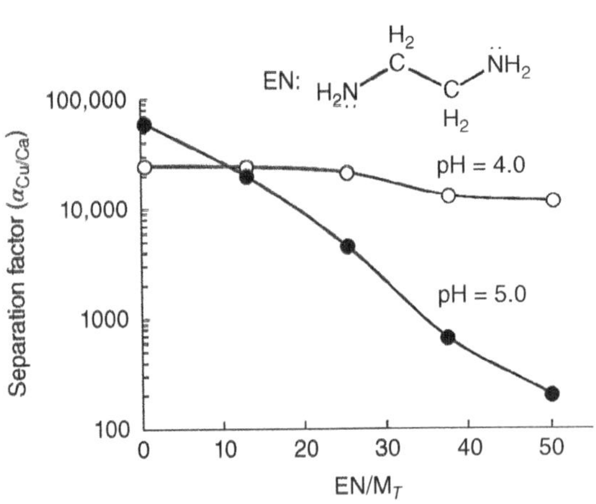

Figure 7.10 Effects of ethylenediamine and pH on copper/calcium separation factor for IRC-718 (iminodiacetate functionality). *Source:* Zhu *et al.* [33]. Reproduced with permission of Elsevier.

Note that at pH 4.0, an increase in EN/M_T had a negligible effect on Cu/Ca separation factor, but at pH 5.0, there was a rapid drop in the separation factor. Mechanistically, the governing factor here is the complexation of EN with H^+ at pH 4.0; EN is protonated and does not complex with Cu(II). But, at pH 5.0, Cu(II) complexes significantly with EN and hence Cu(II) uptake onto IRC-718 is greatly reduced. Quantitative prediction of such an effect is presented in the reference [33]. These examples help illustrate the complexity of a real system where multiple ligands, with varying acid-dissociation constant values, are present simultaneously. It is also pertinent to note that the pH in the presence of appropriate ligands may also be harnessed to separate heavy metals from one another. For further details on the subject, the following reference entitled "Metal Separation by pH-driven Parametric Pumping" is recommended [37].

7.2 Polymeric Ligand Exchange

The field of chelating ion exchange or selective metal sorption has made major strides during the last several decades. It has been possible due to the covalent attachment of chelating functional groups onto the cross-linked polymer matrix. Selective removal of ligands of interest, however, poses insurmountable difficulty because sorption sites must contain transition metal cations or Lewis acids that do not form covalent attachments to a polymer matrix. That is the underlying reason why oxides of polyvalent metals, namely, aluminum, iron, titanium and zirconium, are universally used for sequestration of ligands in preference to polymeric sorbents. It is well recognized that the sorption affinity toward target solutes can be greatly enhanced by modifying and tailoring the interfacial chemistry of the sorbent [38].

To this end, Helfferich [39,40] was the first to conceive the use of Cu(II)- or Ni(II)-loaded weak-acid polymeric cation exchange resins for ligand sorption through LAB interactions. In such a ligand-exchange process, the water molecules (weak ligands) present at the coordination spheres of immobilized Cu(II) or Ni(II) in the cation exchange resins are replaced by relatively strong ligands, such as ammonia or ethylenediamine. The following provides a typical ligand-exchange reaction with ammonia where "M" represents a divalent metal ion, like Ni(II) or Cu(II), with strong Lewis acid properties and the overbar denotes the exchanger phase:

$$\overline{(RCOO^-)_2M^{2+}(H_2O)_n} + nNH_3 \leftrightarrow \overline{(RCOO^-)_2M^{2+}(NH_3)_n} + nH_2O \qquad (7.14)$$

In addition to providing a quantitative approach toward determining ligand-exchange capacity of metal-loaded cation exchangers, Helfferich [40] also unveiled a striking similarity between ion exchange and ligand exchange processes. In heterovalent ion exchange, it is well known that with an increase in electrolyte concentration, the affinity (or binary separation factor) of the counterion with lower valence increases over the counterion with higher valence. In the realm of ion exchange, this phenomenon is popularly known as *electroselectivity reversal*. In a similar vein, Helfferich showed that the affinity of a monodentate ligand (ammonia) toward the exchanger in ligand-exchange processes is enhanced over a bidentate ligand (1,3-diaminopropanol) as the total ligand concentration is increased. This observation provided the basis for efficient regeneration of the bed at the end of the ligand-exchange process. During the last 50 years, much work has been done in applying the concept of ligand exchange in the area of analytical chemistry, separation technology, and pollution control processes with varied degrees of success [41–46].

The investigations in ligand exchange have, however, been confined primarily to non-ionized (uncharged) ligands, namely, various amines and ammonia derivatives. In reality, most of the environmentally significant inorganic and organic ligands are anionic, such as phosphate, selenite, sulfide, arsenate, oxalate, phthalate, phenolate, and naturally occurring fulvates. Ligand-exchange processes using polymeric cation exchangers as metal hosts, as depicted in Eq. (7.14), are unable to sorb any anionic ligand, that is,

$$\overline{(RCOO^-)_2M^{2+}(H_2O)_n} + \left\{ \begin{array}{c} CN^- \\ SeO_3^{2-} \\ HS^- \\ CH_3COO^- \end{array} \right. \rightarrow \text{no reaction} \qquad (7.15)$$

The metal-loaded weak-acid cation exchange resins are electrically neutral and do not have any anion-exchange capacity, and, also, the negatively charged fixed coions (carboxylates in this case) of the polymer will not allow uptake of any anions in accordance with the Donnan coion exclusion principle. Thus, in spite of being strong ligands, the anions in Eq. (7.15) cannot displace water molecules (much weaker ligands) from the coordination spheres of the metal ions (Lewis acids). It is recognized that to sorb anionic ligands selectively, the polymeric substrate upon metal loading must possess fixed positive charges, that is, it should act as an anion exchanger. Obviously, such functional polymers should have high preference toward metal ions, so that the metals do not bleed or bleed only negligibly during the ligand-exchange process.

7.2.1 Conceptualization and Characterization of the Polymeric Ligand Exchanger (PLE)

Conceptually, transition metal cations, say Cu(II), if held firmly onto a solid phase at high concentrations, may act as anion-exchange sites with relatively high affinities toward aqueous phase anions with strong ligand characteristics. Figure 7.11 shows the major constituents of a polymeric ligand exchanger (PLE): first, a cross-linked polymeric template, like other anion exchangers; second, a chelating functional group covalently attached to that template; and, third, a Lewis acid-type metal cation strongly coordinated to the chelating functional group in a manner that its positive charges are not neutralized. The resulting material is essentially polymer anchored metal ions with fixed positive charges, that is, it is essentially an anion exchanger with high affinity for anionic ligands. Note that the coordination requirement of the transition metal cations is satisfied by both the chelating functionality and exchanging anionic ligand.

Specialty chelating polymers with nitrogen donor atoms satisfy the requisite properties to serve as the anchors of the transition metal cations. Chanda *et al.* [44] studied sorption of arsenates and other ligands onto Fe(III)-loaded chelating polymers with nitrogen donor atoms. Iron is a nontoxic, innocuous metal and hence, its bleeding

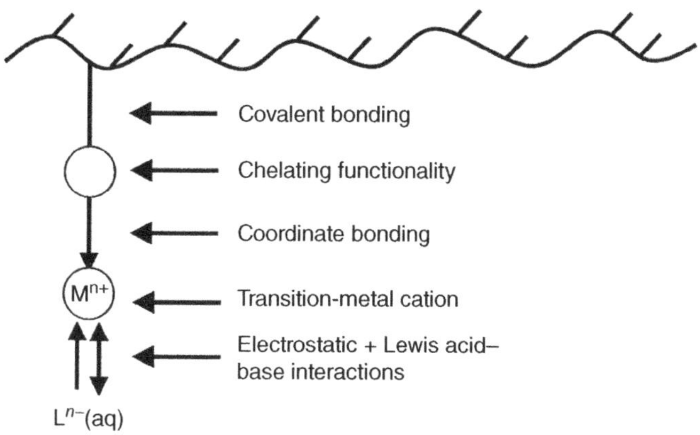

Figure 7.11 Various constituents of a conceptualized polymeric ligand exchanger for selective sorption of an aqueous phase ligand, L^{n-}. *Source:* Zhu and SenGupta 1992 [47]. Reproduced with permission of American Chemical Society.

from the polymer phase does not pose any environmental hazard. However, Fe(III) is a hard Lewis acid with poor affinity toward functional groups containing nitrogen donor atoms. The arsenate removal capacity in the presence of competing sulfate and chloride was rather low.

The primary subject of this section concerns the use of a specific class of commercially available chelating exchangers containing only nitrogen donor atoms (bispicolylamine functional groups). These chelating exchangers are macroporous with polystyrene matrix and divinylbenzene cross-linking, and were first made available from Dow Chemical Co. primarily for copper extraction from dilute wastewaters resulting from mining operations. Currently, many other resin manufacturers also produce similar products. This class of exchangers exhibits extremely high affinities toward Cu(II) ions and is ideally suited for anchoring Cu(II) and acting as a PLE. The exchanger XFS 4195 contains three nitrogen donor atoms per repeating monomer and is referred to as DOW 3N.

7.2.2 Sorption of Polymeric Ligand Exchangers

In order to develop an insight into the underlying sorption mechanism, two strong-base anion exchangers with quaternary ammonium functional groups and one popular chelating exchanger with iminodiacetate functional groups were also included in the study. Table 7.4 includes their salient properties and the names of the manufacturers. DOW 3N or XFS 4195 is converted into copper-loaded forms by passing Cu(II) solution through a column until saturation. For typographical convenience, the PLE will be referred to as DOW 3N-Cu.

Many environmental and process applications require selective separation of trace concentrations (μg/L to mg/L) of anionic ligands (namely, oxalate, phosphate, phthalate, ethylenediaminetetraacetate (EDTA), arsenates, nitrilotriacetate (NTA), selenites, and cyanides) onto a suitable sorbent from the background of high concentrations of competing anions (namely, chloride, and sulfate). As mentioned earlier, phosphate at trace concentration is responsible for eutrophication (algal blooms).

Table 7.4 Background information on ion exchangers.

Composition of the functional group	Characteristic	Matrix, porosity	Manufacturing and trade name
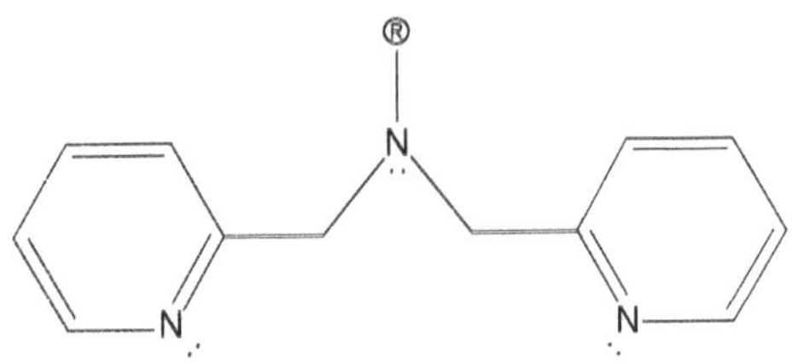	High metal ion affinity (bispicolylamine)	Polystyrene, macroporous	DOW Chemical Co., (Midland, MI): DOW 3N or XFS 4195
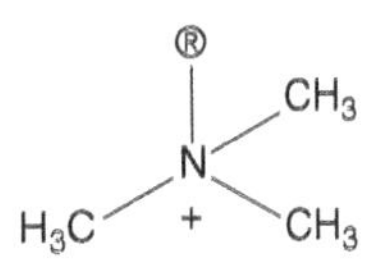	Strong-base anion exchanger	Polystyrene, macroporous Polyacrylic, macroporous	Rohm and Haas, Co.: IRA-900 IRA-958
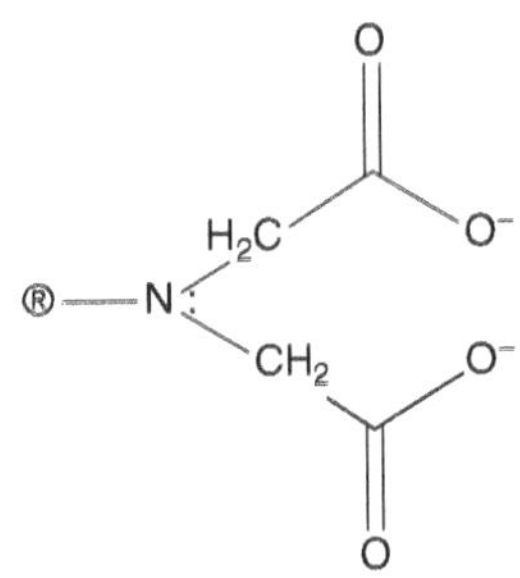	High metal ion affinity (iminodiacetate functionality)	Polystyrene, macroporous	Rohm and Haas, Co.: IRC-718

Note: Circled R denotes the repeating unit of the polymer matrix.

Municipal and industrial wastewater always contains sulfate and chloride anions which will compete with target phosphate for the sorption sites. Of these anions, sulfate possesses higher ionic charges (i.e., divalent) and will offer greater competition through enhanced electrostatic interaction. To assess the competing effects of sulfate on the PLE's phosphate uptake, isotherm tests using the mini-column technique were conducted at two different background sulfate concentrations, namely, 200 and 400 mg/L, with other conditions remaining identical. Figure 7.12 shows the phosphate uptake for these two separate isotherm tests. Note that doubling the concentration of the competing sulfate ion had an insignificant effect on the phosphate removal capacity of the PLE.

Phosphate/sulfate separation factors, as computed from the experimental data, are provided in Figure 7.13. Also superimposed in Figure 7.13 are the experimentally determined separation factors of other anion exchangers studied previously [48,49]. A separation factor, as described in Chapter 2, is a measure of relative selectivity between two competing solutes and is equal to the ratio of their distribution coefficients between the exchanger phase and the aqueous phase. Although both phosphate and sulfate exist as divalent anions at pH 8.3, the average separation factor, $\alpha_{P/S}$, with PLE is well over an order of magnitude greater compared to IRA-958 and other sorbents studied to date.

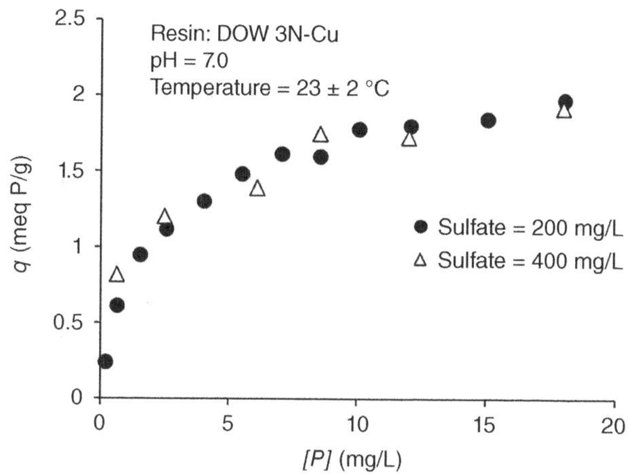

Figure 7.12 Phosphate isotherms for DOW3N-Cu at two different background concentrations (200 and 400 mg/L) of competing sulfate ions. *Source:* Zhao and SenGupta 1998 [48]. Reproduced with permission of Elsevier.

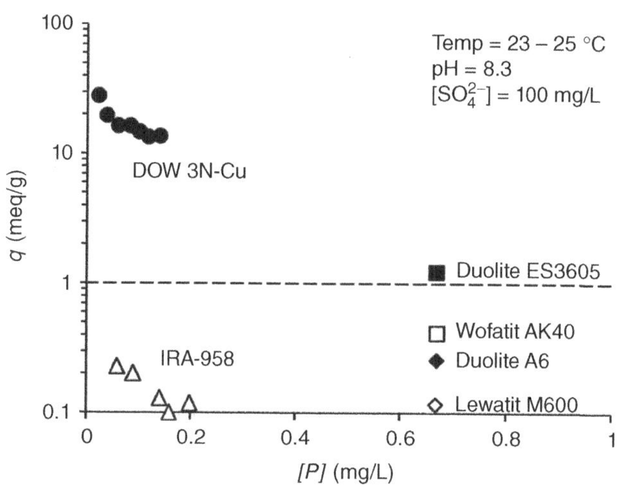

Figure 7.13 Comparison of phosphate/sulfate (P/S) separation factors for various sorbents. *Source:* Zhao and SenGupta 1998 [48]. Reproduced with permission of Elsevier.

Figure 7.14 shows effluent histories of oxalate during separate column runs under identical conditions with four different sorbents: an activated carbon (Filtrasorb 300, Calgon Corp.), a strong-base polymeric anion exchanger (IRA-900, Rohm and Haas, Co.), IRC 718-Cu, and the PLE, DOW 3N-Cu. Compared to other sorbents, oxalate breakthrough for DOW 3N-Cu occurred much later and after 3000 bed volumes, although competing sulfate and chloride were present in the influent at much greater concentrations. In contrast, activated carbon and IRC 718-Cu did not offer practically any oxalate removal capacity, while the strong-base anion exchanger (IRA-900) was completely exhausted in less than 250 bed volumes.

To avoid any possible bleeding of copper from the PLE bed into the exit of the column, a small amount (about 10% of the total bed height) of a virgin chelating ion

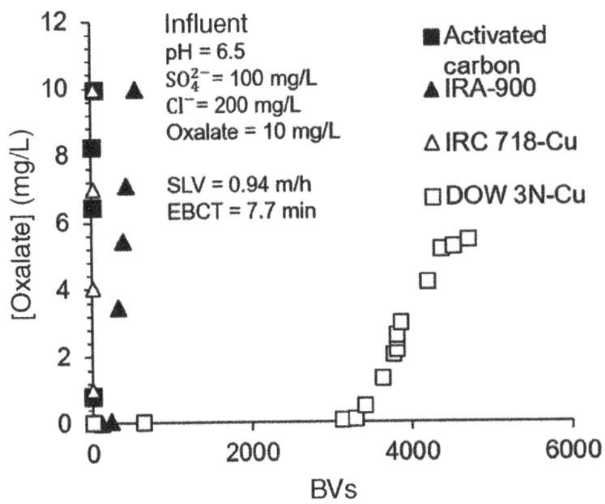

Figure 7.14 Oxalate effluent histories during column runs with four different sorbents under otherwise identical conditions: SLV, superficial liquid-phase velocity, EBCT, empty bed contact time. *Source:* Zhu and SenGupta 1992 [47]. Reproduced with permission of American Chemical Society.

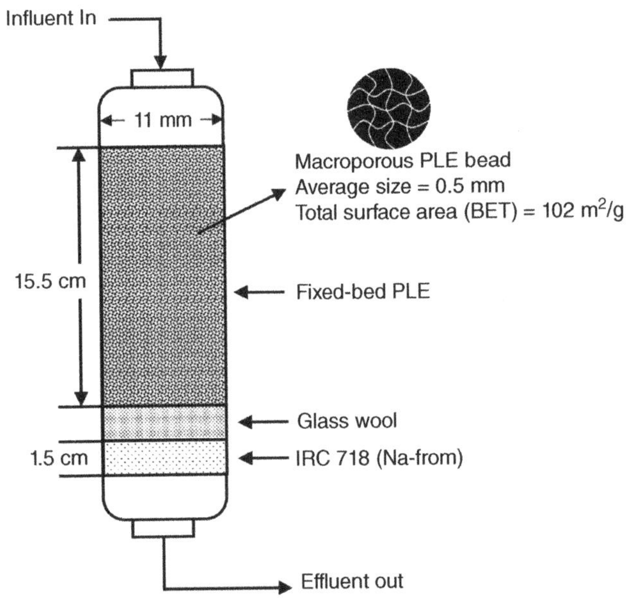

Figure 7.15 General schematic of a fixed-bed column using PLE. *Source:* Sengupta 2001 [38]. Reprinted with permission of Taylor & Francis.

exchanger (IRC-718, Rohm and Haas Co.) in sodium form was kept at the bottom of the column. Figure 7.15 provides a general sketch of the fixed-bed column used in the study. IRC-718 is a chelating cation exchanger with iminodiacetate functional group and polystyrene matrix; its salient properties are provided in Table 7.4.

7.2.3 Validation of Ligand Exchange Mechanism

Column run results and isotherm data clearly demonstrate that the PLE offered much higher phosphate and oxalate removal capacities compared to strong-base anion exchangers (IRA-958 and IRA-900) and IRC 718-Cu. Figure 7.16 provides a schematic presentation of the underlying mechanisms which govern the binding of a bidentate anion (e.g., phosphate or oxalate) onto these three sorbents. Considering the fact

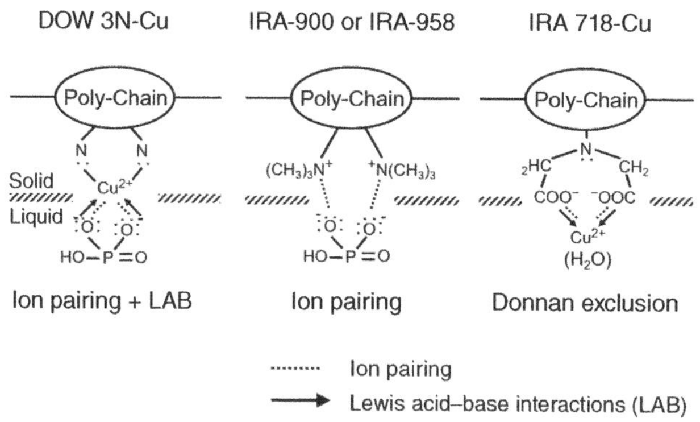

Figure 7.16 Schematic presentation of binding mechanisms of divalent phosphate for different sorbents. *Source:* Sengupta 2001 [38]. Reprinted with permission of Taylor & Francis.

that the Cu(II) ion has four primary coordination numbers, Figure 7.16 illustrates the following:

1. At around neutral to slightly alkaline pH, two coordination numbers for Cu(II) are satisfied by the two nitrogen donor atoms in the polymer phase and, as a result, copper ions are held firmly onto these sites of the PLE. The remaining two coordination numbers and the two residual positive charges of the polymer phase Cu(II) ion are satisfied simultaneously by the divalent phosphate anion (HPO_4^{2-}) with two oxygen donor atoms. Thus, Coulombic interaction (i.e., ion pair formation for maintenance of electroneutrality) is accompanied by LAB interaction where electron deficiencies in the coordination spheres of Cu^{2+} (Lewis acid) are satisfied by donor oxygen atoms of phosphate.
2. For IRA-900 or IRA-958, on the contrary, only Coulombic interaction (ion pair formation) is involved between the positively charged quaternary ammonium group (R_4N^+) and the anion (HPO_4^{2-}) because R_4N^+ does not have any electron-acceptor characteristic. As a result, the sorption affinity of (HPO_4^{2-}) toward IRA-958 or other strong-base anion exchangers cannot be enhanced by LAB interaction.
3. For IRC 718-Cu, the three coordination numbers of Cu(II) are satisfied by the iminodiacetate functional group (two oxygen and one nitrogen donor atoms). Also, its positive charges are neutralized within the polymer phase by the acetate groups. Thus, IRC 718-Cu does not have any available anion-exchange capacity to bind an anion onto it, and the fourth coordination number of Cu(II) is satisfied by a neutral water molecule. In fact, according to the Donnan exclusion principle, anions are rejected by IRC-718 because of its negatively charged (iminodiacetate) functional group. In spite of high Cu(II) affinity, IRC 718-Cu cannot, therefore, sorb anions regardless of their ligand characteristics. However, sorption of nonionized ligands like ammonia and ethylenediamine onto IRC 718-Cu is possible.

The foregoing analysis remains equally valid to explain high affinities of selenite, arsenate, oxalate, and other anionic ligands toward the PLE [38]. It has also been demonstrated that Cr(VI) anions can be very selectively removed at above-neutral pH with the PLE [50].

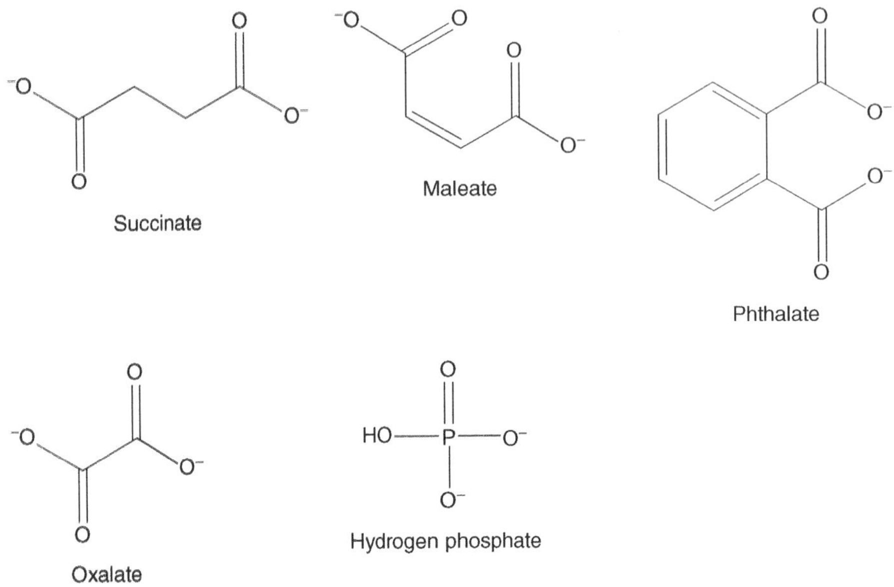

Figure 7.17 Structures of five bidentate anionic ligands: succinate, maleate, phthalate, oxalate, and hydrogen phosphate.

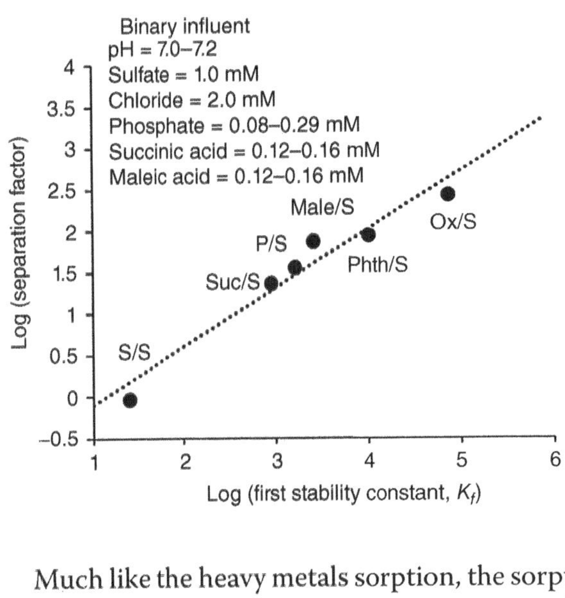

Figure 7.18 Linear free energy relationship (LFER) between the separation factors of DOW 3 N-Cu and the first metal-ligand stability constants. *Source:* Zhao and SenGupta 2000 [15]. Reproduced with permission of American Chemical Society.

Much like the heavy metals sorption, the sorption affinity of anionic ligands onto the PLE may also be predicted from their aqueous phase complexation or stability constant values. For validation, let us consider the first 1 : 1 stability constant of divalent anions for Cu^{2+}, that is,

$$Cu^{2+} + L^{2-} \leftrightarrow CuL \tag{7.16}$$

The divalent anionic ligands in Figure 7.17 were chosen and their first stability constant values are available in the open literature [51,52].

Based on equilibrium isotherm studies, separation factor values of these divalent anions were experimentally determined with respect to sulfate for the PLE DOW-3 N-Cu. Figure 7.18 reveals that a linear relationship exists between the log separation factor values of the anionic ligands and their aqueous-phase log stability constants.

Summary

- Physiologically toxic metals are normally heavy with specific gravity greater than five. They exist in the aqueous phase primarily as transition metal cations and as strong Lewis acids.
- These toxic metal ions, such as Cu(II), Zn(II), Cd(II), etc. form strong complexes with anionic ligands or Lewis bases.
- Metal-selective polymeric ion exchangers, often referred to as chelating ion exchangers, are essentially cation exchangers with covalently attached complexing ligands with oxygen, nitrogen and/or sulfur donor atoms, that is, functional groups are essentially strong Lewis bases.
- Aided by strong LAB interactions, chelating exchangers exhibit unusually high sorption affinity toward toxic metal cations (e.g., Zn^{2+}, Cu^{2+}, Pb^{2+}) over competing alkaline- and alkaline-earth metal cations (e.g., Na^+, Ca^{2+}, Mg^{2+}) that are present in the aqueous phase at much higher concentrations than the toxic metal cations.
- Metal selectivity of chelating exchangers can be predicted from the aqueous-phase metal-ligand stability constants. Regeneration of chelating exchanger with dilute mineral acid is very efficient.
- For chelating ion exchangers, both metal uptake and H^+ regeneration conform to rectangular isotherms. The sorption and desorption kinetics of chelating ion exchangers is often in agreement with the premise of shell progressive kinetics.
- Through intelligent choice of pH, aqueous-phase ligands and chelating exchangers, heavy metals can be separated from one another.
- Weak-base anion exchange resins with soft nitrogen donor atoms (e.g., polyamine, bispicolylamine) have extraordinarily high affinity for Cu(II). Upon copper loading, the nitrogen donor atoms become anion exchange sites with concomitant ability to act as Lewis acids.
- Copper-loaded chelating exchangers containing only nitrogen donor atoms are excellent PLEs with high affinity toward anionic ligands, namely, arsenate, phosphate, oxalate, salicylate and others.
- PLE's sorption affinity for specific anionic ligands can be predicted from the aqueous phase copper-ligand stability constants.

References

1 SenGupta, A.K. (2001) Chapter 1: Principles of heavy metal separation: an introduction, in *Environmental Separation of Heavy Metals: Engineering Processes* (ed. A.K. SenGupta), CRC Press, Boca Raton, FL, pp. 1–14.

2 Bellinger, D.C. (2016) Lead contamination in Flint—an abject failure to protect public health. *New England Journal of Medicine*, **374** (12), 1101–1103.
3 Edwards, M. (2013) Fetal death and reduced birth rates associated with exposure to lead-contaminated drinking water. *Environmental Science & Technology*, **48** (1), 739–746.
4 Edwards, M. Pruden, A. and Falkinham, III, J. (2016) *Flint Water Study*, http://flintwaterstudy.org/ (accessed 15 December 2016).
5 Hanna-Attisha, M., LaChance, J., Sadler, R.C., and Champney, S.A. (2016) Elevated blood lead levels in children associated with the Flint drinking water crisis: a spatial analysis of risk and public health response. *American Journal of Public Health*, **106** (2), 283–290.
6 Henry, W.D., Zhao, D., SenGupta, A.K., and Lange, C. (2004) Preparation and characterization of a new class of polymeric ligand exchangers for selective removal of trace contaminants from water. *Reactive and Functional Polymers*, **60**, 109–120.
7 Lanphear, B.P., Hornung, R., Khoury, J. *et al.* (2005) Low-level environmental lead exposure and children's intellectual function: an international pooled analysis. *Environmental Health Perspectives*, **1**, 894–899.
8 Nelson, R. (2016) Crisis in Flint: lead and Legionnaires' disease. *The Lancet Infectious Diseases*, **16** (3), 298–299.
9 Schwake, D.O., Garner, E., Strom, O.R. *et al.* (2016) Legionella DNA markers in tap water coincident with a spike in Legionnaires' disease in Flint, MI. *Environmental Science & Technology Letters*, **3** (9), 311–315.
10 Pearson, R.G. (1968) Hard and soft acids and bases, HSAB, Part 1: fundamental principles. *Journal of Chemical Education*, **45** (9), 581.
11 Pearson, R.G. (1968) Hard and soft acids and bases, HSAB, Part II: underlying theories. *Journal of Chemical Education*, **45** (10), 643.
12 Förstner, U. and Wittmann, G.T. (1983) *Metal Pollution in the Aquatic Environment*, 2nd edn, Springer-Verlag, Berlin.
13 Nieboer, E. and Richardson, D.H. (1980) The replacement of the nondescript term 'heavy metals' by a biologically and chemically significant classification of metal ions. *Environmental Pollution Series B*, **1** (1), 3–26.
14 Cotton, F.A., Wilkinson, G., Murillo, C.A. *et al.* (1999) *Advanced Inorganic Chemistry*, John Wiley & Sons, New York.
15 Zhao, D. and SenGupta, A.K. (2000) Ligand separation with a copper (II)-loaded polymeric ligand exchanger. *Industrial and Engineering Chemistry Research*, **39** (2), 455–462.
16 Zhu, Y. (1992) *Chelating Polymers with Nitrogen Donor Atoms: Their Unique Properties in Relation to Heavy Metals Sorption and Ligand Exchange*, Lehigh University, Bethlehem, PA.
17 Roy, T.K. (1989) Chelating polymers: their properties and applications in relation to removal, recovery and separation of toxic metals. MS thesis. Lehigh University, Bethlehem, PA.
18 Trujillo, E.M., Spinti, M., and Zhuang, H. (1995) Chapter 6: Immobilized biomass: a new class of heavy-metal-selective ion exchangers, in *Ion Exchange Technology: Advances in Pollution Control Lancaster* (ed. A.K. Sengupta), Technomic Publishing Co., PA, pp. 225–270.

19 Sengupta, S. and SenGupta, A.K. (2002) Trace heavy metals separation by chelating ion exchangers, in *Environmental Separation of Heavy Metals: Engineering Processes* (ed. A.K. SenGupta), Lewis Publishers, Boca Raton, pp. 45–96.
20 Waitz W. (1979) Ion exchange in heavy metals removal and recovery. Amber Hi-Lites. No. 162. Philadelphia, PA: Rohm and Haas; p. 1–12.
21 Bolto, B.A. and Pawłowski, Ł. (1987) *Wastewater Treatment by Ion-Exchange*, E & F.N. Spon, London.
22 Janauer, G., Gibbons, R., Bernier, W. *et al.* (1985) Chapter 2: A systematic approach to reactive ion exchange, in *Ion Exchange and Solvent Extraction*, vol. **9** (ed. J. Marinsky), Marcel Dekker, Inc., New York, pp. 53–174.
23 Urlings, L., Blonk, A., Woelders, J., and Massink, P. (1987) In situ remedial action of cadmium polluted soil, in *Environmental Technology* (ed. M. Streat), Springer, England, pp. 539–548.
24 SenGupta, A.K., Roy, T. and Millan, E. (1988) *Potential of Ion-Exchange Resins and Reactive Polyness in Eliminating/Reducing Hazardous Contaminants.* Proceedings of the 2nd International Conference on 'Physicochemical and Biological Detoxification of Hazardous Wastes, Atlantic City, NJ, 1989 May 3–5, 191–211.
25 Matejka, Z. and Zitkova, Z. (1997) The sorption of heavy-metal cations from EDTA complexes on acrylamide resins having oligo (ethyleneamine) moieties. *Reactive and Functional Polymers*, **35** (1), 81–88.
26 Courduvelis, C., Gallager, G. and Ch, H.N. (1982) *New Developments for the Treatment of wastewater Containing Metal Complexers.* Proceedings AES 4th Conference on Advanced Pollution Control for the Metal Finishing Industry, 77–80.
27 Marton-Schmidt, E., Inczedy, J., Laki, Z., and Szabadka, Ö. (1980) Separation of metal ions on ion-exchange resin with ethylenediamine functional groups. *Journal of Chromatography. A*, **201**, 73–77.
28 Loureiro, J., Costa, C., and Rodrigues, A. (1988) Recovery of copper, zinc and lead from liquid streams by chelating ion exchange resins. *Chemical Engineering Science*, **43** (5), 1115–1123.
29 Burgess, J. (1999) *Ions in Solution: Basic Principles of Chemical Interactions*, 2nd edn, Woodhead Publishing Ltd., Cambridge, UK.
30 Calmon, C. (1984) Impact of improved ion-exchange materials and new techniques on separation performance. *Adsorption and Ion Exchange, AIChE Symp Series No 233*, **80**, 84–95.
31 Helfferich, F.G. (1962) *Ion Exchange*, McGraw-Hill Book Company, Inc., New York.
32 Hudson, M.J. (1986) Coordination chemistry of selective-ion exchange resins, in *Ion Exchange: Science and Technology (NATO AISI Series)* (ed. A. Rodrigues), Martinus Nijhoff Publishers, Boston, pp. 35–66.
33 Zhu, Y., Millan, E., and SenGupta, A.K. (1990) Toward separation of toxic metal (II) cations by chelating polymers: some noteworthy observations. *Reactive Polymers*, **13** (3), 241–253.
34 Kokoszka, L.C. and Flood, J.W. (1990) Chapter 2: Regulatory programs governing toxics management, in *Environmental Management Handbook: Toxic Chemical Materials and Wastes*, Marcel Dekker, Inc., New York, pp. 15–150.
35 Grinstead, R.R. and Paalman, H.H. (1989) Metal ion scavenging from water with fine mesh ion exchangers and microporous membranes. *Environmental Progress*, **8** (1), 35–39.

36 Phelps, D.S. and Ruthven, D.M. (2001) The kinetics of uptake of Cu^{2+} ions in Ionac SR-5 cation exchange resin. *Adsorption*, 7 (3), 221–229.
37 Höll, W., Kiefer, R., Stöhr, C., and Bartosch, C. (1998) Metal separation by pH-driven parametric pumping, in *Ion Exchange and Solvent Extraction: A Series of Advances*, vol. **16** (eds A.K. SenGupta and Y. Marcus), Marcel Dekker, Inc., New York, pp. 228–282.
38 SenGupta, A.K. (2001) Chapter 6: Environmental separation through polymeric ligand exchange, in *Ion Exchange and Solvent Extraction: A Series of Advances*, vol. **14** (eds A.K. SenGupta and Y. Marcus), CRC Press, Boca Raton, FL, pp. 229–258.
39 Helfferich, F. (1961) Ligand exchange: a novel separation technique. *Nature*, **189**, 1001–1002.
40 Helfferich, F. (1962) Ligand exchange. I. Equilibria. *Journal of the American Chemical Society*, **84** (17), 3237–3242.
41 Dobbs, R.A., Uchida, S., Smith, L.M., and Cohen, J.M. (1975) Ammonia removal from wastewater by ligand exchange. *AIChE Symposium Series*, **75**, 157.
42 De Hernandez, C.M. and Walton, H.F. (1972) Ligand exchange chromatography of amphetamine drugs. *Analytical Chemistry*, **44** (6), 890–894.
43 Davankov, V.A. and Semechkin, A.V. (1977) Ligand-exchange chromatography. *Journal of Chromatography. A*, **141** (3), 313–353.
44 Chanda, M., O'Driscoll, K.F., and Rempel, G.L. (1988) Ligand exchange sorption of arsenate and arsenite anions by chelating resins in ferric ion form I. Weak-base chelating resin Dow XFS-4195. Reactive Polymers, Ion Exchangers, Sorbents 7 (2–3), 251–261.
45 Walton, H.F. (1995) Ligand-exchange chromatography: a brief review. *Industrial and Engineering Chemistry Research*, **34** (8), 2553–2554.
46 Groves, F. and White, T. (1984) Mathematical modelling of the ligand exchange process. *AICHE Journal*, **30** (3), 494–496.
47 Zhu, Y. and SenGupta, A.K. (1992) Sorption enhancement of some hydrophilic organic solutes through polymeric ligand exchange. *Environmental Science & Technology*, **26** (10), 1990–1998.
48 Zhao, D. and SenGupta, A.K. (1998) Ultimate removal of phosphate from wastewater using a new class of polymeric ion exchangers. *Water Research*, **32** (5), 1613–1625.
49 Boari, G., Liberti, L., and Passino, R. (1976) Selective renovation of eutrophic wastes—phosphate removal. *Water Research*, **10** (5), 421–428.
50 Zhao, D., SenGupta, A.K., and Stewart, L. (1998) Selective removal of Cr (VI) oxyanions with a new anion exchanger. *Industrial and Engineering Chemistry Research*, **37** (11), 4383–4387.
51 Morel, F.M. and Hering, J.G. (1993) *Principles and Applications of Aquatic Chemistry*, John Wiley & Sons, New York.
52 Sillen, L.G. and Martell, A.E. (1964) *Stability Constants of Metal-Ion Complexes*, The Chemical Society, London.

8

Synergy and Sustainability

Ion exchange science, if appropriately integrated, may help provide synergy and new solutions in areas seemingly unrelated to the process of ion exchange. Sustainable energy generation is a requisite goal for the long-term needs of society. An appropriate approach to achieving this goal is the development of a diverse portfolio of technologies that make use of renewable energy sources. In this chapter, we will present two such possible applications that may benefit through innovative integration of ion exchange:

- Energy recovery from waste acid neutralization.
- Improving stability of methane-generating anaerobic biological reactors.

In addition, a new approach for sustainable softening or hardness removal by ion exchange will be presented.

8.1 Waste Acid Neutralization: An Introduction

Waste acid neutralization is a widely used industrial pollution control process globally. In principle, acid–base neutralization involves association of hydrogen and hydroxyl ions to form neutral water molecules. This is the most thermodynamically favorable aqueous-phase reaction with an equilibrium constant ($K_{1/W}$) value of 1×10^{14} at 25 °C. The Gibbs free energy (ΔG^0) of the reaction at the standard state is −79.85 kJ/mol, whereas the change in enthalpy (ΔH^0) is −55.84 kJ/mol for the reaction as shown below [1].

$$H^+ + OH^- \rightarrow H_2O \tag{8.1}$$

Since the waste acid solution undergoing treatment is often dilute, the significant amount of thermal energy generated in the neutralization reaction causes a very small increase in the temperature of the bulk aqueous phase. As a result, any energy recovery is not possible and has not been reported to date.

According to available data in 2001, the global sulfuric acid production capacity was about 165 million tonnes annually [2]. If one considers that only 1% of this acid appears in the waste stream which needs to be neutralized, that is theoretically equivalent to a generation potential of 522,000 MWh. According to the US Energy Information Administration (EIA), the average carbon dioxide emission per kWh of energy generation by natural gas in the US is 0.55 kg [3]. Thus, energy generation

Ion Exchange in Environmental Processes: Fundamentals, Applications and Sustainable Technology, First Edition. Arup K. SenGupta.

through this route may save the planet from 0.3 million tonnes of carbon dioxide gas every year. Even with 1% efficiency of energy recovery, the impact would be significant. The actual energy generation potential and curbing carbon dioxide emission may go well beyond the above estimate because many industrial and mining processes also generate a huge amount of acidic waste streams, such as sulfur dioxide in the flue gas and naturally produced acid streams in abandoned mines [4,5].

8.1.1 Underlying Scientific Concept

An ion exchange resin may be viewed as a quasi-electrolyte; with the variation in pH, the weak-acid (or weak-base) ion exchanger reversibly acquires and loses its ionic character, thus causing an on–off pattern of osmotic pressure generation inside the polymer. The functional groups are covalently attached to the polymer phase and, hence, cannot diffuse out into the aqueous phase. As a result, due to the concomitant Donnan membrane effect [6–9], these hydrophilic polymers, in consecutive contact with an acid or base, reversibly swell and shrink due to the movement of water in and out of the polymer phase. Thus, acid–base neutralization is accompanied by generation of usable mechanical energy.

Weak-Acid Cation Exchangers

When base solution (NaOH) is added:

$$\overline{R-COOH} + NaOH \leftrightarrow \overline{R-COO^{-}Na^{+}} + H_2O + \text{Swelling} \tag{8.2}$$

When acid solution (HCl) is added:

$$\overline{R-COO^{-}Na^{+}} + HCl \leftrightarrow \overline{R-COOH} + Na^{+} + Cl^{-} + \text{Shrinking} \tag{8.3}$$

$$\text{Overall:}\quad NaOH + HCl \leftrightarrow Na^{+} + Cl^{-} + H_2O + \text{Mechanical Energy} \tag{8.4}$$

Weak-Base Anion Exchangers

When acid solution (HCl) is added:

$$\overline{R_3N} + HCl \leftrightarrow \overline{R_3NH^{+}Cl^{-}} + \text{Swelling} \tag{8.5}$$

When base solution (NaOH) is added:

$$\overline{R_3NH^{+}Cl^{-}} + NaOH \leftrightarrow \overline{R_3N} + Na^{+} + Cl^{-} + \text{Shrinking} \tag{8.6}$$

$$\text{Overall:}\quad NaOH + HCl \leftrightarrow Na^{+} + Cl^{-} + H_2O + \text{Mechanical Energy} \tag{8.7}$$

R_3N and RCOOH represent the weak-base and weak-acid functional groups, respectively and overbar denotes the resin phase.

Figure 8.1a and b illustrates the individual steps leading to the overall acid–base neutralization reaction in the polymer phase for widely available weak-acid cation (WAC) and weak-base anion exchange resins, respectively with the chemical reactions mentioned [10,11].

One single cation-exchange resin bead (C-104, Purolite Inc.) in non-ionized hydrogen form was put in a petri dish under an optical microscope (Testmaster). Salient properties of Purolite C-104 are presented in Table 8.1.

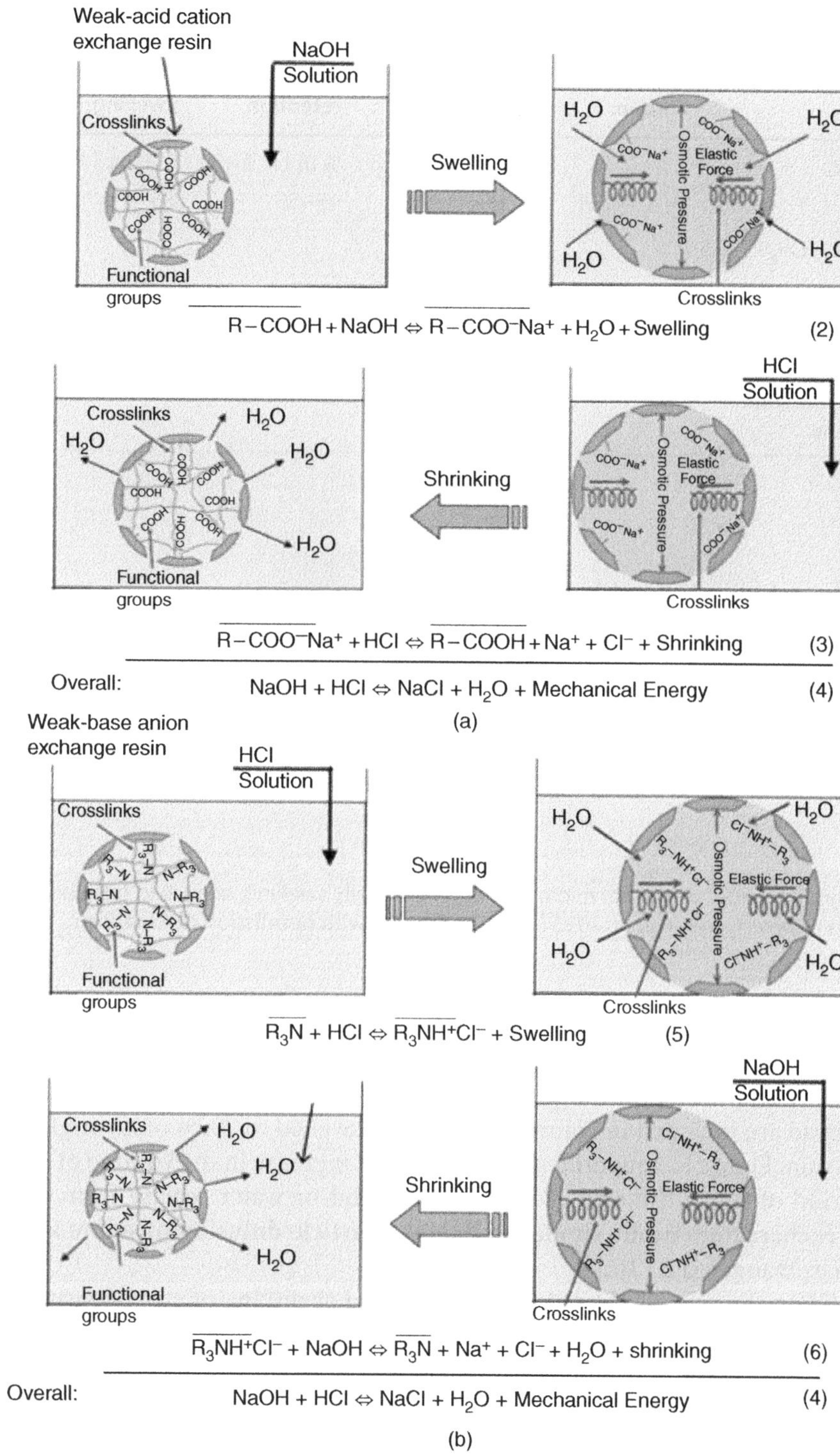

Figure 8.1 Individual steps involved in the process of acid–base neutralization in the polymer phase for: (a) a weak-acid cation exchange resin and (b) a weak-base anion exchange resin. R_3N and RCOOH represent the weak-base and weak-acid functional groups, respectively and overbar denotes the resin phase. *Source:* Sarkar *et al.* 2011 [10]. Reproduced with permission of American Chemical Society.

Table 8.1 Properties of C-104 weak-acid cation exchange resin.

Structure	Functional group	Total capacity	Moisture retention	Reversible swelling
		eq/L in H^+-form	% in H^+ form	% H^+ to Na^+
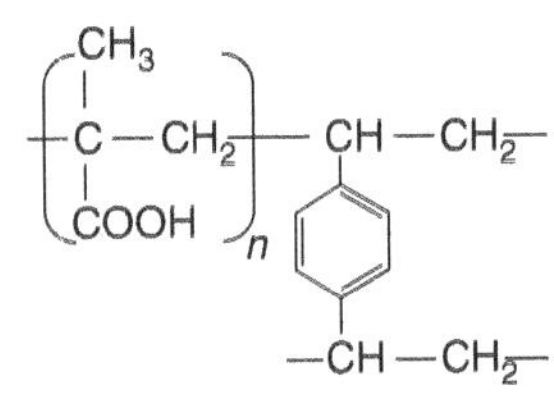 Polymethacrylate crosslinked with divinylbenzene	—COOH	3.4	45–55	85

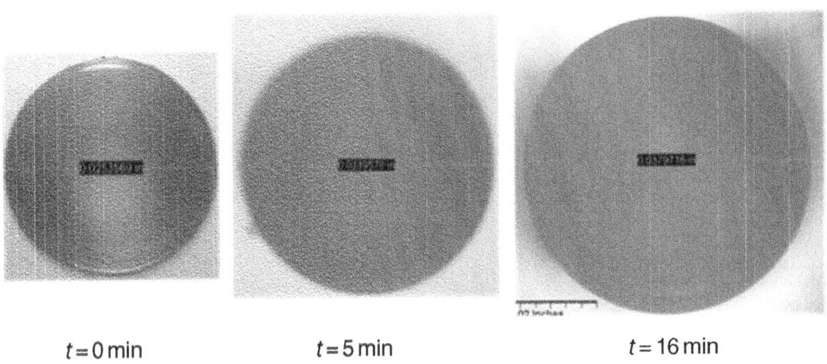

Figure 8.2 Photographs under an optical microscope of a gradually swelling weak-acid resin bead from base addition. *Source:* German *et al.* 2013 [11]. Reproduced with permission of American Chemical Society.

After adding a drop of 4% NaOH solution to the resin, swelling in the volume of the bead was observed in accordance with the reaction (8.2). The diameter of the resin was measured over a length of time, through which it continuously swelled. Photographs of the swelling bead are included in Figure 8.2; the bead swelled to 200% of its original volume after 13 min. Figure 8.3 quantifies the percentage increase in the volume of the bead over a period of time. The rate of water uptake and/or water release from the polymer phase is characteristically similar to the intraparticle diffusion of solute into a polymeric ion exchanger [12,13].

To further validate the reversibility of the swelling and shrinking of a pH-sensitive polymer, 25 mL of Purolite C-104 weak-acid ion exchange resin was placed in a column and subjected to sequential addition of base and acid over three consecutive cycles. The solutions used were 2% mass/volume of NaOH and HCl. The results are presented in Figure 8.4. It may be noted that the bed expanded and contracted to the extent of 50%. Most importantly, the swelling and shrinking for all three cycles were nearly identical, demonstrating the reversibility of the proposed process over multiple cycles.

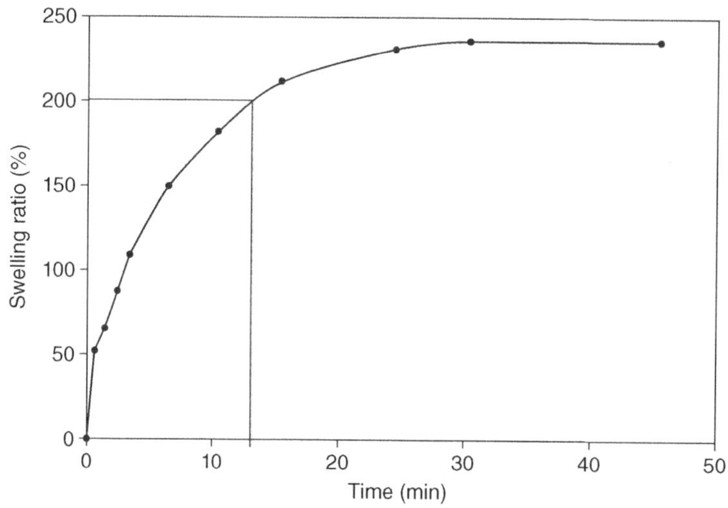

Figure 8.3 Kinetics of swelling for a C-104 resin bead in H-form under the action of NaOH solution. Swelling ratio is defined to be the ratio of the volume of swollen resin bead to the un-swollen one. *Source:* Sarkar *et al.* 2011 [10]. Reproduced with permission of American Chemical Society.

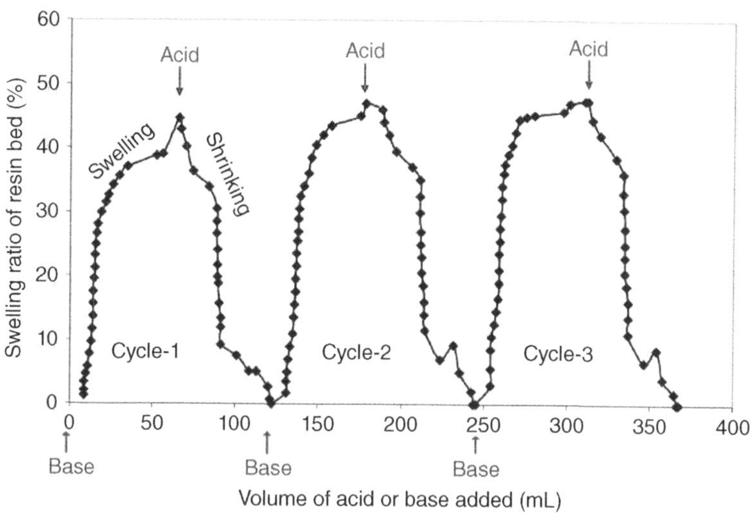

Figure 8.4 Cyclic swelling and shrinking of a weak-acid cation exchange resin bed (Purolite, C-104) when contacted alternately with 2% NaOH and 2% HCl solutions, respectively over multiple numbers of cycles. *Source:* Sarkar *et al.* 2011 [10]. Reproduced with permission of American Chemical Society.

8.1.2 Mechanical Work through a Cyclic Engine

In order to demonstrate the feasibility of transforming cyclic expansion and contraction into mechanical work, an experiment was conducted using a latex-membrane covered container inside which WAC exchange resin beads were allowed to come in consecutive contact with solutions of acid (1% HCl) and base (1% NaOH) during the process of neutralization [10]. Shrinking and swelling of the resin column caused a change in pressure in the air over the resin column, which resulted in deflation

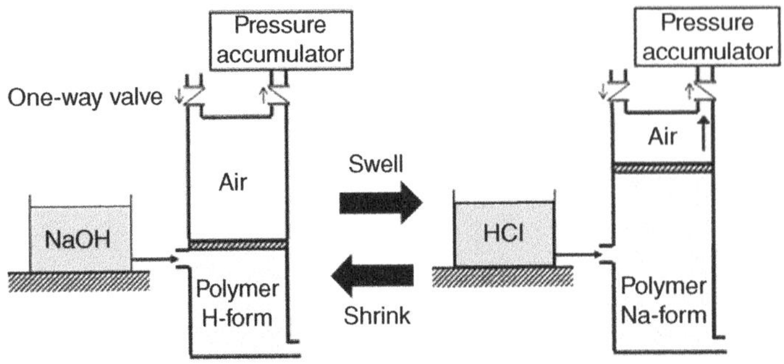

Figure 8.5 Conceptualized reciprocating engine to produce a compressed gas through swelling and shrinking of weak resin. *Source:* German *et al.* 2013 [11]. Reproduced with permission of American Chemical Society.

Figure 8.6 Cyclic deflation (a and c) and inflation (b and d) over two consecutive cycles of a latex membrane covering a container inside which weak-acid cation exchange resin undergoes shrinking and swelling under the action of a solution of acid and base, respectively during a process of neutralization. *Source:* Sarkar *et al.* 2011 [10]. Reproduced with permission of American Chemical Society.

and inflation of the latex membrane. While Figure 8.5 depicts a conceptualized reciprocating engine to provide compressed air as an energy source from acid–base neutralization, Figure 8.6 provides the experimental evidence of inflation and deflation of the latex membrane for two consecutive cycles validating the basic premise of the concept. The external work generated through the expansion and contraction of the chamber can thus be transformed into an engine under isothermal conditions. Obviously, the efficiency of the process directly depends on the swelling/shrinking ratio of the pH-sensitive polymers, which, in turn, intrinsically depend on their degree of cross-linking and hydrophilicity. Many pH-sensitive polymers with a high swelling ratio (over 500%) are available and amenable to synthesis including biorenewable chitosan materials [14]. They are likely to be good candidates for efficient recovery of mechanical energy from acid–base neutralization.

When a bed of a pH sensitive polymer is allowed to swell and shrink under alternate contact with acid and base inside a closed container, the compressed gas (e.g., air) can be stored in a pressure accumulator working like a reciprocating engine under isothermal condition and can be used to produce energy as needed.

However, the mechanical energy recovered from the acid–base neutralization reaction as per the experiment conducted, although noticeable, is below 0.01% of the available enthalpy change. Thus, there is room for improvement. The pressure–volume work is directly correlated to the swelling-shrinking behavior of the pH-sensitive polymer which, in turn, is intrinsically dependent on its degree of cross-linking and hydrophilicity. Many pH-sensitive polymers with high swelling ratio (over 500%) are available and amenable to synthesis [15,16]. In principle, waste acid neutralization offers a scientifically valid opportunity to generate usable energy through use of weak-acid and weak-base ion exchange resins.

8.2 Improving Stability of Anaerobic Biological Reactors

Methane-producing anaerobic biological processes are particularly attractive for treating high-strength organic matters because they produce a gas with high fuel value, do not require any external supply of oxygen, and produce only a small amount of sludge. In spite of these advantages, the anaerobic biological processes are susceptible to operational failures. The major reasons cited for the failure of anaerobic processes are inhibition of the methanogenic bacteria due to (i) buildup of volatile acids and consequent pH reduction, and (ii) toxic overloading of the reactor. An engineered anaerobic biological process for stabilizing dissolved BOD may be viewed as a two-step series reaction where acetogens convert soluble organics to volatile acids which are subsequently converted to methane and carbon dioxide by the methanogens.

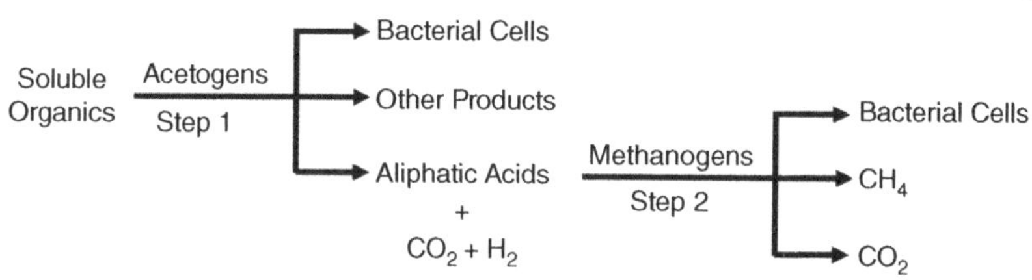

(8.8)

Methanogenesis (i.e., Step 2) is always the rate limiting step in anaerobic biological processes [17]. Inhibition of methanogens starts around pH 6, and is severe below pH 5.5 [18]. According to reaction (8.8), an increase in organic loading results in an accumulation of volatile aliphatic acids, and a consequent lowering of pH. Both low pH [19] and high concentration of unionized aliphatic acids [20,21] inhibit the methanogens and adversely affect the performance of the reactors. To control the pH within neutral range, high alkalinity is maintained in the reactor to resist sudden lowering of pH [22]. But, under steady-state operating conditions, this additional alkalinity is wasted and adds to operational cost. Graef and Andrews [20] suggested feedback control to maintain the optimum pH in the reactor during "acid souring" by scrubbing CO_2 from the gas stream with dilute alkali and then recycling the CO_2-free methane back into the reactor. However, such a feedback control technique is operationally complex and yet to be practiced in commercial applications.

Dissolved heavy metals are toxic to the microorganisms beyond some critical concentrations [23] and have been held responsible for many digester failures. So, heavy metal toxicity for anaerobic processes has been the subject of many investigations. Although most of the cationic heavy metals are least soluble around neutral pH, where anaerobic biological process is the most efficient, an increase in heavy metal concentration may be caused by a sudden increase in influent heavy metal concentration in the feed and/or lowering of pH in the reactor. To control the cationic heavy metal toxicity, Lawrence and McCarty [22] demonstrated the use of sulfide to precipitate heavy metals. But this methodology warrants very precise operation

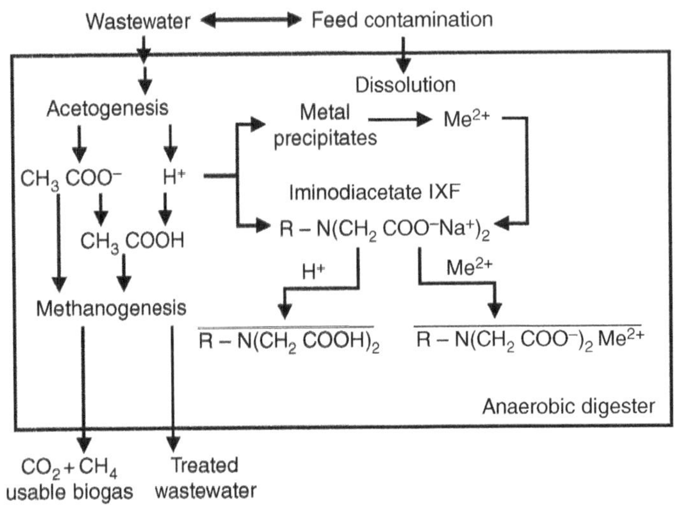

Figure 8.7 A schematic illustrating the dissipation of protons and dissolved cationic metal ions in an anaerobic reactor stabilized by the presence of iminodiacetate functionalized polymers. *Source:* Mitra *et al.* 1998 [27]. Reproduced with permission of Elsevier.

because excess sulfide is toxic to methanogens. Understandably, heavy metal toxicity for anaerobic processes has been the subject of many contemporary studies [24–26], but the details are withheld here for brevity.

8.2.1 Potential Use of Selective Ion Exchanger

It was conceptualized that the instability of anaerobic reactors caused by low pH, high volatile acids and dissolved heavy metals could be controlled by incorporating the appropriate type of polymeric ion exchangers inside the reactor. The incorporation of functionalized polymer using iminodiacetate functionality to stabilize an aerobic biological reactor to both pH fluctuations and toxic metal loading is illustrated in Figure 8.7. Any pH reduction due to organic overloading and/or an increase in toxic metal concentration can be quickly dissipated by selective sorption onto WAC exchanger containing iminodiacetate functional group. The entire process control is passive and requires no external attention. Only during the surge of H^+ and metal ions, the ion exchange materials act to maintain favorable pH and low metals concentration. Once the steady state conditions return, the ion exchangers get slowly desorbed. Thus, this is a self-regenerated, self-adjusting in-situ control process. But, previous studies using ion exchanger beads demonstrated that the process kinetics for dissipation of hydrogen and metal ion is slower than optimum due to poor intraparticle diffusion-controlled kinetics [27,28].

8.2.2 Ion Exchange Fibers: Characterization and Performance

Advances in the development of ion-exchange materials have led to the production of very thin ion-exchange fibers (IXFs), of the order of 20–30 μm diameter, with properties that make them attractive for use in anaerobic biological reactors. These

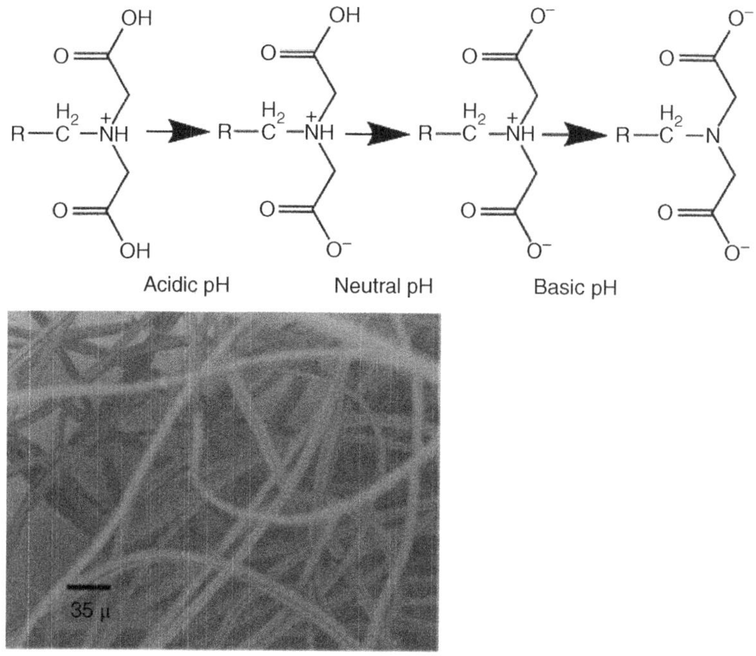

Figure 8.8 Dissociation and high buffer capacity of IXF Fiban X-1 over a wide range of pH. *Source:* Reprinted with permission from Tian [29].

thin cylindrical fibers offer significantly faster sorption kinetics due to their shorter diffusion path length. IXFs also allow practical installation in anaerobic reactors, where they can be suspended from the roof or walls of the reactor as woven sheets or as porous pillows. Fiban X-1, used in the laboratory study, contains the iminodiacetate functional group, which provides pH buffering capacity across a wide pH range, as shown in Figure 8.8 [29].

The iminodiacetate functional group also exhibits high affinity toward metals and commonly encountered cations in the order of $Cu^{2+} > Ni^{2+} > Zn^{2+} > Mn^{2+} > Ca^{2+} > Mg^{2+}$. An example cation-exchange reaction between dissolved calcium and copper ions with iminodiacetate functional groups is

$$\overline{R - CH_2 - N^{+}H(CH_2COO^{-})_2Ca^{2+}} + Cu^{2+} \leftrightarrow \overline{R - CH_2 - N^{+}H(CH_2COO^{-})_2Cu^{2+}} + Ca^{2+} \quad (8.9)$$

A series of laboratory-scale experiments were conducted using multiple 125 mL media bottles modified with gas and liquid sampling ports added to the cap. The reactors were seeded with bacteria from the mother reactor and then step-fed once daily with anaerobic growth media, namely lactose and alkalinity. Figure 8.9 shows the laboratory setup. Two different sets of experiments were conducted, examining the effects of organic overloading and toxic copper addition. Four reactors were set up for each experiment – two reactors without an IXF present, one of which was the control reactor and was not stressed, and two reactors containing different IXF mass loadings (2 and 10 g/L for the organic overloading experiment and 2 and 8 g/L for the copper

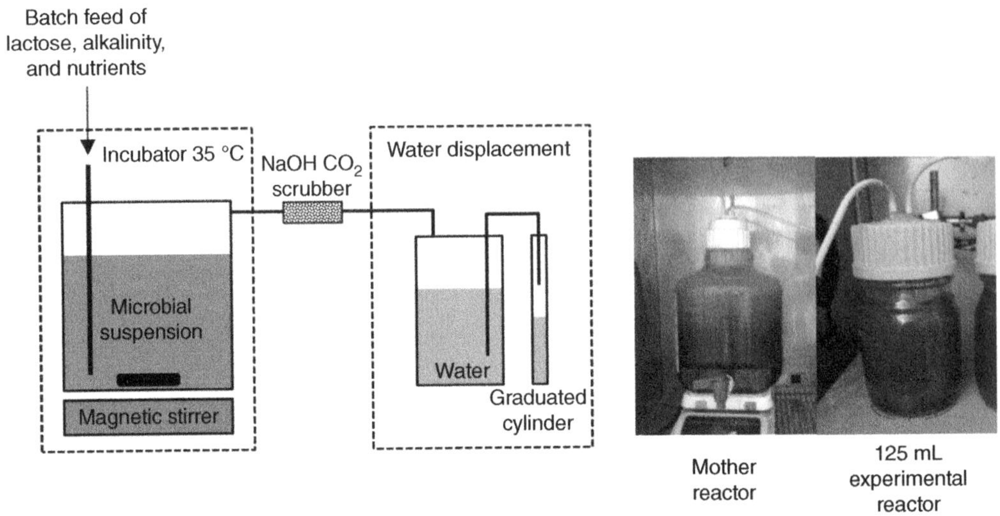

Figure 8.9 (a) Experimental apparatus, showing reactor and water displacement apparatus; (b) mother reactor and experimental reactor. *Source:* Reprinted with permission from Tian [29].

overloading experiment). Various details of the IXF mediated anaerobic reactors are provided by Tian [29].

Results from the organic (lactose) overloading are provided in Figure 8.10, including reactor alkalinity, pH and methane production rate. The results are shown starting at day 31 and the steady-state operation of the reactors is demonstrated by the control reactors (hollow circles). The organic overloading was conducted from days 36 to 46 (depicted by vertical gray shading). All three stressed reactors demonstrated a decrease in alkalinity and pH during the stress period. The key observations are:

1. The reactor with no IXF failed at day 46, that is, methane generation ceased altogether;
2. The methane generation in the reactor containing 2 g/L IXF was moderately impacted during and after the stress period, the reactor with 10 g/L IXF showed no significant adverse impact in alkalinity reduction, pH drop and change in methane generation after the stress period;
3. Upon cessation of the stress period, the two reactors containing Fiban X-1 were able to recover completely, that is, pH, alkalinity and methane production returned to the pre-stress levels. The reactor without IXF could notrecover.

Results from the copper shock-loading experiments are shown in Figure 8.11.

i. The addition of copper on days 32 and 33 results in an immediate decline in methane production and a decrease in pH, indicating that the methanogens were impacted to a greater extent than acetogens. For the reactor without IXF, the pH continued to decrease until the reactor failed near pH 5.6 and methane generation ceased permanently.
ii. For the reactors with IXF, pH rose back to the steady-state value after the copper shock in about three weeks. The methane production in the IXF reactors steadily

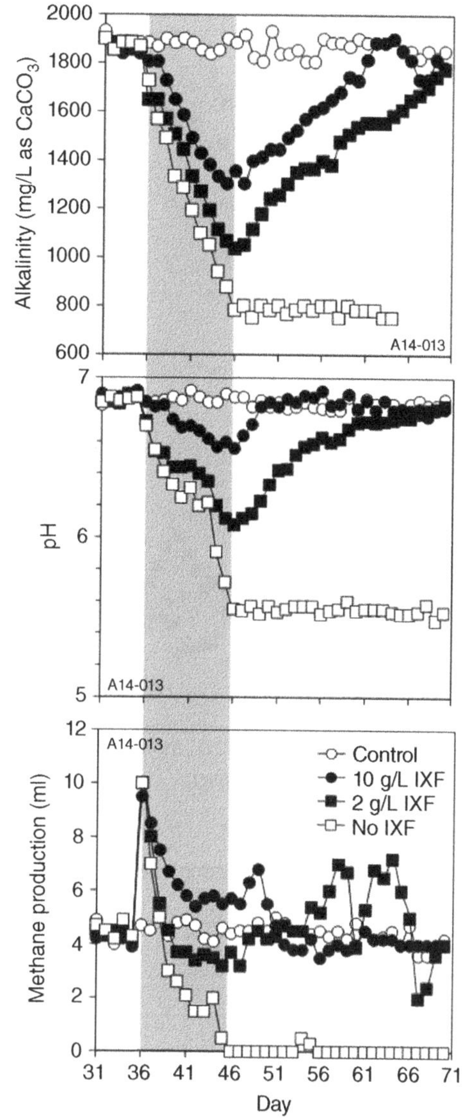

Figure 8.10 Results for organic overloading in the presence of different amounts of iminodiacetate-functionalized IXF. Organic overloading was imparted on the reactors from days 36 to 46 (depicted by the shaded region), with lactose loading increased from the steady-state value of 163 mg/L-d to an overload of 675 mg/L-d and alkalinity loading reduced from a steady-state value of 90.6 mg/L-d as $NaHCO_3$ to zero. The reactors with IXF maintained methane production during the stress, whereas the reactor without IXF failed (methane generation ceased) and was unable to recover. *Source:* Reprinted with permission from Tian [29].

rose after the initial decrease, confirming that the IXF moderated the effects of the copper shock and allowed the reactors to recover.

Figure 8.12 shows nickel exchange capacity of IXF remained practically unchanged after one year of use inside the methane producing anaerobic biological reactor (MPABR).

Use of ion exchange materials in biological reactors containing live bacteria has not been practiced to any significant extent yet. Overall, the study demonstrated that the iminodiacetate-functionalized ion exchange fiber can passively stabilize methane-producing anaerobic biological reactors from failures by (i) buffering pH fluctuations resulting from organic overloading; and (ii) moderating a shock load of dissolved toxic metal.

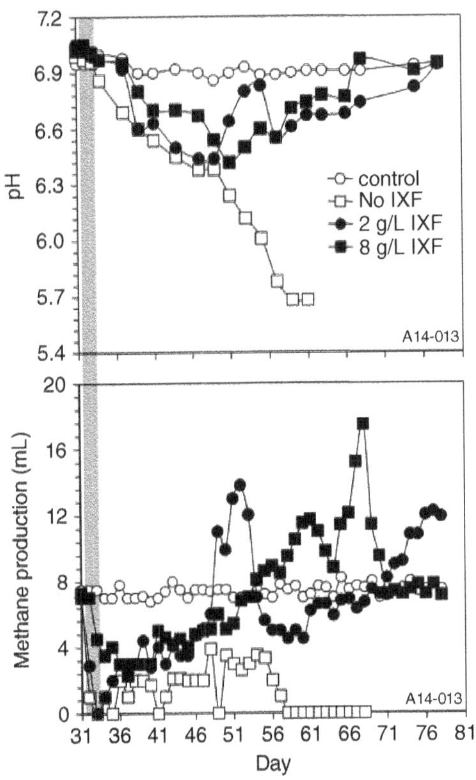

Figure 8.11 Demonstration of copper toxicity prevention with iminodiacetate IXF. Here, 8 mg of $CuCl_2 \cdot 2H_2O$ (47 μmol) was added to 100 mL anaerobic reactors at days 32 and 33 (indicated by gray shading). Methanogenic activity was immediately impacted by the presence of copper, as shown in the lower figure, with a pH decrease continuing afterwards due to a buildup of organic acids. The reactors with the IXF were able to mitigate the effects of the copper addition and these reactors were able to recover. The reactor without IXF failed (methane output went to zero) and was not able to recover. *Source:* Reprinted with permission from Tian [29].

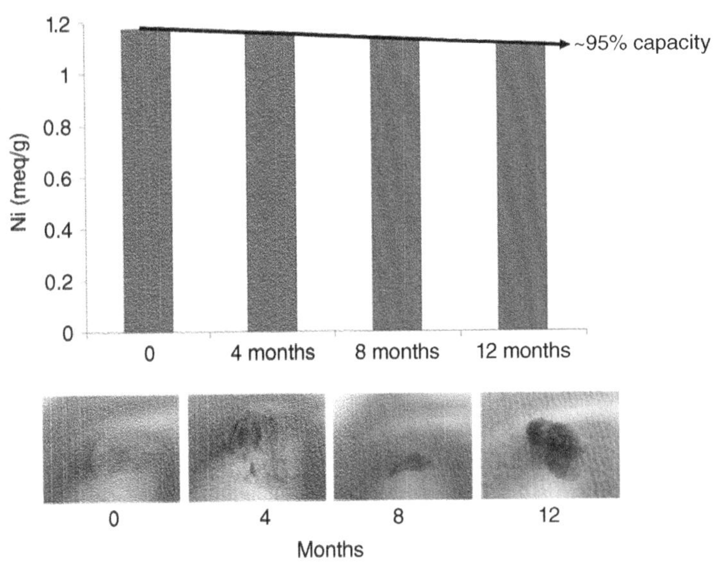

Figure 8.12 FIBAN X-1 Nickel exchange capacity after operation in the MPABR for durations of 4, 8, and 12 months. Initial capacity was 1.16 meq/g-fiber and ~95% capacity remained after one year of operation. Photographs show associated change in fiber color. *Source:* Reprinted with permission from Tian [29].

8.3 Sustainable Aluminum-Cycle Softening for Hardness Removal

8.3.1 Current Status and Challenges

Many industrial and municipal unit operations require removal of hardness, that is, reduction in dissolved calcium and magnesium concentration, to avoid scaling on heat transfer equipment or membranes and to decrease consumption of detergents and chemicals in cooling and washing water. Removal of hardness from water is a treatment or pretreatment commonly referred to as softening. The two universally practiced processes for softening or hardness removal are lime softening and ion exchange. Lime softening produces voluminous sludge to be settled, dewatered and disposed of, and the softened water requires post-treatment for pH adjustment and turbidity removal. Understandably, the development of single-step Na-cycle cation exchange softening that does not produce any sludge or need any post-treatment was hailed as a major breakthrough and adopted universally nearly five decades ago [30–34]. In cation exchange softening, polymeric strong-acid cation exchange resin in Na-form (SAC-Na) replaces one equivalent of hardness (i.e., Ca^{2+}) with one equivalent of Na^+ in accordance with the following reaction:

$$\overline{2(R-SO_3^-)Na^+} + Ca^{2+} \rightarrow \overline{(R-SO_3^-)_2Ca^{2+}} + 2Na^+ \tag{8.10}$$

Thus, the sodium content of the treated water is always higher than in the feed, which poses a health concern when used for drinking or cooking, especially for the growing number of people with hypertension [35–37]. Concentrated brine (5–12% w/v) is used to regenerate the exhausted cation exchange resin back to Na-form, in accordance with the following reaction (where $n = 6–12$):

$$\overline{(R-SO_3^-)_2Ca^{2+}} + nNaCl \rightarrow \overline{2(R-SO_3^-)Na^+} + Ca^{2+} + (n-2)Na^+ + nCl^- \tag{8.11}$$

The consequent environmental impact caused by spent regenerant needing disposal, containing high NaCl concentrations, has surfaced as a major concern in drought-prone and environmentally sensitive ecosystems. More than 25 communities in California face bans or restrictions on the use of SAC-Na softeners because of the risks to agriculture and downstream water quality caused by the spent brine, especially during recent droughts [38,39]. Frequent release of concentrated brine from ion exchange softeners is a global concern, for example, in growing high rise complexes in urban areas.

8.3.2 Sodium-Free Approaches and Alternatives to Na-Cycle Softening

In the mid-1960s, the Sirotherm process was introduced as a novel ion exchange process that relied on thermal regeneration [40,41], but the commercial scale-up was a failure. Similar thermal regeneration processes using hydrogels have been investigated, but currently have limited capacity compared to commercial resins [42]. WAC exchange resin can be a possible alternative to minimize Na-laden spent regenerant, but efficient regeneration of WAC requires mineral acids. From a process safety viewpoint, transport, storage and handling of acid poses safety hazards that often make the softening process uneconomical under the current regulatory environment. Attempts have been made to remove hardness through electrocoagulation also, but the extent of hardness removal is far less than traditional softeners and is pH-sensitive. Scaling on aluminum electrodes and poor performance of iron electrodes also complicate feasibility with low cost electrodes [43,44].

In a bid to identify an environmentally benign regenerant, carbon dioxide-sparged water has been successfully used for removal of temporary hardness using weak-acid IXFs [45–47]. However, high cost and relative unavailability of IXFs are the primary bars against rapid deployment of this softening process. Many novel inorganic, polymeric and hybrid ion exchangers have been developed and commercially deployed during the last two decades to remove contaminating ions selectively, for example, arsenic, boron, chromate, copper, fluoride, nitrate, phosphate, radium, synthetic organics, trace metals, etc. [9,48–61]. However, most of them lack high selectivity, capacity or regeneration efficiency for calcium removal. It is also true that many companies have introduced "salt-free softeners," advertising non-chemical water softening, but these items often have dubious claims, for example, magnetic softeners [62]. An efficient, easy-to-operate hardness removal process with reduced sodium in both treated water and in the spent regenerant is an unmet global need that demands technological innovation to mitigate this water treatment shortcoming.

8.3.3 Underlying Scientific Approach of Al-cycle Cation Exchange

In the proposed new cation exchange softening process, typical monovalent Na^+ would be replaced by a trivalent cation, for example, Al^{3+}, as the pre-saturated form of the resin. Aluminum salts are inexpensive, readily available, exist in the +III oxidation state over a wide range of pH and pe values, and aluminum hydroxide has a very low solubility product ($K_{sp} = 1.3 \times 10^{-33}$). Also, the kinetics of labile aluminum hydroxide precipitate formation are very rapid [63]. Thus, for dilute solutions undergoing hardness removal, exchange of Ca^{2+} with Al^{3+} was expected to proceed with simultaneous precipitation of $Al(OH)_3(s)$ and consequent reduction in alkalinity, as follows:

$$2\overline{(R-SO_3^-)_3Al^{3+}} + 3Ca^{2+} \rightarrow 3\overline{(R-SO_3^-)_2Ca^{2+}} + 2Al^{3+} \quad (8.12)$$

$$2Al^{3+} + 6H_2O \rightarrow 2Al(OH)_3(s)\downarrow + 6H^+ \quad (8.13)$$

$$6H^+ + 6HCO_3^- \rightarrow 6H_2O + 6CO_2(g)\uparrow \quad (8.14)$$

Overall:

$$2\overline{(R-SO_3^-)_3Al^{3+}} + 3Ca^{2+} + 6HCO_3^- \rightarrow 3\overline{(R-SO_3^-)_2Ca^{2+}} + 2Al(OH)_3(s)\downarrow + 6CO_2(g)\uparrow \quad (8.15)$$

Note that in the overall ion exchange reaction, aluminum is precipitated as solid $Al(OH)_3(s)$ and HCO_3^- is partially removed as CO_2 gas. So, according to the LeChatelier Principle, the reaction will proceed to the right, although divalent Ca^{2+} possesses lower affinity than trivalent Al^{3+}. Once the capacity is exhausted, the cation exchanger is regenerated stoichiometrically with dilute $AlCl_3$ on an equivalent basis, as follows:

$$3\overline{(R-SO_3^-)_2Ca^{2+}} + 2Al^{3+} \rightarrow 2\overline{(R-SO_3^-)_3Al^{3+}} + 3Ca^{2+} \quad (8.16)$$

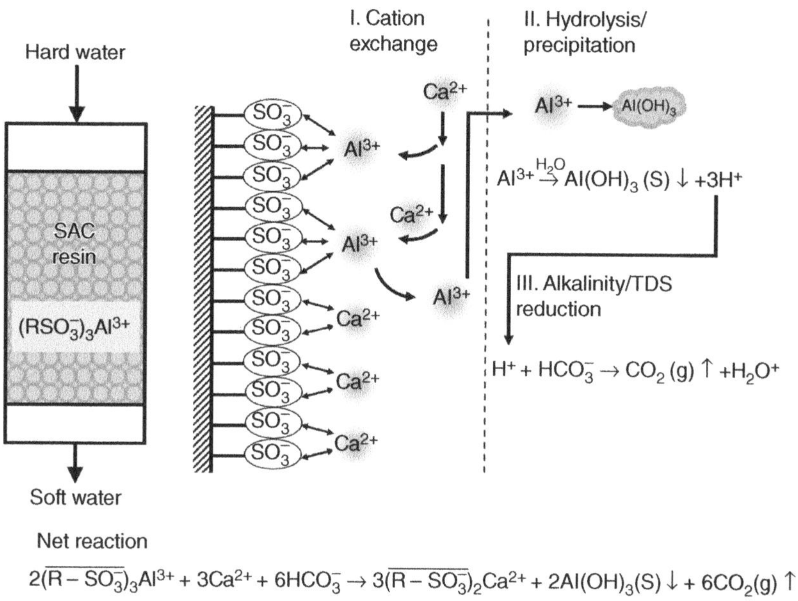

Figure 8.13 A schematic of Al^{3+}-Ca^{2+} exchange followed by $Al(OH)_3$ precipitation with hydrolysis leading to partial TDS reduction. *Source:* Li *et al.* 2016 [64]. Reproduced with permission of American Chemical Society.

Stoichiometric or highly efficient regeneration is possible because of higher Al^{3+} affinity, than Ca^{2+}, for cation exchange sites, that is, the aluminum-calcium separation factor is greater than one, $\alpha_{Al/Ca} > 1$.

Figure 8.13 illustrates individual steps of the conceptualized process and the following points are worth noting:

(i) Spent aluminum regenerant contains no sodium and its disposal is less environmentally stressful than for typical waste sodium brine regenerant.
(ii) Hardness removal can happen on an equivalent basis, that is, one equivalent of Ca^{2+} is removed with an equivalent of Al^{3+} regenerant.
(iii) Sodium concentration in the treated water does not increase.
(iv) There is an overall reduction in TDS of the treated water.

The significant contrast of the proposed concept from traditional practice lies in using a more favorable trivalent cation (Al^{3+}) as the preloaded counterion in the cation exchanger for removing divalent hardness ions, for example, Ca^{2+} and/or Mg^{2+}. From an application viewpoint, the process configuration of the softening process remains the same, except that the cation exchange resin is used in Al^{3+}-form instead of Na^+-form.

Table 8.2 provides the salient properties of the cation exchange resin used in testing. The empty bed contact time (EBCT) and superficial liquid velocity (SLV) were recorded for each experimental column run. Previous studies with polymeric ligand exchangers confirmed that no significant channeling occurs under such operating conditions [65,66].

Table 8.2 Salient properties of Purolite C145, a macroporous strong-acid cation exchange resin.

	Purolite C145[a]
Structure	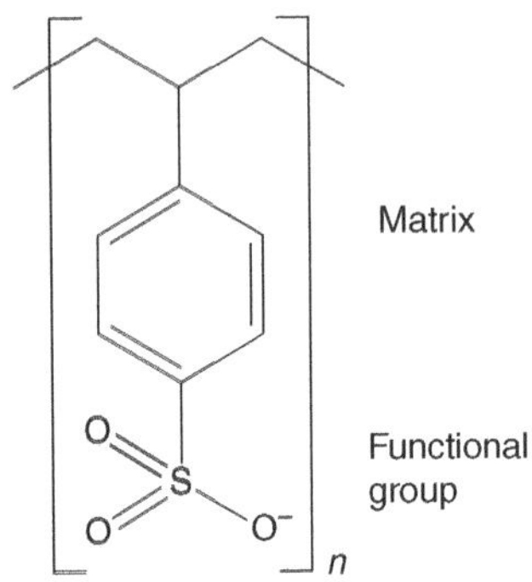
Functional group	Sulfonic acid
Matrix	Macroporous polystyrene crosslinked with divinylbenzene
Capacity (eq/L)	1.5

a) Additional information available from the manufacturer's technical brochure.

Configuration of the ion exchange softening system, for both SAC-Na and SAC-Al operations, remains identical and is shown in Figure 8.14.

8.3.4 Comparison in Performance: Na-Cycle versus Al-Cycle

Softening with SAC-Na Resins

SAC-Na was used to treat synthetic groundwater over two consecutive cycles of exhaustion-regeneration (5% NaCl). Figure 8.15a has the calcium and sodium effluent histories for the SAC-Na softening process. For, say, 300 bed volumes (BVs), calcium was removed and sodium was elevated, that is, reduction in calcium is always accompanied by an equivalent increase in sodium concentration. Two more cycles of operation yielded identical results and confirmed reproducibility of the process. After, say, 350 BVs, effluent concentrations were identical to the influent, signaling capacity exhaustion of the SAC bed.

Figure 8.15b presents calcium elution during regeneration with 5% NaCl, while (c) depicts the stoichiometric regeneration efficiency (R_E), where the dimensionless ratio of the equivalents of regenerant (Na^+) used versus equivalents of Ca^{2+} recovered (Na^+/Ca^{2+}) was plotted against BVs. Note that 86% of removed calcium was eluted after 10 BVs of 5% NaCl regenerant, corresponding to a Na^+/Ca^{2+} recovery ratio of 7.5. It is possible to recover a portion of the spent regenerant with lower calcium concentration and reuse it to attain higher regeneration efficiency. Yet, according to widely used practices in softening, the ratio of regenerant to recovered hardness is rarely less than 4 and often exceeds 6, depending on the treated water quality requirements [30–32].

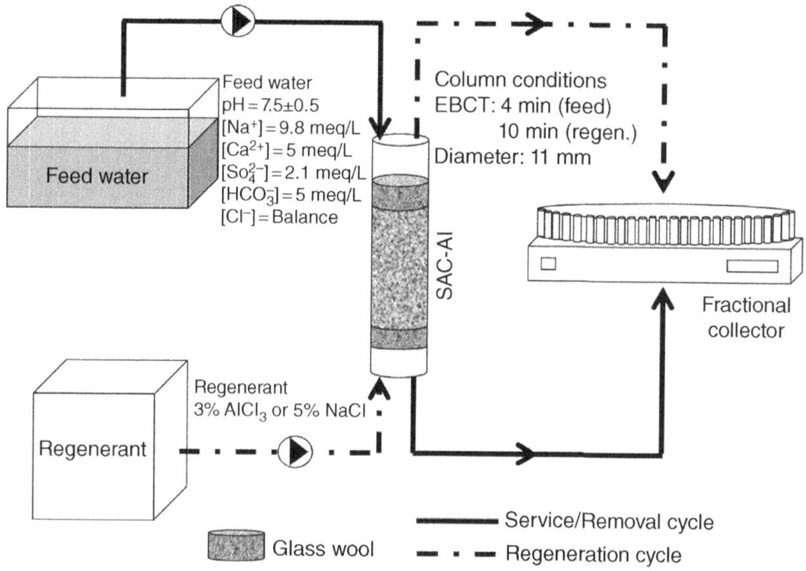

Figure 8.14 A process flow diagram for evaluation of SAC-Al hardness removal and regeneration. *Source:* Li *et al.* 2016 [64]. Reproduced with permission of American Chemical Society.

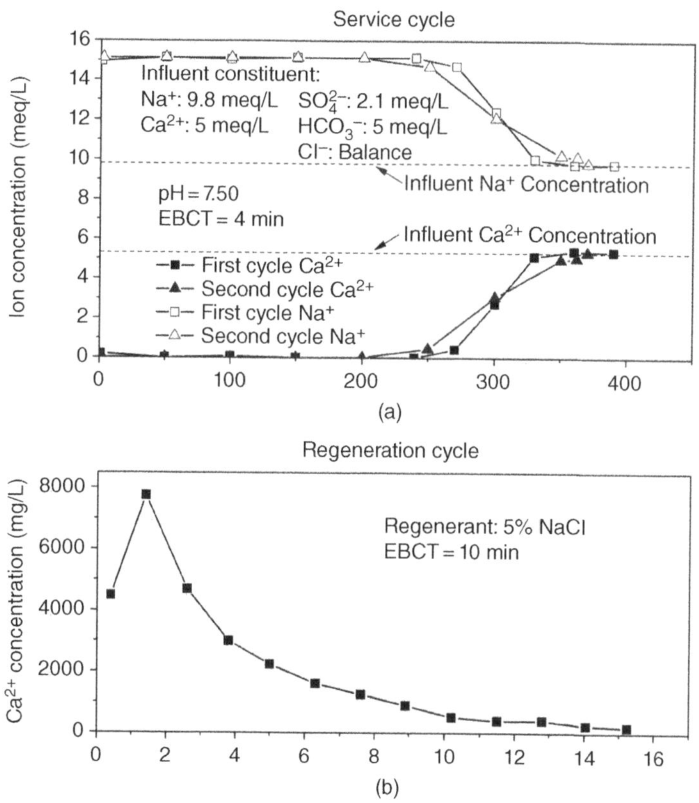

Figure 8.15 Effluent history during operation of SAC-Na resin for (a) sodium and calcium during the service cycle; (b) calcium during the regeneration cycle (5% NaCl); and, (c) regeneration efficiency on an equivalent basis. *Source:* Li *et al.* 2016 [64]. Reproduced with permission of American Chemical Society.

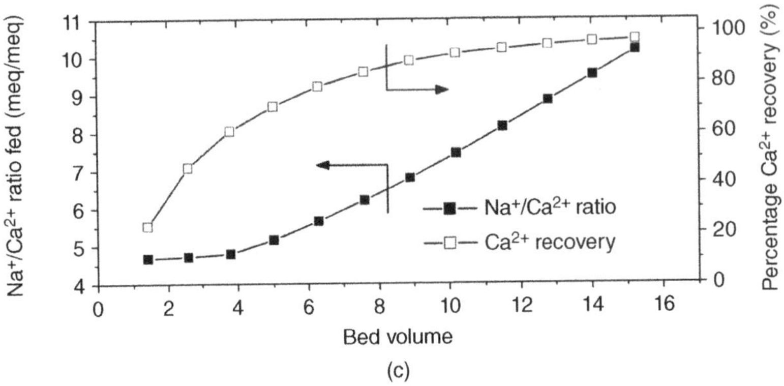

Figure 8.15 (*Continued*)

Softening with SAC-Al Resins

Figure 8.16a shows two consecutive effluent histories during SAC-Al treatment using Purolite C145. All other conditions were identical as the Na-cycle column runs presented in Figure 8.15a, namely, influent composition, EBCT and SLV. Two observations from Figure 8.16a are worth noting: (i) calcium, although removed for nearly 400 BVs, broke through early and gradually; (ii) sodium in the treated water never exceeded the influent concentration. Note that during the first 50 BVs the sodium concentration in the treated water was slightly less than that in the feed water. Figure 8.16b compares the conductivity of the feed and treated water between the SAC-Na and SAC-Al processes under identical conditions, a surrogate parameter for total dissolved solids. Treated water conductivity was noticeably lower after SAC-Al versus the influent water or after SAC-Na treatment. Conductivity decreased from an influent of 1750 μS to an average of 1300 μS in the SAC-Al treated water – a 25% reduction. Two additional column runs, each followed by $AlCl_3$ regeneration, yielded nearly identical effluent or breakthrough histories with respect to calcium, sodium and conductivity.

SAC-Al Regeneration and SEM–EDX Mapping

Figure 8.17a shows elution of calcium during two successive regenerations by 3% $AlCl_3$ for 6 BVs; the bed was then rinsed with 3 BVs of the calcium-containing influent solution. During both regenerations, 86% of calcium on the bed was recovered and the effluent pH followed identical trends. As shown in Figure 8.17b, the pH value dropped to pH 3.5 during regeneration for 6 BVs, followed by an increase during rinsing. Note that when pH reached the minimum value (pH 3.5), regeneration was completed; pH can thus be a surrogate indicator for complete regeneration of SAC-Al.

In a bid to reconcile the experimental results with the underlying scientific hypothesis, slices of both exhausted and regenerated Purolite C145 were characterized by SEM–EDX mapping. Figure 8.18a mapping of the exhausted resin shows Ca throughout the entire resin, while Al predominates in the periphery; this observation was supported by the EDX spectra. In contrast, the regenerated resin in Figure 8.18b shows an absence of calcium in the SEM-EDX mapping and spectra. Ion exchange sites throughout the exchange cation resin were converted into Al-form through Ca^{2+}-Al^{3+}

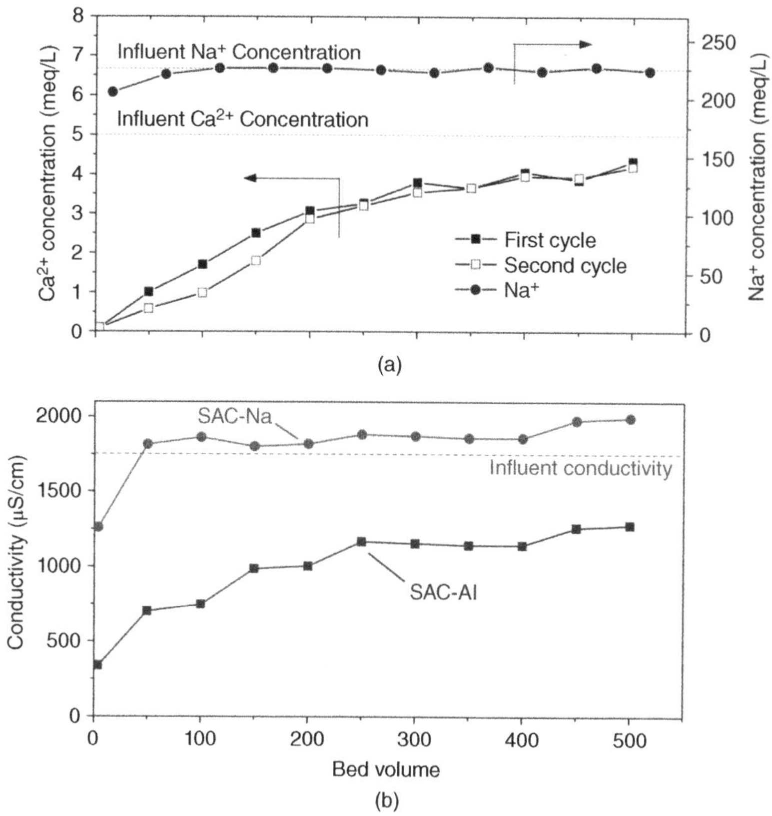

Figure 8.16 Effluent history of (a) sodium and calcium during SAC-Al treatment over two cycles; and (b) conductivity during SAC-Na and SAC-Al treatment. *Source:* Li *et al.* 2016 [64]. Reproduced with permission of American Chemical Society.

Figure 8.17 Effluent history of (a) calcium and (b) pH during SAC-Al regeneration by 3% $AlCl_3$. *Source:* Li *et al.* 2016 [64]. Reproduced with permission of American Chemical Society.

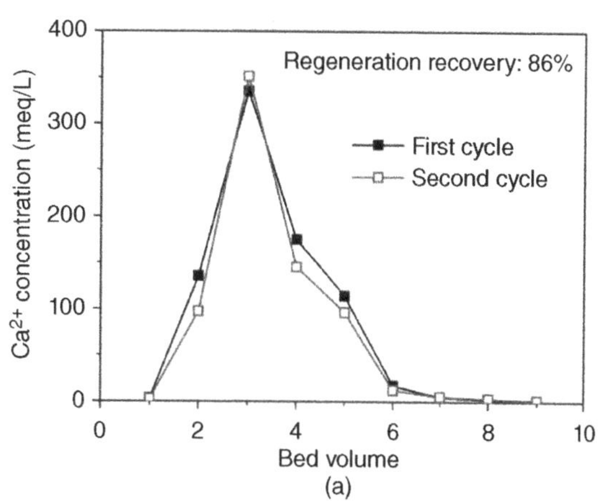

exchange. Aluminum precipitates were present in the macropores of the outer periphery of the cation exchange resin beads in both the exhausted and regenerated SAC-Al. But, aluminum precipitation was never observed in the aqueous phase of the column as white precipitated solids or as impeding the effluent flow. The cyclic process was continued for five sorption–desorption cycles after reaching a steady state.

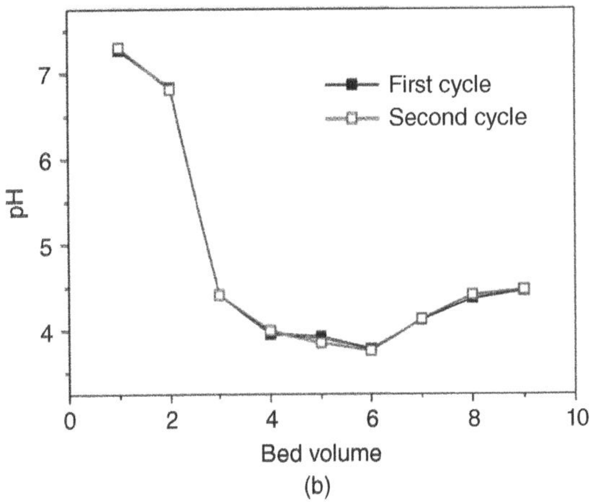

Figure 8.17 (*Continued*)

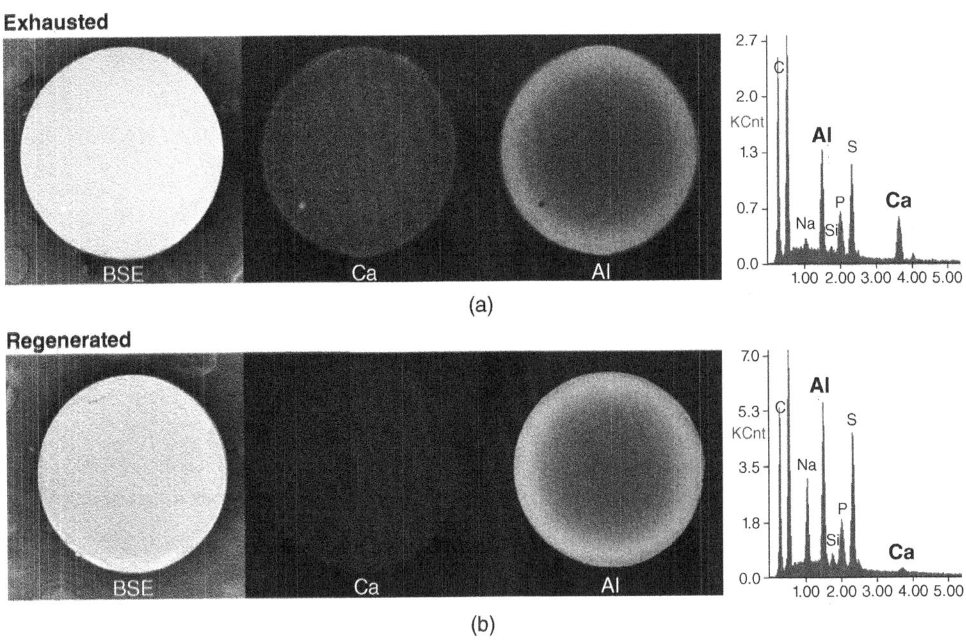

Figure 8.18 SEM-EDX mapping and spectroscopy of a cation exchange resin bead during Al-cycle operations (a) after exhaustion by synthetic hard water and (b) after regeneration by 3% $AlCl_3$. *Source:* Li *et al.* 2016 [64]. Reproduced with permission of American Chemical Society.

8.3.5 Regeneration Efficiency and Calcium Removal Capacity

Figure 8.19a shows the regeneration efficiency, R_E, that is, equivalents of regenerant consumed per equivalent of hardness, for both SAC-Na and SAC-Al softening processes. While R_E was nearly 7.5 for Na-cycle softening, it was close to unity for the Al-cycle. Stoichiometrically, an R_E value of unity is the thermodynamic limit for ion exchange processes and Figure 8.19a validates that no excess regenerant was required during regeneration with $AlCl_3$. Figure 8.19b shows the comparison of calcium removal capacities between the Na- and Al-cycle processes. As anticipated from the column

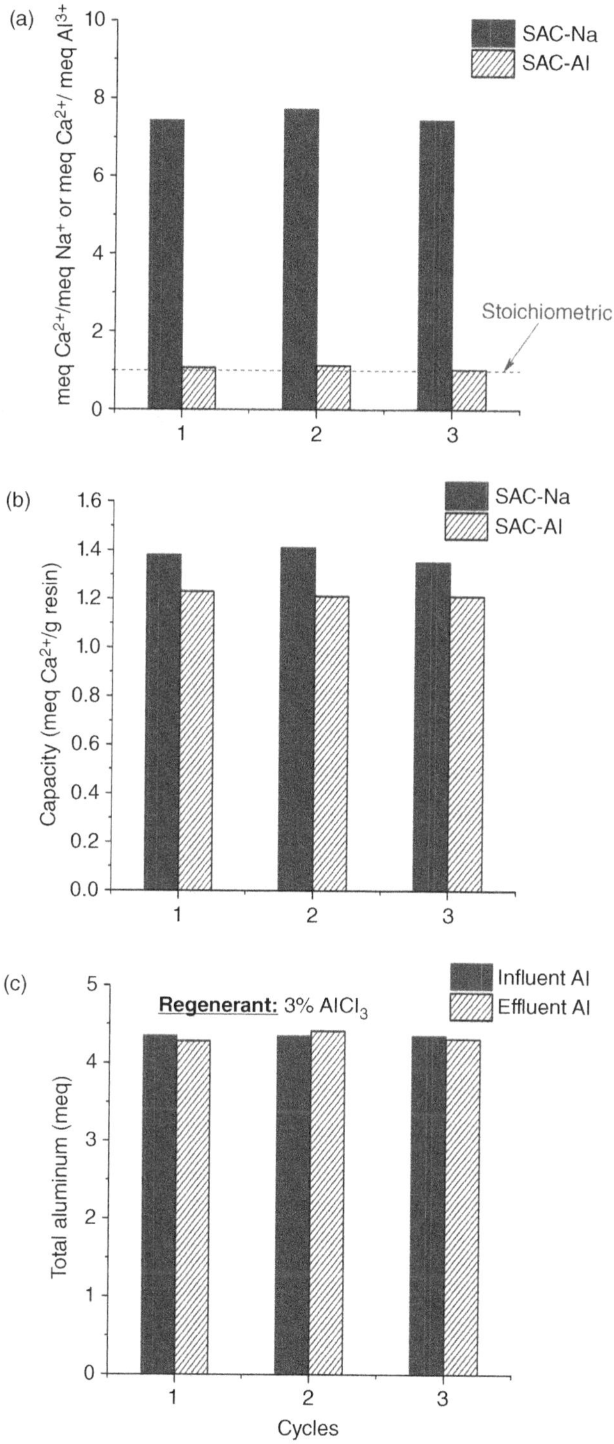

Figure 8.19 Comparison over multiple cycles of operation of (a) regeneration efficiency between SAC-Na and SAC-Al; (b) ion exchange capacity of SAC-Na and SAC-Al; and (c) mass balance of aluminum regenerant influent and effluent. *Source:* Li *et al.* 2016 [64]. Reproduced with permission of American Chemical Society.

effluent histories, the capacity was, say, 20% greater for SAC-Na exchange processes than for SAC-Al. Figure 8.19c shows the mass balance during $AlCl_3$ regeneration: there was an absence of any aluminum accumulation in the bed, that is, the process was at a steady state after several cycles of operation, as there was no buildup of aluminum.

Aluminum chloride regenerant had a pH of 2.9–3.0, thus the free hydrogen ion (H^+) concentration was sufficient to dissolve previously precipitated aluminum hydroxide present in the macropores. Once dissolved, aluminum ions from the precipitates enhanced regeneration by displacing Ca^{2+} from the ion exchange sites:

$$Al(OH)_3(s) + 3H^+ \rightarrow Al^{3+} + 3H_2O \tag{8.17}$$

$$3\overline{(R - SO_3^-)_2Ca^{2+}} + 2Al^{3+} \rightarrow 2\overline{(R - SO_3^-)_3Al^{3+}} + 3Ca^{2+} \tag{8.18}$$

The aluminum content of the cation exchange resins was consistent after every exhaustion cycle (80–85 mg Al/g resin). Comparison of the SEM images shows that colloidal aluminum hydroxide precipitates accumulated in the macropores of the SAC-Al resin beads. Minor aluminum leakage, if any, in the beginning of the service cycle, was completely arrested by having a small cation exchange column in sodium form as post-treatment; the small SAC-Na column also eliminated calcium leakage.

8.3.6 Sustainability Issues and New Opportunities

One of the striking findings of the investigation was that a traditional cation exchange resin, presaturated with a trivalent ion (e.g., Al^{3+}), can remove divalent cations (e.g., Ca^{2+} hardness) and was amenable to efficient regeneration. From a sustainability viewpoint, results of the study are particularly attractive because generation of excess spent regenerant (e.g., SAC-Na processes) or voluminous sludge (e.g., lime softening) was avoided. Equally important, the ion exchange resins, fixed-bed column components and operational protocol are nearly identical to existing SAC-Na processes. Existing systems could be retrofitted with minimal expense and effort [64].

Figure 8.20 provides an overview of the proposed Al-cycle softening process, namely, the absence of sodium in the spent regenerant and the partial reduction of TDS and constant sodium concentration in the treated water.

8.4 Closure

In this chapter, three new possible application opportunities were discussed: recovering energy from waste acid neutralization, passive control of anaerobic biological reactors and sustainable softening without use of NaCl as a regenerant. Although these applications are yet to attain fruition, prospects remain encouraging. Most importantly, the underlying scientific concepts may find acceptance to solve other similar real-life problems; ion exchange remains the unifying commonality in all.

No ingenuity grows in isolation. The true versatility and usefulness of a scientific field lie in the field's ability to connect to others for enhanced synergy. During the six decades following the Second World War, the field of ion exchange has permeated into a wide array of applications ranging from mining to microelectronics, environment to energy, drug delivery to detection, food to fertilizer, chemical cleaning to catalysis,

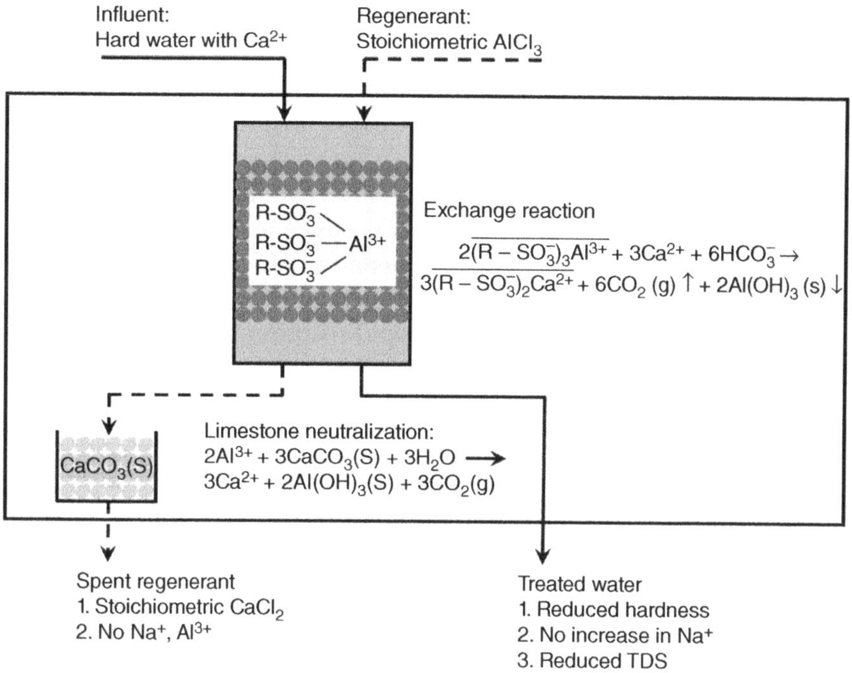

Figure 8.20 Schematic overview of water hardness treatment by SAC-Al and regeneration with $AlCl_3$. *Source:* Li *et al.* 2016 [64]. Reproduced with permission of American Chemical Society.

bioseparation to brackish water, and many others. As environment and energy-related regulations grew, ion exchange took a dominant role in offering solutions to many concurrent problems both in the developed and the developing countries. New materials and new processes are rapidly growing. Ion exchange fundamentals and their application opportunities provide a platform welcoming a diverse group of professionals from chemistry, chemical engineering, environmental engineering, polymer chemistry, physical chemistry, biotechnology, and others to address and offer rational solutions to problems facing human societies globally.

Summary

- Acid–base neutralization (i.e., $H^+ + OH^- \rightarrow H_2O$) is a thermodynamically favorable, exothermic process with the highest equilibrium constant value among all homogeneous, aqueous-phase reactions.
- The heat evolved during waste acid neutralization causes only a little increase in temperature of the bulk aqueous phase. So any energy recovery is not possible and has not been reported to date.
- WAC and weak-base anion exchange resins are pH-sensitive, durable polymers. They swell or shrink a lot when in contact with acid and/or alkali solution due to a major change in the osmotic pressure inside the exchanger phase, that is, swelling/shrinking is reversible.

- A significant amount of mechanical energy can be recovered by carrying out the acid neutralization reaction inside a pH-sensitive, hydrophilic ion exchanger containing covalently attached weak-acid or weak-base functional groups. The process efficiency can be enhanced by tailoring the cross-linking and hydrophilicity of the polymer phase.
- Methane-producing anaerobic biological processes are economically attractive for treating high-strength organic wastes because they produce methane gas with high fuel value and do not require any energy for sustenance. However, they are very susceptible to failure due to organic and toxic overloading.
- Recent advances in the production of IXFs with rapid sorption/desorption kinetics offer unique opportunities to improve the process stability of anaerobic biological reactors through passive control. IXFs with iminodiacetate (IDA) functional groups, if appropriately retained within the reactor, are quite appropriate to protect methanogenic bacteria from sudden surges of acidity and/or toxic metal ions.
- Extensive laboratory studies confirm that IXF greatly improves the stability of anaerobic biological reactors under extreme, unfavorable feed conditions. Also, IXFs do not require any external regeneration and retain capacities for months without being impaired in a biological medium.
- Na-cycle ion exchange softening is universally practiced for hardness removal. However, the regeneration process is inefficient and releases a large amount of brine in the environment. Al-cycle softening, a newly conceived, simple-to-operate hardness removal process, avoids use of brine and greatly improves regeneration efficiency.
- Al-cycle softening can be easily retrofitted into existing Na-cycle systems without any significant changes and is likely to offer ion exchange at the highest thermodynamic efficiency.
- Ion exchange science and advanced ion exchange materials continue to find unexpected novel applications in a wide variety of industries and processes.

References

1 Dean, J.A. and Lange, N.A. (1992) *Lange's Handbook of Chemistry*, 14th edn, McGraw Hill, New York.

2 European Fertilizer Manufacturers' Association (EFMA) (2000) *Production of Sulphuric Acid. Booklet No. 3 of 8*, Ave. E van Nieuwenhuyse 4, B-1160 Brussels, Belgium.

3 US EIA (2016) *How Much Carbon Dioxide is Produced per Kilowatthour when Generating Electricity with Fossil Fuels?* https://www.eia.gov/tools/faqs/faq.cfm?id=74&t=11 (accessed 29 November 2016).

4 Srivastava, R.K. and Jozewicz, W. (2001) Flue gas desulfurization: the state of the art. *Journal of the Air & Waste Management Association*, **51** (12), 1676–1688.

5 Hesketh, A., Broadhurst, J., and Harrison, S. (2010) Mitigating the generation of acid mine drainage from copper sulfide tailings impoundments in perpetuity: a case study for an integrated management strategy. *Minerals Engineering*, **23** (3), 225–229.

6 Donnan, F.G. (1995) Theory of membrane equilibria and membrane potentials in the presence of non-dialysing electrolytes. A contribution to physical-chemical physiology. *Journal of Membrane Science,* **100** (1), 45–55.
7 Sarkar, S., SenGupta, A.K., and Prakash, P. (2010) The Donnan membrane principle: opportunities for sustainable engineered processes and materials. *Environmental Science & Technology,* **44** (4), 1161–1166.
8 Prakash, P. and SenGupta, A. (2005) Coupled transport of trivalentmonovalent ion pair (Al^{3+} and H^+) during Donnan membrane process. *AICHE Journal,* **51** (1), 333–344.
9 Cumbal, L. and SenGupta, A.K. (2005) Arsenic removal using polymer-supported hydrated iron (III) oxide nanoparticles: role of Donnan membrane effect. *Environmental Science & Technology,* **39** (17), 6508–6515.
10 Sarkar, S., SenGupta, A.K., Greenleaf, J.E., and El-Moselhy, M. (2011) Energy recovery from acid–base neutralization process through pH-sensitive polymeric ion exchangers. *Industrial and Engineering Chemistry Research,* **50** (21), 12293–12298.
11 German, M., SenGupta, A.K., and Greenleaf, J. (2013) Hydrogen ion (H^+) in waste acid as a driver for environmentally sustainable processes: opportunities and challenges. *Environmental Science & Technology,* **47** (5), 2145–2150.
12 Streat, M. (1984) Kinetics of slow diffusing species in ion exchangers. *Reactive Polymers, Ion Exchangers, Sorbents,* **2** (1), 79–91.
13 Li, P. and SenGupta, A.K. (2000) Intraparticle diffusion during selective ion exchange with a macroporous exchanger. *Reactive and Functional Polymers,* **44** (3), 273–287.
14 Dragan, E.S. and Valentina, M. (2011) Separations by Multicomponent ionic systems based on natural and synthetic polycations, in *Ion Exchange and Solvent Extraction: A Series of Advances,* vol. **20**, CRC Press, pp. 233–291.
15 Kim, S.J., Lee, C.K., Lee, Y.M. *et al.* (2003) Electrical/pH-sensitive swelling behavior of polyelectrolyte hydrogels prepared with hyaluronic acid–poly (vinyl alcohol) interpenetrating polymer networks. *Reactive and Functional Polymers,* **55** (3), 291–298.
16 Unal, B. and Hedden, R.C. (2009) pH-dependent swelling of hydrogels containing highly branched polyamine macromonomers. *Polymer,* **50** (3), 905–912.
17 Yang, J. and Speece, R. (1985) Effects of engineering controls on methane fermentation toxicity response. *Journal Water Pollution Control Federation,* **57** (12), 1134–1141.
18 Clark, R. and Speece, R. (1970) *The pH Tolerance of Anaerobic Digestion.* 5th International Water Pollution Research Conference, July–August.
19 Speece, R.E. (1983) Anaerobic biotechnology for industrial wastewater treatment. *Environmental Science & Technology,* **17** (9), 416A–427A.
20 Graef, S.P. and Andrews, J.F. (1974) Stability and control of anaerobic digestion. *Journal Water Pollution Control Federation,* **46** (4), 666–683.
21 Hill, D. and Barth, C. (1977) A dynamic model for simulation of animal waste digestion. *Journal Water Pollution Control Federation,* **49** (10), 2129–2143.
22 Lawrence, A.W. and McCarty, P.L. (1965) The role of sulfide in preventing heavy metal toxicity in anaerobic treatment. *Journal Water Pollution Control Federation,* **37** (3), 392–406.
23 Parkin, G.F. and Speece, R.E. (1982) Modeling toxicity in methane fermentation systems. *Journal of Environmental Engineering Division, ASCE; (United States),* **108**, 515–531.

24 Hickey, R.F., Vanderwielen, J., and Switzenbaum, M.S. (1989) The effect of heavy metals on methane production and hydrogen and carbon monoxide levels during batch anaerobic sludge digestion. *Water Research*, **23** (2), 207–218.
25 Mosey, F. and Hughes, D.A. (1975) The toxicity of heavy metal ions to anaerobic digestion. *Water Pollution Control*, **74**, 18–38.
26 Bhattacharya, S.K., Madura, R.L., Uberoi, V., and Haghighi-Podeh, M.R. (1995) Toxic effects of cadmium on methanogenic systems. *Water Research*, **29** (10), 2339–2345.
27 Mitra, I.N., SenGupta, A.K., Kugelman, I.J., and Creighton, R. (1998) Evaluating composite ion exchangers (CIX) for improved stability of anaerobic biological reactors. *Water Research*, **32** (11), 3267–3280.
28 Mitra, I.N., SenGupta, A.K., and Kugelman, I.J. (1998) Improving stability of anaerobic biological reactors using composite ion exchangers. *Water Science and Technology*, **38** (8–9), 369–376.
29 Yu, T. (2016) *Improved Stability of Methane-Producing Anaerobic Biological Reactors through Novel Use of Ion-Exchange Fibers*, Lehigh University, Bethlehem, PA.
30 Clifford, D.A. (1999) Chapter 9: Ion exchange and adsorption of inorganic contaminants, in *Water Quality & Treatment Handbook*, 5th edn (ed. R.D. Letterman), McGraw-Hill Professional, New York, pp. 9.1–9.91.
31 Mihelcic JR, Zimmerman JB. *Environmental Engineering: Fundamentals, Sustainability, Design* : Wiley Global Education; 2014.
32 Clifford, D., Subramonian, S., and Sorg, T.J. (1986) Water treatment processes. III. Removing dissolved inorganic contaminants from water. *Environmental Science & Technology*, **20** (11), 1072–1080.
33 Soday, F.J. (1942) United Gas Improvement Co, assignee. Sulphonated derivatives of polymerized methylstyrene. US patent 2,283,236 A. May 19.
34 Swinton, E. and Weiss, D. (1953) Counter current adsorption separation processes. 1. Equipment. *Australian Journal of Applied Science*, **4** (2), 316–328.
35 Pomeranz, A., Dolfin, T., Korzets, Z. *et al.* (2002) Increased sodium concentrations in drinking water increase blood pressure in neonates. *Journal of Hypertension*, **20** (2), 203–207.
36 Mente, A., O'Donnell, M., Rangarajan, S. *et al.* (2016) Associations of urinary sodium excretion with cardiovascular events in individuals with and without hypertension: a pooled analysis of data from four studies. *The Lancet*, **388** (10043), 465–475.
37 Mills, K.T., Bundy, J.D., Kelly, T.N. *et al.* (2016) Global disparities of hypertension prevalence and control: a systematic analysis of population-based studies from 90 countries. *Circulation*, **134** (6), 441–450.
38 Famiglietti, J. (2014) The global groundwater crisis. *Nature Climate Change*, **4** (11), 945–948.
39 Santa Clarita Sanitation District of Los Angeles County (2008) Santa clara river chloride reduction ordinance of 2008. DOC# 1035050:1–6.
40 Weiss, D., Bolto, B., McNeill, R. *et al.* (1965) The Sirotherm demineralisation process—an ion-exchange process with thermal regeneration. Part I. *Journal of The Institution of Engineers*, **37**, 193.
41 Jackson, M.B. and Bolto, B.A. (1979) Ici Australia Limited, Commonwealth Scientific And Industrial Research Organization, assignees. Amphoteric composite

resins and method of preparing same by polymerization of a two-phase dispersion of monomers. Australian patent US4134815 A. 1979 Jan 16.
42 Custers, J., Stemkens, L.F., Sablong, R. *et al.* (2014) Salt-free softening by thermo-reversible ion-adsorbing hydrogels. *Journal of Applied Polymer Science*, **131** (9), 40216.
43 Liao, Z., Gu, Z., Schulz, M. *et al.* (2009) Treatment of cooling tower blowdown water containing silica, calcium and magnesium by electrocoagulation. *Water Science and Technology*, **60** (9), 2345–2352.
44 Schulz, M., Baygents, J., and Farrell, J. (2009) Laboratory and pilot testing of electrocoagulation for removing scale-forming species from industrial process waters. *International journal of Environmental Science and Technology*, **6** (4), 521–526.
45 Greenleaf, J.E. and SenGupta, A.K. (2006) Environmentally benign hardness removal using ion-exchange fibers and snowmelt. *Environmental Science & Technology*, **40** (1), 370–376.
46 Greenleaf, J.E., Lin, J., and SenGupta, A.K. (2006) Two novel applications of ion exchange fibers: arsenic removal and chemical-free softening of hard water. *Environmental Progress*, **25** (4), 300–311.
47 Padungthon, S., Greenleaf, J., and SenGupta, A.K. (2011) Carbon dioxide sequestration through novel use of ion exchange fibers (IX-fibers). *Chemical Engineering Research and Design*, **89** (9), 1891–1900.
48 Sarkar, S., Blaney, L.M., Gupta, A. *et al.* (2007) Use of ArsenX^{np}, a hybrid anion exchanger, for arsenic removal in remote villages in the Indian subcontinent. *Reactive and Functional Polymers*, **67** (12), 1599–1611.
49 SenGupta, A.K. and Lim, L. (1988) Modeling chromate ion-exchange processes. *AICHE Journal*, **34** (12), 2019–2029.
50 Zhao, D., SenGupta, A.K., and Stewart, L. (1998) Selective removal of Cr (VI) oxyanions with a new anion exchanger. *Industrial and Engineering Chemistry Research*, **37** (11), 4383–4387.
51 Padungthon, S., Li, J., German, M., and SenGupta, A.K. (2014) Hybrid anion exchanger with dispersed zirconium oxide nanoparticles: a durable and reusable fluoride-selective sorbent. *Environmental Engineering Science*, **31** (7), 360–372.
52 Guter, G.A. (1984) inventor. The United States Of America As Represented By The Administrator Environmental Protection Agency, assignee. Removal of nitrate from water supplies using a tributyl amine strong base anion exchange resin. US patent 4479877 A. 1984 Oct 30.
53 Hatch, M.J. (1984) inventor. The Dow Chemical Company, assignee. Removal of metal ions from aqueous medium using a cation-exchange resin having water-insoluble compound dispersed therein. EP0071810A1. 1984 1981-08-03.
54 Zhao, D. and SenGupta, A.K. (1998) Ultimate removal of phosphate from wastewater using a new class of polymeric ion exchangers. *Water Research*, **32** (5), 1613–1625.
55 Blaney, L.M., Cinar, S., and SenGupta, A.K. (2007) Hybrid anion exchanger for trace phosphate removal from water and wastewater. *Water Research*, **41** (7), 1603–1613.
56 Landry, K.A. and Boyer, T.H. (2013) Diclofenac removal in urine using strong-base anion exchange polymer resins. *Water Research*, **47** (17), 6432–6444.
57 Li, P. and SenGupta, A.K. (1998) Genesis of selectivity and reversibility for sorption of synthetic aromatic anions onto polymeric sorbents. *Environmental Science & Technology*, **32** (23), 3756–3766.

58 Li, P. and SenGupta, A.K. (2004) Sorption of hydrophobic ionizable organic compounds (HIOCs) onto polymeric ion exchangers. *Reactive and Functional Polymers*, **60**, 27–39.
59 Zhang, H., Shields, A.J., Jadbabaei, N. *et al.* (2014) Understanding and modeling removal of anionic organic contaminants (AOCs) by anion exchange resins. *Environmental Science & Technology*, **48** (13), 7494–7502.
60 Jadbabaei, N. and Zhang, H. (2014) Sorption mechanism and predictive models for removal of cationic organic contaminants by cation exchange resins. *Environmental Science & Technology*, **48** (24), 14572–14581.
61 Alexandratos, S.D. and Wilson, D.L. (1986) Dual-mechanism bifunctional polymers: polystyrene-based ion-exchange/redox resins. *Macromolecules*, **19** (2), 280–287.
62 Keister, T. (2008) Non-chemical devices: thirty years of myth busting. *Water Conditioning & Purification*, **490**, 11–12.
63 Van Benschoten, J.E. and Edzwald, J.K. (1990) Chemical aspects of coagulation using aluminum salts—I. Hydrolytic reactions of alum and polyaluminum chloride. *Water Research*, **24** (12), 1519–1526.
64 Li, J., Koner, S., German, M., and Sengupta, A.K. (2016) Aluminum-cycle ion exchange process for hardness removal: a new approach for sustainable softening. *Environmental Science & Technology*, **50** (21), 11943–11950.
65 SenGupta, A.K., Zhu, Y., and Hauze, D. (1991) Metal (II) ion binding onto chelating exchangers with nitrogen donor atoms: some new observations and related implications. *Environmental Science & Technology*, **25** (3), 481–488.
66 Zhao, D. and SenGupta, A.K. (2000) Ligand separation with a copper (II)-loaded polymeric ligand exchanger. *Industrial and Engineering Chemistry Research*, **39** (2), 455–462.

A

Commercial Ion Exchangers

For up-to-date information on the high number of ion exchange resins, look at the website of François de Dardel (http://dardel.info/IX/AllResins.php). The website includes an exhaustive list of nearly all current and obsolete ion exchange resins from the year 2000 to present.

The general pH ranges of functionality for each type of ion exchange resin is listed in Table A1. Most metal chelation resins are designed to have higher selectivity versus typical ion exchange resins in challenging conditions (Tables A2A–A20B).

Ion Exchange in Environmental Processes: Fundamentals, Applications and Sustainable Technology,
First Edition. Arup K. SenGupta.

Table A1 General pH ranges of functionality for the major types of ion exchange resins.

Ion exchange resin type	pH range	Notes
Strong acid cation (SAC)	0–14	
Weak acid cation (WAC)	4–14	
Metal chelation		
Amidoxime	6–14	
Aminophophonic	2–6 6–11	H^+-form Na^+-form
Bispicolylamine	0–8	Possibly higher pH, metal-dependent
Epoxypolyamine	<6.5	Cr(VI) → Cr(III) oxidation and precipitation
Hydroxamic acid		Ga-selective from Bayer liquor
Iminodiacetate	2–6 6–11	H^+-form Na^+-form
Methyl glucamate	4–10	Boron-selective
Monophos™	0–4 6–8	Metals, for example, iron, from strong acids Polyvalent metal cations > alkali cations
Polyamine	2–6	Divalent metal ion selective. Selectivity: $Hg^{2+} > Fe^{3+} > Cu^{2+} > Zn^{2+} > Cd^{2+} > Ni^{2+} > Co^{2+} > Ag^{+} > Mn^{2+}$
Thiol	1–13	Hg-selective
Thiourea	1–13	Hg-selective
Strong base anion (SBA)		
Type I	1–13	
Type II	1–13	
Weak base anion (WBA)	1–9 9–12	Cl^--form Free base form
Specialty anion		
Arsenic-selective	4–8.5	Hybrid resin w/Fe(III) nanoparticles, Arsenic-selective
Fluoride-selective	5.5–8.5 (i. Al(III)-based ii. nano ZrO_2-based)	
Nitrate-selective	1–13	
Polyiodide	6.5–8.5	

Notes on companies: Resindion is a subsidiary of Mitsubishi Chemical.

Table A2A Commercial cation exchange resins: strong acid cation (SAC-Na-form), gel-type.

Trade name	Manufacturer	Capacity	
		meq/g dry resin	meq/mL resin bed
C-800 LT UPS	Aldex		1.9
DOWEX Marathon C	DOW		2.0
Indion 225 Na	Ion exchange		2.0
LEWATIT S 1668	Lanxess		2.2
DIAION SK1B	Mitsubishi		2.0
C100	Purolite		2.0
RELITE CF	Resindion		2.0
CG10	ResinTech		2.3
001x8	Suqing	4.5	2.0
Tulsion T-42	Thermax		2.0

Table A2B Commercial cation exchange resins: strong acid cation (SAC-Na-form), MP-type.

Trade name	Manufacturer	Capacity	
		meq/g dry resin	meq/mL resin bed
C-800MP	Aldex		1.8
DOWEX Marathon MSC	DOW		1.7
Indion 730	Ion exchange		1.7
LEWATIT S 2568	Lanxess		1.7
DIAION PK216	Mitsubishi		1.75
C145	Purolite		1.5
RELITE RPS	Resindion		1.8
SACMP	ResinTech		1.65
D001	Suqing	4.35	1.8
Tulsion T-42MP	Thermax		1.7

Table A3A Commercial cation exchange resins: weak acid cation (WAC-H^+-form), gel-type.

Trade name	Manufacturer	Capacity	
		meq/g dry resin	meq/mL resin bed
WAC	Aldex		4.2
AMBERLITE IRC86	DOW		4.1
Indion 236	Ion exchange		4.0
WACG	ResinTech		4.0
112	Suqing	10.0	4.3
Tulsion CXO-12	Thermax		4.2

Table A3B Commercial cation exchange resins: weak acid cation (WAC-H^+-form), MP-type.

Trade name	Manufacturer	Capacity	
		meq/g dry resin	meq/mL resin bed
WAC MP	Aldex		4.3
DOWEX MAC-3	DOW		3.8
Indion 652	Ion exchange		3.5
LEWATIT S 8227 Ca	Lanxess		4.3
WK11	Mitsubishi		2.9
WK40L	Mitsubishi		4.4
C104Plus	Purolite		4.5
C107	Purolite		3.6
RELITE CNS	Resindion		4.3
WACMP	ResinTech		3.8
D113	Suqing	10.8	4.3
Tulsion CXO-12MP	Thermax		3.8

Table A4 Commercial chelating ion exchange resin: amidoxime (free base-form).

Trade name	Manufacturer	Capacity	
		meq/g dry resin	meq/mL resin bed
S910	Purolite		1.25

Table A5 Commercial chelating ion exchange resins: aminophosphonic (Na^+-form; weakly acidic), macroporous-type.

Trade name	Manufacturer	Capacity	
		meq/g dry resin	meq/mL resin bed
CR 2012	Aldex		2.0
AMBERLITE IRC747	DOW		≥1.75
INDION BSR	Ion exchange		2.0
LEWATIT TP-260	Lanxess		2.3
S950	Purolite		1.3 (Ca^{2+})
SIR-500	ResinTech		1.4
D402-II	Suqing	1.45 (Ca^{2+})	0.5 (Ca^{2+})
Tulsion CH-93	Thermax		2.0

Table A6 Commercial chelating ion exchange resins: bispicolylamine (free base/SO_4^{2-}-form).

Trade name	Manufacturer	Capacity	
		meq/g dry resin	meq/mL resin bed
DOWEX M4195 (or 3N)	DOW		1.25 (Cu^{2+})
LEWATIT MDS TP 220	Lanxess		1.13 (Cu^{2+})
S960	Purolite		0.85 (Ni^{2+})
SIR-1000	ResinTech		0.8

Table A7 Commercial chelating ion exchange resins: iminodiacetate (Na^+-form, weak acid), MP-type.

Trade name	Manufacturer	Capacity	
		meq/g dry resin	meq/mL resin bed
CR 29	Aldex		2.0
AMBERLITE IRC78	DOW		1.35
INDION SIR	Ion exchange		2.2
LEWATIT TP 207	Lanxess		2.2
DIAION CR11	Mitsubishi	1.2	0.35
S930/4888	Purolite		1.1
S930Plus	Purolite		2.9
SIR-350	ResinTech		1.6
D401	Suqing	1.95 (Cu^{2+})	0.6 (Cu^{2+})
Tulsion CH-90	Thermax		2.0

Table A8 Commercial chelating ion exchange resins: isothiouronium (H^+-form), MP-type.

Trade name	Manufacturer	Capacity	
		meq/g dry resin	meq Hg^{2+}/mL resin bed
CR 80	Aldex		1.25; 200 g Hg/L
XUS 43600.00	DOW		0.7
LEWATIT TP-214	Lanxess		1.0
S920Plus	Purolite		2.75
SIR-400	ResinTech		1.5
Tulsion CH-95	Thermax		1.25

Table A9 Commercial chelating ion exchange resins: methyl glucamine (Cl^--form, boron-selective).

Trade name	Manufacturer	Capacity	
		meq/mL resin bed	Boron-specific capacity
CR5	ALDEX	0.6	
Amberlite PWA10 (potable)	DOW	≥0.7	
Amberlite IRA743	DOW	0.7	
LEWATIT MonoPlus MK51	Lanxess	0.8	6 mg B/mL
DIAION CRB05	Mitsubishi	0.95	
S108	Purolite	0.6	1–3.5 mg/mL (in operation)
SIR-150	ResinTech	0.7	
D403-II	Suqing	0.9	
Tulsion CH-99	Thermax	0.8	5.7 meq B/mL

Table A10 Commercial chelating ion exchange resin: Monophos™ (sulfonate and phosphonate functional groups);™- EIChroM Industries Inc. (H^+-form).

Trade name	Manufacturer	Capacity	
		meq/g dry resin	meq/mL resin bed
S957	Purolite		1.0

Table A11 Commercial chelating ion exchange resins: thiol (H^+-form), MP-type.

Trade name	Manufacturer	Capacity	
		meq/g dry resin	meq/mL resin bed
AMBERSEP GT74	DOW		1.3 (Hg^{2+})
INDION MSR	Ion exchange	3.6	
S924	Purolite		2.0 (Hg^{2+})
SIR-200	ResinTech		1.2 (Hg^{2+})
D405	Suqing		0.8 (Hg^{2+})
TULSIONCH-97	Thermax		1.5 (Hg^{2+})

Table A12 Commercial chelating ion exchange resins: thiourea, MP-type.

Trade name	Manufacturer	Capacity	
		meq/g dry resin	meq/mL resin bed
XUS 43600.00	DOW		0.7
INDION TCR	Ion exchange		1.6
QuadraPure TU	Johnson Matthey		
LEWATIT MonoPlus TP214	Lanxess		1.0
S914	Purolite		1.0
D405-II	Suqing		1.0
TULSION CH-95	Thermax		150 (g/L as Hg)

Table A13A Commercial anion exchange resins: strong base anion (SBA) Type I (Cl^--form), gel-type.

Trade name	Manufacturer	Capacity	
		meq/g dry resin	meq/mL resin bed
SB-1	Aldex		1.4
DOWEX MARATHON A	DOW		1.3 (Cl^-); 1.0 (OH^-)
Indion GS 300	Ion exchange		1.3
LEWATIT S 6268	Lanxess		1.2
DIAION SA10A	Mitsubishi		1.3
A400	Purolite		1.3
RELITE 3A	Resindion		1.3
SBG1-HP	ResinTech		1.2
201x4	Suqing	3.7	1.2
Tulsion A-23	Thermax		1.3

Table A13B Commercial anion exchange resins: strong base anion (SBA) Type I (Cl^--form), MP-type.

Trade name	Manufacturer	Capacity	
		meq/g dry resin	meq/mL resin bed
SB-1 MP	Aldex		1.2
DOWEX MARATHON MSA	DOW		1.1 (Cl^-)
Indion 810	Ion exchange		1.0
LEWATIT S 6368 A	Lanxess		1.0
LEWATIT VP OC 1074	Lanxess		0.85
DIAION PA316	Mitsubishi		1.3
A500Plus	Purolite		1.15
RELITE 3AS	Resindion		1.2
SBMP1	ResinTech		1.1
D201	Suqing	3.7	1.2
Tulsion A-27MP	Thermax		1.2

Table A14A Commercial anion exchange resins: SBA Type II (Cl^--form), gel-type.

Trade name	Manufacturer	Capacity	
		meq/g dry resin	meq/mL resin bed
SB-2	Aldex		1.4
DOWEX MARATHON A2	DOW		1.2 (Cl^-)
Indion GS 400	Ion exchange		1.2
DIAION SA20A	Mitsubishi		1.3
A300	Purolite		1.4
RELITE 2A	Resindion		1.3
SBG2	ResinTech		1.4
202-II	Suqing	3.5	1.25
Tulsion A-32	Thermax		1.3

Table A14B Commercial anion exchange resins: SBA Type II (Cl^--form), MP-type.

Trade name	Manufacturer	Capacity	
		meq/g dry resin	meq/mL resin bed
DOWEX MARATHON MSA2	DOW		1.0 (Cl^-)
Indion 820	Ion exchange		1.0
LEWATIT S 7468	Lanxess		1.0
DIAION PA418	Mitsubishi		1.3
A510Plus	Purolite		1.15
RELITE A490	Resindion		1.2
D202-II	Suqing	3.6	1.2
Tulsion A-36MP	Thermax		1.2

Table A15A Commercial anion exchange resins: polyacrylic, weak base anion (WBA) (free base-form), gel-type.

Trade name	Manufacturer	Capacity	
		meq/g dry resin	meq/mL resin bed
WB-2	Aldex		1.6
AMBERLITE IRA67	DOW		1.6
LEWATIT S 5228	Lanxess		1.5
DIAION WA10	Mitsubishi		1.2
A845	Purolite		1.6
WBACR	ResinTech		1.6
316	Suqing	5.0	1.6

Table A15B Commercial anion exchange resins: polystyrene, weak base anion (WBA) (free base-form), MP-type.

Trade name	Manufacturer	Capacity	
		meq/g dry resin	meq/mL resin bed
WB1-HC	Aldex		1.6
DOWEX MARATHON WBA2	DOW		1.7
Indion 890	Ion exchange		1.4
QuadraPure DMA	Johnson Matthey		
LEWATIT S 4528	Lanxess		1.7
DIAION WA30	Mitsubishi	3.0	1.5
A100Plus	Purolite		1.3
RELITE RAM1	Resindion		1.6
WBMP-HC	ResinTech		1.7
D306	Suqing	4.8	1.6
Tulsion A-2X MP GP	Thermax		1.5

Table A16A Commercial specialty high capacity anion exchange resin (Cl^--form, strong base), gel-type.

Trade name	Manufacturer	Capacity	
		meq/g dry resin	meq/mL resin bed
AMBERLITE PWA12	DOW		≥1.25
A600E	Purolite		1.6

Table A16B Commercial specialty anion exchange resin: nitrate-selective (Cl^--form, strong base), MP-type.

Trade name	Manufacturer	Capacity	
		meq/g dry resin	meq/mL resin bed
NSR	Aldex		0.9
AMBERLITE PWA5	DOW		≥1.0
Indion NSSR	Ion exchange		0.9
Resinex NR-1	Jacobi Carbons		0.8
Lewatit MonoPlus SR 7	Lanxess		0.6
A520E	Purolite		0.9
RELITE A490 (Type 2)	Resindion		1.2
SIR-100-HP (triethyl)	ResinTech		0.9
SIR-110-HP (tributyl)	ResinTech		0.7
D407; D407-III	Suqing	3.0; 1.6	0.8; 0.5
Tulsion A-62 MP	Thermax		1.0

Table A17A Commercial specialty anion exchange resin: polyamine (free base-form), gel-type.

Trade name	Manufacturer	Capacity	
		meq/g dry resin	meq/mL resin bed
Lewatit A365	Lanxess		3.4
WBG30	ResinTech		2.8

Table A17B Commercial specialty anion exchange resin: polyamine (free base-form), MP-type.

Trade name	Manufacturer	Capacity	
		meq/g dry resin	meq/mL resin bed
DIAION CR20	Mitsubishi		0.8 (Cu)
S985	Purolite		2.3
SQD-96	Suqing	7.5	2.2
TULSION A-10X MP	Thermax		2.5

Table A18 Commercial specialty anion exchange resin: epoxypolyamine (Cl-form), gel-type.

Trade name	Manufacturer	Capacity	
		meq/g resin	meq/mL resin bed
SIR-700	ResinTech		2.1 meq Cl/mL; 10.5 meq Cr/mL

Table A19 Commercial chelating cation exchange resin: hydroxamic acid (Na-form), MP-type.

Trade name	Manufacturer	Capacity	
		meq/g resin	meq/mL resin bed
D409	Suqing	3.5–4.0	

Table A20A Commercial specialty anion-selective resins: trace contaminant selective via non ion-exchange interactions.

Trade name	Manufacturer	Capacity		Physical form
		Selectivity	per volume	
INDION RS-F	Ion exchange	F	1–2 g F/L	HCIX w/ Al(III) nanoparticles
LEWATIT FO36	Lanxess	As	2.5–6 g As/L	HAIX w/ Fe(III) nanoparticles
LEWATIT MonoPlus TP260	Lanxess	F	2.6 g F/L (operating)	Aminophosphonic resin in Al(III)-form
FERRIX A33E	PUROLITE	As	0.5–4 g As/L	HAIX w/ Fe(III) nanoparticles
ASM10HP	ResinTech	As	0.6–1.2 g As/L;40 mg As/g	HAIX w/ Fe(III) nanoparticles
BSM-50	ResinTech	Si, Sb	1.4 eq/L as borate	HAIX w/ Fe(III) nanoparticles in borate form
RSM-50	ResinTech	Ra	1.65 eq Na^+/L	HCIX w/$BaSO_4$ nanoparticles
D406	Suqing	F	0.5 meq/mL	Al(III) nanoparticles
SQ407	Suqing	As		HAIX w/ FeO(OH) nanoparticles
TULSION CH-87	Thermax	F		HCIX w/ Al(III) nanoparticles
HIX2-Nano200	WIST	As, F	10 g F/L	HAIX w/ Zr(IV) nanoparticles

Note: The list of values in the table removal capacity of trace organic contaminants is a significant function of influent conditions, especially pH, contaminant concentration, and competitor concentration. Thus, capacity values for continuous removal of trace contaminants should be determined by rapid small-scale column tests (RSSCTs) using representative influent.
HCIX– hybrid cation exchange resin; HAIX – hybrid anion exchange resin.

Table A20B Commercial specialty anion exchange resin: polyiodide disinfection (I^--form, strong base), MP-type.

Trade name	Manufacturer	Capacity	
		Residual iodine (mg/L)	Capacity (BVs)
Indion SRCD-I	Ion exchange	0.5–1.1	250

B

Different Units of Capacity, Concentration, Mass, and Volume

B.1 Capacity

	meq $CaCO_3$/mL	lb-eq $CaCO_3/ft^3$	kgr (as $CaCO_3$)/ft^3	g CaO/L	g $CaCO_3$/L
1 meq $CaCO_3$/mL	1	0.0624	21.8	28	50
1 lb-eq $CaCO_3/ft^3$	16.0	1	249	449	801
1 kgr (as $CaCO_3$)/ft^3	0.0459	0.00286	1	1.28	2.29
1 g CaO/L	0.0357	0.00223	0.779	1	1.79
1 g $CaCO_3$/L	0.0200	0.00125	0.436	0.560	1

B.2 Concentration (Expressed as $CaCO_3$)

	ppm	meq/L	gr/gal (US)	gr/gal (Imp.)
1 ppm	1	0.02	0.0585	0.0704
1 meq/L	50	1	2.9	3.48
1 gr/gal (US)	17.1	0.34	1	1.20
1 gr/gal (Imp.)	14.2	0.28	0.833	1
1 French degree	10	0.20	0.585	0.704
1 German degree	17.9	0.36	1.05	1.26

Ion Exchange in Environmental Processes: Fundamentals, Applications and Sustainable Technology,
First Edition. Arup K. SenGupta.

B.3 Mass

	Pounds	Grams	Kilograms	Grains	Kilograins
1 lb	1	453.6	0.4536	7000	7
1 g	0.0022	1	0.0001	15.43	0.01543
1 kg	2.205	1000	1	15430	15.43
1 gr	0.000143	0.0648	0.0000648	1	0.001
1 kgr	0.143	64.8	0.0648	1000	1

B.4 Volume

	Cubic foot	Gallon (US)	Gallon (Imp.)	Liter	Cubic meter
1 ft^3	1	7.48	6.23	28.3	0.0283
1 gal (US)	0.134	1	0.833	3.785	0.003785
1 gal (Imp.)	0.161	1.2	1	4.55	0.00455
1 L	0.0353	0.264	0.220	1	0.001
1 m^3	35.3	264	220	1000	1

C

Table of Solubility Product Constants at 25 °C

Anion	Equilibrium reaction	K_{sp}
Carbonates		
$MgCO_3$	$MgCO_3(s) \leftrightarrow Mg^{2+} + CO_3^{2-}$	4.0×10^{-5}
$NiCO_3$	$NiCO_3(s) \leftrightarrow Ni^{2+} + CO_3^{2-}$	1.4×10^{-7}
$CaCO_3$	$CaCO_3(s) \leftrightarrow Ca^{2+} + CO_3^{2-}$	4.7×10^{-9}
$MnCO_3$	$MnCO_3(s) \leftrightarrow Mn^{2+} + CO_3^{2-}$	4.0×10^{-10}
$CuCO_3$	$CuCO_3(s) \leftrightarrow Cu^{2+} + CO_3^{2-}$	2.5×10^{-10}
$FeCO_3$	$FeCO_3(s) \leftrightarrow Fe^{2+} + CO_3^{2-}$	2.0×10^{-11}
$ZnCO_3$	$ZnCO_3(s) \leftrightarrow Zn^{2+} + CO_3^{2-}$	3.0×10^{-11}
$CdCO_3$	$CdCO_3(s) \leftrightarrow Cd^{2+} + CO_3^{2-}$	5.2×10^{-12}
$PbCO_3$	$PbCO_3(s) \leftrightarrow Pb^{2+} + CO_3^{2-}$	1.5×10^{-13}
Chromate		
$CaCrO_4$	$CaCrO_4(s) \leftrightarrow Ca^{2+} + CrO_4^{2-}$	7.1×10^{-4}
$PbCrO_4$	$PbCrO_4(s) \leftrightarrow Pb^{2+} + CrO_4^{2-}$	1.8×10^{-14}
Fluoride		
MgF_2	$MgF_2(s) \leftrightarrow Mg^{2+} + 2F^-$	8×10^{-8}
CaF_2	$CaF_2(s) \leftrightarrow Ca^{2+} + 2F^-$	1.7×10^{-14}
Hydroxide		
$Mg(OH)_2$	$Mg(OH)_2(s) \leftrightarrow Mg^{2+} + 2OH^-$	8.9×10^{-12}
$Mn(OH)_2$	$Mn(OH)_2(s) \leftrightarrow Mn^{2+} + 2OH^-$	2.0×10^{-13}
$Cd(OH)_2$	$Cd(OH)_2(s) \leftrightarrow Cd^{2+} + 2OH^-$	2.0×10^{-14}
$Pb(OH)_2$	$Pb(OH)_2(s) \leftrightarrow Pb^{2+} + 2OH^-$	4.2×10^{-15}
$Fe(OH)_2$	$Fe(OH)_2(s) \leftrightarrow Fe^{2+} + 2OH^-$	1.8×10^{-15}
$Ni(OH)_2$	$Ni(OH)_2(s) \leftrightarrow Ni^{2+} + 2OH^-$	1.6×10^{-16}

Ion Exchange in Environmental Processes: Fundamentals, Applications and Sustainable Technology,
First Edition. Arup K. SenGupta.

Anion	Equilibrium reaction	K_{sp}
$Zn(OH)_2$	$Zn(OH)_2(s) \leftrightarrow Zn^{2+} + 2OH^-$	4.5×10^{-17}
$Cu(OH)_2$	$Cu(OH)_2(s) \leftrightarrow Cu^{2+} + 2OH^-$	1.6×10^{-19}
$Cr(OH)_3$	$Cr(OH)_3(s) \leftrightarrow Cr^{3+} + 3OH^-$	6.7×10^{-31}
$Al(OH)_3$	$Al(OH)_3(s) \leftrightarrow Al^{3+} + 3OH^-$	5.0×10^{-33}
$Fe(OH)_3$	$Fe(OH)_3(s) \leftrightarrow Fe^{3+} + 3OH^-$	6.0×10^{-38}
Phosphate		
$MgNH_4PO_4$	$MgNH_4PO_4(s) \leftrightarrow Mg^{2+} + NH_4^+ + PO_4^{3-}$	2.5×10^{-13}
$AlPO_4$	$AlPO_4(s) \leftrightarrow Al^{3+} + PO_4^{3-}$	6.3×10^{-19}
$Mn_3(PO_4)_2$	$Mn_3(PO_4)_2(s) \leftrightarrow 3Mn^{2+} + 2PO_4^{3-}$	1.0×10^{-22}
$Ca_3(PO_4)_2$	$Ca_3(PO_4)_2(s) \leftrightarrow 3Ca^{2+} + 2PO_4^{3-}$	1.3×10^{-32}
$Mg_3(PO_4)_2$	$Mg_3(PO_4)_2(s) \leftrightarrow 3Mg^{2+} + 2PO_4^{3-}$	1×10^{-32}
$Pb_3(PO_4)_2$	$Pb_3(PO_4)_2(s) \leftrightarrow 3Pb^{2+} + 2PO_4^{3-}$	1.0×10^{-32}
Sulfate		
$CaSO_4$	$CaSO_4(s) \leftrightarrow Ca^{2+} + SO_4^{2-}$	2.5×10^{-5}
$PbSO_4$	$PbSO_4(s) \leftrightarrow Pb^{2+} + SO_4^{2-}$	1.3×10^{-8}
$BaSO_4$	$BaSO_4(s) \leftrightarrow Ba^{2+} + SO_4^{2-}$	1.1×10^{-10}
$RaSO_4$	$RaSO_4(s) \leftrightarrow Ra^{2+} + SO_4^{2-}$	3.7×10^{-11}
Sulfide		
MnS	$MnS(s) \leftrightarrow Mn^{2+} + S^{2-}$	7.0×10^{-16}
FeS	$FeS(s) \leftrightarrow Fe^{2+} + S^{2-}$	4.0×10^{-19}
NiS	$NiS(s) \leftrightarrow Ni^{2+} + S^{2-}$	3.0×10^{-21}
ZnS	$ZnS(s) \leftrightarrow Zn^{2+} + S^{2-}$	1.6×10^{-23}
CdS	$CdS(s) \leftrightarrow Cd^{2+} + S^{2-}$	1.0×10^{-28}
PbS	$PbS(s) \leftrightarrow Pb^{2+} + S^{2-}$	7.0×10^{-29}
CuS	$CuS(s) \leftrightarrow Cu^{2+} + S^{2-}$	8.0×10^{-37}
Cu_2S	$Cu_2S(s) \leftrightarrow 2Cu^+ + S^{2-}$	1.2×10^{-49}
Fe_2S_3	$Fe_2S_3(s) \leftrightarrow 2Fe^{3+} + 3S^{2-}$	1×10^{-88}

D

Acid and Base Dissociation Constants at 25 °C

Acid $-\log K_a = pK_a$			Conjugate base - $\log K_b = pK_b$		
$HClO_4$	Perchloric acid	−7	ClO_4^-	Perchlorate ion	21
HCl	Hydrochloric acid	−3	Cl^-	Chloride ion	17
H_2SO_4	Sulfuric acid	−3	HSO_4^-	Bisulfate ion	17
HNO_3	Nitric acid	0	NO_3^-	Nitrate ion	14
H_3O^+	Hydronium ion	0	H_2O	Water	14
HIO_3	Iodic acid	0.8	IO_3^-	Iodate ion	13.2
HSO_4^-	Bisulfate ion	2	SO_4^{2-}	Sulfate ion	12
H_3PO_4	Phosphoric acid	2.1	$H_2PO_4^-$	Dihydrogen phosphate ion	11.9
$Fe(H_2O)_6^{3+}$	Ferric ion	2.2	$Fe(H_2O)_5OH^{2+}$	Hydroxo iron(III) complex	11.8
HF	Hydrofluoric acid	3.2	F^-	Fluoride ion	10.8
HNO_2	Nitrous acid	4.5	NO_2^-	Nitrite ion	9.5
CH_3COOH	Acetic acid	4.7	CH_3COO^-	Acetate ion	9.3
$Al(H_2O)_6^{3+}$	Aluminum ion	4.9	$Al(H_2O)_5OH^{2+}$	Hydroxo aluminum(III) complex	9.1
$H_2CO_3^*$	Carbon dioxide and carbonic acid	6.3	HCO_3^-	Bicarbonate ion	7.7
$H_2S(aq)$	Hydrogen sulfide	7.1	HS^-	Bisulfide ion	6.9
$H_2PO_4^-$	Dihydrogen phosphate	7.2	HPO_4^{2-}	Monohydrogen phosphate ion	6.8
$HOCl$	Hypochlorous acid	7.5	OCl^-	Hypochlorite ion	6.4
$HCN(aq)$	Hydrocyanic acid	9.3	CN^-	Cyanide ion	4.7
H_3BO_3	Boric acid	9.3	$B(OH)_4^-$	Borate ion	4.7
NH_4^+	Ammonium ion	9.3	$NH_3(aq)$	Ammonia	4.7

Ion Exchange in Environmental Processes: Fundamentals, Applications and Sustainable Technology,
First Edition. Arup K. SenGupta.

Acid $-\log K_a = pK_a$			Conjugate base $-\log K_b = pK_b$		
H_4SiO_4	Orthosilicic acid	9.5	$H_3SiO_4^-$	Trihydrogen silicate ion	4.5
C_6H_5OH	Phenol	9.9	$C_6H_5O^-$	Phenolate ion	4.1
HCO_3^-	Bicarbonate ion	10.3	CO_3^{2-}	Carbonate ion	3.7
HPO_4^{2-}	Monohydrogen phosphate	12.3	PO_4^{3-}	Phosphate ion	1.7
$H_3SiO_4^-$	Trihydrogen silicate	12.6	$H_2SiO_4^{2-}$	Dihydrogen silicate ion	1.4
HS^-	Bisulfide ion	14	S^{2-}	Sulfide ion	0
H_2O	Water	14	OH^-	Hydroxide ion	0
$NH_3(aq)$	Ammonia	~23	NH_2^-	Amide ion	−9
OH^-	Hydroxide ion	~24	O^{2-}	Oxide ion	−10

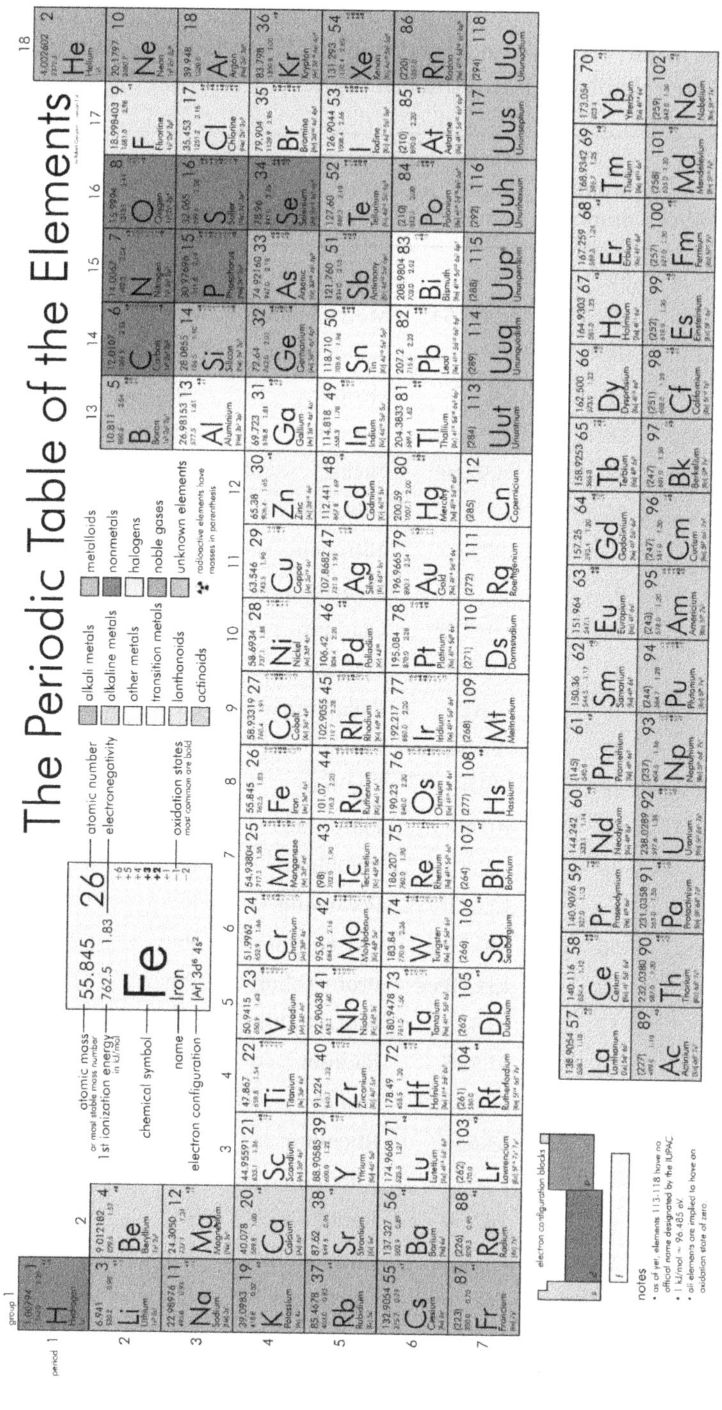

Ion Exchange in Environmental Processes: Fundamentals, Applications and Sustainable Technology, First Edition. Arup K. SenGupta.

ATOMIC WEIGHTS OF THE ELEMENTS

(Adopted by the International Union of Pure and Applied Chemistry in 1973). Based on ^{12}C = 12 atomic mass units (amu). The following values apply to elements as they exist in materials found on earth and to certain artificial elements. When used with the footnotes, they are reliable to ± 1 in the last digit, or ± 3 if that digit is in small type.

	Symbol	Atomic number	Atomic weight (amu)		Symbol	Atomic number	Atomic weight (amu)
Actinium	Ac	89		Mercury	Hg	80	200.59
Aluminum	Al	13	26.98154[a)]	Molybdenum	Mo	42	95.94
Americum	Am	95		Neodymium	Nd	60	144.24
Antimony	Sb	51	121.75	Neon	Ne	10	20.179
Argon	Ar	18	39.948[b), c)]	Neptunium	NP	93	237.0482[b)]
Arsenic	As	33	74.9216[a)]	Nickel	Ni	28	58.71
Astatine	At	85		Niobium	Nb	41	92.9064[a)]
Barium	Ba	56	137.34	Nitrogen	N	7	14.0067[b)]
Berkellum	Bk	97		Nobelium	No	102	
Beryllium	Be	4	9.01218[a)]	Osmium	Os	76	190.2
Bismuth	Bi	83	208.9804[a)]	Oxygen	O	8	15.9994[b), c)]
Boron	B	5	10.81[c)]	Palladium	Pd	46	106.4
Bromine	Br	35	79.904	Phosphorus	P	15	30.97376[a)]
Cadmium	Cd	48	112.40	Platinum	Pt	78	195.09
Calcium	Ca	20	40.08	Plutonium	Pu	94	
Californium	Cl	98		Polonium	Po	84	
Carbon	C	6	12.011[b), c)]	Potassium	K	19	39.0988
Cerium	Ce	58	140.12	Praseodymium	Pr	59	140.9077[a)]
Cesium	Cs	55	132.9054[a)]	Promethlum	Pm	61	
Chlorine	Cl	17	35.453	Protactinium	Pa	91	231.0359[a)]
Chromium	Cr	24	51.996	Radium	Ra	88	226.0254[a), d)]
Cobalt	Co	27	58.9332[a)]	Radon	Rn	86	
Copper	Cu	29	63.546[c)]	Rhenium	Re	75	186.207
Curium	Cm	96		Rhodium	Rh	46	102.9055[a)]
Dysprosium	Dy	66	162.50	Rubidium	Rb	37	85.4678
Einsteinium	Es	99		Ruthenium	Ru	44	101.07
Erbium	Er	68	167.26	Samarium	Sm	62	150.4
Europium	Eu	63	151.96	Scandium	Sc	21	44.9559[a)]
Fermium	Fm	100		Selenium	Se	34	78.96
Fluorine	F	9	18.99840[a)]	Silicon	Si	14	28.086[c)]
Francium	Fr	87		Silver	Ag	47	107.868

	Symbol	Atomic number	Atomic weight (amu)		Symbol	Atomic number	Atomic weight (amu)
Gadolinium	Gd	64	157.255	Sodium	Na	11	22.98977[a)]
Gallium	Ga	31	69.72	Strontium	Sr	38	87.62
Germanium	Ge	32	72.59	Sulfur	S	16	32.06[d)]
Gold	Au	79	196.9665[a)]	Tantalum	Ta	73	180.9479[b)]
Hafnium	Hf	72	178.49	Technelium	Tc	43	98.9062[d)]
Helium	He	2	4.00260[b)]	Tellurium	Te	52	127.60
Hoimium	Ho	67	164.9304[a)]	Terbium	Tb	65	158.9254[a)]
Hydrogen	H	1	1.0079[b), c)]	Thallium	Ti	81	204.37
Indium	In	49	114.82	Thorium	Th	90	232.0381[a)]
Iodine	I	53	126.9045[a)]	Thulium	Tm	69	168.9342[a)]
Iridium	Ir	77	192.22	Tin	Sn	50	118.69
Iron	Fe	26	55.847	Titanium	Ti	22	47.90
Krypton	Kr	36	83.80	Tungsten	W	74	183.85
Lanthanium	La	57	138.9055[b)]	Uranium	U	92	238.029[b)]
Lawrencium	Lr	103		Vanadium	V	23	50.9414[b)]
Lead	Pb	82	207.2[c)]	Xenon	Xe	54	131.30
Lithium	Li	3	6.941[c)]	Ytterbium	Yb	70	173.04
Lutetium	Lu	71	174.97	Yttrium	Y	39	88.9059[a)]
Magnesium	Mg	12	24.305	Zinc	Zn	30	65.38
Manganese	Mn	25	54.9380[a)]	Zirconium	Zr	40	91.22
Mendelevium	Md	101					

a) Elements with no isotopes. These elements exist as one type of atom only.
b) Elements with one predominant isotope (around 99 to 100% one Isotope).
c) Elements for which the variation in Isotopic abundance in earth samples limits the precision of the atomic weight.
d) Most commonly available isotope (radioactive).

Index

Ion Exchange in Environmental Processes: Fundamentals, Applications and Sustainable Technology,
First Edition. Arup K. SenGupta.

i

k

l

Z

Printed and bound by CPI Group (UK) Ltd, Croydon, CR0 4YY

17/09/2026

14964289-0006